Transformers and Motors

Transformers and Motors

A Single-Source Reference for Electricians

George Patrick Shultz

The material in this book has been reviewed by the NJATC (National Joint Apprenticeship and Training Committee for the Electrical Industry) and training directors of the electrical apprenticeship committee.

Newnes

An Imprint of Elsevier

Boston Oxford Johannesburg Melbourne New Delhi Singapore

Newnes is an imprint of Elsevier

 A member of the Reed Elsevier group

 Recognizing the importance of preserving what has been written, Elsevier prints its books on acid-free paper whenever possible.

Elsevier supports the efforts of American Forests and the Global ReLeaf program in its campaign for the betterment of trees, forests, and our environment.

ISBN-13: 978-0-7506-9948-8
ISBN-10: 0-7506-9948-5

The publisher offers special discounts on bulk orders of this book.
For information, please contact:
Manager of Special Sales
Butterworth–Heinemann
An Imprint of Elsevier
225 Wildwood Avenue
Woburn, MA 01801-2041
Tel: 617-928-2500
Fax: 617-928-2620

For information on all Newnes electronics publications available, contact our World Wide Web home page at: http://www.bh.com/newnes

10

Printed in the United States of America

Contents

PART III APPENDICES

Preface

Transformers and Motors was written to assist in the practical training of students of electrical applications. Special consideration was given to the need for increasing the skills of journeyman electricians and for the introduction of these topics into apprentice training programs. *Transformers and motors* can also be used as a primary or supplementary textbook at the high school, vocational-technical school, or junior college level.

Although the basic concepts and the theories of transformers and motors are described, and simple mathematical relationships are explained using examples, I have placed the emphasis on installation and maintenance, and the troubleshooting of faults. I have assumed that the student has studied basic electromagnetic theory and has had an introduction to these concepts.

Because one does not normally remember knowledge which is not applied, you should review electromagnetic theory early in the course and be prepared to ask the instructor for some explanation about any of the concepts you do not fully understand.

This process will not only help the individual become a better student and electrician, but it will also assist others in the class to better understand the principles of electromagnetic theory. At the same time, this process will guide the instructor as to what material to review.

In preparation for writing this text, I made inquiries to many manufacturers, power companies, and maintenance departments in plants throughout the country. Their cooperation in providing the information needed is acknowledged and greatly appreciated.

I also conducted a search of the literature to confirm the theory and practices described in this text. The authors of this technical information are commended for their efforts in preserving, updating, and furthering the knowledge of electrical science.

My special appreciation is extended to the Joint Apprenticeship and Training Committee of National Electrical Contractors Association/International Brotherhood of Electrical Workers (NECA/IBEW) Local 26 in Washington, DC for whom I worked as an instructor for over 20 years under the directorships of Clinton Bearor, John Widener, and Larry Greenhill. The availability to the Clinton Bearor Library extended to me by Mr. Greenhill has accelerated the publication of this text.

I would be remiss in not thanking the members of the National Joint Apprenticeship and Training Committee who took time from their busy schedules to review my efforts. Their criticisms, suggestions, and encouragement have been valuable in eliminating many errors and in providing further clarification of several key points in the text.

The commitment I made to write this book has impressed on me once again the truth that the more one delves into any topic, the less one knows about it. For this reason, this book has no real end. The search for more complete knowledge and understanding of electricity is left to the working electrician.

GEORGE P. SHULTZ

PART

I

Transformers

CHAPTER 1

Fundamental Concepts: Transformers

When power generation was introduced in this country through the efforts of Thomas Edison, the first plants were all direct current (DC) generation facilities. This meant that each power plant had to be close to the consumers because of the inherent losses of DC transmission of power. Efforts to use alternating current (AC) proved to be futile due to hysteresis and eddy current losses. Even with the introduction of AC transmission, many DC plants were still in use in the mid 1900s.

Primarily through the genius of one man, Charles Proteus Steinmetz (1865–1923), the theory of AC transmission was perfected, and the use of AC made possible. Dr. Steinmetz was a physically handicapped refugee from Germany who was almost refused entry into the United States because of his deformity and poverty. His report in 1892 on hysteresis losses gained him an invitation to join a little-known corporation called the General Electric Company (GE). His work at GE is one of the chief contributing factors in the success of that organization and was primary to the company becoming one of the principal manufacturers of transformers and AC motors.

A *transformer* is an electromagnetic device that transfers electrical energy from one circuit to another through mutual inductance. It is one of the most remarkable devices ever conceived. In most cases, it performs its assigned task without supervision year after year and with very little maintenance. It allows the power company to economically supply a single source voltage over a long distance to the customer, where other transformers change the value of the source voltage to the voltages required by the multitude of electrical and electronic devices utilized in the home, office, farm, and industrial plant. Transformers are the single most important apparatus that makes possible our modern electrical distribution system.

Comparing Direct Current to Alternating Current

To illustrate this point, let's assume a power plant is 1000 feet from the user whose equipment requires 300 amperes at 240 volts. For discussion, the current density on the transmission lines will be the same. The DC power plant must supply 300 amperes at 240 volts over

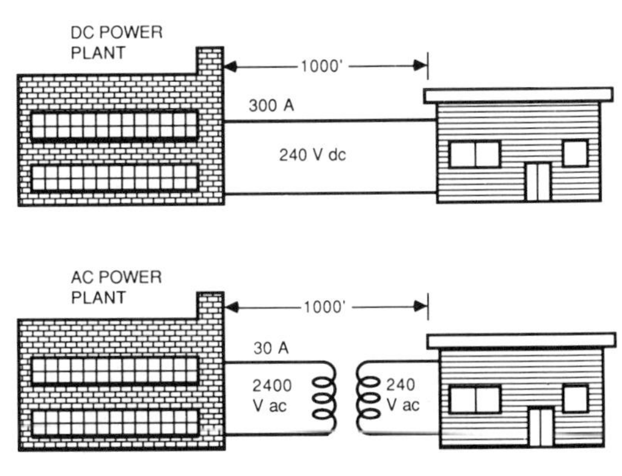

FIGURE 1–1. Comparison of DC and AC power transmission.

the wires. The AC power plant can supply 30 amperes at 2400 volts and deliver the same amount of power to the customer, where it can be transformed to 240 volts, 300 amperes. Figure 1–1 illustrates the two situations.

$$\text{Volt Amperes} = \text{Volts X Amperes}$$
$$72\text{ kVA} = 240 \text{ X } 300$$
$$72\text{ kVA} = 2400 \text{ X } 30$$

In order to have the same current density on both transmission lines, based on the American Standard Wire Gauge ASWG tables, the DC system would require number 0 wire, and the AC system would need number 10 wire. Because the distance involved is the same for both power plants, the volume and weight of the two lines are proportional to the area of the wires. This difference will require the supporting structures (power poles) to have ten times the strength for the number 0 wire with the resulting increased cost of installation and greater maintenance in the future. These costs would be in addition to the wire cost which would be approximately one-tenth as expensive for the AC system.

Number 0 wire has a resistance of 0.1 ohm per 1000 feet. The total resistance for 2000 feet of wire would be 0.2 ohm. Therefore, the voltage drop on the line, current times the resistance, would equal

$$\text{Voltage drop (DC)} = 300\text{ A X } 0.2\ \Omega = 60\text{ V}$$

Number 10 wire has a resistance of 1 ohm per 1000 feet. The total resistance of 2000 feet of wire would be 2 ohms. The voltage drop on the AC line would equal

$$\text{Voltage drop (AC)} = 30\text{ A X } 2\ \Omega = 60\text{ V}$$

Although the voltage drop for both systems is the same, the voltage regulation (VR) for AC transmission is much better than for DC. The DC power plant would have to generate 300 volts, 240 volts plus 60 volts, in order to deliver the required voltage to the customer. The percent of VR would equal

$$\%\text{ VR (DC)} = \frac{60}{240} \text{ X } 100 = 25\%$$

The AC power plant would need to generate 2460 volts to deliver the desired 240 volts. Its VR would equal

$$\%\text{VR} = \left(\frac{60}{2400}\right) \text{ X } 60 = 2.5\%$$

This means that from no load to full load, the customer's voltage would vary from 300 to 240 volts for the DC transmission system. In the case of the AC system, the voltage would vary from 246 to 240 volts from no load to full load.

The difference in the power losses consumed by the two systems is not insignificant. A 60-volt drop at 300 amperes on the DC system would give the following results:

$$\text{Watts loss (DC)} = 60\text{ V X } 300\text{ A}$$
$$= 18{,}000\text{ W}$$

For the AC line the power loss would be

$$\text{Watts loss (AC)} = 60\text{ V X } 30\text{ A}$$
$$= 1800\text{ W}$$

From the customer's point of view, it is easy to see which of these systems will provide the most economical and efficient service. The high voltages on the DC line when going from a light load to full load would shorten the life of all the apparatus being energized. To regulate the DC voltage for the user would incur additional expenses. The AC line can be regulated much more economically and effectively through transformer action.

Direct Current Transmission

DC transmission lines can be economical under certain conditions. For example, when the generation plant is several hundred miles from the center of population. The generator produces AC power which is stepped-up though transformers to very high voltages, often one million volts or more. The AC is then rectified, and the resulting DC is transmitted to the consumers' area.

By using DC for long distance power transmissions, considerable savings can be made. If a good ground return exists between the power plant and the intended destination, only one line is needed instead of three for AC transmission. This reduces the cost of wire by one-third, and the cost of the structures needed to hold the wires will be reduced accordingly. Economies are also realized with savings on losses inherent to AC transmissions such as transformer losses and reactance losses.

These savings are offset by the cost of conversion equipment to change the DC back to AC. Because this equipment is most expensive, the conversion from DC back to AC usually occurs only once.

Principles of Operation

A transformer in its simplest form consists of two windings on an iron core. The winding connected to the source voltage is called the *primary winding*, and the one connected to the load is called the *secondary winding*. Energy is transferred from the primary to the secondary winding through magnetic induction. When AC voltage is applied to the primary, current flows through the windings which creates a constantly changing magnetic field. This varying field cuts through the secondary windings and creates a voltage across the secondary.

Turns Ratio

The relationship between the magnitude of the primary voltage (V_p) to the secondary voltage (V_s) is directly related to the number of turns in the primary (N_p) to the number of turns in the secondary (N_s). This is expressed mathematically as

$$\frac{V_p}{V_s} = \frac{N_p}{N_s}$$

Figure 1–2 depicts a simple transformer. The primary and secondary wires are identified by the standard letter and numbering system. High-voltage (primary) wires are marked with a "H" and low-voltage wires with "X." The turns ratio would be expressed as 2:1, and this would be a step-down transformer. If 480 volts were applied to the primary, the secondary voltage would be 240 volts. If 240 volts were applied to the primary, the output would be 120 volts.

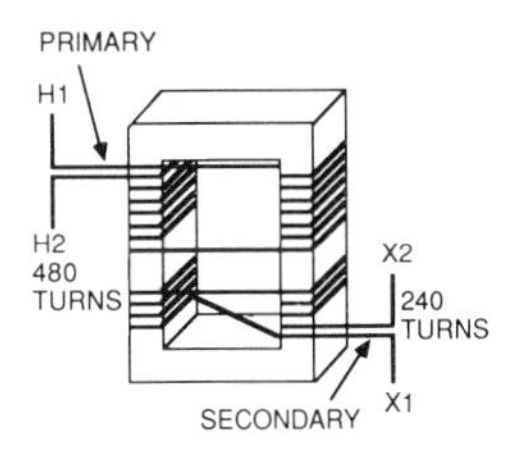

FIGURE 1–2. Simple transformer.

Reversing conditions and having 240 turns on the primary and 480 turns on the secondary would make the device a step-up transformer. Applying 480 volts on the primary would result in 960 volts on the secondary. The turns ratio would then be 1:2.

Most transformers rated above 3 kVA can be used either as step-down or step-up service. Standard transformers below 2 kVA have compensated windings and should not be used in reverse. These transformers have a winding ratio that provides a rated voltage to a rated load.

The source voltage can be connected to either the "H" leads or to the "X" terminals. The primary of the transformer can be either set of terminals, depending on whether the transformer is operated as a step-up or step-down device.

Transformer Rating

Transformers are rated in kilo-volt-amperes (kVA) rather than in watts. The reason for this is that not all loads are purely resistive. Only resistance consumes power, measured in watts. The kVA rating is based on the amount of current a transformer can deliver to a load without exceeding its temperature rise rating.

A large motor load that is running without mechanical load or well below its horsepower capacity will look inductive to the source. This will cause the current to lag the voltage. This inductive current, or lagging current, is doing no work, therefore it is not consuming power. At the same time, the transformer windings must be able to handle the current. The resistance of the windings will use power and cause heating. Under these conditions the circuit is said to have a poor power factor which is stated as a percentage and is equal to the cosine of the angle between the current and voltage.

A motor load will always use some power due to the resistance of its windings and the friction involved with a piece of rotating machinery. For a purely capacitive load, however, very little power or wattage would be consumed outside that used by the resistance of the wires connecting the capacitor to the secondary. A wattmeter connected to this load would indicate zero watts for all practical purposes. The capacitor will consume energy on its charge cycle, and it will return the energy to the circuit when it discharges. At the same time, very high currents could be drawn from the transformer and its kVA rating would need to be sufficient to handle the current or its temperature rating may be exceeded causing damage to the transformer.

Transformer Currents

When calculating the currents of a transformer, the primary current can be determined by dividing the kVA rating by the rated voltage. For example, if the transformer is rated 10 kVA with a primary voltage of

600 volts, then primary current for the ideal transformer (I_p would be

$$I_P = \frac{VA}{V_P}$$

$$I_P = \frac{10{,}000\ VA}{600\ V}$$

$$= 16.67\ A$$

If the 10-kVA transformer has a secondary voltage of 240 volts, the secondary current under full load (I_s) would equal

$$I_S = \frac{VA}{V_S}$$

$$I_S = \frac{10{,}000\ VA}{240\ V}$$

$$= 41.67\ A$$

Primary current is related to secondary current as an inverse relationship to the number of turns in the primary to the number of turns in the secondary. This is expressed mathematically as

$$\frac{N_p}{N_s} = \frac{I_s}{I_p}$$

Primary and Secondary kVA Relationship

The relationships between voltages and currents in a transformer can be confusing at times. One should keep in mind that you will never get something for nothing. The kVA of the primary must equal the kVA of the secondary under the ideal transformer concept. Using the values in the previous example, the following results are computed:

$$kVA_P = kVA_S$$

$$600\ V \times 16.67\ A = 240\ V \times 41.67\ A$$

$$10\ kVA = 10\ kVA$$

Another way of stating this fact is, if the voltage is stepped-down, the current will be stepped-up. Therefore, the relationships that exist between the turns ratio, voltages, and currents of a transformer can be stated as

$$\frac{N_p}{N_s} = \frac{V_p}{V_s} = \frac{I_s}{I_p}$$

Transformer Impedance

Impedance is another factor that needs to be considered when working with transformers. *Impedance* is defined as the total opposition to current flow in an AC circuit.

Transformers are in effect impedance matching devices. In order to deliver maximum power to a load, the impedance of the generator, be it a battery, the secondary of a transformer, the output of an amplifier, or any source of electrical power, must equal the impedance of the load. Figure 1–3 is used to demonstrate this point.

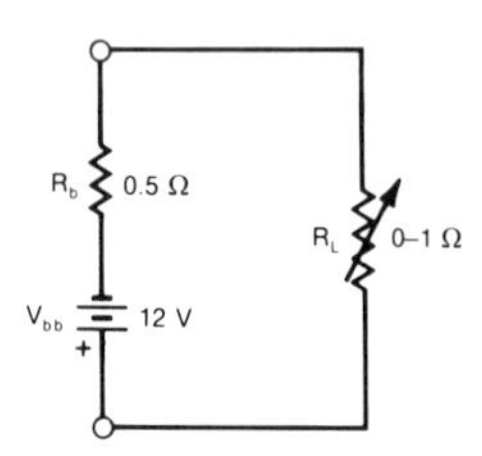

FIGURE 1–3. Maximum transfer of power.

Table 1–1 provides a series of calculations based on the values given in Figure 1–3. A battery is used to simulate a generator. At the two extremes for either a short circuit or an open circuit, no power is consumed by the load. In the first case, the load has no resistance, and only resistance consumes power. In the second case, no current flows when the circuit is opened, and power is calculated using current as a multiplier. Watts equals current times voltage, or substituting for voltage using Ohm's law, current squared (I^2) times resistance (R).

$$\text{Watts} = I^2R$$

TABLE 1–1. Calculations for Maximum Transfer of Power from Generator to Load

R_L	R_T ($R_L + R_b$)	I_T (V_{bb}/R_T)	P_L ($I_T^2 \times R_L$)
0.0	0.5	24	0
0.1	0.6	20	40
0.2	0.7	17	59
0.3	0.8	15	68
0.4	0.9	13	71
0.5	1.0	12	72
0.6	1.1	11	71
0.7	1.2	10	70
0.8	1.3	9	68
0.9	1.4	8.5	66
1.0	1.5	8	64
∞	∞	0	0

Note in Table 1–1 that the power consumed by the load is zero when the value of the load is zero. The total power is consumed within the generator. As load resistance is increased, the power consumed by it increases to a maximum power of 72 watts. This occurs when the resistance of the battery is equal to the resistance of the load. A further increase in the value of the resistance of the load causes the amount of power consumed by it to decrease. The decrease would continue if more resistance were added until zero power would be consumed when the load circuit was open, or undefined.

Phase Relationships

When the secondary of a transformer is open with no load applied, it acts as an inductor. The reactance is very high, and very little current flows in the primary. The primary current lags the applied voltage by nearly 90 degrees. The only power consumed in watts is due to the inherent losses of the transformer. These losses are mostly due to the resistance of the primary windings.

Voltage will be present across the secondary windings of an amplitude corresponding to the turns ratio. The voltage across the secondary will be 180 degrees out of phase with the primary voltage. Figure 1–4 shows this relationship. This is important to understand when connecting a single transformer for additive or subtractive operation. The left-hand rule for electromagnetism can be used to determine the phase relations.

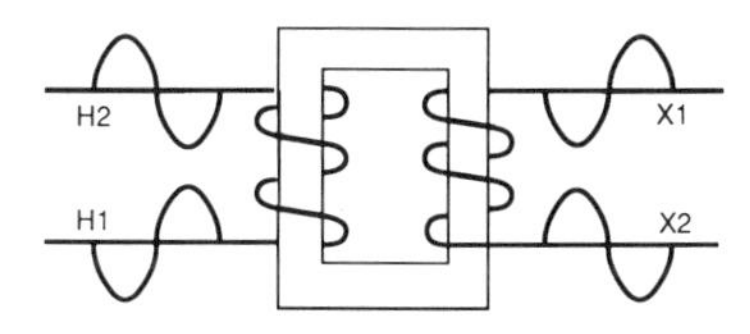

FIGURE 1–4. Phase relationship between primary and secondary voltages.

The terminals of the transformer are marked "H" and "X" which is the common terminology for identifying the leads. "H" indicates the high-voltage leads, and "X" the low-voltage leads. The number 1 indicates the starting point for each winding. For a normally wound transformer, the voltages on H1–X1 and H2–X2 are in phase with each other.

Transformers can be wound so that they have an in-phase relationship. This is accomplished by reverse winding either the primary or the secondary. When this is done, the schematic diagram used in electronic circuits includes a dot on both primary and secondary windings. The solid lines drawn between the windings indicate an iron core. See Figure 1–5.

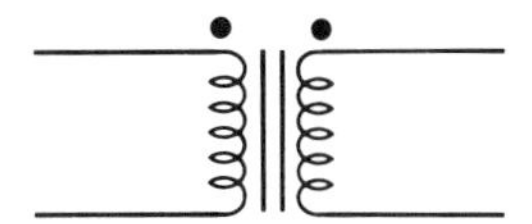

FIGURE 1–5. In-phase transformer.

Figure 1–6 shows a multiple-winding transformer and how the leads are designated for a power transformer. Corresponding numbers between the high-voltage windings and the low-voltage windings will be in phase with each other. For example, H1 will be in phase with X1.

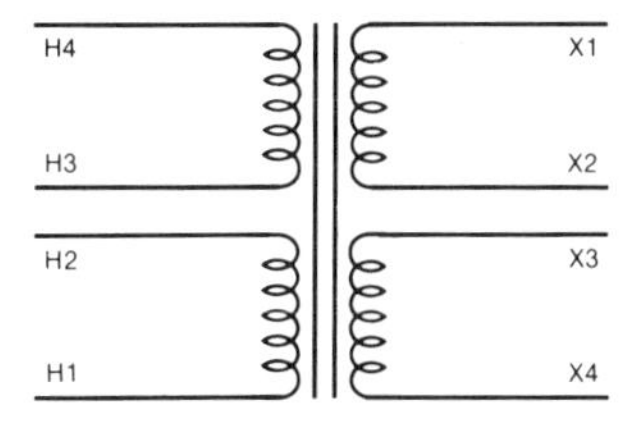

FIGURE 1–6. Multiple windings.

Losses

Transformers when operated within their specifications and temperature range are one of the most efficient devices ever invented by human beings. Efficiencies range from 95% to approximately 99% under full-load conditions. If a transformer is operated under less than full load, the efficiency will decease 1% to 2%.

Losses can be classified into two categories. These are copper losses and core losses. There are several core losses. These include eddy currents, hysteresis, flux leakage, and core saturation.

Copper Losses

Copper losses are due to the resistance of the wire in the primary and secondary windings and the current flowing through them. These losses can be reduced by using wire with large cross-sectional area in the manufacturing of the coils.

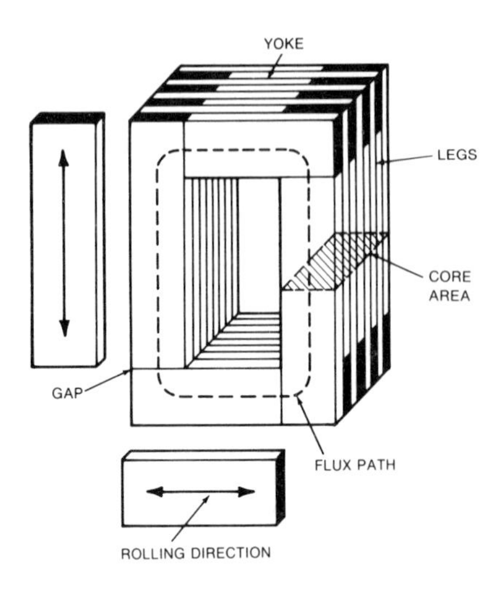

FIGURE 1–7. Butt, wound, and mitered cores. *(Courtesy Sorgel Transformer, Square D Co.)*

Eddy Currents

Eddy currents are those that are introduced into the iron core material of the transformer. They are unwanted currents and consume power which is wasted as heat. A solid iron core looks like a single short circuit winding to the magnetic field. Because of the very low resistance, a very large current can be induced.

This problem is largely overcome by making the core of very thin laminates. See Figure 1–7 for the types of construction. Each lamination is coated on each side with insulating material so that no current can flow between laminates. At the same time, the coating allows the free passage of the flux lines. This process greatly increases the resistance of the core and reduces the amplitude of the eddy currents.

When handling transformers with the core exposed, care should be taken not to break the insulating integrity of the core. For example, dropping a transformer on one of its edges on concrete could short the laminates. Eddy current losses are proportional to the frequency and magnitude of the current in the core of the transformer.

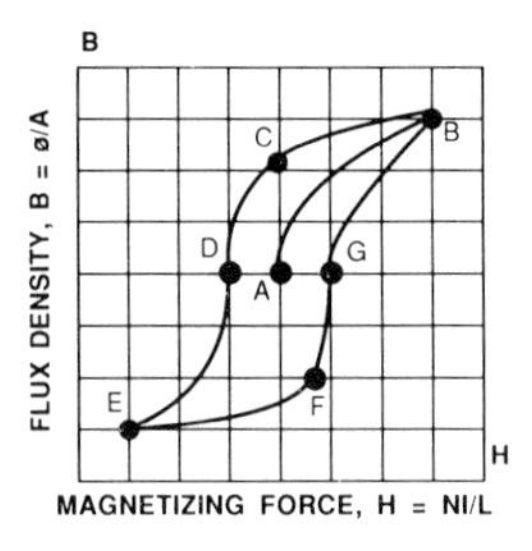

FIGURE 1–8. Hysteresis loop—B/H curve

Hysteresis Losses

Hysteresis losses are due to the magnetic agitation of the molecules in the iron and their resistance to being moved. One theory of magnetism is that in a magnetic material, each molecule has a north and south pole. When the molecules are arranged in a random fashion with north and south poles pointing in every direction, the fields cancel each other and the material is not magnetized. When a loop of wire with a current flowing through it is placed around the core of the transformer, the molecules of the core will align their poles based on the "left-hand rule" of electromagnetism. Because AC is used, the direction of current is constantly changing through the coil, and therefore its magnetic field is constantly changing direction. Therefore, the molecules in the core are constantly moving to align in the proper direction. In their movement, they bump into each other, causing friction and heat.

The molecular alignment lags behind the current change, and a chart of this process is called a hysteresis loop or B/H curve. Figure 1–8 depicts a typical hysteresis loop.

The core begins magnetizing at Point A as the current swings in the positive direction. It continues to magnetize until the increased current of the magnetizing force no longer produces additional flux lines or changes direction. This begins occurring at Point B. At this point, the core is said to be saturated. As the current swings less positive, the core begins to demagnetize. This process lags behind the current, so that when the current passes through its alternation going negative at Point C, the core is still magnetized in the positive direction and passes through the zero base line at Point D when the current is more negative. The core starts to saturate in the negative condition at Point E. As the current becomes less negative and moves back toward its positive alternation, the magnetic field again lags so that it is at Point F when the current is zero. Note that the hysteresis curve does not follow its original curve (Point A to Point B) when the circuit was first energized.

Hysteresis losses will increase with frequency, and they are greatest in materials that have a high retentivity. These materials, once magnetized, tend to retain their magnetism. It requires more energy to demagnetize them than those with low retentivity.

Flux Linkage Loss

Flux linkage loss is small, but one that nevertheless occurs in transformers. For the most part, flux lines will take the path of least reluctance. *Reluctance* is defined as the resistance to the passage of flux lines. These materials are said to have a high permeability.

The *permeability* of a material can be determined by dividing the flux density (B) by the magnetizing force (H). The core material meets this requirement and most of the flux lines will flow though it. Because flux lines going in the same direction repel each other, some of the flux lines are forced into open space and do not cut the secondary windings of the transformer. Because it took energy to create the lines, and they performed no useful work, they become a loss to the system.

Saturation Loss

Saturation losses usually occur when the transformer is being operated beyond its capacity. When the loading in the secondary occurs, the current in the primary must increase. A point is reached when an increase current in the primary produces no additional flux lines. This occurs at the points on each end of the hysteresis curve where the flux density does not increase with increased current (see Figure 1–8).

Loss Reduction

There is little that the maintenance electrician can do to reduce transformer losses other than make sure the transformer is operating within its specifications. The engineers and manufacturers have provided a highly efficient device. To maintain this high efficiency, the electrician should handle transformers with reasonable care when making installations. The installation should be kept clean, and during routine maintenance procedures, all fasteners should be tightened.

Classification

Many different types of transformers are available on the market today. Manufacturers' catalogs list them according to their ratings and construction features and they are often classified according to:

1. service or application.
2. purpose.
3. method of cooling.
4. number of phases.
5. types of insulation.
6. method of mounting.

Application

The most common method of classifying transformers is by their application. For example, *large power* transformers are those above 500 kVA used by the utility company in distribution systems over 67 kilovolts. *Distribution* transformers are used in delivering the appropriate levels of voltages and currents throughout a system. Transformers may be *sign* transformers, used with neon signs, or a *gaseous-discharge lamp* transformer used for outdoor lighting. Transformers may be designated as *control*, *signaling*, *bell*, and so forth. When used in electronic circuits, designations such as *flyback* transformers used in the high-voltage circuit of television sets, *coupling* transformers used to transfer signals from one circuit to another, *pulse* transformers used to develop a signal of proper amplitude and shape are all used.

The list is almost endless, and one needs to be familiar with the field of study to be able to visualize the device. Some transformers may be no larger than a pinhead, whereas others are as large as a room.

Core material changes with frequency. Ferrite and air cores are used with transformers at higher frequency replacing the high permeability steel cores used at the power frequencies. Ferrite cores are often adjustable so that circuits can be tuned by changing the inductance of the coil.

Purpose

General-purpose transformers are usually dry-type transformers 600 volts or less used for stepping voltage down to the utilization voltages for power and lighting. These transformers may be constant or varying voltage or constant or varying current types of transformers depending on the requirements of the loads.

Buck-boost transformers are used to lower or raise the line voltage supplying a power transformer or load. The purpose of the transformer is to step-up or step-down a voltage.

Isolation transformers are used to isolate the ground system of the power line from the area or device being served. For example, operating rooms in hospitals are required to be isolated. When used with electronic devices, it prevents the chassis from becoming energized to the power line ground when the plug on the device is not polarized.

Another purpose might be the use of a transformer to match impedance. The output of a stereo amplifier has a relatively high impedance as opposed to its load, a set of speakers. A transformer with a primary impedance equal to the output of the amplifier, and a secondary impedance equal to the impedance of the speaker, will assure maximum transfer of power.

Cooling Systems

Another way transformers are classified is according to the type of cooling system that is used. The two general classifications would be air or liquid cooled.

Air-cooled transformers are normally small and depend on the circulation of air over or through their enclosures. They may be either ventilated or nonventilated. Forced air provided by fans may be used. The fan(s) may be part of the transformer itself, or installed in a structure to provide general circulation of air for a larger area which includes the transformer(s). The transformers may have smooth surfaces or may be equipped with fins to provide a greater surface area for removing heat from them.

Oil-cooled transformers have the transformer's coils and core submerged in the liquid. The liquid may be mineral oil, silicone fluid, or a synthetic material that has been registered by the particular manufacturer. Natural circulation of the oil due to the heat is used in some of the transformers. Fins are normally provided to dissipate the heat to the surrounding air. Fans may be used to facilitate removing heat from the transformer. At other times, a water jacket with circulating cool water may be inserted inside the transformer housing to cool the oil. Another method would be to pump the oil through the fins or radiator and not depend on the natural circulating currents. Any of these methods or combinations of them may be part of the design of any particular transformer. Figure 1–9 illustrates some of these methods.

An effective cooling system can increase transformer capacity 25% to 50%. Under these circumstances, a 1000-kVA transformer may be operated as high as 1500 kVA without causing damage to the device.

Any steps the maintenance electrician can take to lower the operating temperatures of electrical equipment will assure greater efficiencies and extended operating times without equipment failure. Simple precautions to assure adequate air flow may be all that is needed.

Phases

Transformers are also classified as to the number of phases. Generally, these would include single-phase and three-phase transformers. One may also encounter two-phase or even six-phase transformers.

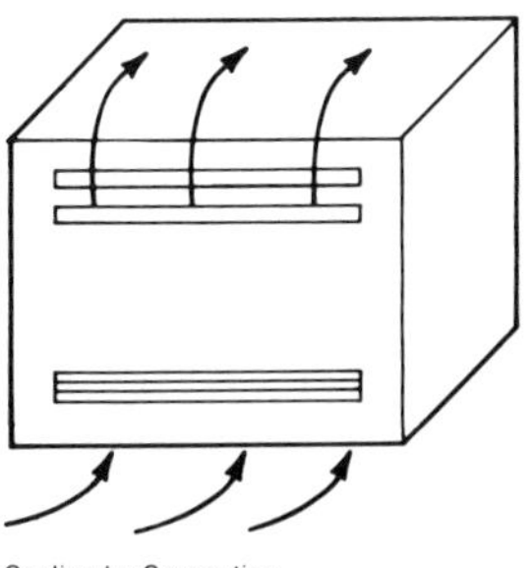

Cooling by Convection. Cool Air Enters Bottom and Exits Top Vents.

Coils Are Immersed in Oil-Filled Tank. Tank Surface May Be Smooth, Corrugated, or Tubular.

Fins Are Attached to Tank to Provide Greater Surface Area for Oil to Circulate.

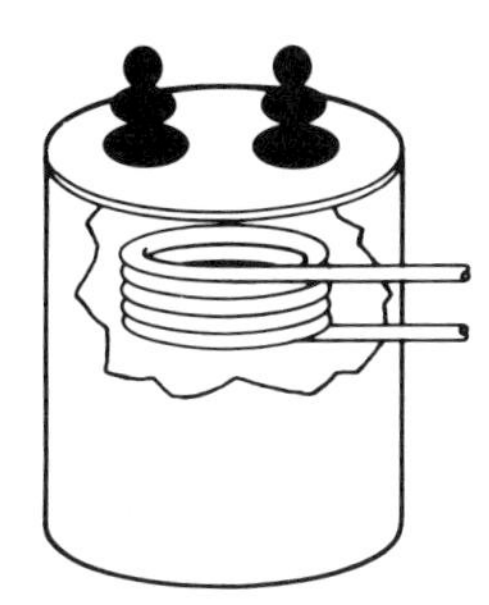

Oil is Cooled by Coils Through Which Cold Water is Circulated.

FIGURE 1–9. Cooling methods for power transformers.

Insulation

The National Electrical Manufacturers Association (NEMA) has designated four classes of insulation with specifications and temperature limits for dry-type transformers. In each case, the temperature base has been set at 40°C or 104°F. Equipment should not be installed in areas with ambient temperatures in excess of this value without having its output rating lowered. The NEMA classifications are

- Class A. Allows for not more than 55°C rise on the coil. This is close to the boiling point of water, but combustible materials may be present in the area with the transformer.
- Class B. Temperature rise may not exceed 80°C rise on the coil. These transformers are smaller than Class A types and weigh about one-half as much. These transformers are becoming less popular than the Class F and H series for distribution systems.
- Class F. This classification allows a temperature rise on the coils up to 115°C. These transformers are smaller than Class B types and are available up to 25 kVA for single- or three-phase applications.
- Class H. A maximum temperature rise of 150°C is allowed on the windings. The insulating materials used with these transformers are high-temperature glass, silicone, and asbestos. These units come in ratings of 30 kVA or greater.

Excessive temperature rise is the primary cause of transformer failure. Transformers are designed to have higher allowable temperature rises. Sorgel Transformers, for example, uses a barrel-type construction on their power transformers that allows for 220°C rise. These transformers are still operated within the NEMA classifications, but this method of construction allows a temperature margin to compensate for any hot spots that may occur.

The half-life of insulation that has been exposed to maximum temperature from the time it is put into operation is approximately 20,000 hours, or 2.3 years. Transformers that are exposed to continuous duty can be designed to withstand these conditions at a cost of approximately 10% above the standard design. This can be accomplished by using larger conductors to reduce copper losses and improving the cooling system.

Mounting

The final method of classifying transformers is the method by which they are mounted. They may be platform mounted, that is, they may stand on their own base on a structure of sufficient strength to hold them. They may be pole mounted, attached to a wall, or installed in a subway or a vault. It is important to specify the method of mounting when ordering a transformer.

Nameplate Nomenclature

Information provided on the nameplate of a transformer (Figure 1–10) is of particular value to the electrician. Areas indicated by numbers 1 and 11 give the name of the manufacturer and where the company is located. Other information that may be included in these blocks are the type of device, country of origin, and if the transformer has been approved by a testing laboratory.

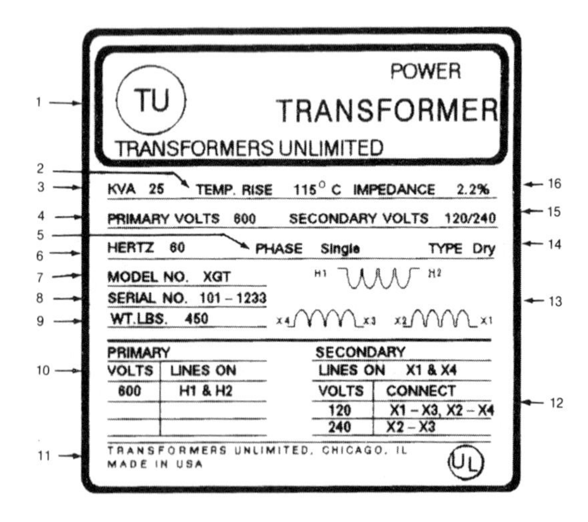

FIGURE 1–10. Transformer nameplate nomenclature.

The kVA rating is given in block 3. As mentioned earlier in this chapter, transformers are rated in kVA rather than watts. All loads are not purely resistive. Even though energy is not consumed by the load, the transformer must still be capable of delivering the required current. The amount of current in amperes can be calculated by dividing the volt-amperes by the secondary voltage (VA/V_s).

Allowable temperature rise is indicated in block 2. This value is usually based on an ambient temperature of 40°C. The combination of ambient and design temperature rise will give the maximum operating value, in this case, 155°C. If the ambient temperature is 30°C, then the transformer rise could be 125°C. If the ambient is 50°C, the allowable temperature rise of the transformer would be limited to 105°C.

Sections 4 and 15 give the voltage ratings for the primary and secondary voltages. Depending on the design, dual primary windings and the number of phases, several values could appear in these blocks. Windings could be connected series or parallel, and for three phase, wye, or delta. These various combinations will be described in detail in Chapter 3.

Block 5 gives the number of phases. This value is usually one or three. Block 6 provides the operating frequency of the transformer. A 50-Hz transformer can be used on a 60-Hz system, but a 60-Hz design usually cannot be utilized on a 50-Hz supply. This combination would result in the transformer drawing approximately 20% greater current.

The model number, serial number, and the weight of the transformer is given in blocks 7, 8, and 9. Manufacturers' design information may be coded in these areas. For example, Westinghouse Corporation uses a 13-digit style number to code information about their distribution transformers. Manufacturers' manuals and catalogs will sometimes provide this information.

Wiring instructions for the primary and secondary windings are given in blocks 10 and 12. The high-voltage leads from the source are connected to H1 and H2, and the load is attached to X1 and X4. For low-voltage operation (120 volts), terminals X1–X3 and terminals X2–X4 are tied together. For high voltage output, terminals X2–X3 are looped together to give 240 volts. If the transformer was to provide both 120 volts and 240 volts, a neutral would be attached to the X2–X3 tie point and brought to the load. This would provide 240 volts between X1 and X4, and 120 volts between both X1 and neutral, and X4 and neutral. A schematic of the transformer is shown in block 13 to aid the electrician in making the proper installation. For three-phase transformers with dual windings, these sections of the nameplate would be more extensive in describing the various combinations.

Although the type (block 14) of transformer in this example is merely described as "Dry," this description can vary with the manufacturer. It may be a coded number similar to that used by Westinghouse Corporation in their Style Number, or a series of letters and numbers that replace the Model Number.

In the case of liquid- or oil-filled transformers, the type of insulating and cooling material may be specified along with the number of gallons needed to fill the transformer. Other designations such as Power, Isolation, Autotransformer, Control, Bell, or other applications may be designated in this section of the nameplate.

The impedance of the transformer is shown in block 16. This value is given as a percentage of change in output voltage from no-load to full-load condition. With a 2.2% impedance transformer, the full-load voltage would equal approximately 234/117 volts for the three-wire system. These values are determined by multiplying the value of the impedance (2.2%) by the no-load voltage (240/120 volts) and subtracting the calculated value from the no-load voltage.

Impedance of the transformer becomes important when two transformers are to be connected in parallel. If both transformers do not have nearly the same value of impedance, the one with the lowest impedance will assume a disproportional amount of the load and could be damaged due to overload.

The short circuit current in the secondary of a transformer can also be determined using the impedance figure. Full-load current multiplied by 100/%Z will give this value. For a single phase transformer the full-load current in the secondary is equal to the kVA/V_s.

Using the information from the nameplate, the current at full-load (I_{FL}) is

$$I_{FL} = \frac{25{,}000 \text{ VA}}{240 \text{ V}} = 104.17 \text{ A}$$

and the absolute maximum secondary current (I_{SC}) under short circuit conditions would equal

$$I_{SC} = \frac{I_{FL}}{\%Z} \times 100 = \frac{104.17}{2.2} = 4734.85$$

This value assumes a sustained primary voltage during fault which implies a zero impedance source. Because the power source in the real world must have some impedance, the actual fault current will be lower than the 1894 amperes calculated in this example.

To determine the full-load current for a three-phase transformer, the following formula would be used:

$$I_{FL} = \frac{\text{kVA} \times 1000}{1.73 \times \text{V}} = \frac{25{,}000}{1.73 \times 240 \text{ V}} = 60.21 \text{ A}$$

Using this value, the short circuit current can be calculated using the same formula as used for the single-phase problem.

$$I_{SC} = \frac{I_{FL}}{\%Z} = \frac{60.21 \text{ A}}{2.2} = 2736.91 \text{ A}$$

Taps

Another item that may appear on the nameplate is information concerning the tap changer. Figure 1–11 shows a typical diagram of this device.

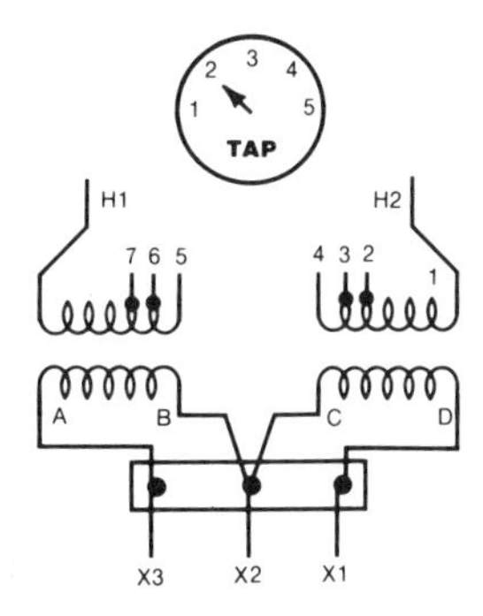

% HIGH VOLTAGE	TAP CHANGER	
	POS	CONNECTS
100.0	1	4 TO 5
97.5	2	3 TO 5
95.0	3	3 TO 6
92.5	4	2 TO 6
90.0	5	2 TO 7

FIGURE 1-11. Tap changer information. (*Courtesy Westinghouse Corp.*)

Because many transformers have a fixed turns ratio, they also have a fixed voltage ratio. If the source voltage is too high or too low, the voltage supplied to the load will be too high or too low. For this reason, taps are often added to the primary windings to provide a means to change the turns ratio as shown in Figure 1–11.

The nameplate always identifies the relationship between the tap positions and the percentage of primary voltage available. The percentage can be determined by dividing the source voltage by the primary voltage given on the nameplate and multiplying by 100.

$$\% = \frac{V_s}{V_p} \times 100$$

On transformers rated 25 kVA and below with secondary voltages of 120/240 volts, the taps are almost always designed to increase the output voltage due to the lighting and appliance loads. The usual arrangement is to have four 2½% taps or two 5% taps. These are known as "full capacity below normal" (FCBN) taps. Tap changing on dry-type transformers, such as the use of a bumper between the taps to complete the primary winding, is a *must*.

For transformers above 30 kVA and/or higher secondary voltages, high-voltage taps are often integrated into the design. General Electric uses a tap combination called universal taps which utilizes four 2½% taps for low voltage and two 2½% taps for high voltage [full capacity above normal (FCAN)]. This combination provides for a 15% voltage difference.

Taps are the cause of approximately 20% of all transformer failures. Westinghouse has an externally operated changer that provides positive sequence line voltage changes. This device features through-type stationary contact studs rigidly supported by a molded plastic channel. The moving contacts are spring-loaded, silver-plated copper which are moved along the stationary studs by a rack and pinion device. Figure 1–12 is an illustration of the Westinghouse device. This design has no rivets, bolts, or nuts, thereby assuring a positive contact on the current carrying connectors. This arrangement has greatly reduced failures due to the tap changer.

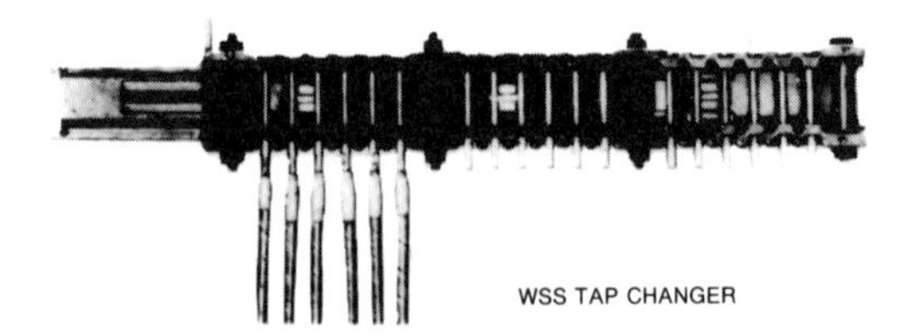

FIGURE 1–12. Tap changer.
(Courtesy Westinghouse Corp.)

Tap changers are normally available to be operated either by removing the cover of the transformer, through a hand hole in the cover, or accessible through switching arrangements brought through the case. *Under no condition* should an attempt be made to the change the tap when the transformer is *hot*. Make sure all power is disconnected from the device.

If the cover must be removed, or a hand hole is used to change the tap, be sure that they are securely closed when replaced. Check that the gasket is in good condition and accurately placed. If moisture and air is allowed to enter the transformer, the need for emergency replacement is sure to follow in a short period of time.

When transformers are operated in parallel, or individual transformers are used for three-phase systems, their taps should all be on the same setting. If not, their impedances will be different, and the transformers with the lowest impedances will take more than their share of the load. Circulating currents will also be set up between the transformers under these conditions, thereby putting unnecessary loads on the transformers and wasting energy.

Protection Devices

Other items that may appear on the nameplate in the schematic are fuses, breakers, and surge arresters. These devices have been designed in the system to protect the transformer, and gear both upstream and downstream from it. Overcurrent protection is designed to be selective and is coordinated with the entire system. This design provides for the isolation of faults without taking out of service those circuits than are operating properly. Figure 1–13 provides some typical high-voltage configurations for transformers along with pictures of fuses and switches.

Fuse links are designed to operate only in the case of winding failure, thereby isolating the transformer from the primary system. The bayonet-type fuse is oil-immersed and is available as either an overload or a fault sensing device. Current limiting fuses are air immersed in drywall canisters and limit both the current magnitude and energy associated with low-impedance faults.

Primary switching may be air- or oil-immersed. The switch shown is oil immersed and hookstick operable. Contacts are ganged together so that they operate as a unit. This switch is available for either radial or loop feed switching.

Figure 1–14 shows the protective devices on and in a canister type transformer. The fuse links are in series with the high-voltage leads, and the circuit breaker is installed in series with the load. In this case, both the fuse link and the circuit breaker operate submerged in oil.

Primary Overcurrent Protection Options

1. Protective Fuse Link

- Internal, oil-immersed, expulsion type
- Replaceable through the handhole
- Sized to operate only in the event of a winding failure, isolating the transformer from the primary system.
- Interrupting rating is 3500 amperes at 8.3 kV

2. D.O. II Bayonet-Type Fuse

- Oil immersed, expulsion type
- Drawout for fuse replacement
- Hookstick operable, loadbreak design
- Available with either overload-sensing or fault sensing
- 3500 AIC at 8.3 kV, 1800 AIC at 15.5 kV

BAYONET-TYPE FUSE ASSEMBLY

WESTINGHOUSE DRAWOUT CURRENT LIMITING FUSE CANNISTER

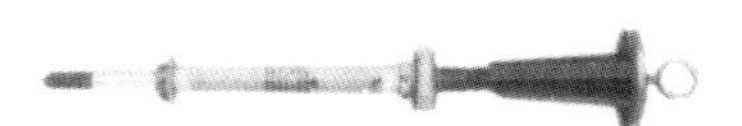

WESTINGHOUSE DRAWOUT CURRENT LIMITING FUSE LOAD BREAK ASSEMBLY

WESTINGHOUSE DRAWOUT CURRENT LIMITING FUSE DEAD BREAK ASSEMBLY

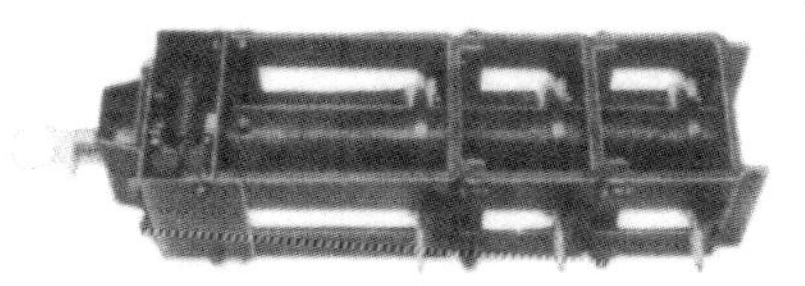

LOADBREAK OIL ROTARY (LBOR)

TYPICAL HIGH VOLTAGE CONFIGURATIONS

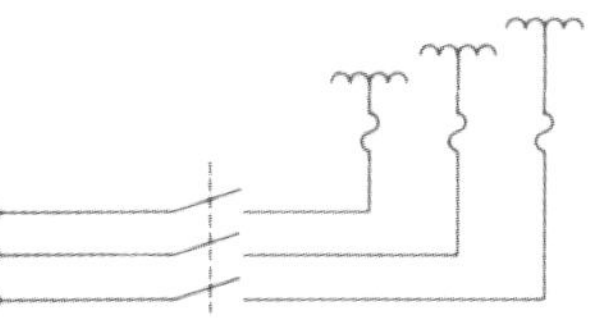

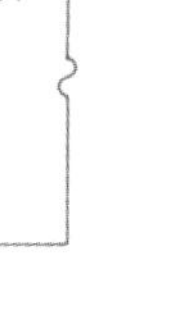

RADIAL FEED, LIVE FRONT WITH INTERNAL PROTECTIVE LINKS AND LBOR SWITCHING

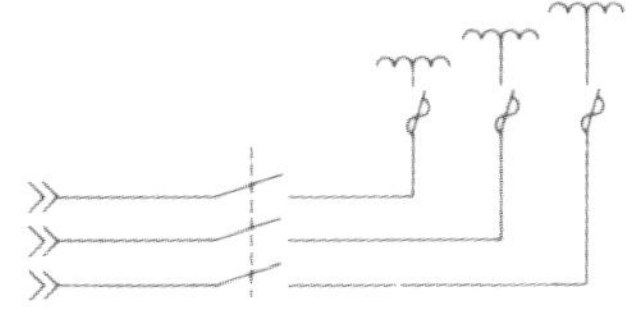

RADIAL FEED, DEAD FRONT WITH DRAWOUT LOADBREAK CURRENT LIMITING FUSES

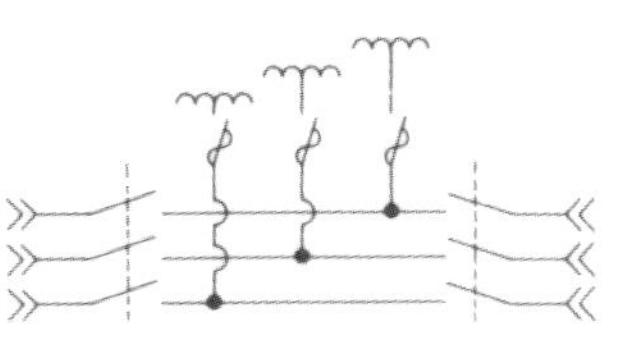

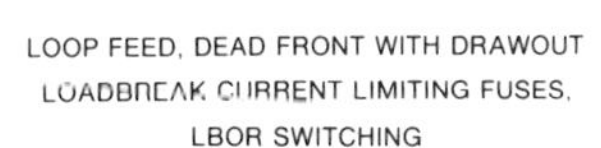

LOOP FEED, DEAD FRONT WITH DRAWOUT LOADBREAK CURRENT LIMITING FUSES, LBOR SWITCHING

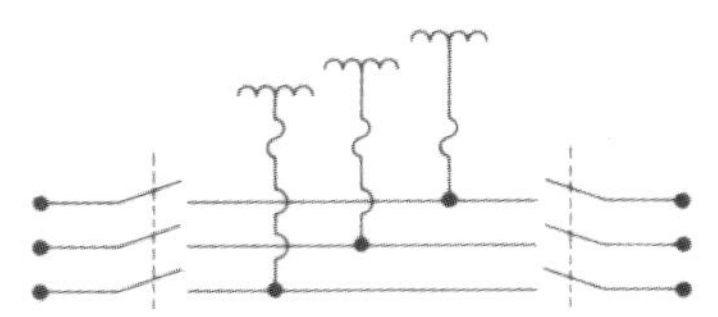

LOOP FEED, LIVE FRONT WITH INTERNAL PROTECTIVE LINKS AND LBOR SWITCHING

FIGURE 1–13. Typical high voltage configurations with components. *(Courtesy Westinghouse Corp.)*

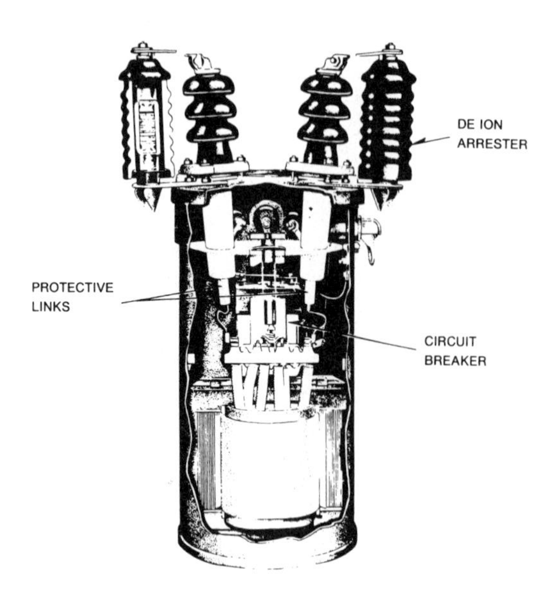

FIGURE 1–14. Protective devices for transformers. *(Courtesy Westinghouse Corp.)*

Lightning arresters are installed to suppress lightning strikes as well as line surges due to equipment being switched. The gap between the high-voltage line and the arrester is normally set at the factory. Size of the gap is normally marked on the arrester. Gaps should be checked when the arresters are installed.

In conclusion, information provided on the nameplate of transformers is important for the electrician to understand. The reversal of primary and secondary leads going to the transformer could cause a small disaster. For example, if the turns ratio of a transformer was 10:1, and the transformer was designed for 2400/240 volt operation, this simple reversal would cause 24,000 volts instead of 240 volts to be supplied to the customer when the system was energized. This would not be good for public relations.

Hardware Identification

Figures 1–15 and 1–16 have been included to help identify the various components on a transformer. Figure 1–15 is used as a secondary unit substation and is fluid filled. Cooling is furnished through flat, tubular fins welded to the tank wall. Standard tank pressure is 5 psi for oil-filled units, and 8 psi for silicone-filled tanks. Transformers over 750 kVA have provision for forced air cooling.

The high- and low-voltage bushings are located on opposite sides of the transformer at a standard height of 55 inches. This provides for "straight-through" line use and ease of coordination with other units. If updating is required at a later date, it can be accomplished with minimum cost.

Figure 1–16 shows some of the standard features available on a distribution transformer used in a substation. Some of the other features not illustrated are the lifting lugs, tank ground pad, and nameplate. The two high-voltage and four low-voltage bushings and radiators can be seen.

Autotransformers

An *autotransformer* is defined as one that has an electrical contact connection between the primary and secondary windings, so that they have part of a winding in common. This arrangement is shown in Figure 1–17. Auto means "self." An autotransformer is a self-induced, electrical-magnetic device.

In Figure 1–17, the autotransformer is wired as a step-down transformer. Under these conditions, the entire winding, x–z, is considered the primary coil. Section x–y is common to the primary and secondary.

Calculations

The same relationships of turns ratio, voltage, and current apply to this type of device as the separate winding transformer. Using "h" for the high-voltage side and "x" for the low-voltage secondary, the following formula is applicable:

$$\frac{V_h}{V_x} = \frac{T_h}{T_x} = \frac{I_x}{I_h}$$

Figure 1–17 provides an example of these relationships. There are 900 turns in the primary with 360 volts impressed. This means that there is 0.4 volt dropped across each turn (360 volts/900 turns). The voltage from x to y would equal:

$$0.4\ \text{V} \times 600\ \text{T} = 240\ \text{V}$$

The windings act as a voltage divider without the losses associated with resistance when it is used for this purpose. If there are 240 volts across the secondary of the autotransformer, then the voltage across the primary, x–z, must equal the source voltage, 360 volts. This voltage is impressed across the primary windings

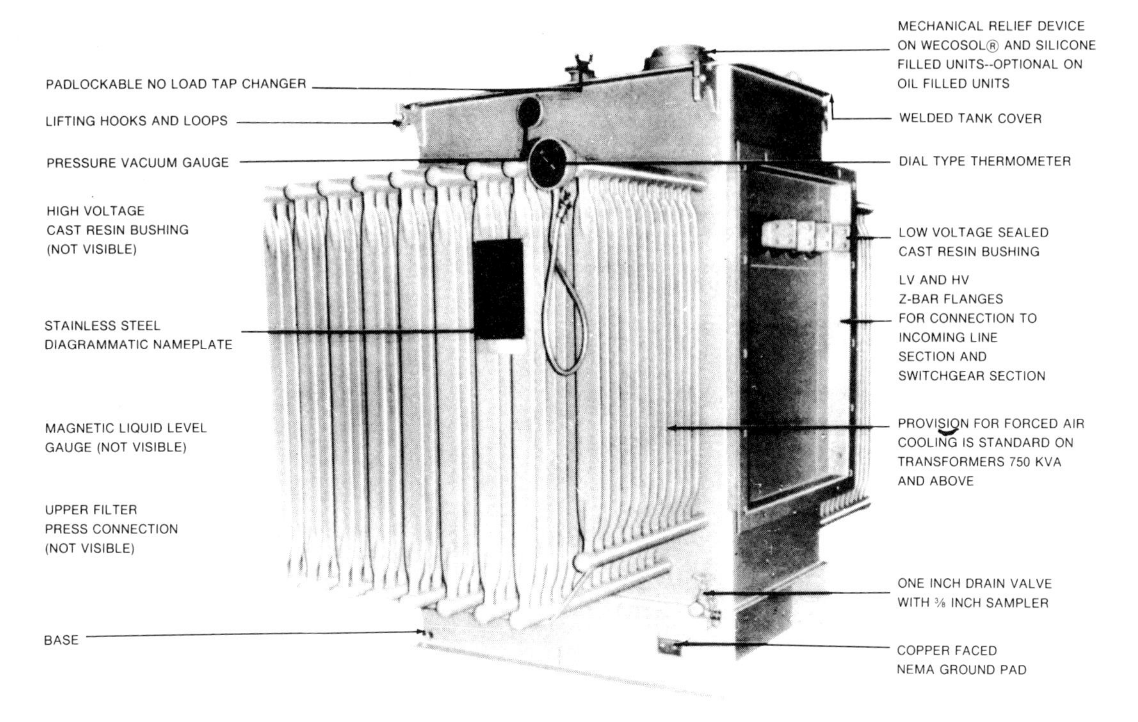

FIGURE 1–15. Standard features and accessories on a Westinghouse substation transformer. *(Courtesy Westinghouse Corp.)*

of 900 turns. Using the voltage and turns ratio formula, and inserting the given values, the following calculation proves the mathematical relationship of using the voltage per turn method.

$$\frac{V_h}{V_x} = \frac{T_h}{T_x}$$

$$\frac{360\ V}{V_x} = \frac{900\ T}{600\ T}$$

$$V_x = \frac{600\ T \times 360\ V}{900\ T}$$

$$V_x = 240\ V$$

The current drawn by the load can be found by applying Ohm's law:

$$I_x = \frac{V}{R}$$

$$20\ A = \frac{240\ V}{12}$$

With this information, the power being consumed by the load is calculated to equal

$$P_x = I_x \times V_x$$

$$4800\ W = 20\ A \times 240\ V$$

If losses and power factor are ignored, the input power must equal the output power. Given this fact, the primary current can be determined by

$$I_h = \frac{P_h}{V_h}$$

$$13.33\ A = \frac{4800\ W}{360\ V}$$

The current relationships of the circuit can now be examined. At point x in Figure 1–17, 20 amperes is flowing to the load. The high-voltage source is providing 13.33 amperes at this junction. According to Kirchhoff's law of currents, the algebraic sum of the currents at any point in a circuit must be equal to zero. Therefore, the remainder of the current must be supplied by the autotransformer. Subtracting 13.33 am-

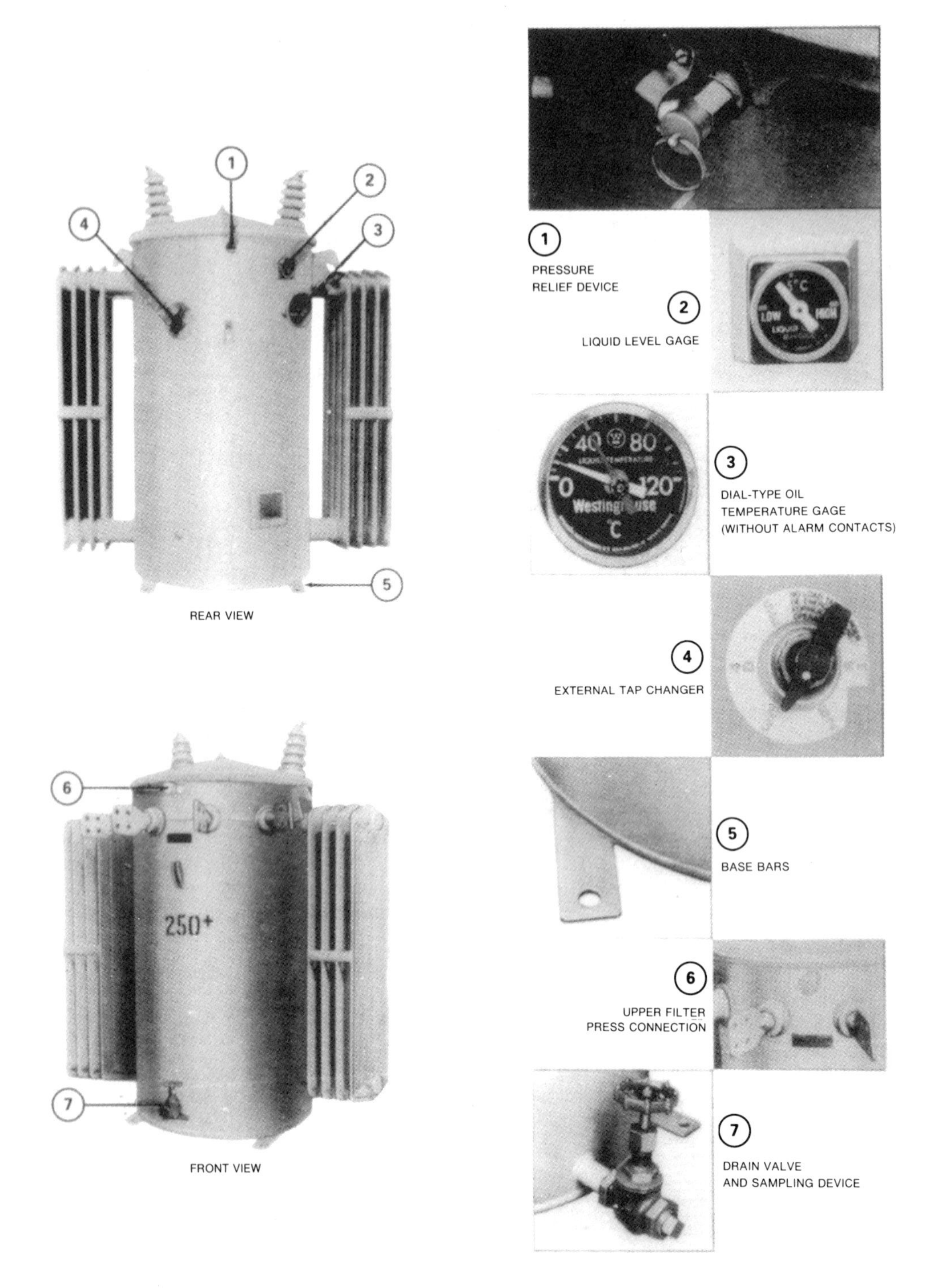

FIGURE 1–16. Standard substation accessories for a tubular type distribution transformer. *(Courtesy Westinghouse Corp.)*

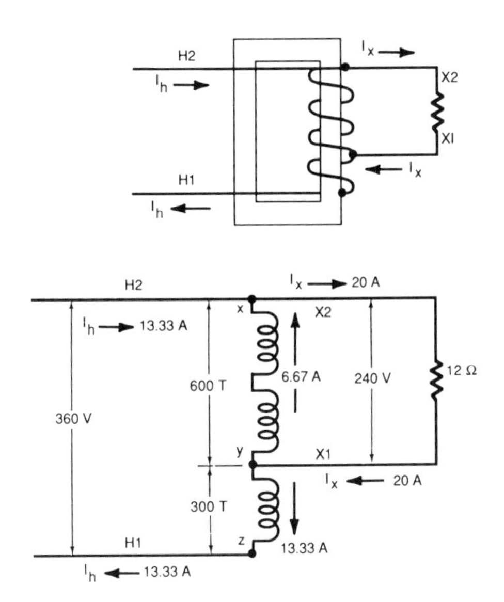

FIGURE 1–17. Step-down autotransformer

peres from 20 amperes gives 6.67 amperes from this source.

Likewise, at point y, 20 amperes is returning from the load. Of this amount, 6.67 amperes is returned to the other side of the autotransformer, and 13.33 amperes goes back to the source. This also complies with Kirchhoff's law in that the amount of current that leaves a power source must also return to that source. Current cannot be stored in a circuit. The winding that is common to both the primary and secondary circuit always carries the difference between load current and source current.

This is what makes the autotransformer more efficient than a two-winding transformer. Because less copper is required for the lower current to achieve the same purpose—to step-down the voltage—the kVA rating can be less with a resulting lower cost.

Co-ratio

The kVA rating of the autotransformer can be approximated by the "co-ratio." The co-ratio is equal to the high voltage minus the low voltage (LV) divided by the high voltage (HV):

$$\text{Co-ratio} = \frac{\text{HV - LV}}{\text{HV}}$$

Using the example in Figure 1–18, the usefulness of the co-ratio can be shown. The high-voltage transformer is rated 75 kVA at 277 volts. The autotransformer will provide the same power to the load at 120 volts.

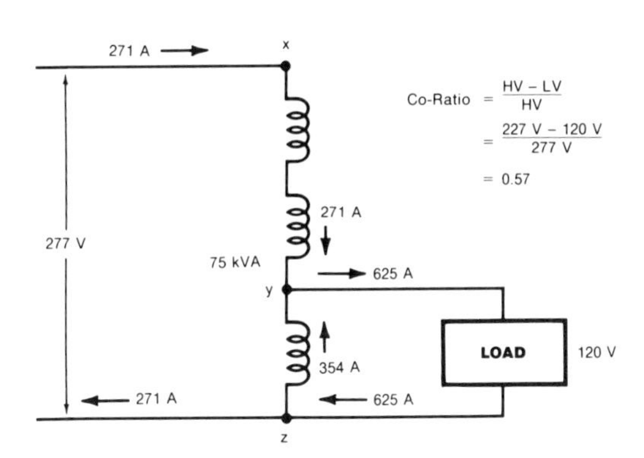

FIGURE 1–18. Proving value of the co-ratio.

The currents in the primary and secondary circuits have been calculated by dividing the 75 kVA by the voltages in each circuit. The currents have been divided at the junctions according to Kirchhoff's law, with 354 amperes flowing through the secondary winding. The kVA of the autotransformer can be calculated by

$$\text{kVA} = \frac{120 \text{ V X } 354 \text{ A}}{1000} = 42.48$$

Dividing the 42.48 kVA of the transformer by the value of source, 75 kVA, gives us 0.57. This is the same as the co-ratio shown in Figure 1–18. If an isolation type of transformer with two separate windings had been used, its value would had to have been 75 kVA, the same as the source.

Advantages

For the same job, autotransformers are superior to isolation-type transformers. They are smaller in size requiring less excitation current, and therefore lower in cost. Overall they have a greater efficiency due to the smaller currents, and they provide better regulation.

Disadvantages

The question might be asked, "Why are not more autotransformers used for these applications?" The answer to this question lies in the disadvantages associated with their use.

The primary concern when using autotransformers is the development of a ground fault. For example, in many installations, the supply voltage may be a delta system that has one leg grounded or a center tap ground. If this system was connected to a wye wired three-phase autotransformer with its neutral grounded, a short circuit would be established between the two. Because of the lower impedance of the autotransformer, the level of fault current will increase in inverse ratio of that of an insulated transformer. The inverse ratio can be determined by dividing the co-ratio into 1. The force between the coils will increase as the square of the current and could cause physical damage.

For example, if the co-ratio is 10%, the fault current using the autotransformer system would be ten times greater. The forces generated between the coils would go up 100 times. Because transformers are not built with safety factors this great, if the source voltage is sustained, the device would be destroyed.

It is mandatory then, either the transformer be built with greater reactance, or that the total system has sufficient impedance to limit the currents to a safe level. This may require additional reactance being added. In most cases, however, the impedance of the source is great enough to accomplish this without further safety features being added. The autotransformer is usually guaranteed to withstand fault currents of 25 times the full-load current. This factor alone is enough to satisfy the safety requirements for all but those few systems with very low co-ratios having very low circuit impedances.

Another disadvantage in the use of autotransformers is the direct contact between the high-voltage and low-voltage circuits. Each circuit is directly affected by the electrical conditions originating in the other. If a ground fault occurs in one, it occurs in both. A ground on the high voltage side may subject the low voltage side to the full high voltage. Although the windings of the autotransformer is designed to withstand the higher voltage, the equipment it supplies is not.

This problem is reduced by grounding the neutral of the autotransformer. A fault on the source under this condition would cause one-phase voltage to collapse, and protective devices would be used to clear the system.

If the neutral of the wye system was not grounded, and a phase of the delta was, the neutral of autotransformer would be free to float at varying potentials depending upon the balance of the load on the three phases. This would be undesirable, and when used, the neutral will always be grounded. This in turn prohibits a grounded phase on the supply side of the system. Some local electrical codes might permit the ground on the delta source and prohibit the use of autotransformers. This should be checked before installing autotransformers.

Because of the low production level and demand for autotransformers, the full economic advantage cannot be taken. The price advantage does not become significant unless the rating is greater than 75 kVA with a voltage ratio of less than 2:1.

As noted above, most problems originate from abnormal conditions or practices not generally in use, such as the grounding of a phase of the delta source. It is beyond the scope of this text to explore all of the abnormal conditions that can happen with the various configurations. When applicable, autotransformers should be considered as alternatives to the insulating, two-windings type.

Variable Autotransformers

Autotransformers are also made to be adjustable. They are rated in kVA and are available in sizes from 0.25 to 200 kVA. Voltage ratings are 120, 240, and 480 volts.

These devices are made as individual units or ganged together from 2 up to 27 units. Ganged units may be made to adjust manually, or they may be motor driven. A schematic diagram of this device is shown in Figure 1–19.

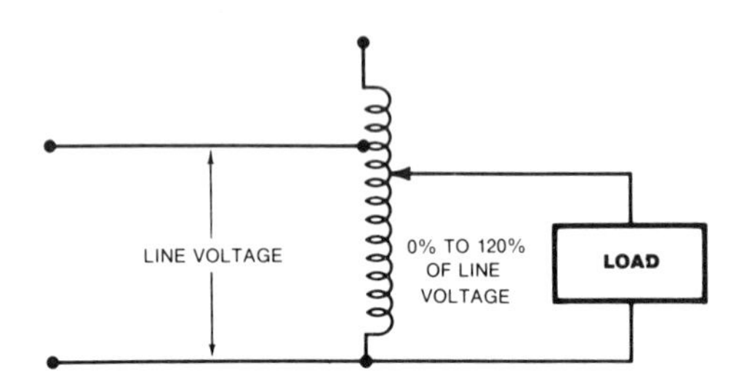

FIGURE 1–19. Variable autotransformer.

Although solid-state silicon-controlled rectifiers are replacing these transformers for such applications as light dimmers and in other high-power control circuits, many of these installations remain in operation today.

Variable autotransformers are particularly useful for test situations where precise values are desired. They are often used in conjunction with DC power supplies to adjust the AC input.

Variable autotransformers are designed to step the

voltage up as well as down. The step-up ratio is usually limited to 20%. They are also manufactured with fixed taps rather than a movable arm.

Instrument Transformers

Because of the dangers involved in measuring the high voltages and currents associated with power distribution systems, transformers are often used to step these values down to safer levels. Besides reducing the current to a level that is safer to measure, the current transformer also isolates the instruments from the main circuit. Figure 1–20 shows a pictorial diagram of two types of current transformers used for this purpose.

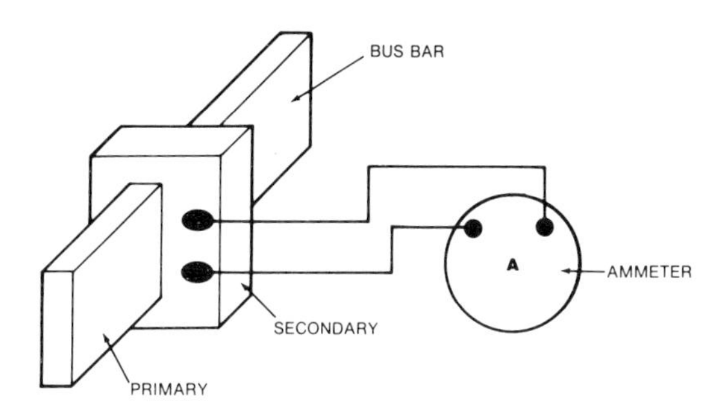

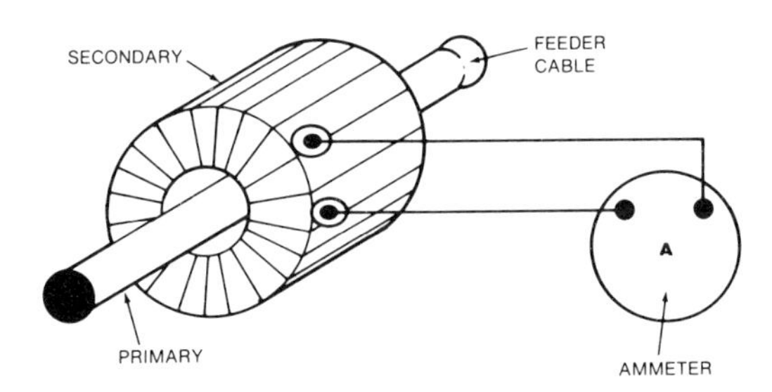

FIGURE 1–20. Current transformers.

The primary of the transformer can be either a bus bar, the feeder cable, or for lower currents, a few turns of wire in series with the line. The impedance of the primary is very low, and the currents very high. The primary current is dependent on the load on the line rather than the load on the secondary circuit. Current drawn by the secondary has little effect on line current.

The secondary of the transformer contains many turns and has a much higher impedance. If the secondary is not loaded, this transformer acts to step-up the voltage to a dangerous level, due to the high turns ratio. Because of this, a current transformer should always have a short-circuit placed across its secondary winding when connecting or removing any device from its output. By heavily loading the secondary, the high voltage is reduced to a safe level.

Current transformers can be used for control circuits as well as for instrumentation. A good example of this would be to apply these devices to detect single-phasing on a three-phase system. If one phase of the power line to a three-phase motor was to become defective, the motor would run under increasing currents and stress. A current transformer on each phase could detect the loss of current in the line and open a control relay to take the motor off the circuit.

Potential Transformers

Potential transformers are used to step-down the high voltages of a distribution system to a lower value. This secondary voltage is always 120 volts when the rated voltage is applied to the primary. Because the load on a potential transformer is very low, the volt-ampere ratings of these devices are in a range of 50 to 200.

Figure 1–21 depicts a schematic diagram using both a potential and a current transformer along with a wattmeter. According to prescribed practices, the voltmeter is connected in parallel with the voltage to be measured, and the ammeter is placed in series with the current. The wattmeter must be supplied with both voltage and current in order to operate properly.

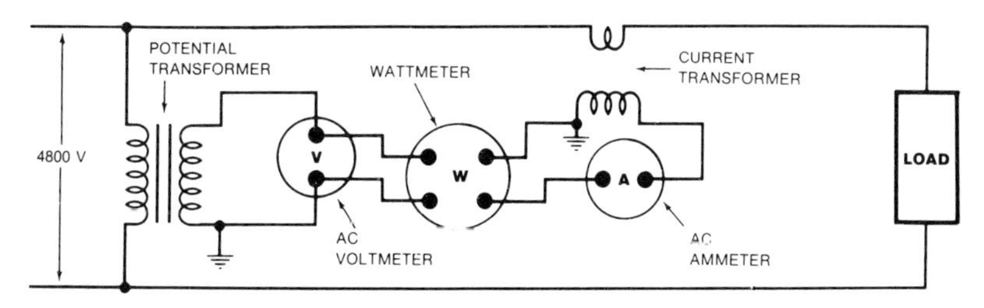

FIGURE 1–21. Transformers used for instrumentation.

Both the voltmeter and the ammeter would be scaled to read the voltage and current of the distribution system. For example, when the voltage across the line is 4800 volts, there will be 120 volts across the meter. The meter scale would indicate 4800 volts. At the same time, the current on the line may be 100 amperes. This may be stepped-down to 1 ampere. The meter scale would indicate 100 amperes when 1 ampere was flowing through it.

Questions

1. Define the term "transformer."
2. Which transmission system, AC or DC, is most prevalent in the electrical industry? Why?
3. Define the terms "primary" and "secondary" when applied to transformers.
4. Explain the terminal markings on a transformer.
5. How are transformers rated? Why?
6. If the secondary voltage is higher than the primary voltage, the transformer is a _______ transformer. Which set of leads, "H" or "X," would be connected to the source?
7. Can a transformer create electrical energy? Explain.
8. Under what conditions will a transformer transfer maximum power to the load?
9. In electronic circuits, the leads which have in-phase voltages are marked with a dot. How can in-phase leads be identified on a power transformer?
10. What are the two major classifications for energy losses for transformers?
11. How are eddy current losses reduced?
12. What is meant when a transformer is said to be saturated?
13. How can an electrician improve the efficiency of a transformer?
14. How are transformers classified?
15. Why is cooling of the transformer important?
16. Which class of insulation allows for the greatest temperature rise on the coils of a transformer?
17. How can the allowable temperature rise on a transformer be determined?
18. The impedance value given on a transformer is given in percentage. What does this mean? Impedance is usually given as an ohmic value.
19. A transformer is rated for a 50-ampere secondary and has an impedance of 2%. What is the maximum current which the secondary can deliver under a fault condition?
20. What is the purpose of the tap changer? What precautions must be taken when changing it?
21. List the types of electrical protection provided to transformers.
22. A 50-kVA transformer has a primary voltage of 4800 volts. The primary windings has 3200 turns, and the secondary has 400 turns. What is the value of the secondary voltage and its rated current?
23. How can an autotransformer be distinguished from other types of transformers?
24. An autotransformer is to be used to replace a regular transformer. How can the size of the autotransformer be approximated?
25. In general, what is the purpose of instrument transformers?

CHAPTER 2

Transformer Connections

Transformer connections and applications have over the years become more varied in their configurations, and in some cases, complicated to understand. The good news is that approximately 90% of all transformer installations are single phase, and with a little study and practice, the theory and techniques of wiring, maintaining, and troubleshooting them can be mastered with ease. It is essential that one understands the single-phase transformer before studying three-phase combinations.

No attempt has been made within the scope of this text to cover all possible three-phase arrangements and how the various voltages and currents are derived. Only the most common connections are discussed in greater detail. This approach has been taken to insure a better understanding as to how to proceed in the proper installation of most transformers with some assurance that the connections are correct before power is applied to the load.

Most electricians will never work on jobs involving some of the more special types of wiring configurations. To study them without having the need to understand them, would only clutter the mind and serve no useful purpose. However, in the event that the electrician is called on to work on one of these systems, he or she needs to be aware that other arrangements are used under certain circumstances. Further study would then be indicated.

Job blueprints and specifications along with the manufacturer's specifications and the nameplate on the transformer should be carefully studied to determine if the device meets the needs of the job. Failure to do this may prove to be very costly, and in some cases, life threatening.

The most important items to check are the voltages involved in the installation and the kVA rating of the transformer. The questions to answer are

- "Does the voltage supplied by the power company match the voltage rating of the primary windings?"
- "Will the secondary voltage match the requirements of the loads?"
- "Is the kVA rating of the transformer greater than or equal to the load to which it will supply power?"

A complete discussion of both of these subjects will be provided later in this chapter.

Grounding connections, an important aspect of transformer connection, are discussed in Chapter 3.

Single-Phase Transformers

A transformer consisting of one primary winding and one secondary winding is the simplest type to install. The high-voltage leads are labeled "H" and the low-voltage leads are marked "X." H1 is found to the left when facing the low-voltage terminals, and H2 is found to the right. X1 is found on the right, and X2 is on the left (Figure 2–1).

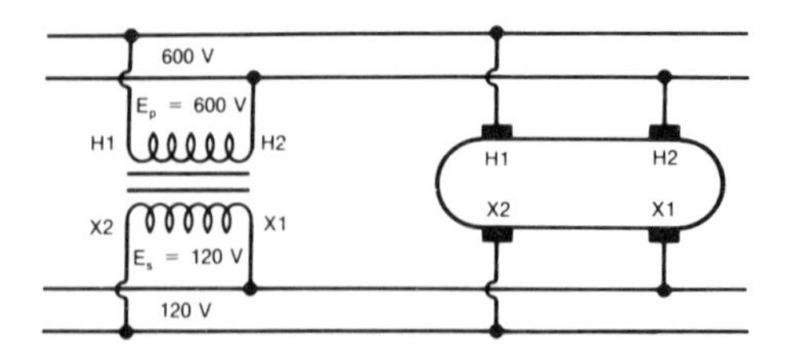

FIGURE 2–1. Connections for a single-phase transformer.

It should be noted that the single-phase supply voltage of 600 volts matches the voltage requirement of the primary winding, and the transformer's output voltage is equal to the voltage required by the load. A pictorial is shown along with the schematic diagram to help the reader visualize this arrangement.

In most cases the transformer will have dual secondary windings. Figure 2–2 provides a schematic and a pictorial diagram of this configuration. With this arrangement, the transformer can deliver 120 volts only, 240 volts only, or both, 240/120 volts to the load. This is the typical type of transformer used to supply power to a residential customer or small business. The most common connection is for 240/120 volts.

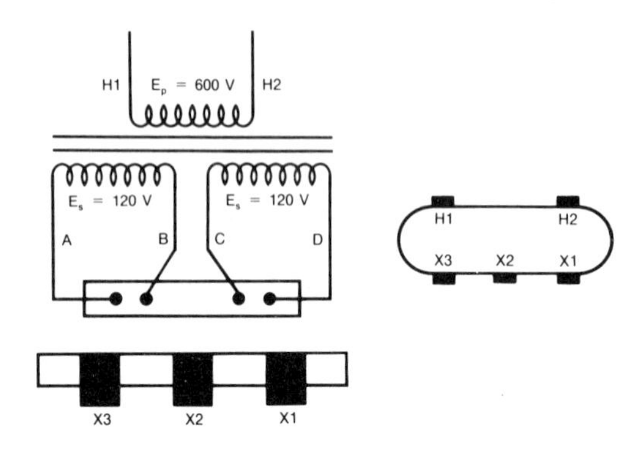

FIGURE 2–2. Typical dual-winding secondary.

120-Volt Load

Figure 2–3 depicts the schematic diagram for wiring the transformer for a single low voltage of 120 volts. Notice that the load requirement is 120 volts only. To accomplish this task, the secondary windings, rated at 120 volts are wired in parallel. On the terminal board, "A" is connected to "C" and "B" is tied to "D." The "A–C" combination is brought to the X2 terminal, and the "B–D" tie is connected to X1. The power lines

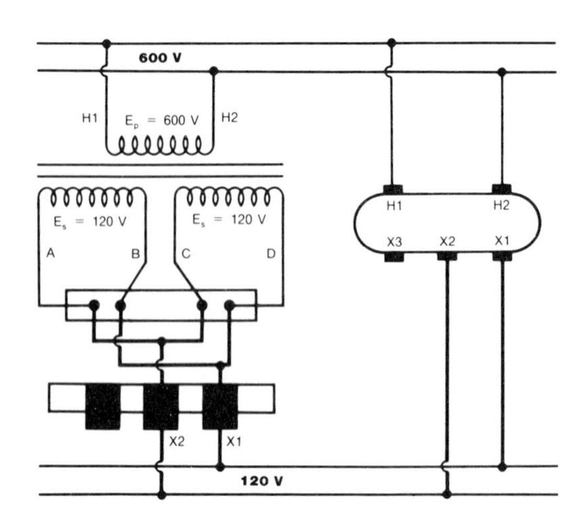

FIGURE 2–3. Typical dual-winding secondary connected for low-voltage output.

going to the load are connected to X1 and X2. H1 and H2 are connected to the high-voltage source supply.

By using both secondary windings in parallel, the load has twice as much current available. Each winding would deliver one-half the kVA used. Because the secondary windings are part of one transformer and have equal ratings, no further considerations need be given when making this type of installation.

240-Volt Loads

Figure 2–4 shows a load requirement of 240 volts. In order to accomplish this, the secondary windings are connected in series with each other. On the terminal board, B and C are wired together. The tie point is connected to X2. A is connected to X3, and D to X1.

Under these conditions, the voltage on the windings of the secondaries will add together to provide the 240 volts. The output current will be one half that of the parallel arrangement in Figure 2–3.

In both cases, the primary voltage is the same, and the primary current will be the same. The maximum kVA available to the loads is equal in either circumstance. The kVA of the primary must equal the kVA of the secondary minus transformer losses.

For example, if the transformer is rated 6 kVA, the primary winding will draw 10 amperes under full load.

$$6 \text{ kVA} = 600 \text{ V} \times 10 \text{ A}$$

The secondary windings at 120 volts can deliver 50 amperes. Each winding would provide 25 amperes.

$$6 \text{ kVA} = 120 \text{ V} \times 50 \text{ A}$$

or

$$6 \text{ kVA} = 2(120 \text{ V} \times 25 \text{ A})$$

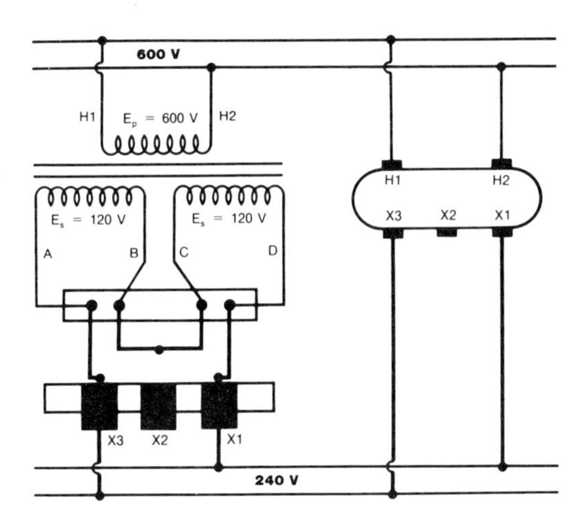

FIGURE 2–4. Typical dual-winding secondary connected for high-voltage output.

When 240 volts are delivered to the load by wiring the secondary windings in series, the transformer can deliver 25 amperes of utilization current.

$$6 \text{ kVA} = 240 \text{ V} \times 25 \text{ A}$$

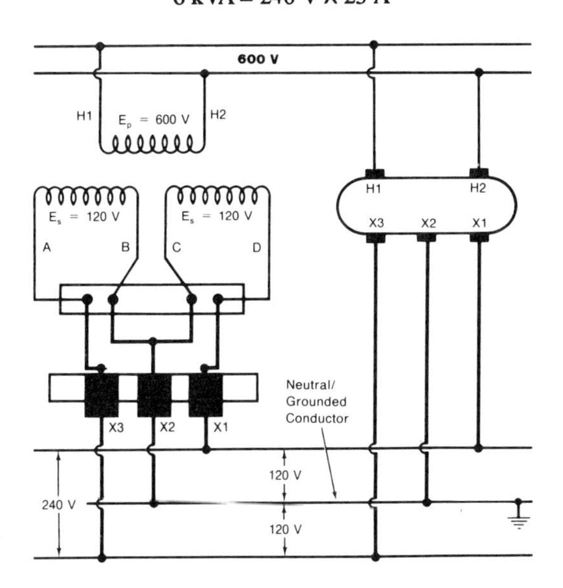

FIGURE 2–5. Typical dual-winding secondary connected for dual-voltage output.

In either case, the current rating of the secondary windings is 25 amperes, and the current rating of the primary winding is 10 amperes.

240 Volt/120 Volt System

Figure 2–5 shows the typical wiring system for the single-phase transformer with dual 120-volt secondaries. By adding a third wire at the tie between B and C, and terminating it at X2, the transformer can provide a voltage of 240/120 volts to the load.

Terminals A and D are wired to X3 and X1, respectively. The voltage between X1 and X2 and X3 and X2 is 120 volts. X2 is the grounded neutral in this system. The voltage between X1 and X3 is 240 volts.

In earlier days when power requirements of homes and small commercial plants were limited, only two-wire, 120-volt systems were used. With increased power demand, the three-wire, 240/120 volt system was introduced. The addition of the third wire provides a 100% increase in capacity with only a 50% increase in the cost of the conductors.

The loads on the 120-volt circuits should be as balanced as possible. Referring to our example before, each winding was capable of delivering 25 amperes to the load. This requirement remains—one secondary winding cannot deliver 35 amperes while the other provides 15 amperes, even though the total kVA requirement is still the same.

Dual Primaries

Transformers may have dual primaries as well as dual secondaries. Figure 2–6 depicts the wiring arrangement for a low-voltage connection, and Figure 2–7 shows the high-voltage system.

In each of the cases above, the rating of the primary windings is 600 volts. When the source voltage is 600 volts, the windings are wired in parallel to each other. When connected to the line, each winding will have 600 volts across it, or its required value. When the source voltage is 1200 volts, the windings are connected in series. Because the windings have equal impedances, and the current in a series circuit is the same, the voltage drop across each of the windings will be 600 volts, or the required voltage rating.

Summary

Keep in mind that the connection diagram on the transformer nameplate will provide the necessary wiring information to make the proper series or parallel connections. One must still make certain that the

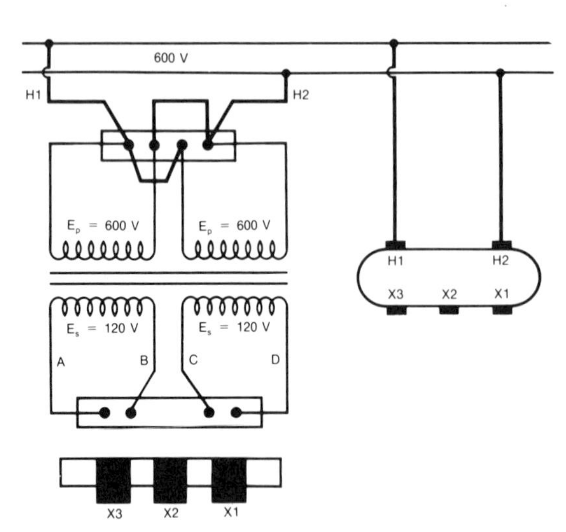

FIGURE 2–6. Dual-winding primary connected for low-voltage input.

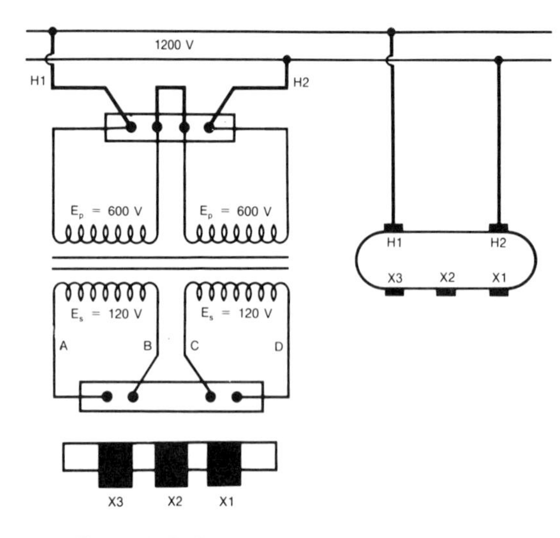

FIGURE 2–7. Dual-winding primary connected for high-voltage input.

source voltage will match the voltage rating of the windings, and the secondary voltages will match the requirements of the user.

Some transformers may have only one high-voltage bushing, H1. In these cases, the other side of the winding is connected to the tank of the transformer. The second line supplying power to the transformer must be a grounded conductor. This wire is connected to the grounding lug on the transformer's case and taken to earth ground.

Transformer Polarity

Transformers do not have a fixed polarity like that of a DC battery. They do, however, have relative polarities that must be considered when connecting them together. Because the voltages across the windings are constantly changing, we normally describe these voltages in terms of vectors.

Vectors provide a graphic representation of each voltage in terms of its amplitude or size, and the angular displacement, the phase relationship, between the voltages. In the case of a single-phase transformer, these voltage relationships are either in phase, or 180 degrees out of phase. Therefore, when the windings are wired together, the output voltage can be either additive or subtractive.

Battery Polarity

Figure 2–8 depicts these relationships using DC batteries. When batteries are connected in series with opposite polarities wired together, their voltages add. This is represented by vectors with the arrowhead of one vector attached to the tail of the second. A common device using batteries in this manner is a flashlight having two 1½-volt batteries connected positive to negative. If the polarity on one of the batteries is reversed, their voltages will cancel each other, and the flashlight will not function. This arrangement is represented by the vectors with the arrowheads opposing each other.

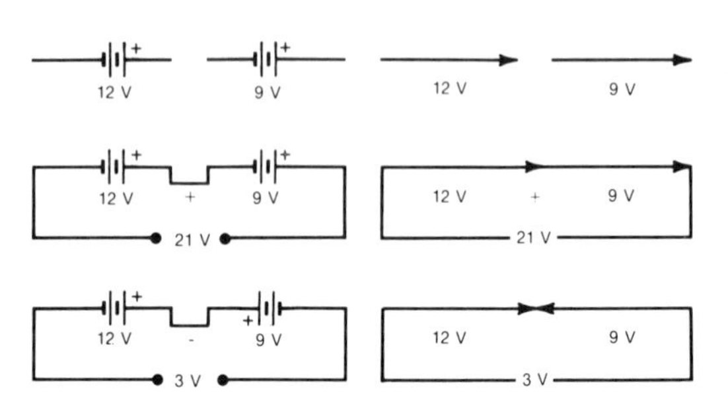

FIGURE 2–8. Additive and subtractive polarities using batteries.

Identifying Transformer Polarity Using Bushings Numbers

Figure 2–9 illustrates a similar situation using transformers. Although the voltages of the primary and secondary are constantly changing, their relative phase angle to each other is constant. The vector is a stop-

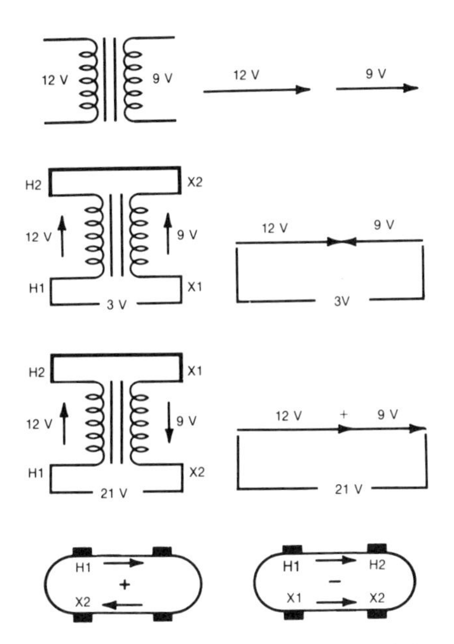

FIGURE 2–9. Additive and subtractive polarity connections for transformers.

motion concept that allows us to visualize the relationships. This is difficult to do if one thinks in terms of the constantly changing voltage sine waves in the primary and secondary.

The standard practice for wiring a single-phase transformer for additive polarity, or high-voltage output, is to connect the high-voltage bushing H2 to the low-voltage bushing opposite it, X1. When facing the low-voltage side of the transformer, H1 will be on your left, and H2 will be on the right. The output voltage is taken from H1 to X2 terminals. The primary voltage of 12 volts and the secondary voltage of 9 volts are in phase with each other as represented by the vectors. The output voltage will be 21 volts.

Transformers are also manufactured with subtractive polarities. This is accomplished by having the secondary winding wound in the reverse direction of the primary. With this type of arrangement, the low-voltage bushing X1 is opposite H1. The position of the bushings is shown in the pictorial in Figure 2–9.

Regardless of how the secondary is wound, the current flow in the secondary will always be in such a direction as to create a magnetic field that will oppose the primary field. This concept is known as Lenz's law. If the opposite was true, the transformer would become a perpetual motion device and create its own energy.

Standard practice for wiring the single-phase transformer for low-voltage output is to connect H2 to the bushing opposite it. In this case, X2 would be connected to H2. Because the primary and secondary voltages are opposed to each other, 180 degrees out of phase, the low voltage will subtract from the high voltage. A 3-volt output would be taken from H1–X1.

A joint committee of NEMA and The Edison Electric Institute recommended

1. Transformers ≦ 200 kVA with voltage ratings ≦ 9 kilovolts will be additive.
2. Transformers manufactured above 200 kVA will have subtractive polarities.
3. Transformers with voltage ratings above 9 kilovolts, regardless of the kVA rating, will have subtractive polarities.

Determining Unknown Polarity

There may be cases with older transformers, some manufacturers not adhering to this recommendation, or where the markings for the bushings no longer exist, that some doubt may exist whether a transformer has additive or subtractive polarity. Figure 2–10 provides two test conditions for determining the polarity of the device.

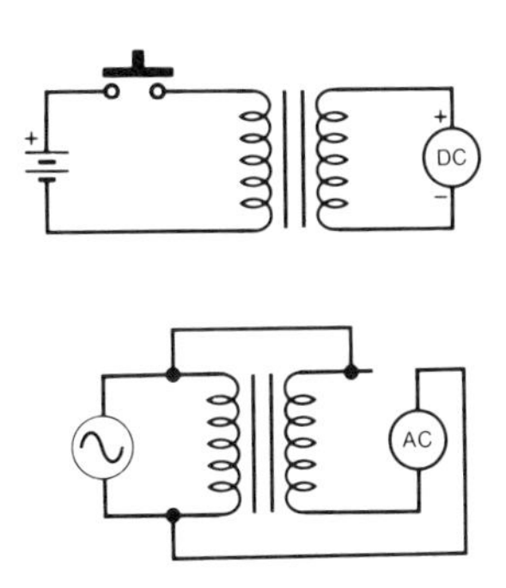

FIGURE 2–10. Determining the polarity of a transformer.

The first step in determining the correct markings for the bushings on a transformer is to designate the high-voltage bushing on the left-hand side as "H1." This is done while facing the device from the low-voltage side.

A small DC source is then applied to the primary windings of the transformer. A DC voltmeter is then connected across the secondary windings. The polarity of the voltmeter should match the polarity of the battery on the high-voltage bushings, and the meter should be set on its highest voltage range.

Next, break the battery's connection while observing the meter. If the deflection is up-scale, the transformer is additive, and the low-voltage bushing with the positive lead of the meter is X1. A deflection downscale would indicate that the transformer is subtractive, and the low-voltage bushing would be designated X2, with X1 opposite of H1. In the event that no deflection is observed, the meter multiplier should be lowered to the next lowest range, and the procedure should be repeated. Continue the process until a deflection occurs on the meter.

If no deflection occurs on the meter even at its lowest range, the windings of the transformer should be checked using an ohmmeter to determine if one of the windings is open. If both windings are found to have continuity, the DC voltage should be increased on the primary and the test repeated. Under no circumstance should a high DC voltage be applied. Current is limited only by the resistance of the windings, and damage to the transformer may occur.

A second method of testing for additive or subtractive polarity is to apply AC to the primary. This voltage need not be the same as the rated primary voltage of the transformer, and a much lower voltage can be used for safety purposes. For example, the primary may be rated for 600 volts, 1200 volts, 2400 volts or higher, and 120 V AC can be used for this test. The secondary voltage will be reduced according to the turns ratio. For example, if the transformer voltage rating is 1200/120 volts, 120 volts applied to the primary would produce 12 volts on the secondary. These values will be used to describe this method.

Before voltage is applied to the primary, connect H2 to the low-voltage bushing opposite it. An AC voltmeter is attached between H1 and the other low voltage bushing. AC voltage is then applied.

If the meter indicates 132 volts, 120 volts plus 12 volts, the transformer is additive. Otherwise, the meter would read 108 volts, 120 volts minus 12 volts, and the transformer would be subtractive.

More often than not, the transformers that need to be tested will have multiple windings. The test procedures would still be the same, but the leads to individual windings would need to be identified. These pairs can be found by using an ohmmeter. The polarity of each pair can then be identified by using the designated H1 and its corresponding lead to apply the test voltage.

Wiring Transformers in Parallel

Although a single transformer is more efficient than two or more transformers connected in parallel to supply a load, there are times that parallel connections may be necessary. A larger transformer may not be immediately available, or the higher kVA is needed only on a temporary basis.

When connecting transformers in parallel, the polarity of the transformers must be considered. Windings with the same polarity must be connected to the same line. Regardless if the transformers being connected are additive or subtractive, polarity will be observed if all "H" and "X" terminals with the same numbers are connected to the same line. If the transformers are not marked, the procedures for determining polarity should be followed (Figure 2–11).

Voltage

Transformers being connected in parallel should also have the same voltage ratings and the same ratio of transformation. If the low voltages are different, the winding with the higher voltage will constantly circulate current to the lower voltage winding causing a reduction in efficiency. Under these circumstances, transformers should not be wired together.

Care should be taken to insure that wire size is the same for the installation, and precautions taken to reduce to minimum any resistance in the joints of the interconnections. An unbalance may be created, and one transformer may carry more than its share of the load.

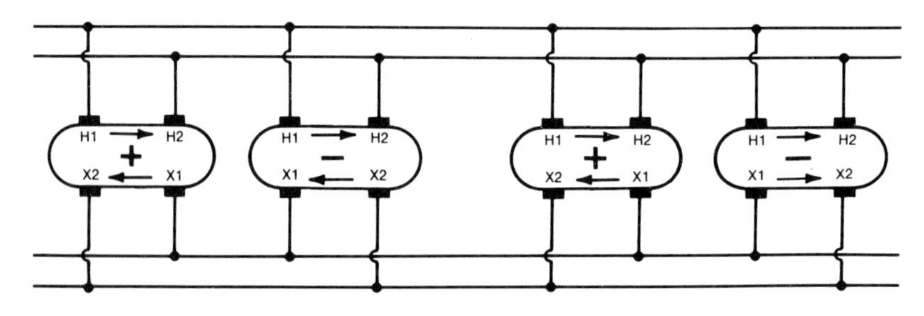

FIGURE 2–11. Matching polarities when connecting transformers in parallel.

Impedance

Impedance of the transformers should be nearly the same. Figure 1–10, block 16, showing the name plate nomenclature, provides an example of this value.

Impedance is given in terms of percent of regulation from no-load to full-load. The maximum difference allowable is approximately 7.5%. This translates into an exception of only a few one-hundredths of a percent range. When this value is not observed, the transformer with the lowest impedance will carry a disproportional amount of the current and may become overloaded. Characteristic impedance does not usually vary greatly among major manufacturers, however, and the losses encountered are often offset by the practicality of the problem to provide added power to the load.

Limitations

It is not considered good practice to operate transformers in parallel under the following conditions:

1. When the total load is equal to the combined kVA of the transformers, the total current supplied by either transformer shall not exceed 110% of its rating.
2. The no-load circulating current in either transformer shall not exceed 10% of the full-load rated current.
3. The arithmetical sum of the circulating and load currents shall not be greater than 110% of the full-load current.

Circulating current is the current flowing at no-load in the primary and secondary windings minus the excitation current. The load current is equal to the amount of current going to the load. This would be the total current minus the circulating and excitation currents.

Parallel Wiring Transformers with Different Impedances

Under certain circumstances, a transformer can be wired so that an equal division of load can be accomplished. Figure 2–12 provides one example of how this can be done with a transformer having dual primary and secondary windings.

The parameters of the circuit are a source voltage of 4800 volts with a load requirement of 240 volts. The primary windings are rated at 2400 volts and the secondaries at 120 volts.

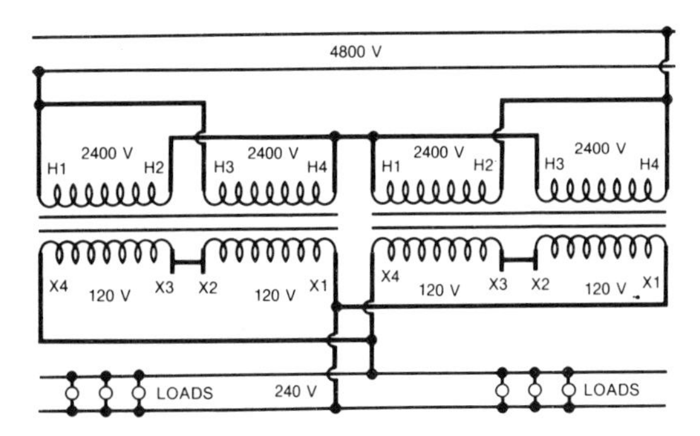

FIGURE 2–12. Wiring diagram for parallel operation of two transformers with different impedances.

In order to drop 2400 volts across each primary winding with 4800 volts impressed by the source, each pair of windings will need to be wired in series. In this case, one primary winding of each transformer is wired in series with one primary winding of the second transformer. H2 of T1 is connected to H3 of T2, and H4 of T1 is connected to H1 of T2 to provide the series connections. This provides two sets of windings with equal impedances which are then connected in parallel with each other and wired to the source.

To obtain the 240 volts required by the load, X2 and X3 of each individual transformer is connected to put the secondary windings of each in series to provide the needed voltage. These two sets of windings are then connected in parallel and attached to the feeder to supply the load.

Because the primary windings now have equal currents flowing through them, the currents in the secondary windings will also be equal. The kVA of the primaries and secondaries must be equal. The two transformers in parallel now look like a single transformer with a different impedance than either of the individual transformers.

The voltage drop across each of the series primary windings will be slightly different due to the differences of the impedances in series. Therefore, the induced voltages across each of the secondary windings will be unequal. However, the voltages across the two sets of secondary windings will be equal, 240 volts, because equal values are added in each series combination.

If the load requirement in the above example had been 120 volts instead of 240 volts, the paralleling of all of the secondary windings would have set up a circulating current. The 10% rules given previously would need to be invoked to determine if the transformers could be used in this manner.

A similar problem would have occurred if the source voltage had been 2400 volts. All four primary windings would have been connected in parallel instead of the series-parallel combination. In this case, it might have been possible to connect an inductor in series with the primaries having the lowest impedance to make their total impedance equal to the higher value.

There are many different conditions that may exist when connecting transformers in parallel. Each set of circumstances must be examined independently in order to determine the feasibility of meeting load requirements with this arrangement.

Three-Phase Transformers

In the preceding section, our discussion of transformers was limited to single-phase systems. Transformer voltages were either in phase, or 180 degrees out of phase. Voltages, therefore, either added or subtracted arithmetically from each other.

Although single-phase systems are the most frequently used by most utilization equipment, they are not economical for large power distribution or for supplying large loads. Three-phase systems are the most popular in this country for these purposes. They require approximately 25% less copper than single-phase systems to deliver the same amount of power to a load. In addition, three-phase transformers are smaller and simpler to manufacture, and they have better operating characteristics than the single-phase types. Motors operated on three-phase have approximately 50% greater rating than a comparable motor wired for single-phase operation.

Three-Phase Power Generation

Three-phase voltages are generated by rotating three coils of conductors counterclockwise in a magnetic field. The start of each coil is designated with a "1," and the end with a "2." Each coil is 120 electrical degrees apart. Figure 2–13 is a pictorial of the generator.

The ends of the coils can be brought out of the generator through slip rings and brushes. On large generators, the field magnet is rotated, and the armature is stationary, thereby eliminating the brushes and slip rings for the high-power circuits.

Each coil will provide one single-phase voltage source. Usually, the coils are interconnected, and three-phase power is taken from the generator rather than three, single-phase voltages. The waveshape of the voltages is a sine wave whose value is zero at zero degrees and has a maximum value of 100% (1 X E_{max}) at 90 degrees.

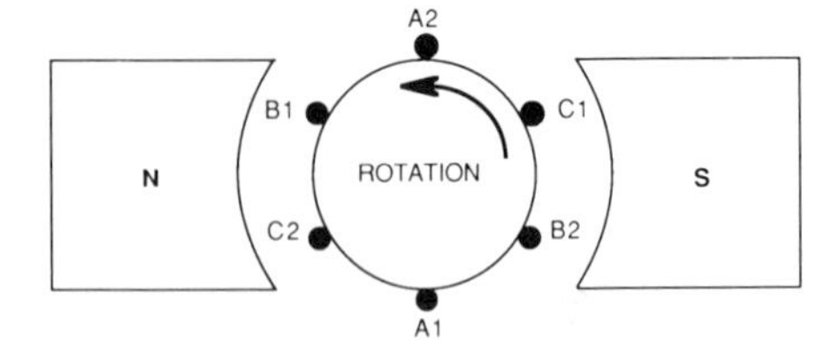

FIGURE 2–13. Three-phase power generation.

The sine waves representing the three voltages are shown in Figure 2–14. The amplitudes of the voltages correspond to the position of the coils as shown in Figure 2–13.

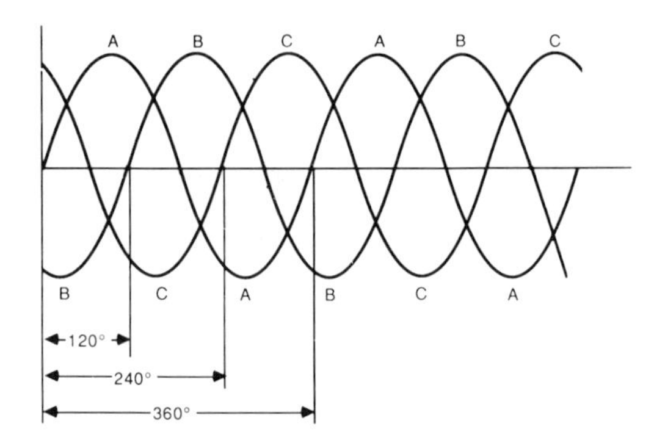

FIGURE 2–14. Three-phase voltage waves.

As represented, the start of coil "A" is at zero degrees. Because it would be moving parallel to the magnetic flux lines at this point, no voltage would be induced into it.

Coil B is moving down through the field, and Coil C is moving up through the field. Since both coils are 120 degrees from the start point, they are cutting the magnetic field at 60 degrees, and 86.7% of the voltage is being induced into them. A trigonometric table will give the values for any degree based on the laws of the sine wave.

Vectors

Figure 2–15 provides a vectorial representation of the voltages across the three coils. Their start points are at the center, and the end points are at the arrow heads.

At any degree on the rotation, their algebraic sum would be equal to zero. For example, at zero degrees, the voltage across A is zero, the voltage across B is –0.867 X E_{max}, and the voltage across C is +0.867 X

E_{max}. Figure 2–14 shows the polarities and amplitudes of the three phases.

Figure 2–15 also shows the vector resolution for determining the voltage between any two phases. In this case, if the maximum voltage for each phase was 120 volts, and they are 120 degrees apart, connecting the vectors head to tail will give a resultant of 208 volts. Because the values are constant in this case, the mathematical multiplier can be determined by dividing 208 by 120. This gives the value of 1.73, or the square root of 3. To reinforce this concept, 480/277 volt systems are also common. Dividing 480 by 277 gives the same result, 1.73. These values should be familiar to any working electrician.

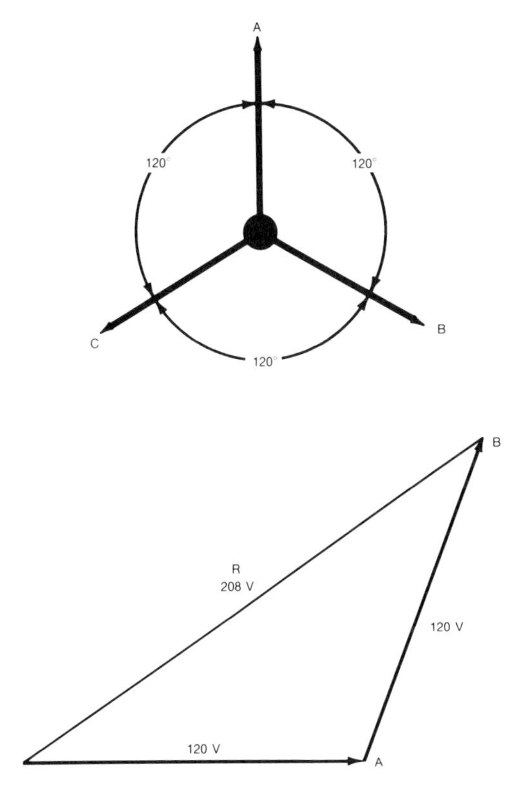

FIGURE 2–15. Vectorial representation of three-phase sine waves.

The phase sequence is the order in which each phase becomes maximum. As coil A rotates counterclockwise though the magnetic field from zero degrees, its value rises to maximum when it is opposite the south pole of the magnet, or at 90 degrees (see Figures 2–13 and 2–14). Coil A is followed by coil B and then by coil C. Maximum negative voltage is induced in each coil as they pass through 270 degrees.

Transformers are used to change the power generated at the power plant to values used for transmission and again to the voltages utilized by the customers. The most frequent type of wiring configurations encountered with these transformers are the wye and delta connections. These will be discussed in detail later in this chapter.

Three-Phase Transformer Construction

A three-phase transformer will usually be internally prewired. The nameplate will provide the information if it is wye or delta connected.

If wye connected, the end of each coil will be internally connected and brought to the ground lug of the case. The other side of each coil will be brought to each high tension terminal in proper phase and marked H1, H2, and H3, respectively. Figure 2–16 depicts this situation.

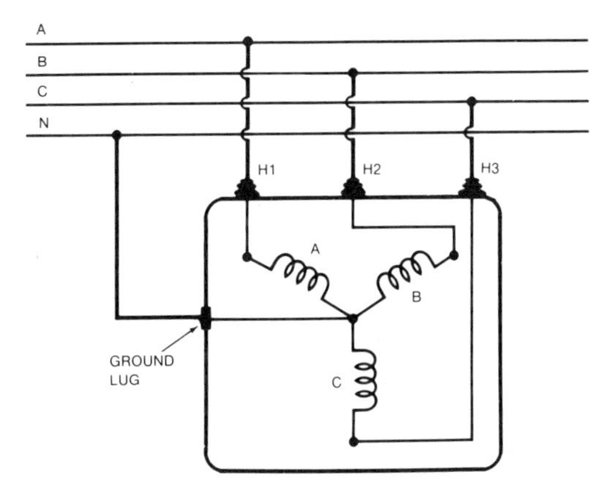

FIGURE 2–16. Wye wired three-phase transformer.

Advantages of Three-Phase Transformers

Three-phase transformers are easier to install due to being prewired and ready for service. They have lower cost than three single-phase units, and their efficiency is higher. These devices require less core material to provide the same kVA when individual units are used. Therefore, they are lighter and smaller. This in turn requires less space for installation and reduces the costs of transportation.

Disadvantages of Three-Phase Transformers

The primary disadvantage of three-phase transformers is in the case of fault. Loss of one phase results in the entire unit being shut down. Although provisions could be made in the design to remove the bad section and allow the other two sections to operate when three single-phase transformers are used, this is not possible with a three-phase transformer. The reason for this is that all three units share a common core. The core of the defective unit would quickly saturate magnetically due to the lack of an opposing magnetic field. Flux lines would escape the core and begin passing though the metal enclosure causing severe heating, the temperature generated in some cases being enough to blister the paint and cause fires. Because of this, when one phase is out, the entire unit must be shut down.

Cost of repair is greater for three-phase transformers, and spare units cost more than one single-phase transformer which could be used in that system to restore service. Three-phase units also have reduced capacity when they are self-cooled.

Even with these disadvantages, three-phase transformers are becoming more popular. With improved designs these transformers are becoming more reliable, and the frequency of failure is becoming more rare.

Wye Connection

Wye connected transformers, sometimes referred to as "star" systems, are normally used when the source voltage comes from a wye connected secondary. The wiring system for wye connections is shown in Figure 2–17.

Four separate representations are presented so that the reader can better understand the connections. All four are used at times by various writers to illustrate their works. This writer prefers, for simplicity, to think in terms of the letter "Y." When making this installation, if the electrician draws a simple "Y," labels each leg, and uses this representation to check the work, a mistake will seldom be made.

Three single-phase transformers are used in this illustration. When making the connections, the first step is to wire all H2 terminals together. The H2 terminals are then connected to the neutral.

A neutral is required for wye connections to carry any unbalance in load between the three phases of the transformer. If the unbalanced currents were forced back through the windings, the voltages across the loads would change. Wye connected transformers require four wires, unless the transformer is feeding a balanced load, such as a three-phase motor.

The rest of the installation is simple. The H1 high-tension terminal of each single-phase transformer is attached to separate phases of the source. Use the "Y" diagram to check the accuracy of the work.

Currents

For the wye connected transformer, the winding is in series with the line, and there is only one current path. Therefore, the current carried by the winding is equal to the line current of the phase to which it is attached. The current that enters a junction, must leave the junction, as expressed by Kirchhoff's law of currents:

$$I_{winding} = I_{phase}$$

Voltages

Voltage relationships for the wye connected transformer are different. Two windings, attached at a common point, are connected to two separate phases of the power line. The voltage drops across the two windings must be algebraically equal to the voltage impressed across them as expressed by Kirchhoff's law of voltage drops around a closed loop.

The voltages across the windings, however, are 120 degrees out of phase with each other. Their algebraic sums must be added together at this angle to equal the voltage between phases.

For example, if the phase-to-phase voltage of the line is equal to 4160 volts, what must the voltage rating of the windings of the transformer be in order to obtain proper transformation using the wye system? The

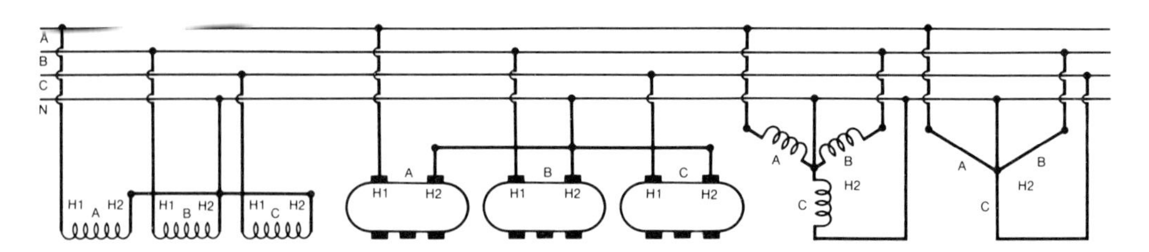

FIGURE 2–17. Three single-phase transformers wired for three-phase wye.

correct answer would be 2400 volts. The calculations below show how this value was determined.

$$V_{winding} = \frac{V_{a-b}}{1.73}$$

$$2400\text{ V} = \frac{4160\text{ V}}{1.73}$$

or

$$2400\text{ V} = 4160\text{ V X } 0.58$$

The multiplier, 0.58, is obtained by dividing 1 by 1.73. If you are good with numbers, it is a handy value to know, but not necessary. If you will remember the value of the square root of 3, 1.73, and know how it is used, that is all that is required.

To illustrate this point further, suppose you have a three-phase, wye connected transformer, with primary windings rated at 20 kilovolts. What would be the correct voltage between phases of the source voltage? If your answer is 34.6 kilovolts, your calculations are correct.

$$V_{a-b} = V_{winding} \text{ X } 1.73$$

$$34.6\text{ kV} = 20\text{ kV X } 1.73$$

When to Use Wye Connection

Another indicator in determining when a transformer must be wired with the wye configuration is the numbers involved. If the voltage rating of the windings is not equal to the phase-to-phase voltage, or equal when the voltage rating is added for dual windings, a wye connection would be necessary. You would still need to use the multiplier of 1.73 times the winding voltage to determine if the match is compatible. Remember 1.73, it is important on your job. Wye connections are normally used when the voltage is 100 kV or more.

Delta Connection

The other common method of wiring three-phase transformers is the delta system. Figure 2–18 illustrates the connections for this configuration. Once again, four separate representations are provided to assist the reader in understanding the proper wiring methods.

To make a delta installation using three single-phase transformers, H2 of Coil A is tied to H1 of Coil B, H2 of Coil B is connected to H1 of Coil C, and H2 of Coil C is tied back to H1 of Coil A. An examination of the above wiring instructions in Figure 2–18 shows that in effect, the windings have been connected in series with each other. These connections should be made first. The line connections are now simple. H1 of each coil is wired to its respective phase. Delta connected transformers require only three wires.

Voltage

This arrangement places each coil across the source voltage. Therefore, the voltage across the winding will equal the line voltage. This is expressed mathematically as

$$V_{winding} = V_{line}$$

In the case of dual windings, the source voltage may be twice the winding voltage. In this case, the windings would first be hooked in series with each other before making the delta connection. When dual windings exist, and the line voltage equals the winding voltage, then the windings would be wired in parallel.

Current

The line current is split between the two phases it feeds. These two currents flowing through the windings are 120 degrees out of phase with each other. Their algebraic sums must equal the line current, because Kirchhoff's law of current states that the algebraic sum of the current at any junction must be equal to zero. The relationship between the line current and the phase currents can be expressed mathematically as

$$I_{line} = I_{winding} \text{ X } 1.73$$

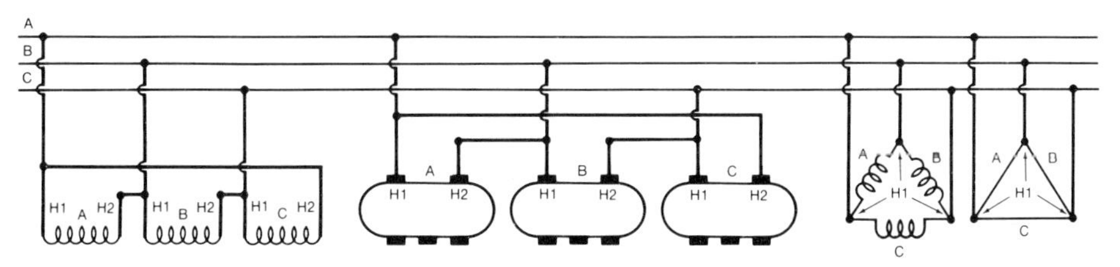

FIGURE 2–18. Three single-phase transformers wired for three-phase delta.

If the line current is known, then the winding currents can be determined by dividing line current by 1.73 or multiplying by 0.58.

$$I_{winding} = \frac{I_{line}}{1.73}$$

or

$$= I_{line} \times 0.58$$

When to Use Delta Connection

Usually, delta primary connections are made when the source voltage is delta supplied, or only a single voltage is available, and the distribution system requires only a short distance. Once again, the numbers attached to the voltages are the key to help determine this. If the source voltage is equal to the voltage specified for the primary winding, then a delta connection is indicated. In the case of dual windings, the source voltage can either be equal to or twice the value of the winding voltage.

Ground

When the delta system is operated without a ground, an artificial ground tends to exist vectorially in the center of the delta. Potential between each phase and the artificial ground under balanced conditions will equal the line voltage divided by the square root of 3. When the phases are unbalanced, the artificial ground will float and shift with corresponding differences in the currents in each phase. This voltage cannot be measured because any connection between a phase and ground effectively brings that phase to ground potential as a reference. A voltmeter placed between any phase and ground, therefore, will read zero volts.

A single ground fault can occur with this system without any noticeable effect on the operation. The second fault is a short circuit. For this reason, some jurisdictions require grounds and/or ground fault detectors with ungrounded delta configuration.

Summary

A quick determination on how to wire a three-phase transformer can be accomplished by comparing the voltages involved. If the source voltage is equal to the voltage required by the primary windings of the transformer(s), delta connections are required. If the source voltage is greater than the voltage required by the primary windings (1.73 E_p), then the primary will be wired wye.

When the secondary voltage of the transformer is equal to the required utilization voltage, then the transformer's secondary would be connected delta. If the utilization equipment requires a higher voltage, 1.73 times as great, as the secondary voltage, the secondary would be wired wye.

Common Transformer Connections

In this section the most common connections for three-phase systems will be explored. These include

Primary	Secondary
Delta	Delta
Delta	Wye (star)
Wye	Wye
Wye	Delta

Each of these arrangements has advantages and disadvantages. Many of these have already been discussed in the previous two sections of this chapter, where only the primary circuits were investigated. With the addition of the secondary windings, further information about the systems is needed.

Figure 2–19 shows the wiring for a delta-delta, three-phase transformer system. Both the primary and secondary windings are connected in series. H1 of each coil is then wired to its respective phase on the feeder, and X1 of each secondary coil is likewise connected to the load circuit.

The delta diagrams opposite the schematic representations illustrate the voltage phase relationships between the primary and secondary windings. Lines drawn parallel to each other are in phase. These lines

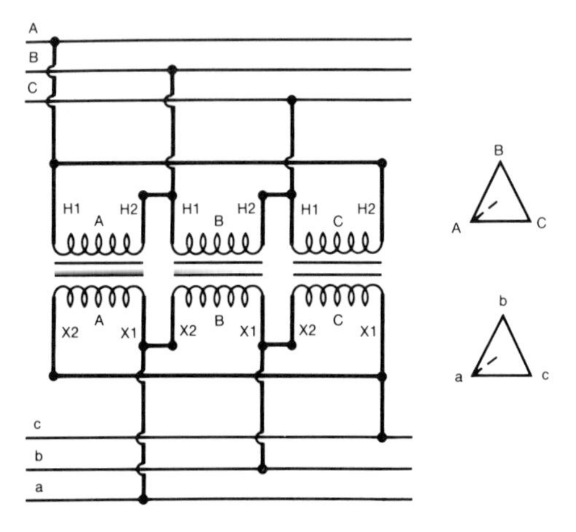

FIGURE 2–19. Primary and secondary of transformer delta connected.

are still vectors, but the arrows have been replaced with letters designating the phase at H1 or X1 on the transformer(s). Positive direction is in the direction of the ascending alphabetical order. The dotted line indicates the neutral point of the diagram.

The load should be divided equally between each transformer. If the total load was 75 kVA, each transformer would be rated at 25 kVA, or one third of the total load. Total power delivered to the circuit can be calculated by multiplying the phase voltage with line current times the square root of 3. Dividing the value by 1000 will put the units in kilovolt amperes. The formula is

$$\text{kVA} = \frac{V_{line} \times L_{line} \times 1.73}{1000}$$

Voltage across the primary will equal the source voltage, and the voltage across the load will equal the secondary voltage of the transformer. Feeder current will be equal to 1.73 times the current in the primary winding, and load current will equal 1.73 times the current in the secondary windings. Because of the high source and load currents, both the power source and the load should be close to the transformer.

Each of the delta drawings showing the wiring arrangements of the primary and secondary windings has a dotted line extending from the "A" phase at 30 degrees. This line represents the phase relationship between the primary and secondary voltages. When paralleling banks of transformers, this phase relation must be observed. If the transformers are out of phase, voltage would exist between parallel connections, and a current would flow between the two points. Even two banks of transformers wired delta-delta may have phase differences due to the interconnections of their windings. Under these conditions, they may not be connected in parallel with each other without modifying the internal wiring of one to make it equivalent to the other.

A transformer bank connected delta-delta can never be paralleled with a bank wired delta-Y of Y-delta. The diagrams in Figures 2–20 and 2–22 depicts the phase differences that exist. All the other rules regarding the parallel connection of single-phase transformers would be applicable for three-phase systems.

The delta primary is normally used when the supply source is delta connected. The delta secondary is normally used only when the load requires a single voltage and a high current. The load should be balanced so that an equal amount of current is supplied by each phase. A three-phase motor would meet these requirements.

Delta-Wye

Figure 2–20 shows the wiring for the delta-wye system. The previous description about delta primaries would also apply to this arrangement.

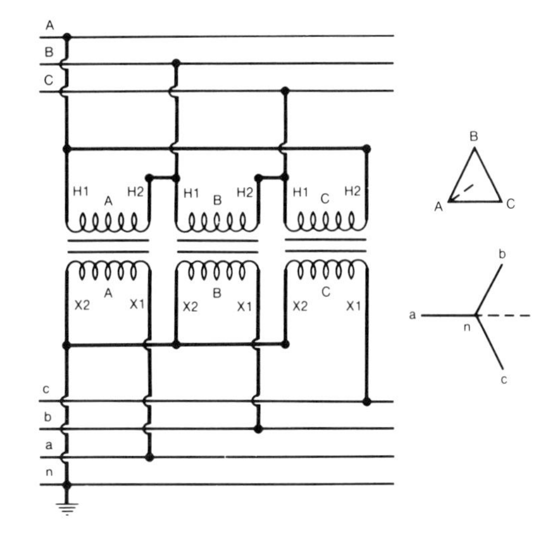

FIGURE 2–20. Transformer connected for delta primary, wye secondary.

The wye connected secondary is accomplished by first wiring the X2 terminals of the three transformers together and tying them to a neutral ground. Each of the X1 terminals of the coils is then wired to its respective phase.

Secondary voltage to the load will be 1.73 times the voltage of a delta connected secondary. Load current and secondary winding current will be the same, because the winding is in series with the load.

This arrangement of the secondary provides dual voltages to the load. Typical values would be 120/208 volts for small installations, or 277/480 volts for larger areas of distribution. The system will provide three single-phase circuits at both the higher and lower voltages, and one three-phase circuit at the higher voltages.

The 120-volt and 277-volt single-phase circuits are obtained by wiring between any phase and the neutral. The higher single-phase voltages are obtained by wiring between any two phases, and the three-phase voltage for the load would have all three phases connected to the load.

Besides the advantages of the dual voltages and having available both single-phase and three-phase power, the loads on the single-phase circuits can be balanced more easily on a wye secondary, so that each transformer in the bank is carrying approximately one third of the load. In addition, if the load increases beyond the bank's rating, this delta-wye system allows for the paralleling of the secondaries of another bank of transformers that is wired in the same manner.

When transformers are wired delta-wye or wye-delta, a phase shift between the phases of the source voltage and the utilization voltage is introduced. The dotted lines in the graphic representation of the delta-wye arrangement depicts this shift as being 30 degrees.

Wye-Wye

Figure 2–21 illustrates a three-phase bank of single-phase transformers that has been connected wye-wye. In this case the currents of both the primary and secondary windings will be equal to the currents on the lines to which they are connected. The voltages between line phases will be equal to 1.73 times the winding's voltage. If the source voltage is a given, the voltage rating necessary for the primary can be calculated by dividing the source voltage by 1.73 or multiplying it by 0.58.

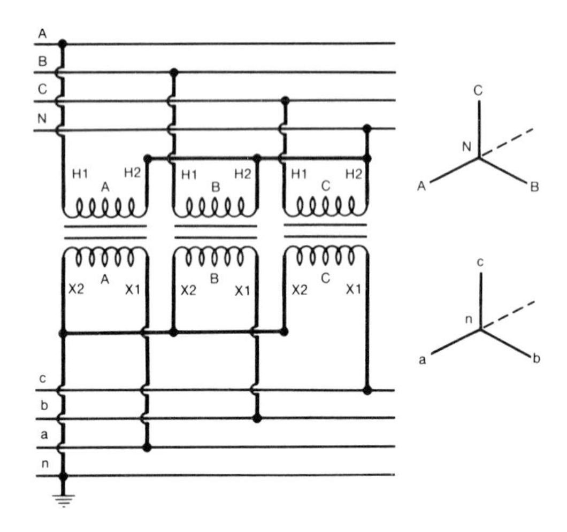

FIGURE 2–21. Transformer connected for wye primary, wye secondary.

The load should be divided equally between each transformer. kVA rating for each transformer would equal the source voltage times source current times 1.73, divided by 3 times 1000:

$$\text{kVA} = \frac{I_{source} \times V_{source} \times 1.73}{3 \times 1000}$$

or

$$\text{kVA} = \frac{I_{source} \times V_{source} \times 0.58}{1000}$$

Wye-wye have been relatively unpopular due to the inherent neutral instability of this system. This is due to the fact that currents flowing in the branches of the wye connections are not independent of each other. Usually, the neutral of each of the primary windings must be tied together if the system is operating with unbalanced loads. When this is not done, the voltage across the largest load with drop, and the voltage across the other two single-phase loads will increase. When three single-phase transformers are used, this system is incapable of supplying an appreciable single-phase load from line to neutral without a serious shift in the neutral. An exception to this is single-unit three-phase transformers with a common core which can give good results.

Without the neutral tie, large third harmonic voltages (3 X 60 Hz = 180 Hz) would appear both in the primary and secondary windings. These voltage are often excessive and could stress the transformer's insulation beyond its capacity, causing a breakdown of the system.

As mentioned earlier in the text, if a primary source voltage was to fail, current would still flow through the secondary winding of that phase due to the load. In the case of a single-unit, three-phase transformer, the core would quickly saturate. This is due to the lack of current flowing through the primary winding to cause a counter magnetic field to oppose the field of the secondary. The flux lines would leak to protective covers and metals surrounding the transformer. The heat generated by the currents in these parts can often be high enough to cause damage to the transformer or start a fire.

Wye-wye systems generate considerable interference with communications systems. The level of this noise is often so high that telephone lines cannot be run parallel with this configuration. Some jurisdictions prohibit the use of wye-wye transformer connections for this reason, and the disadvantages given above. The system is seldom used even in those areas where it is permitted.

Wye-Delta

Figure 2–22 depicts the wye-delta system. Transformers with primaries of 2400 volts can easily be changed to handle 4160 volts from a four-wire distribution

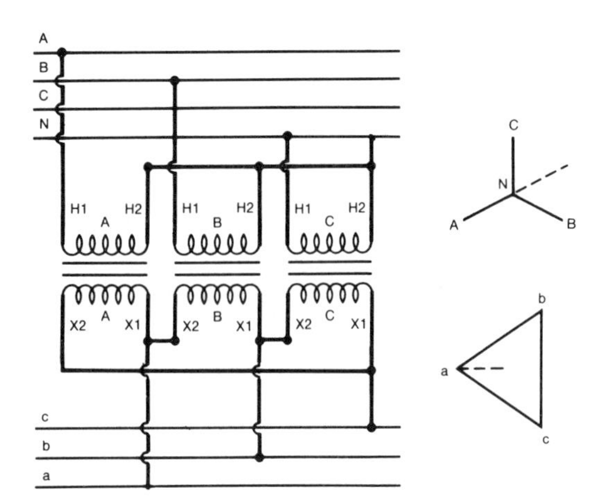

FIGURE 2–22. Transformer connected for wye primary, delta secondary.

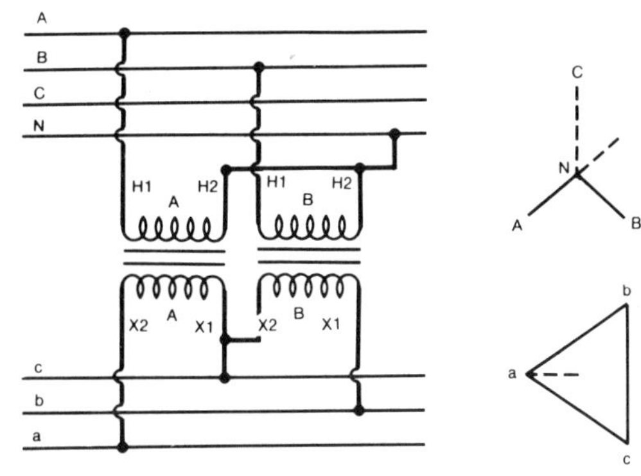

FIGURE 2–23. Two-unit wye-delta system.

system by connecting them wye instead of delta. The tendency of utilities companies is to use the higher voltage, for distribution purposes.

The neutrals of the primaries of the transformers in this Y-Δ system are not tied back to the neutral of the source. This allows for a three-wire distribution system, as used with the lower-voltage delta-connected primary, to now be used with a higher voltage.

In the event of failure of one of the transformers in the bank, two transformers can be connected wye-open delta to continue service. Figure 2–23 shows this arrangement.

In order to wire the two primaries as wye connected, the neutral wire of the supply side must be present. The two transformers are now connected between phase and the neutral, which will provide the required 2400 volts across the primaries. With the three-phase banks, two primary windings were in series across 4160 volts of the two phases. The secondary will supply all these phases operating open delta. This is discussed later in this chapter.

Dual Windings

Transformers are sometimes manufactured with dual windings. More often than not, the secondaries have this feature, whereas the primaries do not. For discussion purposes in this section, the possibilities of the different wiring arrangements will be explored.

Figure 2–24 shows three identical transformers having dual windings on both the primary and secondary sides. Each primary winding is rated at 1200 volts, and each secondary winding is rated at 120 volts. These values will be used in the examples described in this section.

This arrangement presents the possibility of several different wiring schemes. For example, windings can be connected in series for higher voltages and low current, or in parallel for lower voltages and high current. These combinations can then be connected either wye for high voltage and low current, or delta for low voltage and high current. Some of these will be explored based on different transmission voltages and the load requirements.

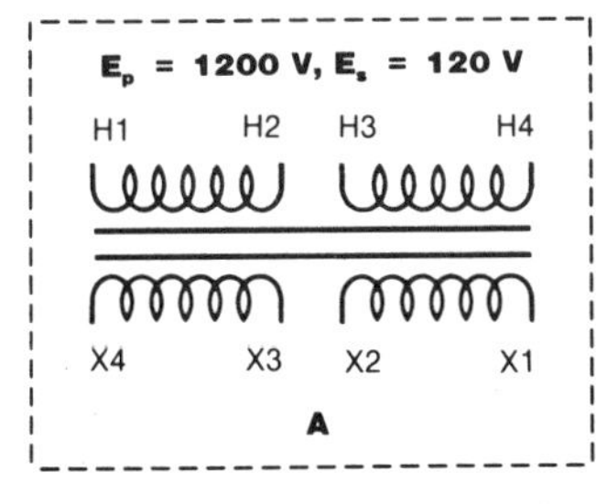

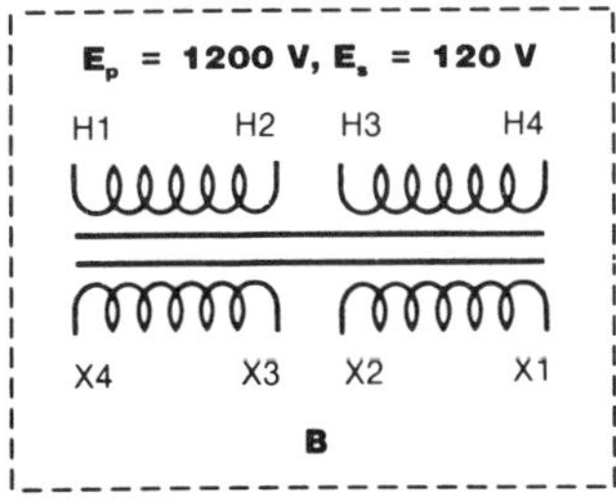

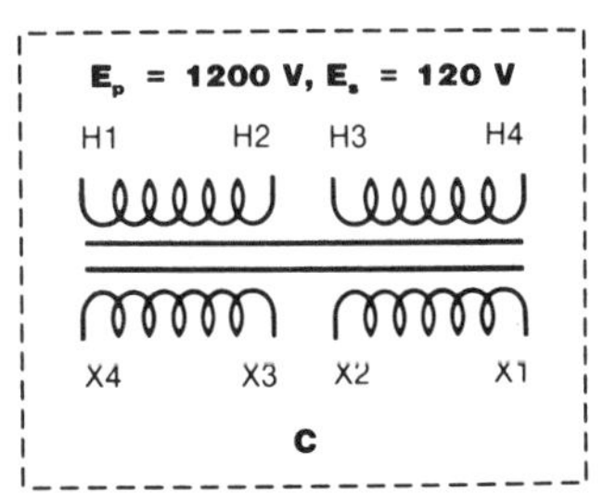

FIGURE 2–24. Three, single-phase transformers having dual-primaries and dual-secondaries.

Delta Parallel: Delta Parallel

Example 1. Source Voltage: 1200 Volts; Load Requirements: 120 Volts—Figure 2–25 meets the requirements of the parameters that are given. With a source voltage equal to the voltage ratings of the primary windings, the windings on each phase must be connected in parallel. With this arrangement the voltage impressed will equal the required voltage across the primary windings.

The load requirement of 120 volts is equal to the voltage across each of the secondary windings. To obtain the 120 volt load requirement, each of the secondary windings on each phase must be wired in parallel to each other.

When the transformer's windings of either the primary or secondary are wired in such a manner that the voltage across the windings equals the line voltage, then the three-phase set of coils must have the delta configuration. This arrangement is illustrated in Figures 2–25 and 2–26. In Figure 2–26, all wiring is shown as vertical connections to the lines in order to simplify the drawing and make it easier to understand. Only three wires are required to connect the bank of transformers to the lines.

In order to wire the primary coils of each phase in parallel with each other, the following connections are made:

H1–H3; H2–H4

Likewise, the dual windings of the secondary would be wired

X1–X3; X2–X4

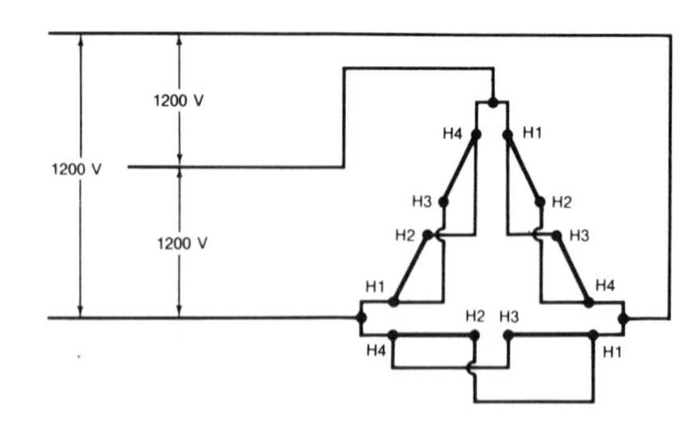

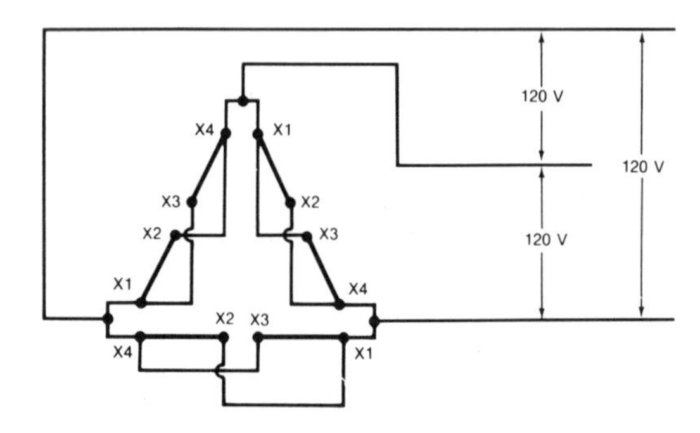

FIGURE 2–25. Transformer wired for 1200-volt source, 120-volt load.

To complete the delta primary, the parallel windings of each phase are connected in series with each other. The following terminals are wired:

H4A–H1B; H4B–H1C; H4C–H1A

The delta secondary can then be completed by wiring

X4a–X1b; X4b–X1c; X4c–X1a

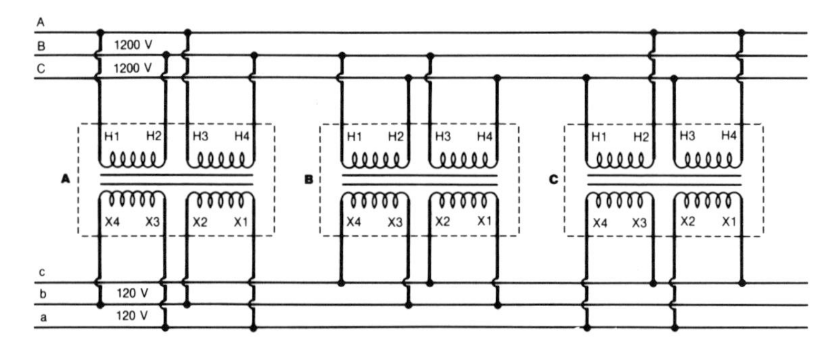

FIGURE 2–26. Dual-primary, dual-secondary transformer rated 1200/120 volts, wired for 1200-volt source with 120-volt load requirements.

H1 of each phase of the primary can now be connected to its corresponding source voltage. X1 of each of the three secondary windings is connected to the load.

Delta Series: Delta Series

Example 2. Source Voltage: 2400 Volts; Load Requirements: 240 Volts/120 Volts—In this case the voltage being delivered to the transformer is twice the voltage rating of the primary windings. In order for each winding to have its rated voltage of 1200 volts, the primary windings of each phase will need to be connected in series.

H2–H3

The voltage drop across each pair of windings must now equal the source voltage. When source voltage equals the voltage across the phase of the transformer, the wiring choice is delta. The same connections made in Example 1 would apply here (Figure 2–27). H1 of each winding would be wired to H4 of the next phase.

H4A–H1B; H4B–H1C; H4C–H1A

To provide the required 240 volts, the 120-volt secondary windings must be connected in series. X2 and X3 of each transformer would be tied together to accomplish this.

The three transformers' secondaries would also be wired in series, X4 of one phase connected to X1 of the next phase, to provide a delta system. The three-phase load will have twice the voltage as that in Example 1, but only one half the current.

In this case, the load also requires 120 volts. This can be accomplished by using a center tap on one of the phases. This center tap becomes the neutral of the secondary, and it is grounded. Between A phase and the neutral, and C phase and the neutral, 120 volts are available for lighting and small appliances.

Only one of the three phases can be grounded. Potential exists between the center taps of the three phases (120 volts). If they were connected to a common ground, the bank would be shorted. This would result in excessive current which could destroy the transformer and cause other property damage.

This arrangement also has a third voltage (208 volts) available between the neutral and the high leg of the delta. The high leg is B phase in Figures 2–27 and 2–28.

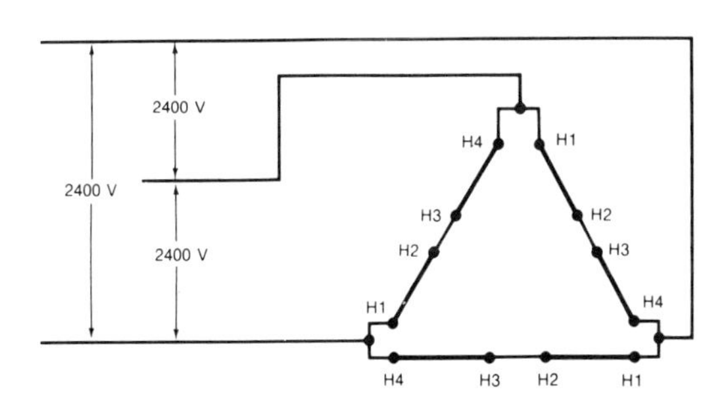

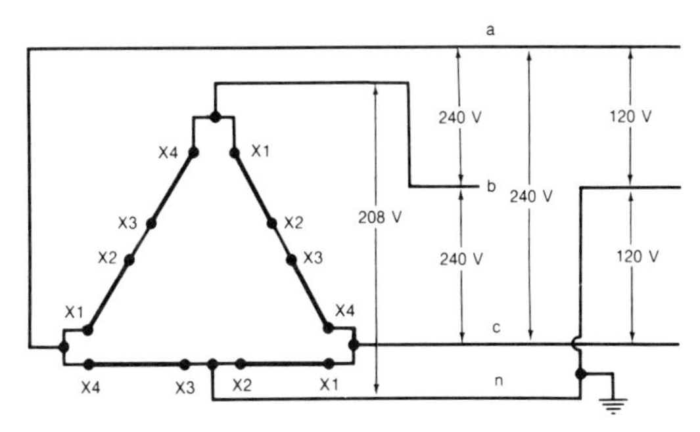

FIGURE 2–27. Transformers wired for 2400-volt source, 240/120-volt load.

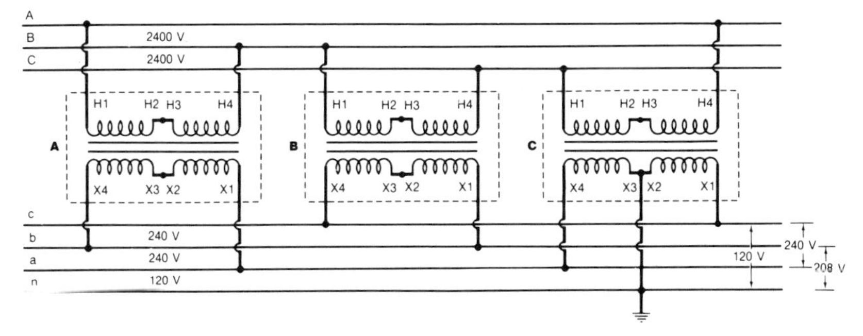

FIGURE 2–28. Dual-primary, dual-secondary transformer rated for 1200/120 volts, wired for 2400-volt source with 240/120-volt load requirement.

Although delta secondaries are normally used to provide high currents to the load, and the 120-volt circuits are usually small, the opposite may be true. If the requirements were for a small three-phase load and large load for the 120-volt phase, two small transformers could be used in the delta to supply the three-phase power. A much larger transformer would be used to supply the lighting and small appliance load at 120 volts and the third leg of the three-phase power at 240 volts.

Wye Series; Wye Parallel

Example 3. Source Voltage: 4160 Volts; Load Requirements: 208/120 Volts—The tendency is to use 4160 volts for distribution rather than 2400 volts. This change does not require a change in the transformers used to step-down the voltage for utilization purposes. In this case, the transformers used in these examples can still perform the job.

In Example 2, the primary windings were connected in series which required a 2400-volt source so that 1200 volts would be dropped across each winding. The three primaries were then wired delta so that the line voltage would equal the required voltage across the primaries for proper transformation to 120 volts on the secondary windings.

To obtain the same results with 4160 volts, the primary windings would once again have to be in series with each other. H2 and H3 on each transformer would be connected.

The transformers would then have to be wired in a wye configuration rather than delta. If this change is not made, each of the 1200-volt windings would have 2080 volts across them. This would result in a transformation of 208 volts across the 120-volt secondary windings. Some customers might complain about this inconvenience.

When the value of the supply voltage is greater than the voltage required by the windings, the transformers will have to be wired wye. The supply voltage should be divided by 1.73 or multiplied by 0.58 to assure that the voltage across the primary windings will be correct. The nameplate on the transformer should be checked to confirm the calculations.

To complete the installation, H4 of each of the transformers is connected together to form the neutral. H1 of each transformer is then wired to its respective phase of the service. Figures 2–29 and 2–30 depict this arrangement.

In Figure 2–29, the voltages shown on the primary are representative for all phases. 2400 volts will be found from each phase and the neutral. 4160 volts will be measured between the phases A–B, B–C, and A–C.

The same is true for the secondary. There are 208 volts between each phase AB, BC, and AC. Between each phase and neutral, there are 120 volts.

Figure 2–30 depicts each of these values. The windings in the secondary were put in parallel by drawing vertical lines to the load phases. This was to simplify the schematic so that it would be easier to understand. Only four wires would be necessary to connect the transformer to the load.

With the 208/120-volt secondary arrangement, three single-phase 120-volt circuits are available. This makes it easier to equalize the 120-volt loads on each phase, providing a balanced three-phase system. This is not possible for the three-phase delta secondary, because the entire single-phase load is on one phase.

The delta system does have an advantage over the wye system. If a phase is lost with the wye secondary, all three phases must be shut down until repairs are made. On the delta system, the two remaining banks can continue to operate with all three phases present, provided the remaining two transformers can carry the load.

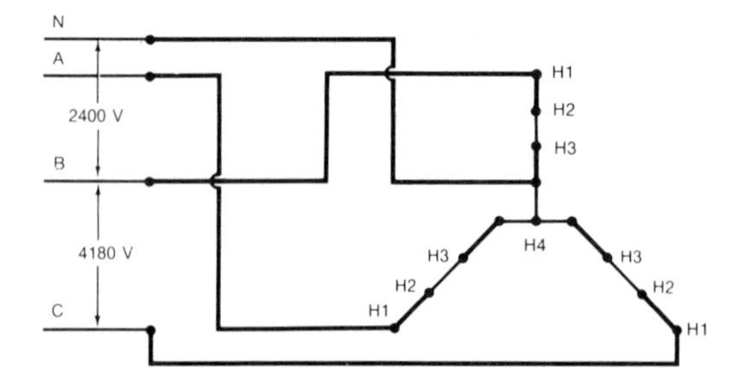

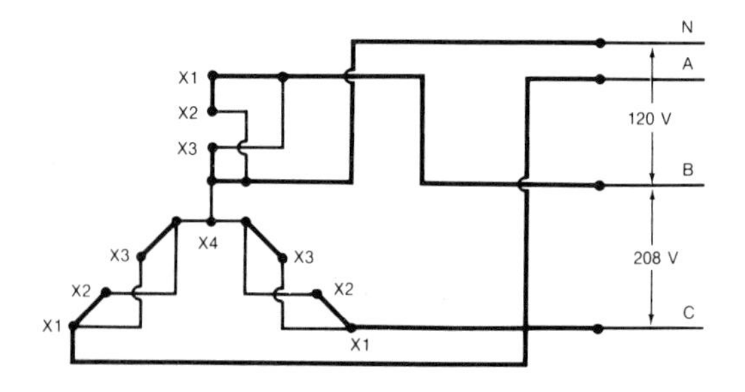

FIGURE 2–29. Transformers wired for 4160-volt source, 120/208-volt load.

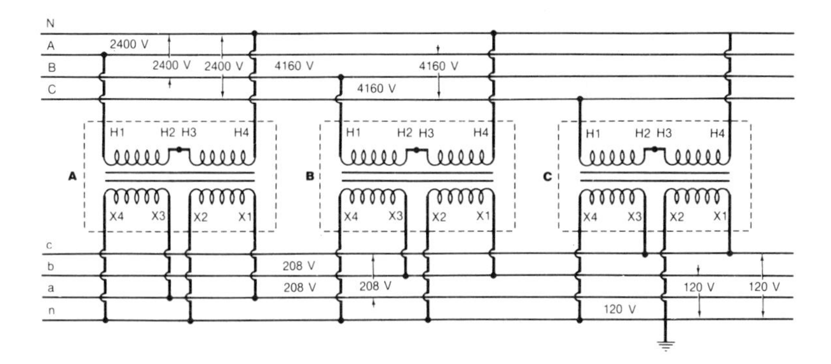

FIGURE 2–30. Dual-primary, dual-secondary transformer rated for 1200/120 volts, wired for 4160-volt source with 208/120-volt load.

Open Delta

When a three-transformer bank is converted from a closed-delta to an open-delta (V) system, the load cannot be greater than 58% of the kVA capacity of the three-phase system. This value is derived from the fact that for the closed-delta system, each winding provides 58% (1/1.73) of the current for each phase. With open-delta, each secondary winding must supply 100% of the current. The winding and the line are now in series with each other and must be equal. Figure 2–31 shows these relationships. Figure 2–23 provides the schematic diagram for a three-phase circuit using two transformers which are wired wye-open delta.

Each bank is providing three-phase power to the load. The three-phase bank has 100 amperes flowing through its secondary windings, with 173 amperes flowing to the load. The two-transformer open delta has 173 amperes flowing through the secondary windings with the same value being delivered to the load. The kVA of the load is equal to

$$
\begin{aligned}
\text{kVA} &= (E \times I \times 1.73)/1000 \\
&= (480\ \text{V} \times 173 \times 1.73)/1000 \\
&= 143.659\ \text{kVA}
\end{aligned}
$$

Assuming that each transformer in the bank is rated at 50 kVA, it would be impossible for only two of them to carry the 143.7 kVA load. In order to continue service with the original equipment, the load would need to be reduced to 58% of its original value.

If the delta system were to be replaced by an open delta, the replacement transformers would need to be at least 15.5% larger than the total kVA rating of the delta. In this case, if the three 50 kVA transformers were replaced with two 75 kVA transformers, each of them would need to carry 173 amperes on their secondary windings.

$$
\begin{aligned}
\text{kVA} &= (480\ \text{V} \times 173\ \text{A})/1000 \\
&= 83.040\ \text{kVA}
\end{aligned}
$$

As can be seen, the transformers will need to provide at least 83 kVA, and they would not be large enough to handle the existing load. Adding the 15.5% capacity to the 75 kVA transformers would increase their size to 86.6 kVA. The next largest standard size of transformer would be selected.

The open delta can be used to supply a high level of single-phase power for lighting and small appliances as well as a small three-phase load. When this is done, the single-phase source can be much larger than the second transformer needed to obtain the three phases.

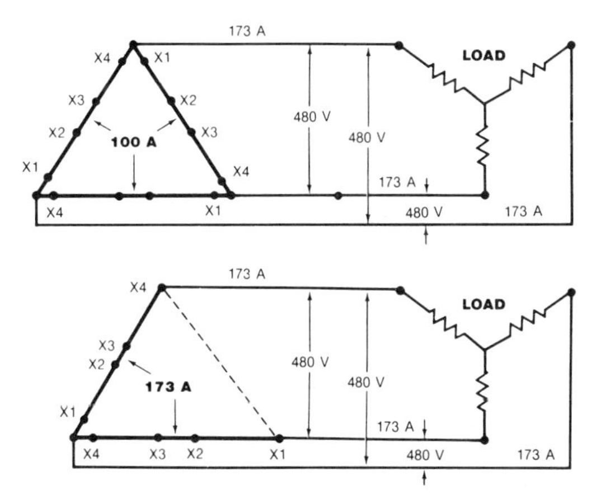

FIGURE 2–31. Comparison of delta and open-delta.

An open delta system may be selected for use in an area where projections for greater power consumption are predicted. When the demand is reached, the addition of the third transformer will add almost 75% capacity to the bank. This is also due to the fact that each pair of transformers in the three-transformer bank is providing only 58% (1/1.73) of the line current on each phase with the installation of the full delta arrangement.

Special Connections

Other than standard wye and delta type of connections, there exist many other designs for wiring transformers. These systems are usually installed for their economic advantage, or to overcome some deficiency in the operation of the present system. Some of these will be discussed.

Buck-Boost Transformers

Because transformer voltages have relative polarity, they can be wired as autotransformers so the secondary voltage either adds to the line voltage or subtracts from it. When the voltage adds, it is a "boost" transformer, and when it subtracts, it is called a "buck" transformer.

Standard-size transformers when wired for either bucking or boosting applications offer the electrician great flexibility and versatility in providing the proper voltage to the utilization of equipment with different voltage requirements than that of the power supplied in a given building.

Transformers with dual-winding primaries and secondaries are the most popular for this purpose. These transformers may be transmission types using 1200/2400-V primaries and 120/240-V secondaries, or transformers rated with voltages under 600 V using 120/240-V primaries with 6/12-, 12/24-, and 16/32-volt secondaries. Figure 2–32 shows this arrangement of the dual-primary, dual-secondary transformer.

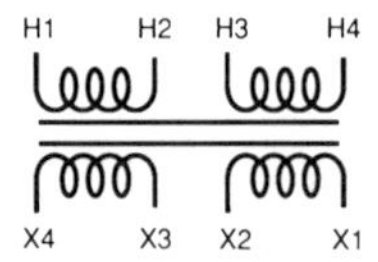

FIGURE 2–32. Typical dual-primary, dual-secondary transformer used for buck-boost applications.

A good example of this application is an air conditioner requiring 230-volts, single-phase in a small commercial building supplied with 120/208 V. In this case a transformer with voltage ratings of 120/240:12/24 volts could be used. This transformer would have a turns ratio of 10:1 (see Figure 2–33). In this case the primary would be connected in series by connecting H3–H2 together. The secondary would likewise be in series with X3–X2 wired together.

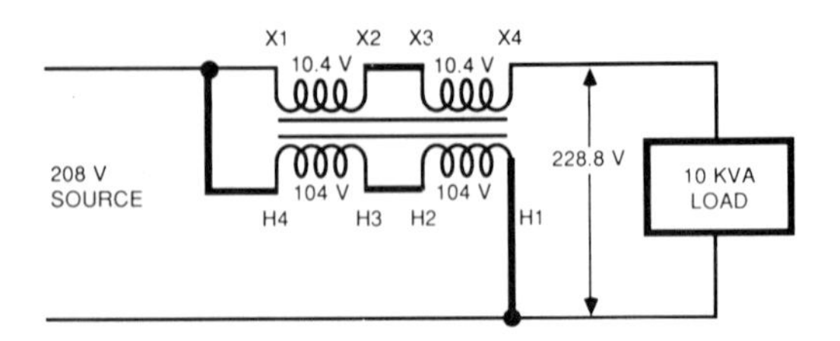

FIGURE 2–33. Boost transformer used to raise a 208-volt source to supply a 230-volt load.

The primary leads, H1 and H4, are connected to the 208-volt source. This will impress 104 volts across each of the primary windings. The 10:1 turns ratio of the transformer will provide 10.4 volts across each secondary winding, or 20.8 volts total with this wiring arrangement. X1 and X4 are wired in series with the power line. The 208-volt source when added to the secondary boost voltage of 20.8 volts will provide a voltage of 228.8 volts for the 230-volt air conditioner. The 228.3 volts is close enough to 230 volts that the compressor will operate properly.

Now that we know how to boost the voltage, how do we determine the size of the transformer to use to accomplish this purpose? Let's assume that the single-phase load is 10 kVA. What size boost transformer do we need to raise the voltage?

The total current of the load will flow through the secondary of the boost transformer. This current can be calculated by:

$$\begin{aligned} I_L &= 10000/228.8 \\ &= 43.71 \text{ A} \end{aligned}$$

This current when multiplied with the 24-volt secondary of the boost transformer gives:

$$VA = 43.71 \text{ X } 24 = 1049 \text{ VA}$$

A standard rated 1 kVA transformer may be sufficient to supply the higher voltage to the 10 kVA load. However, a transformer rated at 1 kVA with a 24-volt secondary would have rated current for the secondary of 41.67 A. This means the secondary winding is exposed to about 4.8% higher current than its rating. Because the heating effect of this current varies as the current squared, total heating effect will increase by nearly 10%. Most transformers are designed to take this overload.

A question needs to be answered at this point. Is this enough difference to select a larger rating for the boost transformer? This would depend upon the conditions involved. If there is a chance the source voltage of 208 volts will drop to a lower value, the current will increase even more. Current for induction or synchronous motors used with air conditioners will increase at about the same rate that the voltage drops. This would not be a consideration with a resistive load; here the current will decrease proportionately with the voltage.

Another question involves the load on the air conditioner. Will it continuously operate in an overload condition? If so, one could expect an increased current. In most cases, air conditioners do not operate continuously and proper design takes care of this problem. Consult the manufacturer's catalog to determine the load limit on the transformer used to boost the voltage. When a serious doubt exists, err on the side of the next larger transformer.

Current supplied by the 208-volt source will also increase when the boost transformer is introduced to the circuit. This is due to the current flow through the primary of the transformer. In this case it would be 1/10 (turns ratio) the current in the secondary, or 4.37 A. Total line current is 43.71 A + 4.37 A or about 48 A. This should be checked to determine if the ampacity of the conductors needs to be increased or if the overcurrent protection meets the requirements of the code.

There are a couple of factors the electrician needs to be aware of when wiring buck-boost transformer circuits. When a standard transformer is used, the application is different from that which the manufacturer intended. The transformer when wired as an autotransformer may be subjected to voltage stresses which never occur when operated as a conventional transformer. If the primary of the transformer opens, for example the fuse blows, the secondary in series with the line acts like a choke which may have from two to five times its normal voltage impressed across it. This may result in insulation breakdown and destroy the transformer. The higher the voltages involved in the boost operation, the greater the chance of breakdown when the system malfunctions. Isolating the case of the transformer from ground is one way of preventing this from occurring. The use of transformers designed for buck-boost operation is better.

Another factor involves the familiarity that the electrician has for a given task. Assume the boost transformer for the air conditioner is several hundred feet from the 208-volt source and the air conditioner also requires 120 volts for its control circuits. Instinctively the electrician might think why run a neutral several hundred feet to the source transformer just to get 120 volts. Why not ground the midpoint (G) of the 228.8 volts on the autotransformer and use a ground return to obtain the 120 volts? (Figure 2–34).

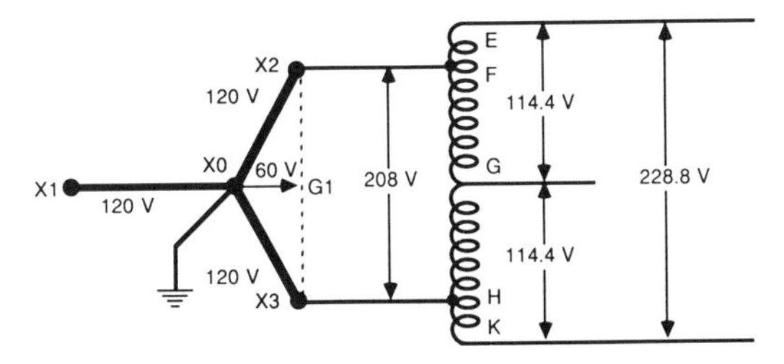

FIGURE 2–34. Effects of grounding the midpoint of the boost transformers. *(Courtesy Songel Transformer.)*

The average electrician is accustomed to connecting the midpoint of a 120/240-volt secondary ground. This is when the trouble begins. G and X0 are not the same electrically in this circuit. A potential of 60 volts exist between the two points in this case. This is shown pictorially by the dotted line between X0 and G1. G and G1 are electrically the same point with a zero potential between them.

Connecting X0 to G causes a short circuit. The current of the short is limited only by the impedance of the transformer and the resistance of the ground return. If the source transformer is capable of sustaining the additional load, and the resistance of the ground is very low, the system will continue to operate. However, there is now about 120 volts between F and G and 120 volts between G and H instead of 104 volts (½ of 208 volts). The primary voltage of the boost transformer has been raised with the secondary voltage rising accordingly. The worse possible case would be that 264 volts is across the load instead of the nearly 230 volts the air conditioner is rated.

The results of using this ground return are:

1. The ground short circuit will cause increased current flows which will tend to cause premature failure of the transformers.
2. The overvoltage on the air conditioner will shorten its life and may cause failure of auxiliary components such as fans.

The solution to obtaining the 120 volts is to run the neutral from the transformer to the air conditioner. *Do not ground* point G on the boost transformer.

Instead of single-phase power, the air conditioner may require three-phase power at 230 volts. In this event, three, single-phase transformers would be used with the same voltage ratings used in the example for the single-phase requirement. Figure 2–35 illustrates the wiring for the three-phase system.

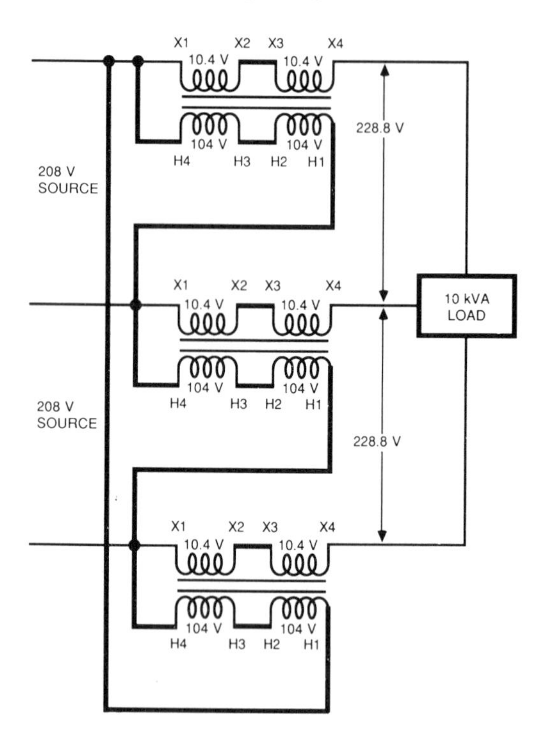

FIGURE 2–35. Using boost transformers in a three-phase power system to raise voltage to the load.

The VA rating of the three single-phase transformers can be reduced due to the decrease in line current. The line current for three-phase operation will equal:

$$\begin{aligned} I_L &= VA/V_L \text{ X } 1.73 \\ &= 10000/228.8 \text{ X } 1.73 \\ &= 25.26 \text{ A} \end{aligned}$$

The VA rating for each transformer would equal:

$$\begin{aligned} VA &= 24 \text{ X } 25.26 \\ &= 606.24 \text{ VA} \end{aligned}$$

There are times when the line voltage may be too high. A standard rated transformer can be connected to buck the line voltage, thereby lowering the voltage to the load. Figure 2–36 illustrates this arrangement.

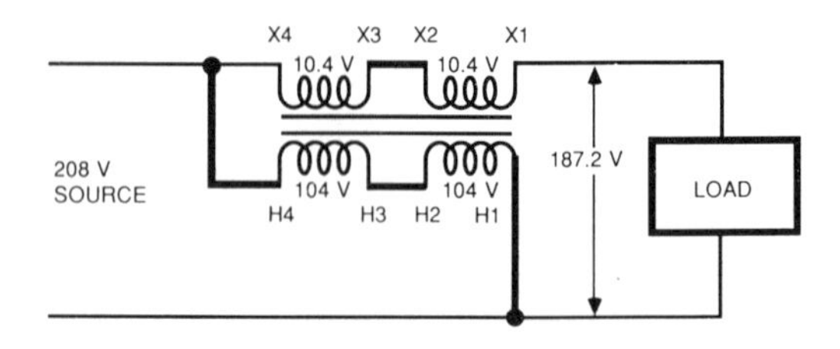

FIGURE 2–36. Buck transformers used to lower the voltage to the load.

Under this condition, the connections on the secondary of the back transformer are reversed. X4 is now connected to H4. The secondary of the transformer is now acting like a choke which results in 187.2 volts (208 volts–20.8 volts) being delivered to the load.

Figure 2–37 shows an application of a booster amplifier used to raise the voltage on a 2400-volt distribution line. In this case, line voltage has fallen to 2200 volts. Use of the booster transformer raises the voltage back to 2420 volts.

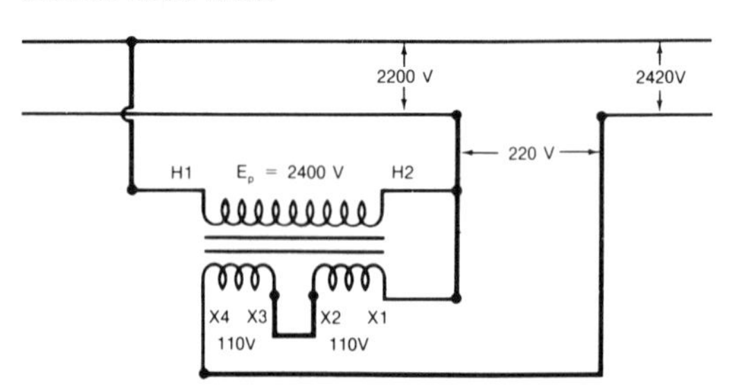

FIGURE 2–37. Boost transformer used to raise 2200 volts to 2420 volts.

Tee-Connected Transformers

It is well known to the manufacturers of transformers that, as the size of transformers goes down, the cost per kVA increases. Given the high competition within the industry, anything that can be done to reduce costs without sacrificing performance should be attempted. The Tee-connected transformers is one attempt to achieve this goal. Both the manufacturer and the

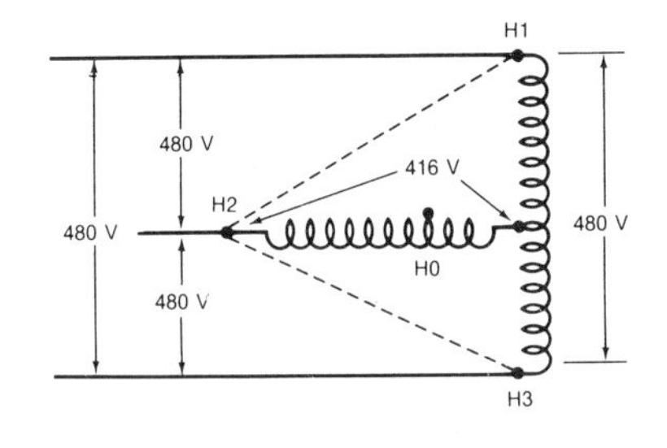

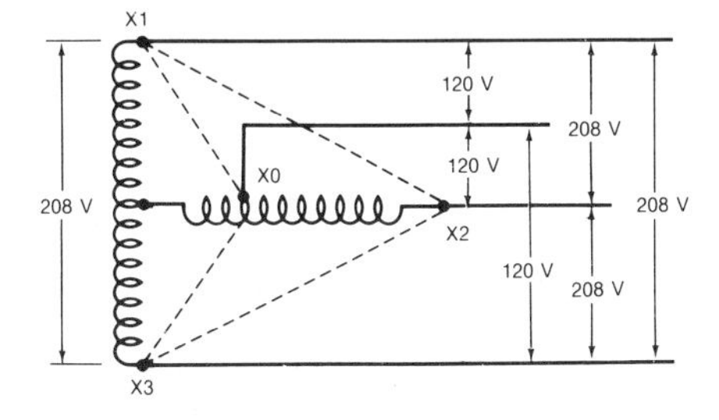

FIGURE 2–38. Tee-connected transformer.

customer save money when this arrangement is selected. Tee-connected transformers are also called Scott transformers.

When a three-phase, delta-wye transformer is purchased, it usually has three primary and three secondary windings. The Tee-type transformer has a two-coil primary and a two-coil secondary (Figure 2–38). By reducing the number of coils and achieving the same purpose, the Tee-connected transformer can be sold for less money than its typical three-phase equivalent.

The connections for the primary are the same as for any three-phase transformer. Each high-voltage terminal is wired to the appropriate phase of the supply. Between the center tap and H2, 416 volts are impressed across the winding. The turns ratio of the transformer is adjusted to compensate for the lower voltage. Between H0 and any of the high-voltage terminals, 277 volts will be present.

The secondary terminals, X1, X2, and X3, will have 208 volts between them. From X0 and any of the low-voltage terminals, 120 volts are available. X0 is the electrical neutral for the system and it may be grounded.

Another way of obtaining 208/120 volts from a 480/277 volt distribution system is to use autotransformers. Figure 2–39 depicts this arrangement. There are 208 volts between the lines attached to the taps. Between the lines and the neutral/ground, there are 120 volts.

Zigzag Connection

The zigzag connection of transformers is used to provide a neutral for a three-phase system when the neutral is not supplied through the system. This arrangement is particularly useful where a wye-wye connected system is to be used. Wye-wye systems using three single-phase transformers with a common ground may resonate with the capacitance in the system and cause high stress voltages at the third harmonic on the system. This problem is greatly reduced through the use of the zigzag connection which reduces third harmonic voltages to acceptable levels. Figure 2–40 illustrates this arrangement. The kVA ratings of the transformers are based on the amount of unbalance current and the voltage of the system.

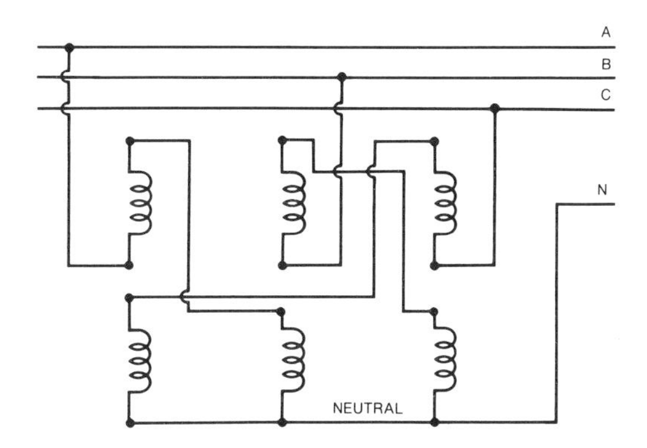

FIGURE 2–40. Zigzag connection using six autotransformers

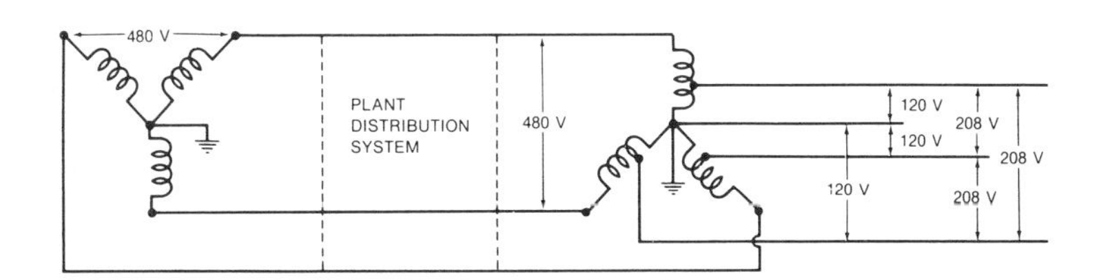

FIGURE 2–39. Obtaining 208/120 volts for plant distribution using autotransformers.

Questions

1. When installing a transformer, which are the most important specifications to be confirmed?
2. How are high- and low-voltage leads identified on a transformer?
3. Draw the schematic diagram for a dual-winding secondary having 120 volts on each winding to produce a 240-volt output.
4. Does the kVA calculation of the primary circuit equal the kVA rating of the secondary? Explain.
5. A transformer is rated at 15 kVA, 2400/600 volts. What is the primary current? How much secondary current is available?
6. A transformer has dual primaries rated at 600 volts. Connect them for a 1200-volt source.
7. If two 1½-volt batteries are connected positive to positive, how much voltage will be available at their other terminals?
8. A transformer is rated 2400/240 volts. H1 is wired to X1. What is the voltage between H2 and X2?
9. Will a 12-kilovolt transformer have an additive or subtractive polarity?
10. How can the continuity of a winding be determined?
11. Draw a circuit for determining the polarity of a transformer using an AC voltage source.
12. If you have two transformers, one additive and the other subtractive, which are to be connected in parallel to each other to supply a load, which terminals on each transformer would be connected to the same line?
13. Can transformers having different voltage ratings be connected in parallel? Explain.
14. What is the effect of connecting two transformers that do not have the same impedances in parallel with each other?
15. Are there any wiring methods that would allow transformers of different impedance to be connected in parallel?
16. How many degrees separate the phases of a three-phase system?
17. The amount of copper in a single-phase system can be reduced by ______ % if a three-phase system is used to deliver the same amount of power.
18. In which direction does the rotor of a generator turn?
19. What is the algebraic sum of the voltages in a three-phase system?
20. In Figure 2–15, if the phase voltages were 277 volts, what would be the voltage between B and C?
21. How are the H2 terminals of the transformers connected for a wye system?
22. List the advantages of using a three-phase transformer rather than three separate single-phase transformers.
23. What is the relationship between the line current and the winding current of a wye connected transformer?
24. What is the relationship between the line voltage and the winding voltage of a wye connected transformer?
25. What is the relationship between the line current and the winding current of a delta connected transformer?
26. What is the relationship between the line voltage and the winding voltage of a delta connected transformer?
27. On which system, wye or delta, is it easier to balance the single-phase loads?
28. Which system, wye or delta, is more likely to be used with a high three-phase current requirement with a minimal single-phase load?
29. Can three-phase banks of transformers be wired in parallel with each other? Are there any exceptions?
30. Draw the schematic diagram for a simple delta-wye wired transformer.
31. Calculate the kVA of a three-phase transformer with 2400-volt, 15-ampere primary windings.
32. Which of the three-phase wiring systems is capable of generating excessive voltages at the third harmonic?

33. Is it possible to obtain three phases for a load while using only two single-unit transformers? Explain.
34. Using dual secondaries rated at 120 volts, what is the maximum number of voltages available to the load? What are their values?
35. Draw the secondary circuit that would provide the voltages in the question above.
36. In the event of failure of a single transformer in a delta system, how much of the full load will the open-delta system carry?
37. When replacing a delta system with an open-delta one, how much larger must the replacement transformers be in comparison to the original ones?
38. What is the purpose of buck-boost transformers?
39. Give another name for Tee-connected transformers.
40. Give an application for zigzag-connected transformers.

CHAPTER 3

Installation

When a transformer is received at the worksite for installation, it should be inspected. If any apparent damage is evident, this should be reported immediately by filing a claim form to the responsible party, carrier, or manufacturer.

General Considerations

The transformer should be examined for any damage to the case in the form of excessive dents. Protruding parts such as insulators or indicating meters should be given individual attention to detect cracks in porcelain or broken glass.

All bolts, nuts, and fittings are checked for proper tightness. If the transformer is opened for inspection or installation, take care that foreign materials do not fall into the transformer. Dropping tools, bolts, nuts, and other objects into the transformer could damage the insulation on the coils.

Coolants

If the transformer is oil filled, a check is made of the liquid level and for any leaks. The transformer may have an oil level gauge, or the level will be marked inside the tank (Figure 3–1). Larger transformers have a gauge, and smaller types have a level mark.

If the coolant level is low and a leak is suspected, locate the exact point of the leak by cleaning the surface of the transformer with a solvent. When the surface is dry, dust the area with a dry powder such as talcum, lime, or cement dust. The exact point of the leak can be readily located in this manner.

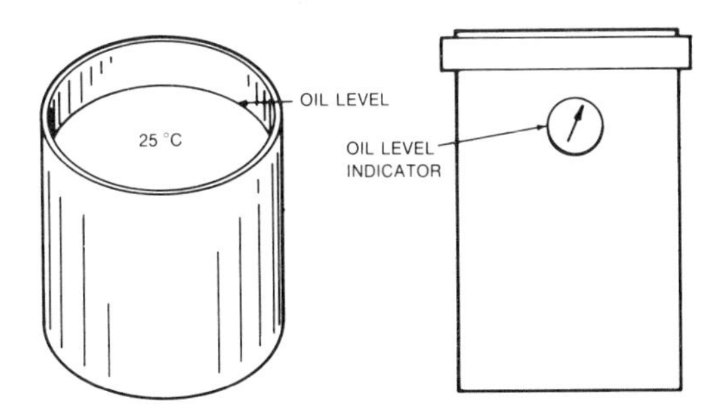

FIGURE 3–1. Transformer oil level indicators.

If the leak is above the oil level, an inert gas such as nitrogen can be inserted at 3 to 5 psi. Use a solution of soapy water on all joints and welds. The location of the leak will show up as a fine stream of bubbles. Leaks above the oil-level mark can occur because the coolant rises in the tank when it is heated during normal operation.

Once the leak has been located, it can be repaired if the transformer has already been accepted. Small leaks in welded joints can usually be repaired by "peening" them with a ball peen hammer. Larger leaks in joints need to be welded, brazed, or soldered. Leaks around gaskets can normally be stopped by tightening the bolts. If not, the gasket will need to be replaced. Leaks around threaded joints can be repaired by applying a joint compound that will not dissolve in oil. In the event this does not work, the joints need to be re-threaded.

Once repairs have been effected, the oil level must be replenished. Be sure the replacement oil is the same as that in the transformer. If mineral oil is mixed with any of the synthetic oils, it is almost impossible to separate them. The entire transformer may have to be drained of the contaminated oil and refilled with the recommended coolant. A 1000 kVA transformer may hold from 500 to 1000 gallons of oil, depending on its voltage rating.

Transformers should not be overfilled. In operation the oil becomes hot and expands, building up pressure inside sealed transformers. If the transformer is not sealed, the oil may spill out under normal operating conditions.

Testing Coolant

If the transformer is oil filled, the oil will be checked for moisture. A moisture content of 0.06% reduces the dielectric strength of the oil by about 50%. If the transformer is not going to be installed immediately, a quick test can be conducted with a red-hot nail. If the nail is placed in the oil, and the oil "crackles," moisture is present, and the transformer will need to be dried out. A dielectric test should be performed on the oil if the "hot nail" test indicates that water is present in the transformer.

Figure 3–2 depicts a receptacle for the dielectric testing of the oil. The standard is to use electrodes 1 inch in diameter that are separated by one-tenth of an inch. The complete procedure for conducting this test will be described later in this chapter.

Removing Moisture from Coolant

Moisture can be removed from the oil by raising the temperature of the oil above the boiling point of water. This is normally accomplished by applying power to the transformer using the short-circuit method. The short-circuit method will also be discussed later in this chapter when electrical testing methods are described.

When the water is boiled, the temperature is not to exceed the maximum temperature ratings shown on the nameplate of the transformer. Good ventilation is needed to take the water vapor away from the transformer. Because it may take several days to remove the water in this manner, checking the transformer on receipt may provide adequate time to dry it out before it is scheduled to be installed.

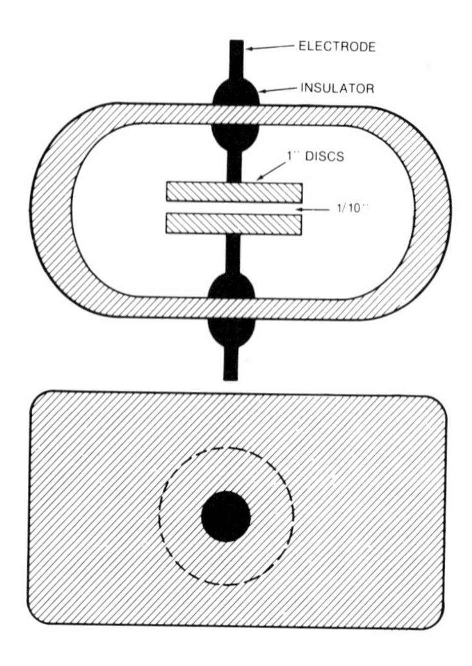

FIGURE 3–2. Receptacle for dielectric testing of transformer oil.

Storing Coolants

Oil for large transformers is normally shipped separately to reduce the weight of the transformer. For a 1000 kVA transformer rated at 4160 volts primary, the oil adds approximately 1 ton to the total weight. The oil comes in 55-gallon drums. It should be stored inside and on its side (Figure 3–3). The bung on the barrel should be checked for any leaks, and the drums remain sealed until the oil is needed.

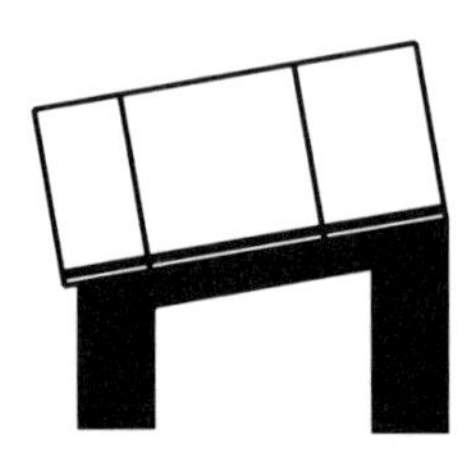

FIGURE 3–3. Drum storage for transformer oil.

Outdoor transformers usually use mineral oil, and transformers for inside use are filled with a synthetic liquid (askarel) that is nonflammable and nonexplosive. (Pyranol® is GE's tradename and Intereen® is Westinghouse's tradename for askarel.) Care should be taken to keep the mineral oil and askarel separate so that they are not accidentally mixed together when filling a transformer. The mineral oil will ignite at a much lower temperature, thereby affecting the characteristics of the synthetic oil.

Health Hazards

Synthetic oils sometimes cause skin irritation, and one must be careful not to become exposed to them. Some types of synthetic oil used in past years contain polychlorinated biphenyls (PCBs) which are known to cause cancer. These coolants are being replaced, but many transformers still contain them. Contact with them should be avoided, and in the case of a transformer fire, the area should be evacuated. Fire fighting personnel use special breathing apparatus when fighting these types of fires.

Moving Transformers

When the transformer is to be moved, it should never be dragged along the floor. Use the lifting eye(s) or other designated welded fixtures on the tank to lift it onto casters or rollers. Operational fixtures on the transformer should never be used as hand-holds or lifting points. If a hoist or other rigging is used to lift the transformer, check that the ropes or cables have the capacity for the job.

If a forklift is available on the job, it is an excellent tool within its rating for this purpose. Always move the transformer in an upright position. Be sure that any apparatus used to move the transformer does not come in contact with parts, such as insulators or gauges, that might be broken if any force is applied to them.

Verify Order

The nameplate on the transformer should be checked to see if the correct device has been received. It is important to check the kVA rating and the voltage ratings.

It is customary for the manufacturer to ship transformers wired for their highest voltage rating. This is true for the secondary as well as the primary.

Storage

When the transformer is to be stored for a period of time before installation, some precautions should be taken. Distribution transformers designed for outside use need no special handling. However, they are not to be kept in an area where they are standing in water or corrosive chemicals or where gases are present. Transformers designed for inside use are to be stored in a dry, clean area where the temperature of the surrounding air is higher than the case of the transformer. Otherwise, condensation will form on and in the transformer. This may increase the drying out period. Any entrance to the inside of the transformer must be blocked to prevent rodents, insects, and excessive amounts of dust and dirt from entering the transformer.

Follow the manufacturer's recommendations when installing the transformer. Check local codes and the specifications of the job along with any addendum for compliance and proper procedures.

Environmental Considerations

Several environmental factors must be considered when installing a transformer. The general rule is that the transformer must be accessible to qualified personnel for maintenance and repair while access is restricted to laypersons. Some of these factors are mandatory as prescribed by the National Electrical Code (NEC), codes of local jurisdictions, manufacturer's specifications, and common sense.

Guarding

Safety is the most important factor that must be considered. When possible, transformers with exposed live parts should be placed in a locked room or vault with only qualified personnel allowed to have a key. If this is not possible, transformers rated over 50 volts and not over 600 volts may be elevated so that their live parts are a minimum of 8 feet above floor level.

In areas where the transformer may be exposed to physical damage, a permanent barrier of sufficient strength to prevent damage is permitted. Only qualified personnel are allowed access to the area. Warning signs which forbid entrance are posted at the entrances of guarded locations. (Figure 3–4).

If the transformer carries over 600 volts, elevation is not permitted as a means of guarding. The equipment is to be enclosed in a locked room or vault to which only qualified personnel have access. Other requirements are specified for the room or vault depending on the nature of the equipment being installed. If a barrier or fence is used for this purpose, it shall have a minimum height of 8 feet.

Metal-enclosed transformers are permitted to be installed in areas open to the general public. Appropriate caution signs are required. Any openings to the device are designed in such a manner that foreign objects cannot be forced into the enclosure.

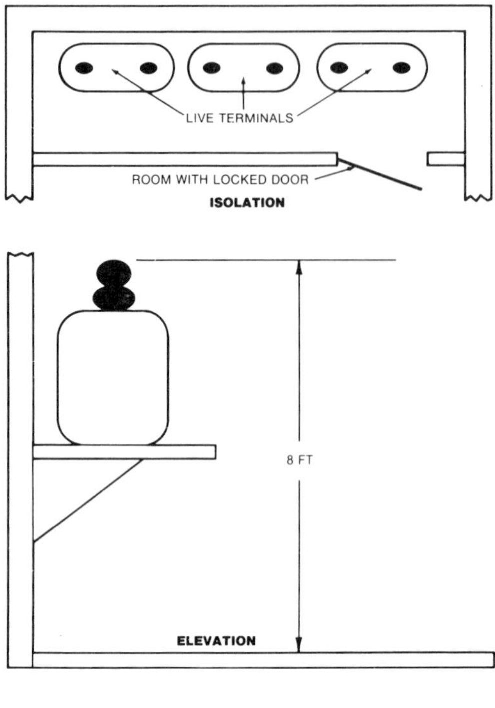

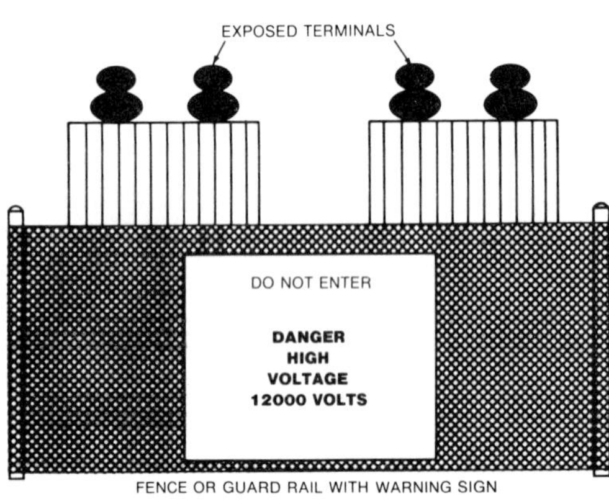

FIGURE 3–4. Guarding live parts of a transformer.

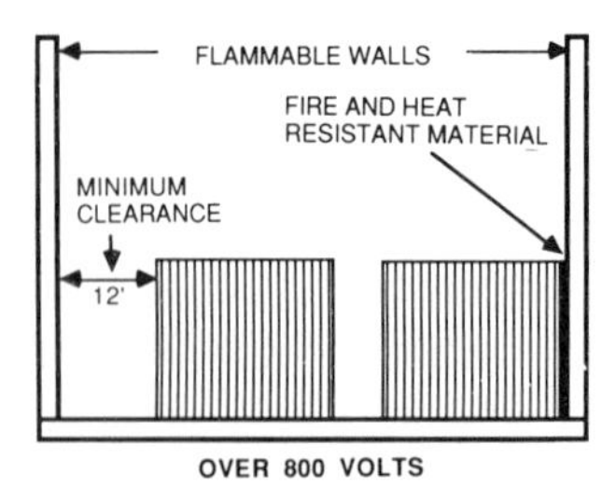

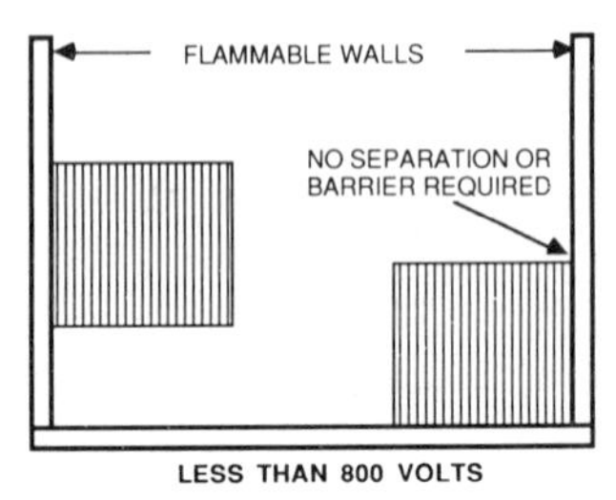

FIGURE 3–5. NEC requirements for dry-type transformers of 112.5 kVA or less.

Transformer Rooms and Vaults

The NEC in Article 450 gives specific provisions about the environment in which transformers shall be installed. These vary depending on the kVA, voltage, and the type of coolants used.

Dry-type transformers rated over 600 volts at not more than 112.5 kVA when installed indoors shall be separated from combustible materials by a minimum of 12 inches. A fire-resistant, heat-insulating barrier must be installed between the two if this requirement cannot be met. If the transformer is rated at 600 volts or less, no barrier is necessary (Figure 3–5).

Dry-type transformers over 112.5 kVA or with voltage ratings above 35 kV that are installed inside a building shall be in a fire-resistant room. No special spacing is required when this condition of the code is met. An exception to this rule allows these transformers to be installed in a room made of combustible materials provided there is 6 feet clearance horizontal and 12 feet vertical. Also under these conditions, the insulation class must be Class B or H and allow for an 80°C temperature rise or greater. Figure 3–6 depicts these conditions.

Transformers insulated and cooled with less-flammable liquids (fire point 300°C, 572°F) and less than 35 kV may be installed indoors. They shall be located in noncombustible areas of noncombustible buildings. Minimum clearance required for the heat release of the liquid used shall be maintained. A liquid confinement area around the transformer is required. The minimum height of the sill is set at 4 inches, but it must be high enough to contain the liquid. If these transformers are not in a vault, an automatic fire extinguishing system shall be incorporated. Drains should be installed in vaults, if practical, to dispose of any accumulation of water and oil. Liquid-cooled transformers rated above 35 kV shall be installed in a vault. Figure 3–7 shows these requirements.

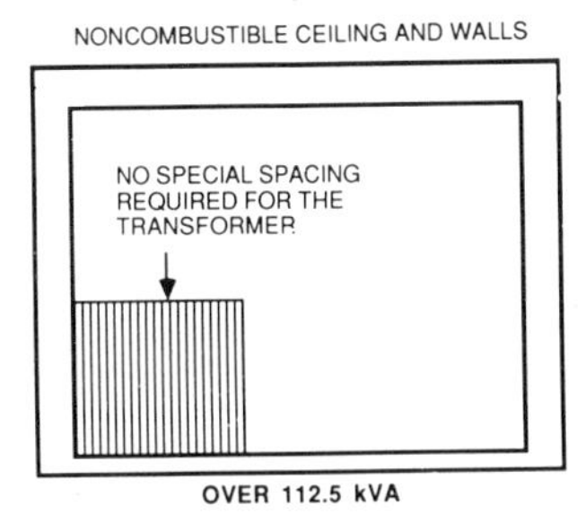

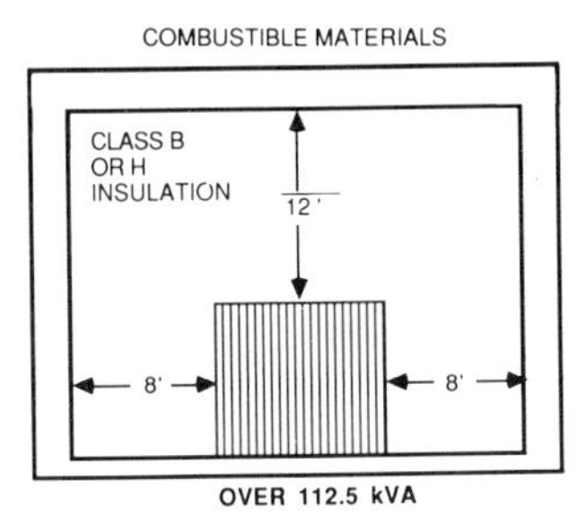

FIGURE 3–6. Requirements for dry-type transformers over 112.5 kVA when installed indoors.

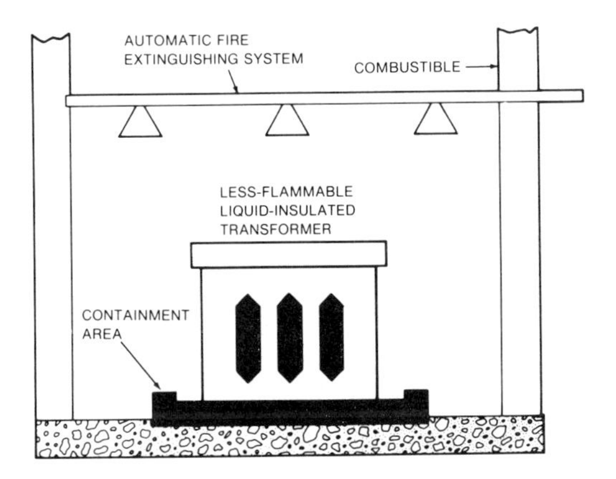

FIGURE 3–7. Installing less-flammable liquid type transformer near combustible materials.

Floors of vaults that are in contact with the earth shall be at least 4 inches thick. If the vault is above the ground level, the floor shall be so constructed as to be fire resistant for a minimum of 3 hours. The roof, walls, and doors of a vault must also meet the 3-hour requirement. The roof and walls must be constructed with a minimum of 6 inches of reinforced concrete. A 1-hour requirement is placed on those vaults that are equipped with automatic water, carbon dioxide, or Halon fire protection systems. No other piping or duct system is allowed to pass through the vault other than that used for cooling the transformers.

Vaults shall be used only for their intended purpose of housing the transformers in a safe manner. No other foreign materials of any nature are permitted to be stored in them (Figure 3–8).

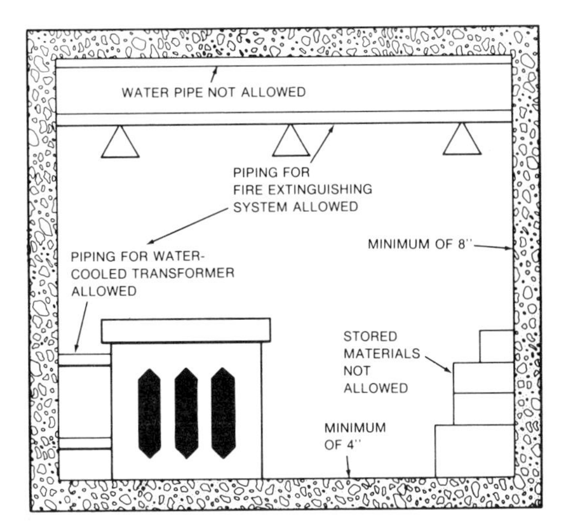

FIGURE 3–8. Requirements for transformer vaults.

Ventilation

When some synthetic coolants that do not ignite at relatively low temperatures are used, they give off nonflammable gases due to internal arcing in the transformer. In poorly ventilated rooms, means of absorbing or disposing of the gases must be included in the design. If the transformer is rated over 25 kVA, a pressure release value is required on the transformer. Duct work may be used to carry the gases to the outside (Figure 3–9).

An adequate ventilation system is defined as one that is capable of carrying off the heat losses of the transformer without causing a temperature rise that would be in excess of the transformer rating. Generally this requires the free exchange of 100 cubic feet per minute of air for each kilowatt of loss. Vaults, whenever possible, should be located in areas where access to outside air is readily available.

Ventilation openings are not to be installed in areas where people are likely to be present, nor should the vents be exhausted into any area where combustible materials may be present. When the natural circulation

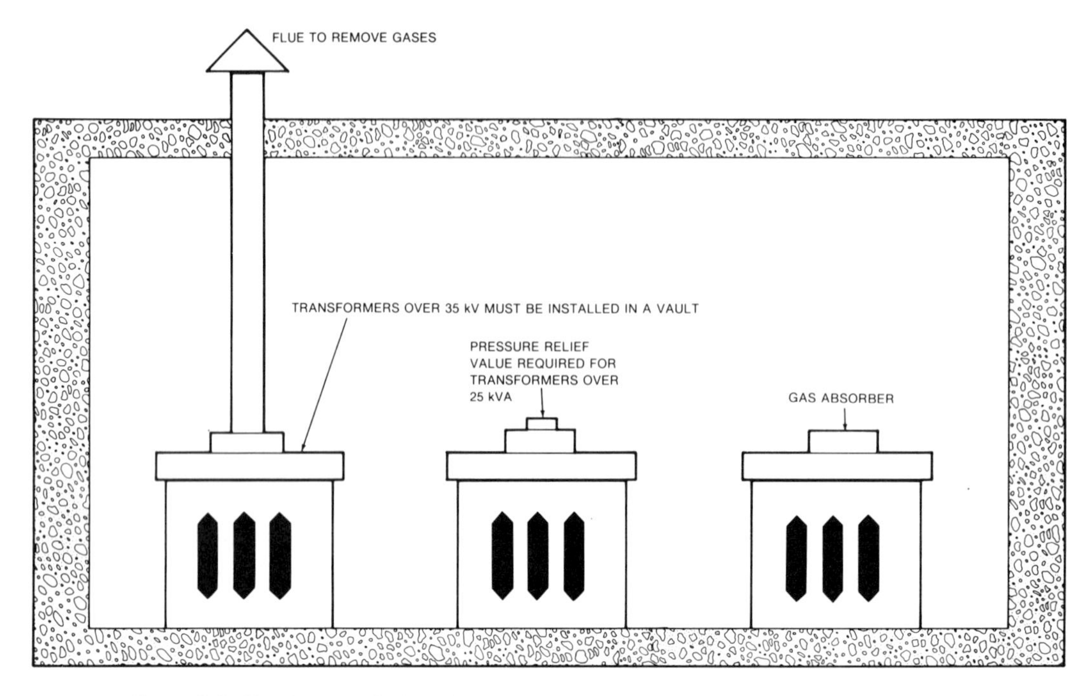

FIGURE 3–9. Requirements for installing synthetic oil-cooled and installed transformers indoors.

of air is used, one-half the total area of the vents is to be installed near floor level and the other half in the roof or near the ceiling. Three square inches of vent opening are required for each kVA. Under no condition shall the vent be less than 1 square foot for transformer capacity under 50 kVA.

Vents shall have automatic fire dampers in their design that will close in the event of a fire. They must be covered with durable screens, louvers, or grates to prevent entrance into the transformer by birds or animals and to avoid other unsafe conditions.

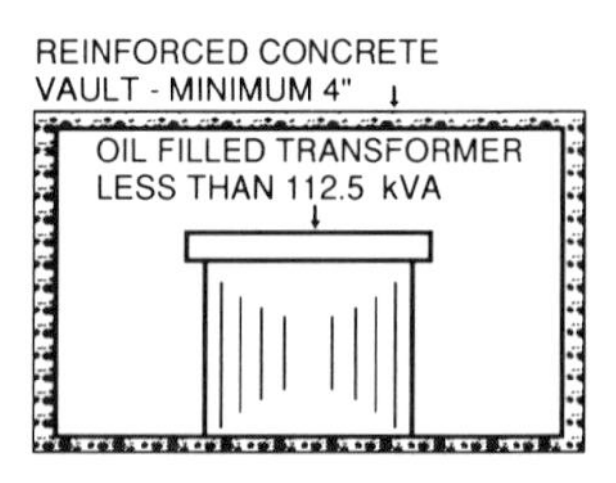

FIGURE 3–10. Requirements for oil-filled transformers installed indoors.

Flammable Coolants

Oil-insulated transformers whose coolant is flammable must be in a vault that has a minimum of 4 inches of concrete on all sides when installed indoors. The total capacity of the transformers under these conditions must not exceed 112.5 kVA. There are exceptions to this requirement, and the local and national codes should be consulted if this installation is proposed. Figure 3–10 illustrates this requirement.

Noise Levels

Another consideration when installing transformers is the level of noise that they transmit. All transformers make noise due to the vibration of the iron core as the magnetic field builds and collapses. The frequency of this noise is twice that of the frequency applied to the transformer. Transformers should be chosen that have sound levels below the ambient noise level in the area in which they are to be installed. Figure 3–11 provides typical sound levels in decibels for various locations. Most manufacturers of transformers usually provide the sound levels of their products. The measurements are normally made based on American National Standards Institute (ANSI) standards.

The first rule in installing transformers so that their noise level does not interfere with people is to select

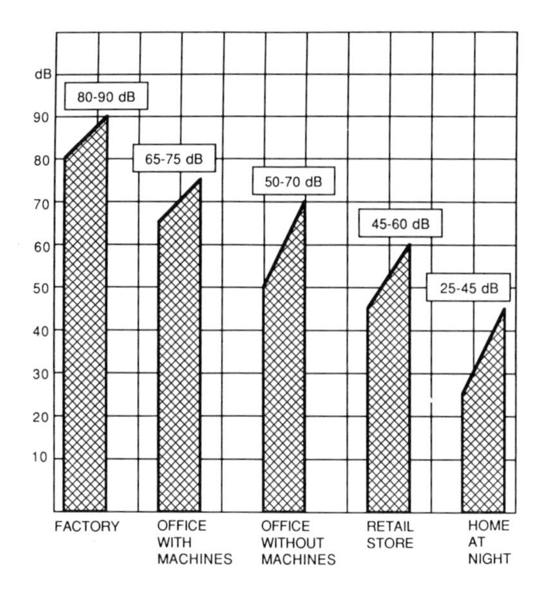

FIGURE 3–11. Ambient noise level in decibels.

one that is quiet. It must have a sound level below that of the area in which it is going to be installed if noise interference is to be avoided. The ambient sound level of the area will mask the noise of the transformer. The noise from the transformer, however, will be heard when the normal ambient noise level drops. When possible, install the transformer as far away from the activities of people as feasible.

Locate transformers where the normal acoustics of the room do not amplify their sound. The least desirable location is in a corner near the ceiling. The corner area of a room acts as a megaphone. Likewise, narrow corridors or stairwells are to be avoided since these areas tend to amplify the noise.

Mount the transformer so that its vibration is not transmitted to the structural parts of the building. Flexible mounts can be be used when the structure of the building is used. The mounts must provide isolation from the structure itself, and they must be properly loaded.

Solid mounting of small transformers to reinforced concrete walls or floors is usually not objectionable. Their mass is not sufficient to cause vibrations in the structure.

The use of flexible couplings between wiring troughs and conduits will reduce the chance of noise transfer through these devices. Flexible wire for the hook-up is also recommended.

Following these procedures will allow most transformers to be installed in areas where people are working. Avoid trying to reduce the noise level after the job is completed. Curtains and screens have not proven to be effective in reducing transformer noises. The use of acoustic tile has minimal effect at transformer frequencies. Construction of sound barriers after the fact is expensive and may not prove effective.

Most manufacturers provide low noise level types of transformers. General Electric and Westinghouse sell a quiet, dry-type transformer that averages 40 decibels in the 0 to 9 kVA range, and 65 decibels for the 1001 to 1500 kVA devices. It should be kept in mind that a 3-decibel rise in noise results in a doubling of amplitude. On the other hand, a reduction of 3 decibels will cut the noise level in half.

For example, if the noise level of a 5 kVA transformer changed from 40 to 43 decibels, its vibrations would be twice as loud. If the noise level of a 1500 kVA transformer was reduced from 65 to 62 decibels, it would be only one-half as noisy as before.

Location

Whenever possible, open ventilated transformers should be installed in areas that are clean and dry. If the location does not meet these specifications, suitable protection is needed. Dry-type transformers depend on the free flow of clean air. They are not to be subjected to areas containing excessive dust and corrosive chemicals. In extreme cases, filtering of the air may be necessary.

Ambient temperature of the room in which transformers are to be installed should be checked. If the temperature is above the normal value of 40°C, the transformer may have to be derated, or the ventilation increased to reduce the ambient temperature.

Transformers are normally designed according to industry standards to operate below altitudes of 3300 feet. Because the air becomes thinner with increase in altitude, less heat is removed from the device. For each 330 feet above 3300 feet, the transformer is to be derated 0.3%. Transformers operated 6600 feet or more above sea level should have their outputs reduced to 97% of their kVA rating. In most cases, this should not be a problem. The majority of installations have a safety factor sufficient to offset any needed altitude corrections.

Grounding

Grounding is probably the most misunderstood of all the topics regarding electrical systems. Several books and large sections of others have been written to help explain the topic. An examination of Article 250 of the NEC reveals that grounding requirements are associated with almost every other other article of the code. A complete discussion of this topic is far beyond the scope of this text and will be limited to a few basic considerations and explanations.

Purpose

The purpose of grounding systems is threefold:

1. to keep noncurrent-carrying metallic parts of electrical systems at zero potential of the earth to protect personnel who come in contact with them.
2. to limit excessive voltages caused by lightning, transients, and line faults with higher-voltage systems to protect the equipment and associated systems from damage and to avoid life-threatening situations due to these accidental happenings.
3. to stabilize the voltage in respect to ground and to facilitate the operation of overcurrent protection devices when a ground fault occurs.

Definitions

Before discussing this topic, a few definitions are in order. *Ground* is generally defined as being earth potential. In some cases, where the earth is not very conductive, such as in very dry regions, artificial grounds can be established for an electrical system. This may include the introduction of chemicals into the earth, or a counterpoise, consisting of conductors laid out in a radial fashion and buried in the earth. When a system is said to be *grounded*, this means that it is brought to the reference potential of the earth or the established ground.

A *grounding conductor* is the electrical lead that connects the system or equipment to earth. This lead is normally connected to the grounded electrode which generally is a metal stake driven into the earth. Copper or aluminum conductors are required for the grounding conductor. It may be bare or insulated. When insulated, the color shall be green if the wire gauge is number 6 or smaller. If the wire is larger than number 6, the grounding conductor must be identified in some manner.

Bonding is the permanent connection between metallic parts of a system to provide electrical conductivity between them. It assures that all related parts of the system are at the same ground potential. Methods of acceptable bonding usually involve the creation of good mechanical connections between metallic parts. Pressure connections and clamps are used for this purpose. Soldering the parts together is not permitted.

System Grounds

Preference for establishing the system ground is shown in Figure 3–12. The first choice would be the steel frame of the building. Where this is not possible, a metallic cold water pipe is the next best source of obtaining a good ground. Finally, if a ground cannot be established in one of these ways, a rod is driven into the ground as close as possible to the equipment.

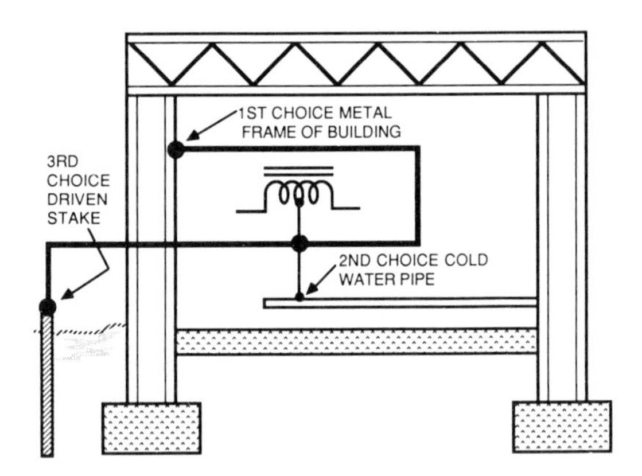

FIGURE 3–12. Sources for grounding systems.

Equipment Grounding

In general, all covers, conduits, cable trays, and other noncurrent-carrying metallic parts of electrical systems must be grounded as provided in Article 250 of the NEC. There are exceptions to this general rule, and they should be checked.

Care must be taken that the integrity of the grounding system is not broken. Grounding conductors and those providing bonding between parts shall not be wired through a switching device, which thereby creates a means of breaking the continuity of the system.

An exception also exists to this rule. If a circuit breaker is used that simultaneously breaks all live conductors when the ground breaker opens, then it is permitted. Further study of the requirements for grounding is necessary for the electrician.

I had the unique "pleasure" of experiencing a situation where article 250 of the code had not been followed. The case in point took place in Japan.

Additional new wiring for an interoffice communication system was being installed in a dropped ceiling which had existing wiring for light and power. The crawl space in the ceiling was approximately 2½ feet, and the ceiling itself was capable of supporting the weight of a man. The dropped ceiling was approximately 12 feet from the deck, and required a 10 foot stepladder to enter it.

Work had progressed in an orderly manner for several hours. The wiring had been taken to a point in the ceiling where a supporting beam prevented passage to another area of the room where the wiring was to be terminated. There was a space of approximately 2 to 3 inches between the beam and the dropped ceiling through which the wire could pass. It was necessary to climb down from the ceiling and move the ladder to the other side of the beam.

Immediately on entering the ceiling on the other side of the beam, the planets and all the stars in heaven whirled around in my head, and I became unconscious for a short period of time. On regaining consciousness, I found my head at the opening in the ceiling. Perspiration ran profusely from my body. To the best of my recollection, I did not know what had happened, and it took several seconds for the realization of my situation to register. At this point, pure fear took hold of me. I was afraid to move in anticipation of a recurrence of the shock. My flashlight had gone out, and I could see very little of what surrounded me.

It took several minutes before I summoned up enough courage to wiggle my toes. Finally I begin working myself inch by inch backward by pushing with my heels. I came out of the hole in the ceiling head first down the ladder.

When I reported the incident, a local electrician was summoned to determine the cause of the fault. The electrician was a Japanese national who informed me, "It never happens in Japan."

My integrity being at stake, I asked him what he meant that it could never happen in Japan. He proceeded at that point to explain grounding systems to me and why in Japan such an accident could never happen.

"You see," he said, "In Japan all conduits are bonded together and taken to the ground. It never happens that you get shocked."

Being frustrated, young, inexperienced, and having survived this dangerous experience, I suggested that perhaps he would like to go up into the ceiling to prove his point. Now, being older, more experienced, and more tolerant of people, I would not take this approach.

Without hesitation the local electrician climbed the ladder into the ceiling. There was a very loud scream and the man came falling from the ceiling. I caught him in my arms as he came head-first down the ladder. The Japanese electrician looked up at me and said, "It happened in Japan."

To make a long story short, a fluorescent fixture ballast had shorted. The conduit supplying the fixture had a break in continuity and had not been grounded as required by article 250 of the code. Once the necessary repairs had been made, the intercommunications job was completed without further incident, but not without some trepidation on my part.

There are several lessons to be learned from this incident. The most important is, take the time to do the job right in the beginning. You may be setting a booby trap for someone who will work on the equipment in the future. Second, believe it if someone tells you a noncurrent-carrying part of an electrical system is hot. Test it before you touch it.

Rules for Grounding Transformers

In the case of a transformer guarded by a chain link fence, not only must the cover of the transformer be grounded, but also the fence itself. Figure 3—13 illustrates this requirement. The grounding conductor must be of sufficient gauge to provide a return path for ground faults without burning out. The size of this conductor has been established in the NEC, Article 250, based on the rating of the protective device upstream from the equipment.

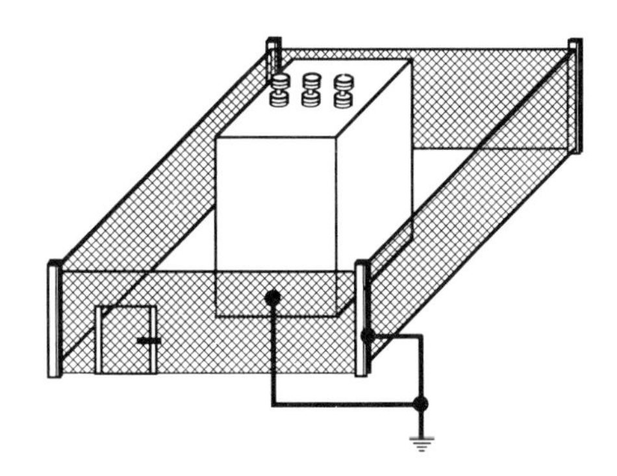

FIGURE 3–13. Grounding requirements for a transformer guarded by a chain-link fence.

There are some general requirements for the grounding of AC systems. These rules are divided according to the secondary and primary voltages of the transformer.

Transformers having secondary voltages of less than 50 volts must be grounded if the primary voltage is greater than 150 volts to ground, or when the transformer's source voltage is ungrounded. Also, this system requires a ground if the conductor carrying a voltage of less than 50 volts is installed overhead outside the building.

Figure 3–14 depicts the requirements for grounds of systems from 50 to 1000 volts AC (NEC 250-5b). The general rule is that circuits with voltages less than 150 volts shall be grounded systems. In addition, in 277/480 volts, three-phase, four-wire wye systems, where the neutral is used as a circuit conductor, the neutral must be grounded. Also, 240/120 volt single-phase, three-wire systems must have the neutral grounded. There are exceptions to these regulations. When in doubt, consult the appropriate code.

Systems over 1 kilovolt need not be grounded, but it is permitted. When systems over 1 kilovolt supply portable equipment, they shall be grounded.

Separately derived systems such as a transformer on each floor of a high rise building have the same grounding requirements as specified previously. Exceptions to this rule would include electrical cranes operating over combustible fibers and the anesthetizing locations such as the operating rooms of hospitals.

Electrical Protection

Transformers are designed for a particular voltage and a maximum kVA rating. When these values are exceeded, the insulation quality of the transformer is reduced, and serious damage or destruction can occur. Both overvoltages and excessive currents attack the insulation quality of the device. Overvoltages often exceed the dielectric strength of the insulator and arc-over occurs. Excessive currents cause the temperature of the transformer to exceed its rating. These high

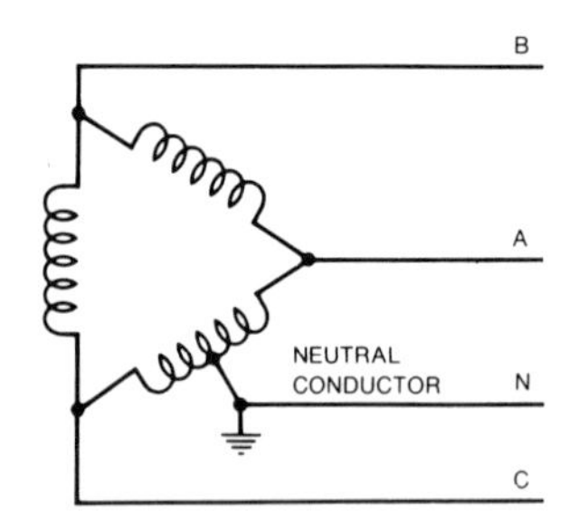

(A) Center-tap grounded, 3-phase, 4-wire delta.

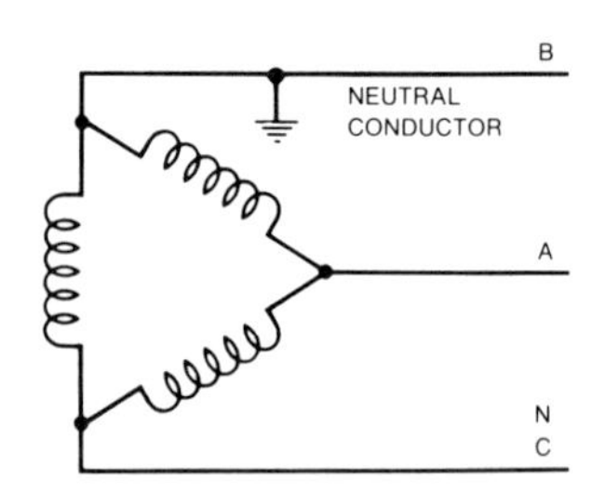

(B) Corner grounded, 3-phase, 3-wire delta.

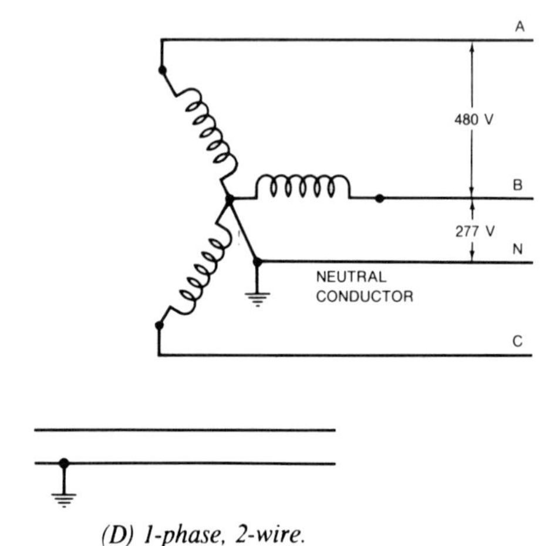

(C) 3-phase, 4-wire wye.

(D) 1-phase, 2-wire.

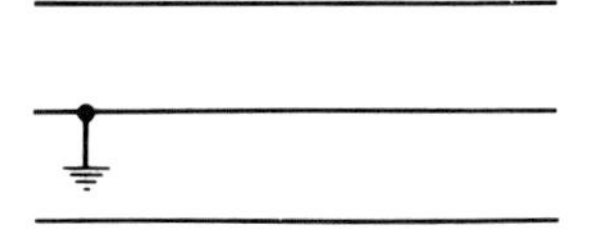

(E) 1-phase, 3-wire.

FIGURE 3–14. System grounds.

currents cause deterioration of the dielectric. Therefore, the transformer must be protected from both overvoltage and excessive currents.

The most common causes of overvoltages is lightning strikes and surges due to switching on the line. The conventional protective device used to alleviate this problem is the *surge arrester*.

Arresters

Arresters are normally installed before the high-voltage bushings of the transformer. They must be attached to each hot line feeding the transformer. A separate device is needed for each line protected. The voltage rating of the arrester is usually selected based on the voltage present on the line. Its breakdown voltage is slightly higher than the rated voltage. When this voltage is exceeded, the arrester shorts the excessively high voltage to ground preventing it from reaching the primary of the transformer.

Wire gauge of the grounding conductor must be at least as large as the feeder, nor shall it be smaller than number 14 copper wire. The grounding conductor should be routed for the shortest possible path to ground. Article 280 of the NEC describes the requirements for the installation of these devices.

Arresters may also be of an ungrounded type. These are wired between the phases of the incoming line. They effectively short-circuit surges on the line and prevent the higher than normal voltages from being impressed across the primary of the transformer.

Manufacturer's recommendations should be followed when selecting and installing arresters.

Some transformers come equipped with lightning arresters already installed. Some are connected directly to the line, while others depend on an adjustable air gap for the high voltages to jump. These gaps have been set at the factory, but they should be checked. Figure 3–15 shows this arrangement.

The size of the gap is marked on the arrester. Feeler gauges are used to check and set the gap. These adjustments should be made before putting the transformer in operation. Never attempt this adjustment when the transformer is hot.

Lightning arresters are designed so that the line current does not follow the overvoltage current to ground. This would be a considerable waste of power and would increase the outage upstream by taking out fuses and breakers. When the arrester is activated by high voltages, fibrous materials inside the arrester release a deionization gas, creating a nonconductive gap, thereby forcing the line current through the lower impedance of the primary of the transformer.

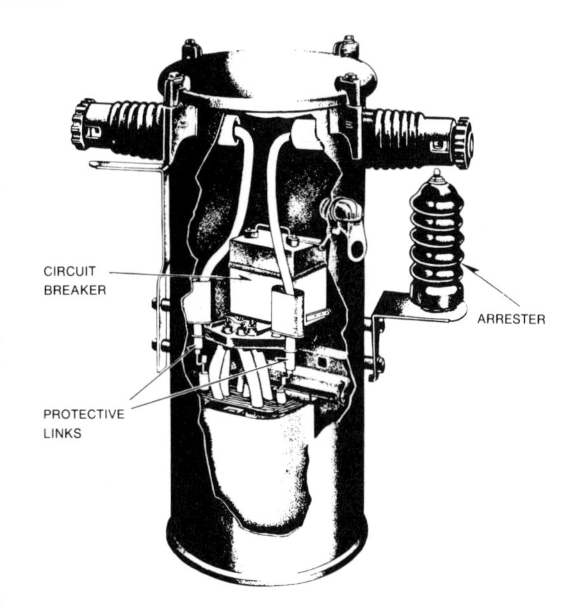

FIGURE 3–15. Transformer protected by lightning arrester. (*Courtesy Westinghouse Corp.*)

Another useful feature of the lightning arrester is that it can do its job over and over without being replaced. Credit for the basic research and initial development on this device is given to Dr. Charles Steinmetz, the father of alternating current technology.

Fuses and Breakers

There are two classifications of overcurrents. The first is a condition where the transformer is delivering from one to six times its normal current rating. The current is confined to its normal circuit path, and a temperature rise takes place in the transformer. Protective devices must react within seconds to this condition, which is generally called an "overload."

The second condition that can exist which will cause overcurrents is when the circuit is operating under abnormal conditions. In this case, currents are no longer forced through the normal wiring channels to the loads. Under abnormal conditions currents may be hundreds of times the normal operating currents and may reach levels of 50,000 amperes or greater. Tremendous magnetic forces are generated that can completely destroy bus bars, switching gear, and other related electrical equipment, if not the building housing the electrical equipment itself. Electrical arcing and fires are common outcomes under abnormal conditions, with the possibility of toxic fumes and danger to property and human life. Protective devices must react in milliseconds to prevent the effects of short circuits.

Protection against overcurrents is provided by fuses and circuit breakers. These devices are rated according to their voltage and current handling capabilities.

The voltage rating of the protective devices must be at least as high as the voltage in the circuit it is to protect. A fuse or circuit rated for a higher voltage breaker may be installed, but a lower rating must never be used. The voltage rating is based on dielectric strength, the size of the gap created when the device opens, and its ability to prevent arc-over. If the voltage rating is too low, the protective device may not clear the fault, thereby causing severe damage.

Fuses and circuit breakers are also rated according to their ampacity. Generally, this protection should not exceed the current carrying capacity of the circuit it is protecting. If the conductors of the circuit are rated at 50 amperes, then the protective device should not be greater than 50 amperes. However, there are times when codes permit a higher rating than the full-load current of the circuit.

Even when the voltage ratings and recommended ampacity of circuit protection are scrupulously followed, this may not provide protection against a short circuit fault and the resulting destructive forces that may result. Fuses and circuit breakers are designed to withstand these tremendous currents without rupturing and causing further damage. It is important that they are selected based on their amperes interrupting capacity (AIC) rating which should be greater than the maximum current that can possibly be delivered by the source. The level of this current will differ at various points in the circuit. The further downstream from the source that the fault occurs, the smaller will be the fault current.

The lesson to be learned from this is that it is not sufficient to replace a 600 volt/100 ampere fuse with another 600 volt/100 ampere fuse. This may not assure the integrity of the circuit under fault conditions. The AIC of the replacement fuse must be checked to see if it is as large or greater than that of the defective fuse. To do otherwise would be very foolish, and may cause a fusebox to blow up in your face. This might be disastrous. Figure 3–16 shows two fuses rated at 200,000 AIC that are manufactured by Bussman Division of McGraw-Edison Company.

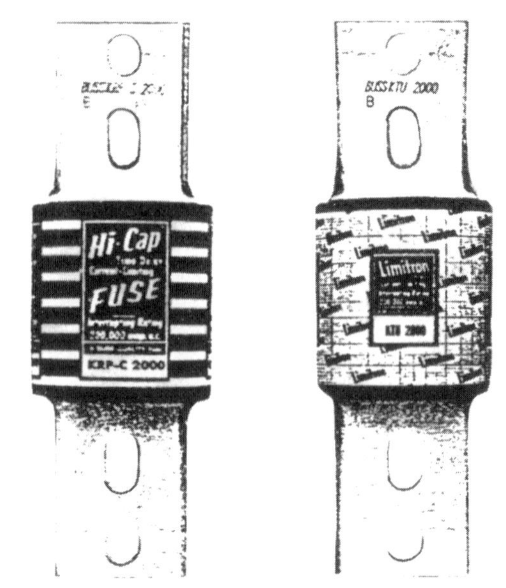

FIGURE 3–16. Fuses providing protection up to 200,000 AIC.
(Courtesy Bussman Division, McGraw-Edison Co.)

Transformer Protection

To protect against overloads and short-circuit conditions, some transformers have a fuse cutout in the primary circuit. This fuse is installed after the lightning arrester and before the primary windings inside the transformer. It is sensitive to heat rise in the transformer. Its primary purpose is to protect equipment upstream due to defects in the transformer, such as coil to core shorts or shorted windings. Figure 3–17 shows how to identify the protective link.

The secondary of the transformer is also protected by fuses or circuit breakers. Their primary purpose is to protect the transformer from overloads or shorts in the secondary circuit. Their size must be selected so that they will blow out or open before a temperature rise due to overloads and shorts can cause damage to the transformer. Yet, they must be large enough that they do not open under normal operating conditions which may place temporary overloads on the system. Such overloads occur regularly, for example, when large motors are started. Figure 3–18 depicts the circuit breaker and circuit.

CSP Transformers

Many distribution transformers come equipped with all three of the devices listed above: surge arresters, primary fuse links, and secondary breakers. These

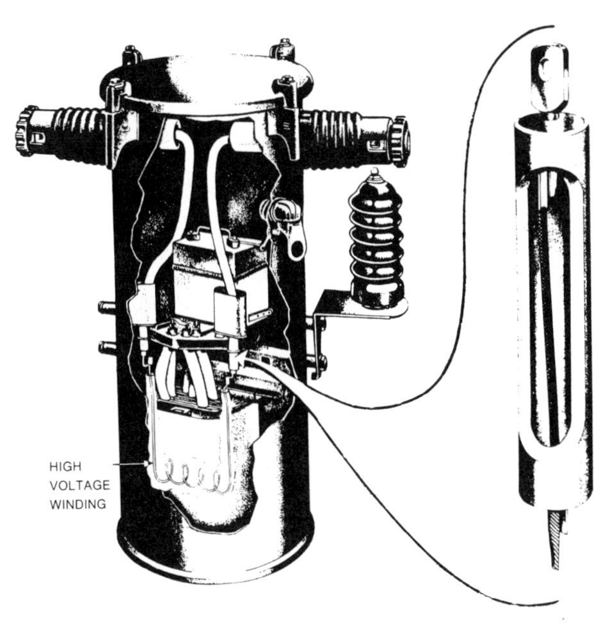

FIGURE 3–17. Internal protective link. (*Courtesy Westinghouse Corp.*)

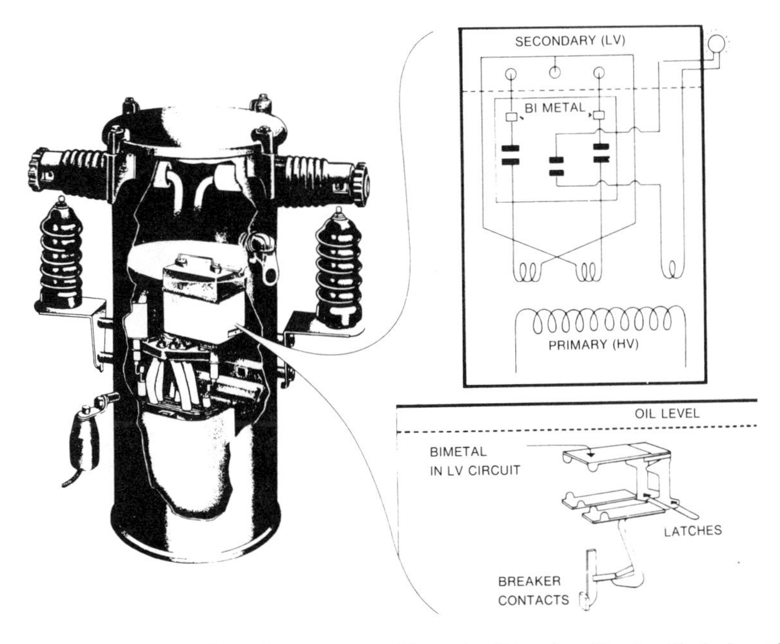

FIGURE 3–18. Secondary of transformer protected by a circuit breaker. (*Courtesy Westinghouse Corp.*)

transformers are known as "completely self-protected" (CSP) transformers.

CSP transformers have a red signal light that comes on prior to the breaker opening on the transformer. This light warns of pending overload so that the conditions can be remedied prior to the secondary circuits being dropped from the system. There is a reset handle for the breaker on the exterior of the tank that allows the breaker to be quickly reset. Figure 3–19 shows the handle and the warning light.

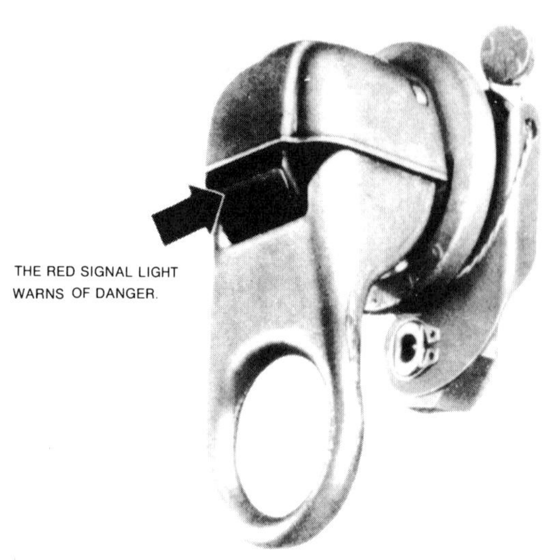

FIGURE 3–19. Breaker reset handle and warning light. (*Courtesy Westinghouse Corp.*)

These transformers also have reserve capacity that can be used in the event of continuous overloads. This reserve can be activated by recalibrating the circuit breaker. This is readily accomplished by breaking a seal on the circuit breaker reset handle and turning it to the higher current rating. This feature reduces the need for emergency changes of the transformer. However, once the handle has been set for the higher current rating, the transformer should be replaced within a few days with one of higher kVA rating. Figure 3–20 illustrates the reset and recalibration handle.

Once the smaller transformer has been taken out of service, the circuit breaker is to be recalibrated for the proper current capacity before using it on another job. At the time of recalibration, the seal is replaced on the handle.

Transformers that have been in service for long periods of time and are being replaced often fail at this point. The probable cause is the physical shock to the insulation when moving the transformer. The insulation, through use, may have deteriorated and become brittle. When the transformer is moved, the impact on the insulation causes it to break and flake resulting in shorts between windings and the core. Before putting a transformer back in service, it should be tested completely.

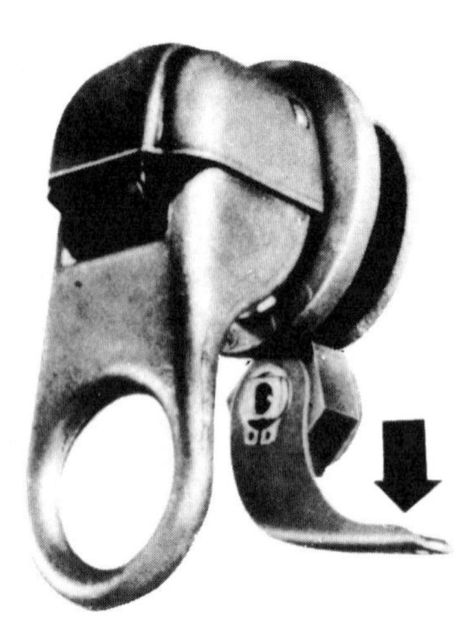

FIGURE 3–20. Recalibrate circuit breaker for higher current. (*Courtesy Westinghouse Corp.*)

Requirements for Transformer Protection

Article 450 of the NEC provides the requirements for protecting transformers from overloads. The transformers are first separated into those handling voltages over 600 volts nominal, and those handling 600 volts nominal or less. The NEC Code divides transformers rated over 600 volts, into two main branches: those with primary protection only and those with both primary and secondary protection.

When only primary protection is provided, the maximum current rating of the fuse is 250% of the full-load current (FLA) of the primary. If this calculation does not equal a standard size fuse, the next larger size may be used. If a circuit breaker is used for protection, its maximum size is limited to 300% of FLA. When this value is not equal to a standard size circuit breaker, the next larger size may be used. Prior to the 1987 NEC, the size of the circuit breaker would have been reduced to the next lower standard size. Figure 3–21 depicts these conditions. The secondary of the transformer must still be protected under these circumstances by providing individual overload protection for the loads.

For example, let us assume you are installing a 1200/240-volt, single-phase, 100-kVA transformer. What size fuse or circuit breaker is permitted to protect the primary?

The first step is to determine the primary current

$$I_P = \frac{kVA}{V_P} = \frac{100 \times 1000}{1200\ V} = 83.3\ A$$

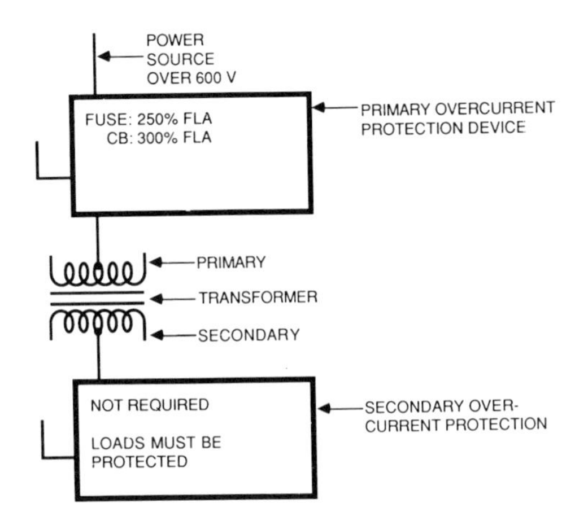

FIGURE 3–21. Primary protection only for circuit over 600 volts.

Fuse size would equal

$$250\% \times 83.3 = 208.25 \text{ A}$$

The next larger standard size fuse would be

$$225 \text{ A}$$

Circuit breaker size would equal

$$300\% \times 83.3 \text{ A} = 250 \text{ A}$$

A 250-ampere circuit breaker would be a standard size. In this example, the size would be the same as the calculation. If the value had not been a standard size, the circuit breaker ampacity may be increased to the next higher value.

If the problem above was changed from a single-phase transformer to a three-phase circuit, the formula for calculating the current would change to

$$I_P = \frac{100 \times 1000}{1200 \times 1.73} = 48.17 \text{ A}$$

Required fuse size would be 250% X 48.15 = 120.4 amperes, with the next larger standard size being 125 amperes. Circuit breaker size would equal 300% X 48.15 = 144.45 amperes with the next higher standard value being 150 amperes.

When both primary and secondary protection is provided the transformer, the impedance of the transformer is considered. With both the primary and secondary voltages over 600 volts, and the transformer having an impedance of 6% or less, the primary may be fused at a maximum of 300% of primary current. The secondary may have a maximum fuse size of 250% of the secondary current. If circuit breakers are used for protection, the primary device may be 600% of the current, and the secondary has a maximum value of 300%. Figure 3–22 shows these requirements.

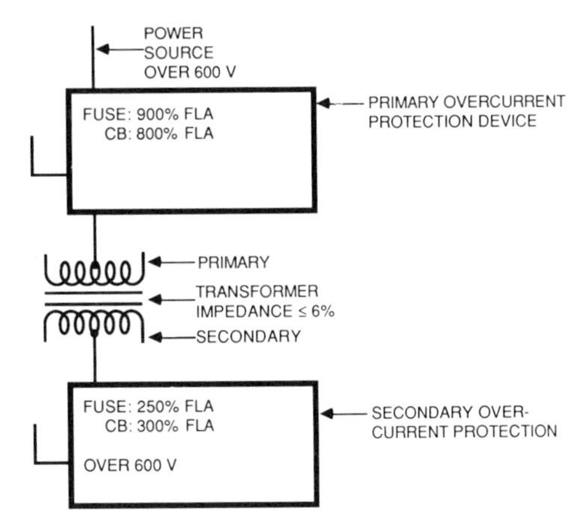

FIGURE 3–22. Overcurrent protection for transformers over 600 volts having 6% or less impedance.

If the transformer has an impedance greater than 6% but equal to or less than 10%, the value of the multiple for the primary fuse remains at 300%. The secondary fuse, however, is reduced to 225%. If circuit breakers are used, both the multiple for the primary and secondary devices are reduced. The primary breaker shall not exceed 400% of primary current, and the secondary breaker shall not exceed 250% of the secondary current. Figure 3–23 illustrates these conditions.

When secondary voltages are 600 volts or less, then the overcurrent device, fuse or circuit breaker, is set at a maximum 125% of the secondary currents, regardless of the impedance. For supervised transformers, this value is increased to 250%. Primary protection remains the same as for unsupervised transformers. Where the conditions for secondary protection are met, or the transformer is thermal protected, primary protection is not required if the feeder is protected by a fuse or circuit breaker that does not exceed the percentages for primary protection (Figure 3–24).

Transformers rated 600 volts nominal or less are divided into two main categories for overcurrent protection. These are "Primary Protection Only" and "Primary and Secondary Protection."

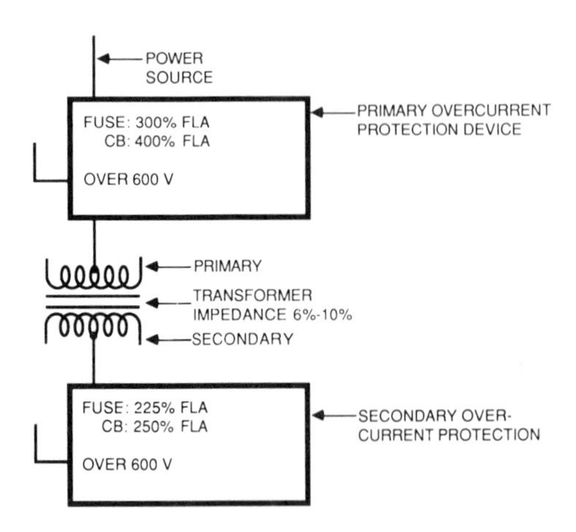

FIGURE 3–23. Overcurrent protection for transformers over 600 volts having an impedance between 6% and 10%.

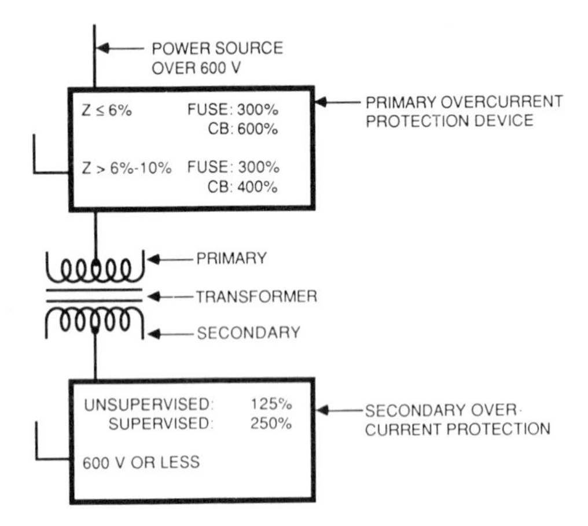

FIGURE 3–24. Overcurrent protection for transformers with primary voltage over 600 volts and secondary voltage of 600 volts or less.

Where only primary protection is provided, the general rule is that the fuse or circuit breaker shall not exceed 125%. This would be the optimum value using dual element fuses.

Exceptions are made to this rule for two-wire secondaries based on the amount of maximum primary current. These values are: less than 2 amperes; 2 amperes but less than 9 amperes; and more than 9 amperes. The flow chart in Figure 3–25 depicts these conditions.

The maximum values set by the NEC may vary depending on the type of load the transformer is supplying. Those sections pertaining to overcurrent protection for the particular device should be checked. Loads supplied by the transformer must have overcurrent protection when the secondary of the transformer is not protected.

When both primary and secondary protection is provided for the transformer, the NEC again divides the requirements for the protective devices into two categories. These are for those transformers with and without thermal protection.

In both cases, the optimum protection is to use dual-element fuses rated at 125% of full-load current. This block is not shown in Figures 3–26 through 3–28 where the maximum allowable values for the various conditions are given. The best and safest policy is to protect the transformer at the lowest possible rating at which proper operation of the different types of loads is obtained. Do not exceed the maximum values provided by the code.

Figure 3–26 shows the maximum allowable values under the NEC when both primary and secondary protection is provided for a transformer without thermal protection. Figures 3–27 and 3–28 provide the same values where thermal overload protection is part of the transformer design. Thermal-protected transformers are subdivided into those that have impedances of 6% or less and that have impedances greater

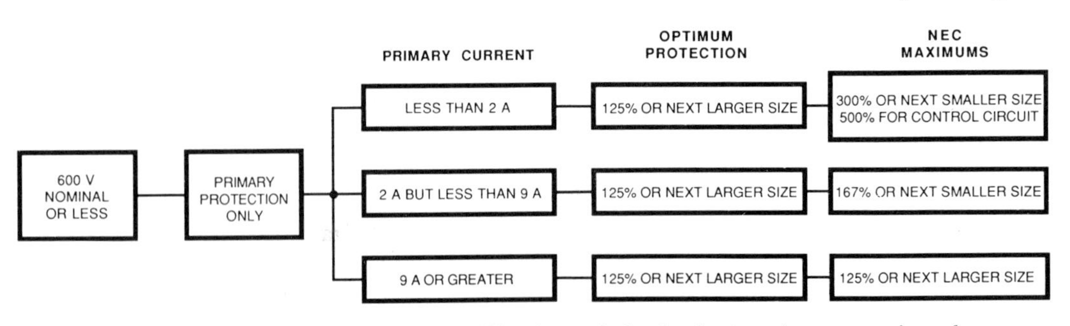

FIGURE 3–25. Requirements for transformer 600 volts nominal or less having primary protection only.

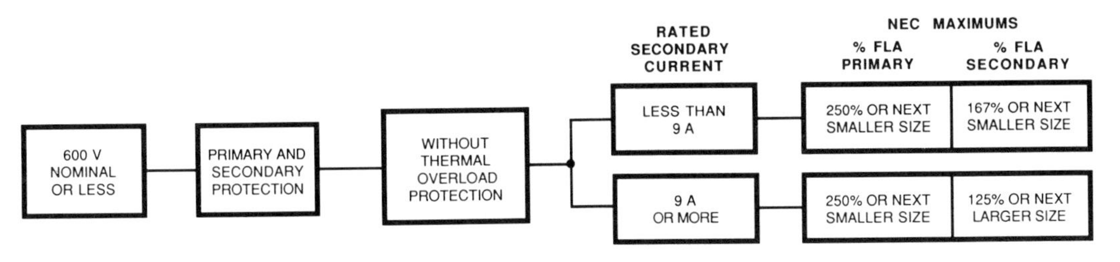

FIGURE 3–26. NEC maximum ratings for protective devices for transformers without thermal protection.

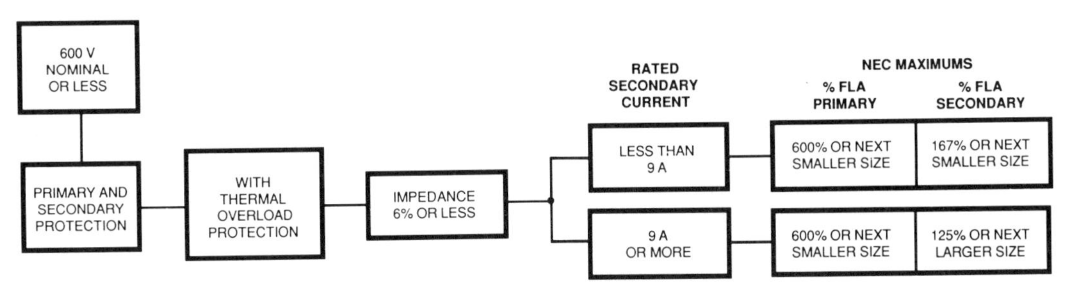

FIGURE 3–27. NEC maximum ratings for protective devices for thermally protected transformers with impedances of 6% or less.

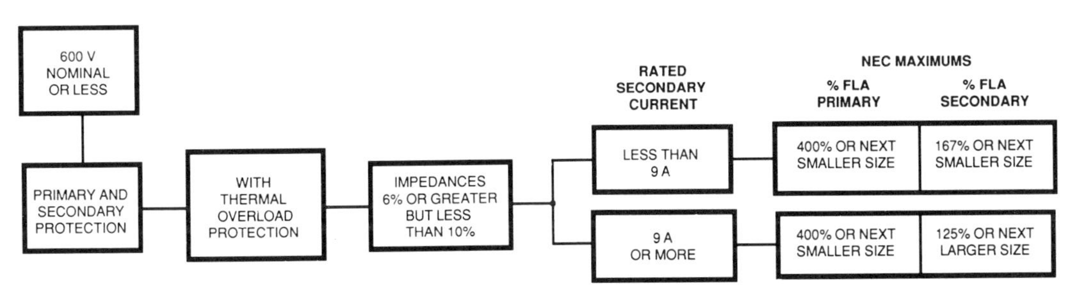

FIGURE 3–28. NEC maximum ratings for protective devices for thermally protected transformers with impedances greater than 6% but less than 10%.

than 6% but less than 10%. Each of these divisions is again divided into transformers having primary currents of less than 9 amperes and those of 9 amperes or greater.

Interrupting Capacity

Besides determining the voltage and ampere rating of a fuse, it is also necessary that the interrupting capacity (AIC) is also calculated. As discussed previously, if the AIC is not large enough, shorts and ground fault currents can cause considerable damage.

The impedance rating of the transformer, which is given as a percentage, is used to make this calculation. The following example demonstrates how this is done.

Given a 50-kVA single-phase transformer with a 240/120-volt secondary and an impedance of 1.7%, what value of interrupting capacity should the fuse have?

$$\text{Full-load current} = \frac{\text{kVA X 1000}}{\text{V}}$$

$$= \frac{50 \text{ X } 1000}{240} = 208.33 \text{ A}$$

$$\text{Interrupting capacity} = \frac{\text{FLA}}{\%Z} \times 100$$

$$= \frac{208.33 \text{ A}}{1.7} \times 100$$

$$= 12{,}255 \text{ A}$$

This would be the absolute maximum amount of current the transformer could deliver if the fault was directly at the output terminals of the transformer. The further downstream the fault occurred, the less current would flow. This is due to the impedance of the conductors and devices between the transformer and the fault. This value should always be considered when replacing or installing fuses.

It is desirable when a fault occurs in a transformer or a load downstream that the malfunction does not cause the feeder systems upstream to also drop off the line. This generally can be accomplished by providing protection on a 2:1 ratio. For example, if the feeder is protected at 1200 amperes, the maximum fusing on each of its loads must not exceed 600 amperes. If the 600-ampere circuit is supplying other loads, the maximum fusing of those loads should be 300 amperes or less. These ratios may vary with different circumstances.

Cooling Systems

All transformers must be equipped with some method for cooling. Heat decreases the transformer's efficiency and reduces the unit's life expectancy. Operating the transformer below rated temperature allows for a greater kVA capacity to be delivered to the load.

Dry-Type Transformers

For dry-type transformers, no particular check is necessary if conditions for ventilation have been provided for the area in which the transformer is to be installed. In this event, the ventilation system of the room should be inspected to make sure it is operating properly. Transformer radiator vents should be clear of any obstructions that may prevent the heat from escaping.

Forced Air

If the cooling system consists of forced air for maintaining the temperature of the transformer, the motors should be checked for proper lubrication and operation. These motors will often be controlled by thermostats which also need to be inspected and tested to determine if the motors operate within the preset temperature ranges. Manufacturers' specifications are to be followed for making these tests. When located in closed vaults, means should be provided to exchange the air in the vaults.

Water-Cooled Systems

Water-cooled systems must be tested for leaks and proper operation. To make these tests, the cooling coils should be removed from the transformer if water is to be used to provide the pressure. Care must be taken not to damage the coils when removing or replacing them. If the coils are to be checked in the transformer, air or the type of coolant oil used by the transformer can be used to make the pressure checks.

If the coils are removed from the transformer, the total water cooling system can be checked. The coils are filled with water under pressure of 80 to 100 psi and they are checked for leaks. The coils remain under pressure for at least 1 hour to assure that no leaks will develop. A drop in pressure would indicate a leak.

This time is well spent. As mentioned before, a very small amount of moisture, 0.06%, will reduce the dielectric strength of the oil by 50%. Moisture content beyond this point does not cause the dielectric strength of the oil to deteriorate much more.

Associated equipment with the water-cooled system also needs to be checked, including the water pump, pressure gauges, temperature gauges, and alarm system. In addition, make sure you have a water source with an adequate flow and pressure.

Liquid Coolants

Oil coolants are specially prepared and dehydrated. They are processed to be free of acids, alkali, and sulfur. To ensure good circulation, they have a low viscosity.

Need for Dielectric Testing

When a liquid coolant is used, its dielectric strength must be tested. An oil-cooled transformer should never be placed in service without this test.

This writer is aware of one disastrous incident where this test was neglected on a very large transformer which required two railroad flatcars to deliver it to an ammunition factory in central Illinois. When the transformer was energized, not only was the transformer destroyed, but also a part of the factory went with it.

Cautions

Exposed areas of the body must be protected when working with askarel. Askarel may cause severe irritation of the skin, especially of the lips, nose, and eyes. A coating of cold cream can help prevent injury to these areas. Eye protection is essential. In case askarel comes in contact with the skin, the affected area should be thoroughly washed with soap and water.

Older types of askarel contain PCBs that have proven to be cancer causing. Therefore, contact with these materials and breathing vapors they produce should be avoided. When a transformer is operated at 80°C or above, these vapors may be emitted.

Taking Oil Samples

When putting oil in a transformer or removing samples for testing, care should be taken not to introduce impurities. The proper results of the dielectric tests are dependent on how clean the sampling and test equipment is maintained.

Canning jars with two-piece lids are good for collecting the sample. Glass jars allow you to inspect the liquid, and if a high content of water is present, it can be seen.

The jars should be thoroughly cleaned with an oil-disolving solvent and washed in a strong detergent and water. Following this, they are dried in an oven for several hours at 230°F. The jars should be stored open to prevent condensation in a dust-free environment which is maintained around 100°F. A small, closed cabinet with a 60-watt light bulb burning inside it will maintain this requirement.

If the oil is stored in drums, allow at least 8 hours to lapse after the drums have been moved and before a sample is taken. This time permits the oil to settle. In the case of a large oil-filled transformer, several days may be required after the coolant is added. This is especially true if the oil is cold.

A "thief," a long glass or brass tube, is used to remove the oil for testing. The thief should have the same cleaning, drying, and storing procedures performed on it as the jars. A rubber hose must never be used to obtain the sample. Rubber has sulfur in it, and this element will attack the copper windings of the transformer.

Mineral oil samples are taken from the bottom of a drum, and askarel samples taken from the top. When taking oil from a transformer with a drain valve, allow the oil to run out for a few seconds before taking the sample. This will clean out any sludge or water in the valve. Rinse the container several times with the oil from the transformer and drain it from the jar before taking the sample for testing.

Approximately 1 pint to 1 quart of oil will be needed for the test. Leave room for the sample to expand. Mark the source of each sample on the bottle and whether it came from the bottom or the top of the drum or transformer.

Dielectric Testing

Even though water is not visible in the sample, moisture may still be suspended in the oil. The sample should be tested for its dielectric strength. A test receptacle similar to that shown in Figure 3–2 must be used in order to obtain accurate results based on the established standard. The receptacle requires the same cleanliness as the sample bottle and the thief. Standard procedure after cleaning is to use the oil in the sample to wash out the receptacle before using more oil for the test.

Tests on the oil are conducted at room temperature. Agitate the oil in the sample each time before pouring it into the receptacle. This will help to prevent any variations in test results. Be sure that the oil covers the 1-inch electrodes. To eliminate air bubbles in the oil, rock the receptacle gently. This should be done prior to applying the voltage.

A standard oil dielectric test set is needed. Voltage is applied in increasing amounts at 3 kilovolts per second. Each sample of mineral oil is tested for not more than five breakdowns. Allow approximately 1 minute after each breakdown before resuming testing. This process is usually repeated on three samples for a total of 15 breakdowns. Tests must be repeated if the results are not consistent. If the average breakdown voltage is less than 20 kilovolts, the oil will need to be filtered until a minimum average breakdown of 25 kilovolts is achieved.

When testing askarel, only one breakdown per sample is permitted. An average of three samples is the normal procedure, but the tests should be repeated if the results are not consistent. If the askarel tests less than 25 kilovolts, it is to be filtered until 30 kilovolts or greater is achieved.

The above recommendations are general in nature, and compliance with the operating instructions for the particular tester used in testing must be followed to obtain reliable results. Also the dielectric strength of each type of coolant may be slightly different, and manufacturer's specifications as to the minimum dielectric strength acceptable should be used for the tests.

When oil is removed from a transformer for testing, it must be replaced. A good way of doing this, and at the same time assuring that a clean bottle is used to collect the sample, is to obtain samples of the oil from the supplier. Pour the sample oil into the transformer just prior to refilling the jar with oil from the transformer.

Filling Transformer with Coolant

Prior to filling the transformer with the coolant, the inside of the tank of the transformer must be checked for cleanliness and lack of moisture. If the inside shows signs of condensation, dry it. These conditions would become evident when electrical testing is performed on the transformer prior to putting it into operation, but there is no reason to compound the problem and increase the amount of time to get the transformer on line.

It is good practice to filter the oil when filling a transformer prior to putting it into service. If a filter press is available, use it. If not, the oil can be strained through cheesecloth.

Large transformers that have pressure release valves are filled through the drain valve. This helps prevent air mixing with the oil and causing the relief diaphragm to rupture as pressure builds during the filling process. Reducing the rate of fill when the transformer is near full also helps avoid this problem.

After the oil has had adequate time to settle, which may be 12 hours or more for a large transformer, check the level gauge to determine if the proper amount of oil is present. Do not overfill. At this time, conduct another series of tests on the dielectric strength of the oil.

Problems with Coolants

Oxidation of the coolant is another problem encountered with oil and askarel-filled transformers. Oxidation causes the buildup of sludge inside the transformer and results in reduction in the insulation quality of the coolant.

On very high-voltage transformers, a closed system has been developed. Air is driven out of the transformer and replaced by an inert gas, normally nitrogen. This prevents the discoloration of the coolant and stops oxidation.

Electrical Testing

Many possible tests may be conducted on a transformer when it is manufactured, before it is put in service, during the initial start-up, and after it has been in operation for various periods of time. Before performing any electrical tests, be familiar with the manufacturer's recommendations and the proper operation of the test equipment you will use. Interpretation of various measurements may require special training and may be the responsibility of a group other than the electricians making the installation. Know the specifications of the job.

Hipot (High-Potential) Testing

Before the transformer is filled with coolant, and after it has been filled, measure the resistance between windings and the resistance between windings and ground. Tests before and after adding oil to the transformer are necessary to insure safe operation.

Either an AC or DC high-resistance tester (megger) can be used for this purpose. However, some authorities believe that DC testers should not be used when the transformer is filled with coolant. Direct current testers will polarize particles in the liquid, thereby lowering its insulation quality. The megohmmeter usually provides a standard 500 volts for testing, but such devices are available ranging from 50 to 2500 volts.

Insulation resistance is not directly related to the dielectric strength of a material. Any material will conduct if enough potential is applied to it. The results from these instruments will not tell if the insulation is going to break down when higher potentials are applied. The insulation test will, however, indicate if it is feasible to make a high-voltage dielectric strength test. If the resistance of the insulation is below the required value, high-voltage tests would not be attempted until corrective action is taken.

A resistance test before the transformer is filled with coolant can determine if moisture is present in the coils. This test may be misleading because at low temperature the resistance may be very high and within tolerance. However, at operating temperature, the transformer may not meet the insulation levels required for safe operation. Insulation resistance will be reduced by one-half for each increase in temperature of 10°C. If the resistance of the insulation was 100 megohms at 20°C, it would be 50 megohms at 30°C, and 25 megohms at 40°C.

Allis-Chalmers Manufacturing Company recommends that the minimum resistance in megohms to ground of each winding at 85°C in air and at 40°C in the coolant be equal to at least that calculated by the following formula:

$$R\text{ (megohms)} = \frac{\text{kV X 30}}{\sqrt{\text{kVA/Hz}}}$$

Using a 100 kVA transformer rated at 2.4 kilovolts, 60 Hz as an example, the recommendation would be approximately 55 megohms.

$$R\ (\text{megohms}) = \frac{2.4 \times 30}{\sqrt{100/60}} = \frac{72}{\sqrt{1.67}} = \frac{72}{1.29} = 55.8$$

To make the insulation resistance test, all windings should be grounded except the winding being tested (Figure 3–29). Each winding on the transformer should be tested in this manner.

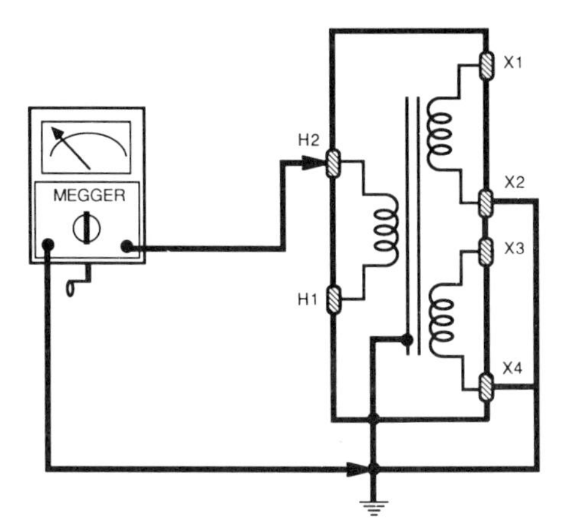

FIGURE 3–29. Resistance insulation to ground test.

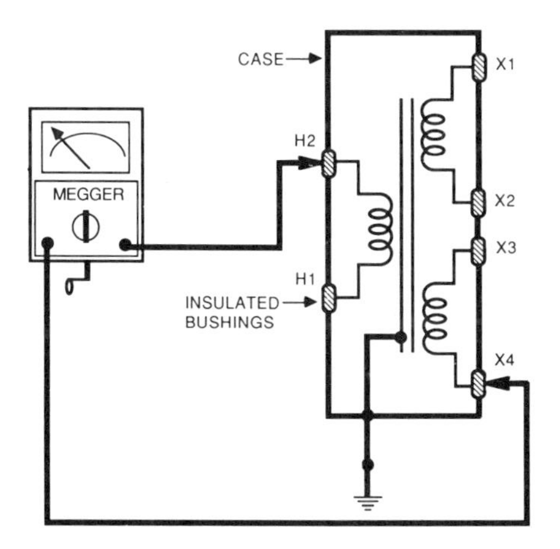

FIGURE 3–30. Resistance of insulation between coils.

After conducting the ground test, insulation between windings are tested. Figure 3–30 illustrates this test. Conduct tests on all possible combinations. The coils not being tested are grounded.

Short-Circuit Test

If the transformer passes the insulation test at room temperature, it may not do so at a higher temperature. In order to raise the temperature of the transformer to its operating value, a short-circuit test may be performed. Figure 3–31 shows the circuit for this test. This test can also prove the proper transfer of power from primary to secondary.

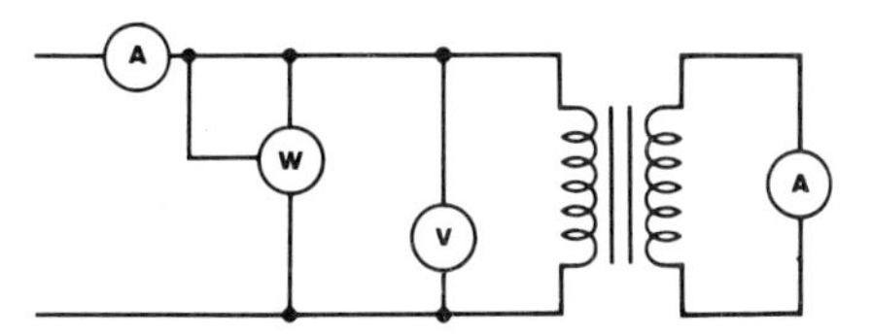

FIGURE 3–31. Short-circuit test.

The secondaries of the transformer are shorted when conducting this test. A controlled voltage from 3% to 15% of rated voltage at the specified frequency is applied to the primary windings. This voltage is carefully adjusted until the rated primary current is indicated on the ammeter. The ammeter across the secondary is for all practical purpose a short circuit due to its very low resistance.

The ratio of the currents will be the inverse of the turns ratio of the transformer. If the transformer is rated 2400/240 volts and the primary current meter indicates 3 amperes, then the secondary short circuit meter would indicate approximately 30 amperes.

This test can be run in the reverse direction with voltage applied to the secondary windings with the primary windings shorted. Caution needs to be exercised because the primary voltage in this example will be ten times the voltage applied to the secondary, due to the turns ratio.

Results should be the same in either direction. When these tests under the above circumstances are conducted on the transformer without coolant, the voltage must be reduced immediately after taking the readings to prevent a temperature rise above the rating of the transformer.

The voltage is then adjusted to a level that will maintain the recommended operating temperature of the transformer. This is approximately in the range of 20% of the full-load current. Only 0.5% to 1.5% of normal voltage at the rated frequency is necessary. Power is applied to the end terminals of the coils and not to tap connections so that current passes through all of the turns.

Several thermometers are placed in and around the transformer coils and near the top. As many as possible of the thermometers should be located so that they can be read without moving them. They must be guarded against air currents which may cause lower values than that of the transformer coil.

It cannot be stressed enough that temperatures must not exceed 85°C under any circumstance. Above this value, the insulation will begin to break down, and permanent damage can be caused.

As the temperature rises, the process is stopped at regular intervals, and the high-resistance tests are run on the transformer. Power must be removed during testing. The results of the tests for each temperature are recorded.

Drying Out the Transformer

Resistance levels will decrease with the increase in temperature. If the ohmic values fall below the recommended levels for the transformer, it is necessary to dry out the transformer. This can be accomplished by continuing to use the short-circuit method to heat the transformer. Temperatures should be maintained between 75 and 80°C. Drying will continue until the transformer meets the specifications for the insulation resistance. Frequent temperature checks are taken during heating with resistance measurements made at 4-hour intervals. The transformer must not be left unattended while it is drying.

This process of drying out the transformer is very slow and may take several weeks to accomplish. To accelerate the drying, external heat may be applied to the transformer. The transformer should be enclosed with adequate ventilation to take off the moisture. Heat is applied at the bottom of the enclosure and allowed to rise. Temperature of the air entering the enclosure must not be allowed to rise above 85 to 90°C.

It is recommended that electric heat be the source for this process. Using combustible material will make it difficult to control the temperature and gases may cause further contamination of the transformer.

If the transformer has already been filled with oil when it fails the insulation test, the same processes for drying can be used. Oil temperature at the top of the tank must not exceed manufacturer's specifications. The range should be between 75 and 85°C.

The standard dielectric strength test on the oil is performed at 4-hour intervals. Samples for testing are taken from the top and the bottom of the tank. The oil must test 22 kilovolts or higher on seven consecutive tries before drying is discontinued. Values lower than this would indicate that moisture is still being transferred from the transformer into the coolant, and the drying process would be continued.

High-Voltage Dielectric Test

Once the transformer has passed the insulation test, it is possible to conduct a high-voltage dielectric test of the system. This test determines the ability of the transformer to withstand abnormal conditions such as voltage surges and switching transients. Such a test also detects the presence of faulty insulators and bushings. Surge protectors are disconnected during the test.

Dielectric testing can be either destructive or nondestructive. In the case of the transformer, nondestructive testing is used. This is accomplished by increasing the voltage at a fixed rate to a value just below the breakdown value of the transformer. The normal rate of increase is on the order of 3000 volts per second. This may vary with the type of test instrument used and the technician needs to be skilled to make this measurement. Mistakes with this test can cause permanent damage to the transformer.

If the transformer fails the high-potential test, obvious causes need to be examined before rejecting the transformer. Examine the insulators and clean them with an electrical solvent. Replace any broken insulators or damaged seals.

The transformer was tested at the factory, so some damage has occurred since then. If the transformer comes assembled, there is little chance that the insulation on the coils has been damaged unless someone has dropped an object on them in preparation of putting the device into service. Be sure that the surge arresters have been removed from the circuit during testing. Don't forget to reconnect them after testing.

In the great majority of cases, if a new transformer passes the insulation resistance test, it will also pass the high-potential test. Various authorities differ on how often to test for high-voltage breakdown. Some say it is safe to repeat the tests if properly performed as many times as necessary. Others believe that the test should be performed only at prescribed times, and that these should be limited. In any event, the test should not be repeated continuously without allowing time for the dielectric to depolarize after being stressed.

Voltage and Loss Tests

The transformer may also be tested for its rated voltage regulation and losses. This can be performed using the open-circuit test. See Figure 3–32 for the schematic diagram of the test set-up.

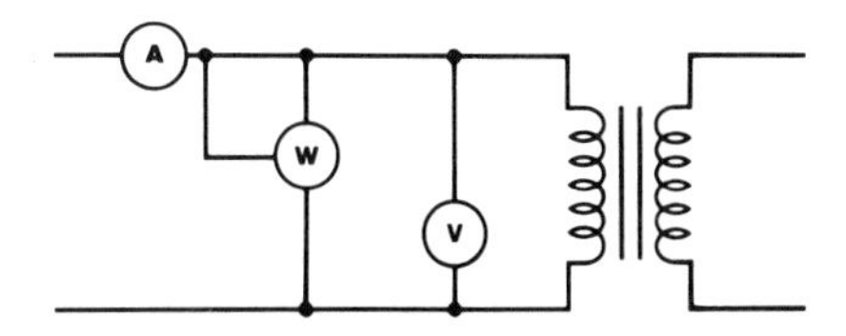

FIGURE 3–32. Open-circuit test.

This test may be run on either the primary or the secondary circuit. Available voltage usually determines which is used. With rated voltage at the specified frequency applied to the primary as shown in Figure 3–32, current of less than 10% of maximum will flow. The wattmeter will indicate a wattage in the order of 1% to 1.5% of the rated kVA. Because no power is being used by the secondary, open circuit, this wattage represents the inherent losses of a transformer as described in Chapter 1.

Wiring and Putting the Transformer in Service

After conducting the prewiring tests with satisfactory results, you are ready to begin connecting the transformer according to the specifications of the job. The neutrals and grounds for the system are installed first.

Preparing to Wire

Use the nameplate as your primary guide in wiring the transformer along with the explanations provided you in Chapter 2 of this text. Check the source voltage to determine if it is the value specified for the transformer. If the source is slightly high or low, the tap of the transformer should be set to accommodate the deviation.

The general rule for the type of wiring connection remains that if the line voltage is equal to the voltage rating of the windings, the windings will be connected delta. If the line voltage is 1.73 times the voltage ratings of the windings, the transformer will be connected wye. Where dual windings are involved, a determination will need to be made if they are to be put in a parallel or series configuration.

Testing Delta Systems

When in doubt, some additional electrical tests may be made prior to final wiring of the transformer or before putting it on line. In the case of a delta wired secondary, a voltmeter, lamp, or piece of fuse wire can be connected across the final junction before connecting them together. Figure 3–33 illustrates this test.

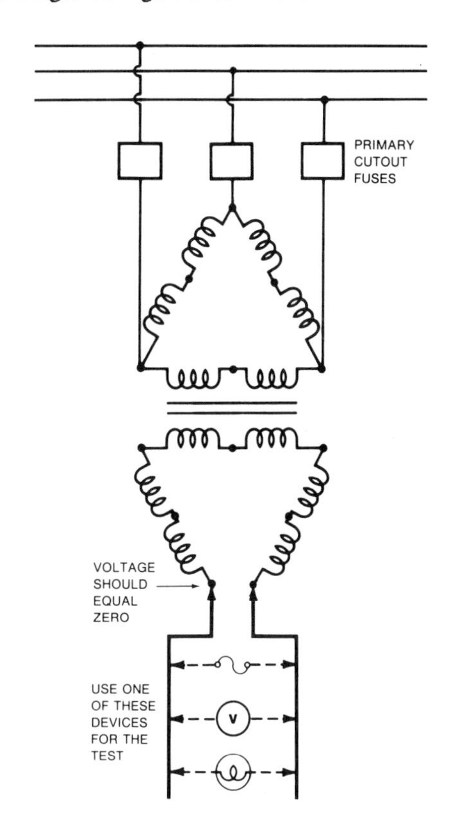

FIGURE 3–33. Testing delta wiring.

Apply power to the primary. If the meter reads approximately twice the secondary winding voltage, the lamp burns brightly, or the fuse blows, one of the secondary windings has been reversed. Connecting the junction will cause a short circuit under these conditions.

To correct the situation, the connections on one of the windings is reversed and the test repeated. Continue reversing the windings one at a time until zero potential exists between the open terminals.

After making the connection, measure the voltage(s) across the output terminals to determine if the voltage(s) is equal to that required by the load. If not, either the connections in the primary circuit or the secondary circuit can be at fault and should be corrected before the transformer is put on line. Review Chapter 2 regarding transformer connections if the need arises.

Testing for Phase and Paralleled Transformers

If the transformer is going to be connected to a secondary main to which power has already been applied, certain precautions need to be taken. Because the transformer may be energized from either the primary or the secondary, feedback voltage may be present. Protective clothing should be worn when making the connections. Rubber gloves tested to withstand the available voltages are a must.

Be sure that the voltage present on the main and the transformer voltage are the same. For a two-wire system, connect one lead of the secondary to one line of the main. Energize the transformer and test for voltage between the other secondary lead and the other line before making a connection. A fuse, voltmeter, or test lamp may be used. If a lamp is used, test it before using. If the fuse blows, the meter indicates voltage, or if the lamp lights, the two leads to the main will have to be interchanged. Otherwise a short circuit will result. Figure 3–34 illustrates this test.

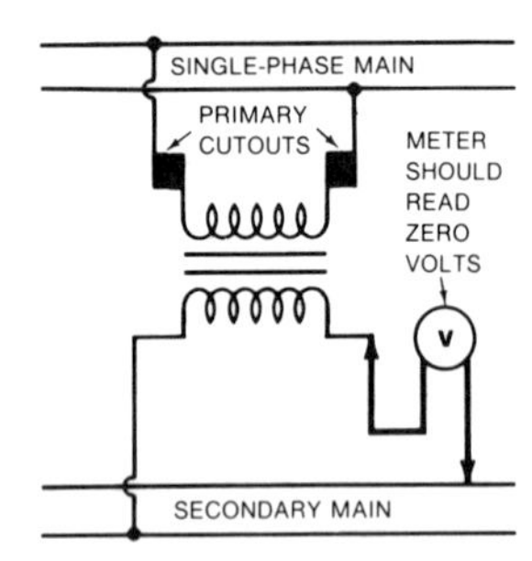

FIGURE 3–34. Testing before connecting transformer to live secondary mains.

This same test can be used when paralleling transformers. As in the previous example, the voltage between the lines to be connected must read zero or nearly zero volts before they are connected. Otherwise, the leads will need to be interchanged.

In the case of a three-phase transformer, proper phasing can be determined also using the above method. If the phases are the same, the voltage between the secondary lead and the line will be zero.

Start-Up

It is always good practice to energize the primary of a transformer without load on the secondary. If possible, this voltage should be brought up slowly so that any defects can be detected before full-voltage is applied, and thus perhaps prevent damage to the transformer in case a defect is present.

After the rated voltage is applied, the transformer is to be observed for several minutes before applying the load. If possible, the load should be divided, and each part applied one at a time while allowing adequate time after each increase to detect any abnormal conditions. When approximately 75% of the total load is being carried by the transformer, it should be operated several hours under these conditions. Temperature checks should be made frequently.

If the transformer appears to be operating normally under these conditions, full-load can be applied, and the transformer brought up to normal operating temperature. After operating under these conditions for several hours, in the case of oil-filled transformers, samples should be taken from the bottom and top of the tank and its dielectric strength tested. This is to ensure that the transformer has thoroughly dried out and is no longer giving off moisture.

Once the transformer is in operation at full load for several days, and all tests have proven good, no further testing is required. Provided loading does not increase beyond the capacity of the transformer, and environmental conditions do not deteriorate substantially, the device will operate year in and out with only periodic inspection and maintenance.

Questions

1. What should be done when a transformer is received at the worksite for installation?
2. How would you know if an oil-filled transformer is properly filled?
3. If a transformer has a leak above the oil level, how can it be located?
4. Why should transformers not be overfilled with coolant?
5. What percent of moisture content in the coolant will reduce its dielectric strength by 50%?
6. Explain how moisture can be removed from the coolant without removing the oil from the transformer.
7. What type coolant does the outdoor type of transformer use?
8. Askarels, which are synthetic coolants, may be the cause of ______ ______, and they contain the chemical ______ which is known to cause cancer.

9. It is customary for the manufacturer to ship the transformer with its windings connected for the ______ voltage rating.
10. What is the minimum height for the live parts of a transformer that is guarded by elevation?
11. What are the requirements for a chain link fence provided as a guard around a transformer?
12. How must a transformer rated over 600 volts be guarded when installed indoors?
13. What is the minimum height of the sill of a vault which contains an oil-filled transformer?
14. May fire protection be provided by a water sprinkler system in a transformer vault?
15. Define an adequate ventilation system.
16. A noise level meter reads 65 decibels before the installation is modified and the level of noise is reduced by one-half. What is the new meter reading?
17. State the first rule in reducing the noise level of a transformer in a new installation.
18. Define ground; grounding conductor; bonding.
19. List the two primary purposes of grounding electrical equipment.
20. What two electrical conditions will quickly attack the insulation quality of a transformer?
21. How are transformers protected from large voltage spikes on the power line? Overcurrents?
22. Overcurrents have two classifications. Define them.
23. How are fuses and breakers rated?
24. How can an electrician determine if a fuse has opened due to an overload or a short circuit?
25. Define what is meant by a CSP transformer.
26. A 200 kVA/2400 volt transformer has primary protection only. What size fuse or breaker would you select?
27. In a commercial structure, what is the first choice for establishing a ground for the system?
28. The red pilot light on a CSP transformer indicates that the transformer is ______.
29. List the electrical factors taken into consideration by the NEC in prescribing the sizes of fuses and breakers for a transformer.
30. Name the type of instrument used to measure the insulation resistance of a transformer.
31. A 500 kVA/4800 volt/480 volt, three-phase transformer with an impedance of 3% will deliver how much current under a short-circuit condition?
32. How much voltage would be applied to the primary of a transformer when a short-circuit test is run on it?

CHAPTER 4

Maintenance, Troubleshooting, and Repair

Anyone who has owned a home or worked in business or industry realizes that maintenance is a fact of life. My father-in-law, while he lived with us, had a policy of fixing one item each day. He said that in this way, 365 things would be in working order during the year and would not have to be thrown away. This added spice to his life and made our home a more efficient and cost effective operation. Unfortunately, since his death, this policy has not been continued, with ensuing emergency repairs, inconveniences, interrupted work schedules, and increased costs.

If this is true of an individual's home, it is even more important that these functions be a part of every business and industry where neglect of the maintenance and repair functions can cost thousands of dollars. They may be performed by the owner or the boss, or there may be an entire department for operations and maintenance for a large plant, building, or institution. These departments include all the services necessary for the continued operation of the facility. Among these are the major trades which include carpenters, plumbers, and electricians.

Preventive Maintenance Program

When an institution, business, or plant facility is of sufficient size to warrant the employment of a repair crew, management often finds that people are just sitting and waiting until something happens. Repair crews are inherently inefficient, and yet they are a necessary cost factor in the operation of any enterprise.

To overcome some of these deficiencies, it makes good sense to initiate a routine maintenance program. Not only will this more fully occupy the time of the repair crew, it will more than likely make them better and happier workers. Idle time on the job is difficult for most people. In addition, routine maintenance in most cases will extend the life of the equipment and head off outages and down-time before they occur. This, in turn, will change the repair crew into a money savings operation rather than increasing the cost of doing business. The adage, "If it isn't broken, don't fix it," does not hold true when it comes to preventive maintenance.

Institutional Maintenance Program

For example, at the University of Illinois, Champaign, there are approximately 100 engineers, technicians, mechanics, and helpers in the Electrical Section of the O & M Division. Forty-five members of this group are assigned to the Construction and the Electrician Maintenance Shop. They work on every electrical item from high-voltage power generation down to changing the light bulbs in 200 major buildings on the campus. Their objective is to provide and maintain a comfortable working and learning environment for the 6000 employees and 35,000 students of the university, which in itself constitutes a small city.

The 119-year-old university has over 300 distribution transformers, 5100 air conditioners, 10,000 electrical motors, and 372,000 light fixtures to maintain and keep in operation. Some of these items were installed as long ago as 1939 and are still in operation. As the equipment becomes older, it requires more maintenance than when it was first installed.

Although the Construction and Maintenance Shop is always involved with new facilities, there is flexibility, and the maintenance of existing equipment is scheduled on a regular basis. When the preventive maintenance program was first established, it was planned on a 1-year rotation basis. Every piece of electrical equipment was scheduled. Management found that all equipment could not be checked in this period, so the time was extended to 18 months rotation.

A daily log book is keep by the supervisor of this department. Each piece of equipment has its own file sheet which shows the maintenance work performed. The date when the work was done is recorded. The file sheets are kept updated from the work orders which are used to assign the daily work schedule of the electrician.

Transformer Maintenance

Records for the maintenance on a transformer begin when the transformer is installed. Data concerning the operation of the device should be recorded when the transformer is put on-line and operating normally. This would include the voltages present, the kVA delivered to the load, and the appropriate temperature readings for the type of cooling system associated with the device. Other information may also be included such as "hipot" (high-potential) tests of the transformer and the conductors, along with any dielectric strength tests performed.

This information serves as a basis for comparing data collected during routine maintenance and for determining changes in conditions that may lead to the damage of the transformer. It is also useful for troubleshooting the system when faults occur.

There is normally space on the form to make notations about the work performed on the transformer. Materials used can be recorded in this area. It is also useful to note any changes in the environment or increasing loads on the transformer.

Another factor that may be part of the maintenance record is cost. The original installation cost may be included on this sheet with a running record of labor and material costs to keep the device in operation. This information is useful to the accounting and purchasing departments of the company in comparing different devices from various manufacturers. Long-range economies can be realized from these records. Figure 4–1 shows a suggested maintenance information sheet.

Need for Transformer Maintenance

Transformers properly selected for the loads usually operate for years without giving any trouble or needing to be replaced. Because of this, and because they have no moving parts, it might be construed to mean that they never have to be given attention until they fail.

This is not the case. Many of the factors that were acceptable when the transformer was installed can change drastically over the years. Any of these changes that cause moisture to be introduced into the transformer or result in the rise of temperatures to unacceptable levels need to be corrected before any damage can occur. Transformers that are operated in areas where corrosive atmospheres and electrically conductive dusts are present need frequent inspections. Moisture, temperature, and detrimental environments are the primary enemies that can destroy a transformer.

Even dry-type transformers that are hermetically sealed may need frequent inspections. Let us take the case of three transformers rated at 500 kVA with a primary voltage of 600 volts and secondary voltages of 240/120 volts. These transformers provided power and lighting in a plant where metallic dust and shavings were present. The system failed within a year. Failure occurred during peak production time idling 70 workers for several hours.

Figure 4–2 illustrates the arc paths between the terminals of the transformer that failed. The dust had accumulated over the several months after the transformers had been put into service. Arcing had occurred many times in this period, clearing the fault before a good conductive path was established across the 240-volt secondary terminals, effectively shorting out the transformer.

When a company keeps good records of failures like the one described above, they provide a guide to how frequent maintenance should be scheduled. In this case, routine maintenance was scheduled on a bi-weekly basis, and later adjusted to once a month. A shield was also installed to prevent metal shavings from falling onto the terminal board.

Transformer manufacturers provide guidelines to begin a preventive maintenance program for the various types they sell. These guidelines should be followed in the beginning of a new program where records of failures have not been kept. Once experience has been gained for your particular situation, these time intervals can be increased or decreased depending on the particular circumstances under which the transformers are operated. Usually the frequency of inspection and maintenance is going to depend on

INSTALLATION, MAINTENANCE, REPAIR, AND REPLACEMENT FORM FOR TRANSFORMERS

TYPE: MODEL NO: SERIAL NO: MFGR: PURCHASED FROM: P.O. NO: COST: DATE:

KVA: PRI. VOLTS: SEC. VOLTS: HERTZ: PHASE: TEMP.RISE: IMPEDANCE: COOLANT: WEIGHT SIZE:

INSTALLATION INFORMATION

DATE:

COSTS

MATERIALS:
LABOR:
TRANSPORATION:
MISCELLANEOUS:

TOTAL COST:

TEST DATA

PRIMARY VOLTS:
SECONDARY VOLTS:
KVA OF LOAD:
AMB. TEMP:
TEMP. RISE:
INSULATION TEST:
(RESISTANCE TO GND.)
WINDINGS
PRIMARY:
SECONDARY:
CONDUCTORS:
PRIMARY:
SECONDARY:
DIELECTRIC TEST
OIL:
TRANSFORMER:

DATE TAKEN OUT OF
SERVICE:
LOCATION:

DATE:	**NOTATIONS**	**MAINTENANCE RECORD**		
		DATE	LABOR	MATERIALS
	CONTINUE ON THE BACK			

FIGURE 4–1. Transformer maintenance, repair, and replacement form.

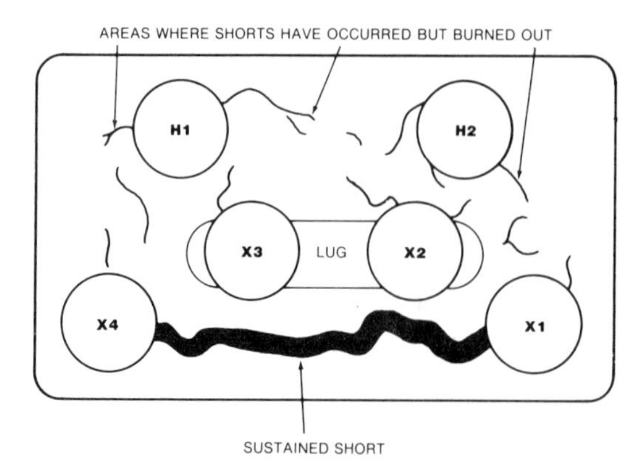

FIGURE 4–2. Arc paths on transformer terminals.

how critical it is to maintain service and the environment in which the transformer must operate.

Inspection and Maintenance

A good transformer maintenance program begins with a thorough inspection of the transformer. Operations that have many transformers to maintain will have checklists for the particular transformer scheduled for maintenance. The work order or job ticket may have this information on it.

Sensory Inspection

Do not limit the number of senses you use when performing this function. The tendency of many people is to use only the sense of sight, and this can cause you to miss many problems.

Noise and vibration can be detected by listening to the device and by touching it. If these levels have increased, it can indicate future problems with the system. Hardware may have loosened and needs to be tightened. A heavier load may have been put on the system which may account for the problems. In any event, do not ignore these clues; the cause needs to be investigated and corrected.

The sense of smell is a good means of detecting burned insulation and arcing. High-voltage arcing will cause ozone to form, which also can be detected by the taste buds. Smells and tastes are difficult to describe, and therefore, it will be left to the individual to gain the necessary experience in order to use these senses to identify various electrical problems.

Dry-Type Transformers

Dry-type transformers and small distribution transformers need far less attention than large power transformers or those that have special provisions for cooling or moisture prevention measures using inert gases. Inspections normally would occur once a year or even longer. They would, however, be subject to more frequent maintenance when continuity of service was critical, or where they are operated under adverse conditions.

The exterior of the transformer is examined for any rust or corrosion. Any deterioration of the transformer finish should be sanded and repainted. This may require taking the transformer out of service.

When inspecting the exterior of the transformer, particular attention should be given to the condition of the conductors, connectors, and the high- and low-voltage bushings. Look for signs of discoloration where an arc may have occurred. Check for broken insulators or cracks and frayed insulation on the conductors. Damaged conductors are most often found around the pothead or where they are attached to the transformer.

Electrical connections may have loosened and will need to be tightened. Loose connections cause hot spots due to the increased resistance and may cause the connector and insulation to break down under electrical stress.

Lightning strikes and line surges may have caused damage around the bushings or pothead. The surge arrester needs to be examined for any obvious damage and proper gap. Vibration may have caused it to loosen.

All gaskets on the bushings, hand hole, and cover are to be examined for any deterioration. These will need to be replaced if they are not in good condition. In most cases their primary purpose is to prevent moisture from entering the transformer. As mentioned before, a little moisture can cause big problems. All covers and bushings should be tightly secured to prevent this problem.

Hardware on the transformer may have loosened due to vibrations. All hardware should be checked for this condition and tightened if necessary.

Ventilation ducts need to be examined and cleared of any obstructions. The surfaces of the core, coils, and terminal boards should be examined for the accumulation of dust and other debris. This buildup can cause the transformer not to cool properly.

Compressed air of 25 to 50 psi can be used to clear the vents and blow the dust off the surfaces of the transformer. While performing this task, the terminal boards and coils can be inspected for any deterioration.

Some dry-type transformers may also include forced air cooling using fans. Fan motors and related control devices, fuses, and wiring need to be checked at least once a month. Use compressed air to clean these areas of the transformer.

The wiper arm on the voltage ratio adjuster should be rotated through all the positions several times to clean off any oxidation and to assure good contact. The contact presents a higher resistance to current flow when oxidized, thereby creating a hot spot. Transformers often fail due to this problem. If the oxidation is fairly heavy, an electrical contact cleaner needs to be used. Be sure to return the tap back to its original setting before returning it to service. If the tap changer is left between contacts, the transformer will not operate when it is put back on the line. In the event the voltages have changed, the tap may be set at a different position to compensate for the change.

Coolant-Filled Transformers

Examine coolant-filled transformers to ascertain the level of the coolant and look for any possible oil leaks. Most coolants have a low evaporation rate at their operating temperatures. If the level is low, more than likely, oil is leaking from the transformer. This may be due to a rupture in the tank, or the transformer becoming so hot that oil was boiled out of it.

Where oil leaks are evident, they will require immediate attention. When the cause of the leak has been remedied, the oil level needs to be restored. Coolants containing PCBs should be replaced if the transformer has to be taken out of service. The replacement liquids developed for this purpose do not propagate the flame and have a fire point of not less than 300°C. If ignited by an arc, the flames will not spread. These coolants may be used either indoors or outdoors. When the transformers are rated over 35,000 volts, they shall be installed in a vault. Care should be taken to meet federal and local regulations when replacing PCB coolants.

Dielectric Test

Discolored coolant may indicate a high degree of oxidation and the presence of sludge. A dielectric test should be performed on the coolant. Use the procedures outlined in Chapter 3 for making these tests. Mineral oil that tests much lower than 22 kilovolts and askarel below 25 kilovolts will need to be filtered using a filter press (Figure 4–3). After filtering, mineral oil should test for 25 kilovolts and askarel for 30 kilovolts.

Filter Press

It is important that the filter press is clean before it is used. Old filters need to be replaced. Use the same type of care cleaning the filter press as you would use on sample bottles and the thief. All equipment associated

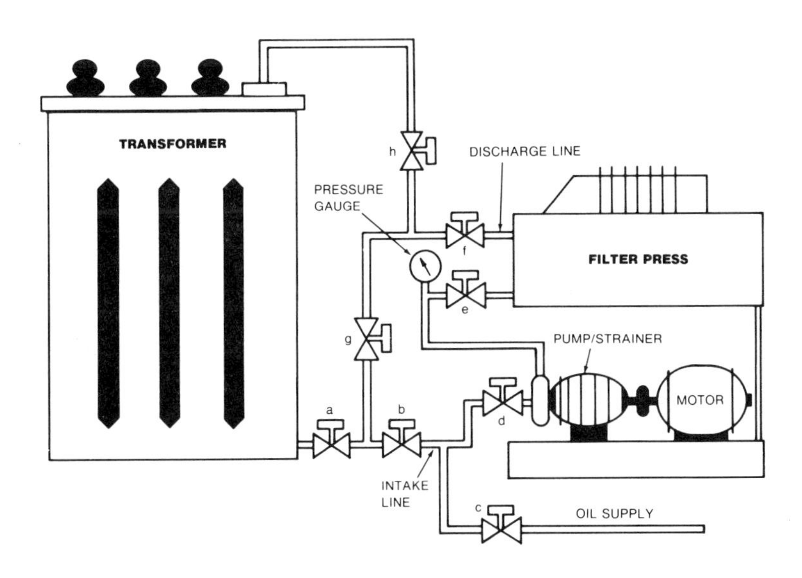

FIGURE 4–3. Filtering coolant using a filter press.

with working with the coolant needs special care to be sure that they are free from any contaminates. Remember that rubber hoses are never to be used due to the presence of sulfur in the rubber.

The filter press shown in Figure 4–3 provides for the filtering of the oil already in the transformer or for the filtering of oil from the supply. This is accomplished by opening and closing the valves.

To filter the oil from the tank, valves c and g would be closed with the other valves open. Filtering the new oil supply would require closing valve b and opening valve c. All other valves should be opened, and the oil would be supplied to the top and bottom of the tank at the same time.

Closing valve h would cause the oil to be inserted into the bottom of the tank. If valve g were closed instead, the oil would be supplied to the top of the tank.

The pressure gauge provides a means of determining when the filters in the filter press need to be changed. As the press takes contaminants out of the oil, the screens become clogged. This results in an increased resistance to the flow of oil, and the pressure on the gauge increases. Manufacturer's recommendations should be followed as to what pressure the filters need to be changed.

A decreased pressure would indicate that the strainer on the pump has become clogged. Under this condition the process would have to be stopped, and the strainer cleared of any obstructions.

If a filter press is not available, the coolant will have to be changed. After the old coolant has been removed, the interior of the transformer will need to be cleaned and dried thoroughly before the new coolant is installed. This will prevent the contamination of the new coolant with the residue that may remain in the tank.

The supplier will more than likely have a filter press. The contaminated coolant can be saved and shipped to the supplier for filtering. The old coolant should never be dumped, particularly if PCBs are present in it. The Environmental Protection Agency has levied substantial fines for this practice.

Temperature Check

Recording temperature on each inspection is a requirement for a good maintenance program. Most large installations will have a thermometer as a standard accessory, and making these readings is relatively simple. If not, a thermometer should be placed in the coolant and a reading taken. Ascertaining the ambient temperature of the installation is also useful so that the temperature rise of the transformer can be compared to that at the time of installation.

Sudden rise in temperature is one of the best indicators available for predicting the need for corrective action. Under supervised conditions, these readings may be taken hourly. In other cases, they may only be performed when regular maintenance is scheduled.

Gas Test

Another test to be performed on coolant-filled transformers is for the presence of combustible gases. Some coolants, when heated, give off gases which will ignite and explode. An arc inside the transformer may cause this to happen. Proper ventilation is the solution to this problem. Where proper ventilation cannot be achieved, gas absorbers on the transformers would be used.

The coolants are combined chemically so that they do not usually give off enough explosive gases until their temperatures exceed 180°C. Only Class H–rated transformers are designed to operate with temperature this high on the windings. The chances are that the transformer will be damaged if the temperatures go above the rating of the insulation for any long period of time.

Specific Tests

Transformers are cooled in a variety of ways. Other types of transformers protect the coolant from oxidation by sealing the transformer and inserting inert gas in the air space. These differences in design require some specific tests and maintenance procedures.

Water-Cooled Transformers

The temperatures of the ingoing and outgoing water should be taken at least weekly. Records of these readings are kept to determine the efficiency of the cooling system and to detect any changes in the operating conditions of the transformer.

When the ratio of these two temperatures starts to rise, the cause needs to be determined. If the increase cannot be explained due to additional loading or by other maintenance checks, the water pressure and flow would need to be checked. These values would have been recorded when the transformer was put in service.

A pressure meter on the input line indicating high pressure may mean the coils are full of scale and are in need of maintenance. The input and output pipes would be disconnected from the coils. Examine the insides of the supply pipes as well as the coils for scale and sediment. A bluish-green deposit (copper sulfate) would indicate that the water being used is acidic and is eating at the copper tubes.

Some authorities state that the coil does not have to be removed from the transformer for cleaning. This is probably true in most cases. However, if the copper sulfate deposits are present, and the tubing is showing signs of deterioration, then it should be removed from the transformer. The tubing may be damaged further when the following procedure using a solution containing acid is used to clean it. If this were to happen, water and acid will contaminate the transformer oil if the cooling coil is not removed.

To clean the pipes of scale and sediment, a solution containing equal parts of hydrochloric acid and water having a specific gravity of 1.10 is used. When mixing the solution, the acid is to be poured into the water. Pouring water on acid will cause the solution to boil up violently and can cause severe burns on the skin or possible eye damage.

The solution is then poured into the coil. If the coils have not been removed, it may be necessary to add pipes to the input fittings and arrange them in a vertical position so that the solution can be poured. After the solution is inserted into the coil, it should be allowed to stand for about 1 hour and then flushed out with clean water. If a single treatment does not clear the coil, the process needs to be repeated. Use a new solution of acid and water each time. Usually, one or two cleanings will be sufficient to clear the coils.

If the transformer is going to be taken out of operation for a period of time and be subject to freezing weather, the water should be blown out of the pipes and coils. They need to be thoroughly dried with hot air to remove all moisture, or they may be filled with transformer oil. Oil used for this purpose will need to be dried after it is removed from the coils, because it will contain the water that remained in the coils.

When scaling and other sediment problems are frequent, the water may need treatment before it can be used. A chemical feeder can be added to the water system for this purpose. The acidity can be corrected by inserting a base chemical such as sodium hydroxide or soda ash. Other chemicals or filtering will be needed to prevent the scaling and sediments. An analysis will have to be performed on the water to determine precisely which chemicals would be effective.

If purified water is used with a closed system for cooling the transformer, a pump is needed. The motor and pump should be checked for proper operation. Motor maintenance will be discussed in Chapter 10.

Sealed Transformers

Conservator transformers have a pressure relief diaphragm. This should be checked for any cracks or breaks. Damaged diaphragms are to be replaced immediately. Loss of pressure, or if pressure does not change as the temperature of the transformer increases, may indicate a defective diaphragm. The pressure in the transformer can be read on the meter on the mercury relief valve.

If the diaphragm is not at fault, there is probably a leak in the tank above the oil level. A soapy suds test can be conducted on all seams and around the fittings of any meters or pipes to locate the leak. Figure 4–4 depicts a gas-sealed power transformer.

The alarm on this system should be checked with each inspection. This can be accomplished by closing the cutoff valve on the return line and allowing gas to escape slowly from the exhaust valve. The low-pressure value at which the alarm sounds can be read on the pressure meter.

Air Blast Transformers

Air blast transformers that are operated indoors, and rely on the forced air blast provided by electrical fans, should have their fans running at all times that they are in operation. When these transformers are operated outdoors, the fans are controlled by a thermostat to conserve energy in cold weather.

Each fan needs to be checked for proper operation. If the fans are controlled by a thermostat, a check needs to be made to determine if the fans are turning on at the proper temperature.

The temperature of the air going into and exiting the cooling duct is to be taken and compared to previous readings. Corrective action would be indicated if there is a sudden increase in the differences of the two readings.

As part of the preventive maintenance program, the duct works, fans, and cooling coils are to be cleaned with dry compressed air. Accumulation of dirt in these areas and on the transformer will reduce the cooling efficiency of the system and may cause high-voltage leakage around the bushings. Air filters may be part of the forced air system. They need to be inspected and changed if dirty. Frequency of cleaning will depend on the type of environment in which the transformer is operating.

These transformers are normally equipped with a high-temperature alarm system. The thermostat for the alarm should be turned up until the alarm sounds. Check that the thermostat level for this test is set at a

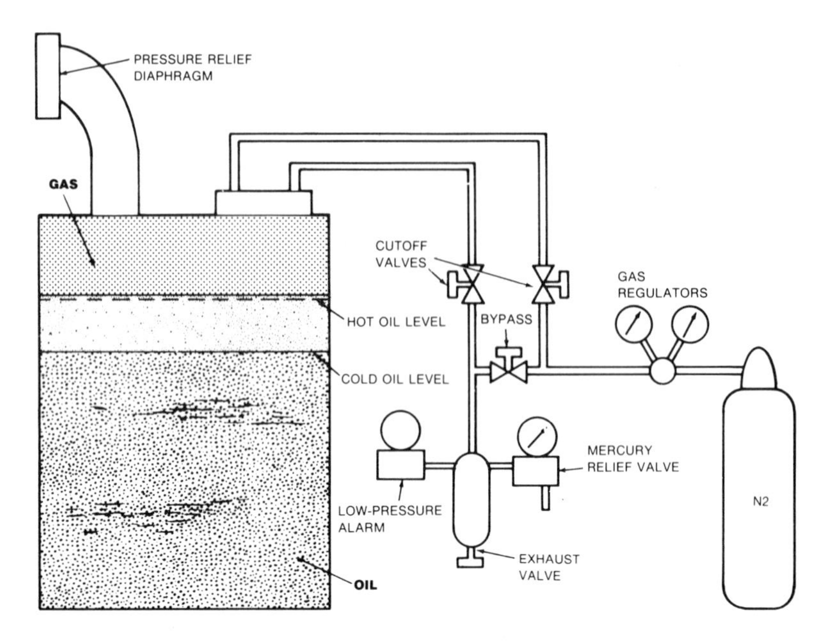

FIGURE 4–4. Gas sealed power transformer.

level where the heat generated by the transformer is at or below the manufacturer's recommendation.

Forced-Oil–Cooled Transformers

Temperatures of the ingoing and outgoing oil is to be measured. The difference in temperature can be compared to the readings taken previously. A sudden increase in this value indicates a possible problem.

These transformers are equipped with an oil strainer. If there is a large difference between the ingoing and outgoing pressure across the oil strainer, the strainer will need to be cleaned of any foreign material which has collected in it.

Oil-Immersed, Air-Pressure–Cooled Transformers

Measure the temperature of the oil in the transformer to determine if the fans should be running. If not turn the thermostat to the prescribed level to start the fans. The motors, fuses, conductors, and related control devices should be checked for any deterioration and proper operation. Accumulation of dust and dirt should be blown off the various parts of the installation with compressed air.

Gas Absorbers

Transformers installed in poorly ventilated areas may have a gas absorber. These devices need to be checked for deterioration evidenced by dampness, cracking, and flaking. A new absorber should be installed when the need is indicated.

Summary of Transformer Maintenance Program

A well-planned transformer maintenance program begins when a transformer is first installed and a record is made concerning all tests and procedures performed at that time. This record is updated whenever any future inspection, maintenance, or repair is performed.

This record is a guide to any electrician that works on the installation in the future. It provides the means for the more efficient servicing of the device and reduces the amount of time the transformer must be off-line during maintenance, or whenever a fault occurs. It takes the guess work out of the operation.

Frequency of inspection and maintenance is dependent on such factors as cleanliness, atmosphere, shock and vibration, temperature, and operating conditions. If no problems are found on several routine inspections, the chances are inspections are too often. If the system fails before routine maintenance, then the chances are that the inspections are performed at intervals that are too great. The record of the transformer's history is the key to setting the schedule.

Where applicable, the transformer inspection should include the following procedures.

1. Check temperature of
   a. Coolant.
   b. Incoming and outgoing water on a water-cooled transformer.
   c. Incoming and outgoing air on air blast transformers.
   d. Incoming and outgoing oil on a forced oil-cooled transformer.
2. Make sure the air vents are working for the enclosure in which the transformer is located.
3. Vents on the transformer should be free of any obstructions.
4. Fan motors need to be checked.
5. Check water and oil pumps.
6. Clean strainers and filters.
7. Check pressure and flow of oil and water.
8. Check pressure and temperature gauges.
9. Test automatic alarm systems.
10. Test automatic controls on motors and pumps.
11. Look for deterioration of wiring, bushings, insulators, fuses, and circuit breakers.
12. Blow dust out of ventilators and off insulators, core, coils, and terminal boards. The transformer should be kept as clean as possible.
13. Test the dielectric strength of the coolant. Filter if necessary.
14. Check the level of the coolant. Add additional coolant if the level is low. Do not fill beyond the level mark. Use proper procedures and protection when working with askarel.
15. Measure the primary and secondary voltages of the transformer. Adjust the voltage tap if necessary. Be sure the transformer is disconnected from the source voltage when this adjustment is made. Be aware that if the secondary is not disconnected, a backfeed may be possible on some systems. The policy of some power companies is that the secondary is always disconnected along with the primary.
16. Measure the kVA output of the transformer. Compare it with previous recorded data.

Anyone experienced in servicing transformers will come to the conclusion that *heat, dirt, and moisture in the coolant* are the primary enemies of transformers. Every effort should be made to reduce these hazards in a good maintenance program.

Excessive heat may be caused by

1. Poor ventilation in the area in which the transformer is installed.
2. Improper circulation of the air around the transformer or the coolant through it.
3. A low level of coolant.
4. Accumulation of dirt and dust on the trans former and in any cooling paths.
5. Electrical overload on the circuit.

Water may get into the coolant by various methods.

1. It may be introduced by "breathing." This happens when the transformer is cooler than the surrounding air, and moisture condenses on the metallic parts of the transformer. Usually this only occurs when the transformer is not in service.
2. Defective gaskets and seals on the device may be the cause.
3. Persons working on the transformer have not taken proper precautions with the equipment used to install, test, or replace coolants. Other foreign substances detrimental to the proper operation of the transformer can also be introduced by the careless worker.

Troubleshooting

The time to start troubleshooting any piece of equipment or system is when it is installed, not when it fails. If you were not involved in the initial installation, become familiar with the system when you are assigned its care. Hopefully, this will not happen simultaneously with the fault.

Study the equipment record. Pay special attention to previous reports of failure. A pattern may become obvious. Read any notations made by the person who serviced the transformer on the last few occasions. There may be a clue that may save considerable time in restoring service.

Be familiar with the power source supplying the transformer. Know where the disconnect is and the value of the voltage(s) along with the kVA that is available.

Know the locations of the panels served by the transformer and the loads supplied. Different types of loads have different current requirements. Some loads are more likely to cause problems than others. You may want to take these off-line before energizing the transformer if any of these have been proven to be a problem in the past.

Learn to use all the test equipment associated with troubleshooting the transformer before an outage occurs. When you are under stress while trying to repair the device is a poor time to be reading a manual on how to use the test equipment. Attempting to make an insulation test on a live circuit would prove rather embarrassing under the circumstances.

There is no substitute for previous hands-on experience when troubleshooting a system. Previous knowledge about the system and all related equipment, however, is an important supplement to becoming a great troubleshooter. Learning by experience alone can be very expensive.

What to Do

Be alert! Make use of all of your senses to determine what has happened. To often we look, but do not see; listen, and do not hear; touch, but do not feel; smell and taste, but do not comprehend. Someone once said, "You can see a lot by looking."

When this information gathered by your sensory perceptions is integrated with your previous experiences and knowledge about the system, the problem will usually be simple to determine. The one thing about power systems is that physical evidence of the malfunction is normally present. Be prepared to understand what the evidence is telling you. The time spent analyzing and diagnosing the fault will in itself save time in getting the transformer back on-line. Haphazard methods of troubleshooting are time consuming and should be avoided. Use a logical sequence.

If the transformer is still on-line when you arrive, it should be disconnected immediately if any smoke or liquid is coming from the tank. The same procedure should be followed if there is any unusual noise coming from the transformer. Do not re-energize until the cause of the malfunction has been established. Activate any ventilation systems to clear dangerous gases from the area and to lower the ambient temperature.

Take a temperature reading of the coolant, if liquid cooled, as soon as possible. This reading will be a good indicator if damage has occurred to the insulating quality of the transformer. If feasible, allow any cooling system to continue to operate until the temperature of the transformer is reduced to an acceptable level.

In the event the transformer is not in immediate danger of being destroyed or exploding, and can continue to be operated, make ampere readings. If the system is not metered, a clamp-on ammeter can be used for this purpose.

Feeder loads can be disconnected one at a time in order to determine which feeder has the problem. Individual loads can then be disconnected in a like manner. If the problem is not isolated in this way, then one of the conductors is at fault. Normal current readings on the feeders indicate that the problem is in the transformer itself.

With the transformer off-line and with the loads disconnected, check for any external damage to the transformer. This may be either electrical or mechanical. Examine the potheads, conductors, insulators, bushings, and surge arresters. Look for evidence that the coolant is leaking from the tank.

If some of the fuses on the transformer have blown, they can also provide valuable information. If the fuses are in the secondary circuit, the problem is more than likely being caused by the load. If the primary fuses are out, the transformer itself is most likely at fault.

How the fuse blows can also tell you something about the fault. Figure 4–5 shows a good single-element fuse linkage. Figure 4–6 depicts a fuse that has opened due to an overload—notice that only one segment of the element has opened. Figure 4–7 illustrates a fuse that has opened under short-circuit conditions. In this case all three segments of the fuse have melted.

FIGURE 4–5. Typical single-element fuse.
(Courtesy Bussman Division, McGraw-Edison Co.)

FIGURE 4–6. Single-element fuse opened under overload conditions.
(Courtesy Bussman Division, McGraw-Edison Co.)

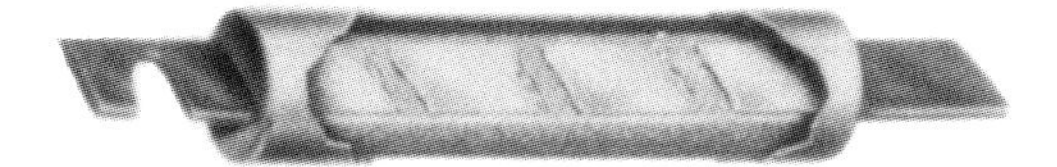

FIGURE 4–7. Single-element fuse opened under short-circuit conditions.
(*Courtesy Bussman Division, McGraw-Edison Co.*)

Figures 4–8 through 4–10 show similar conditions when a dual-element fuse is used. The spring-loaded element is shown open in Figure 4–9. This indicates an overload on the circuit. In Figure 4–10, the short-circuit element is shown opened. Note that the spring-loaded contact did not have time to open under this condition.

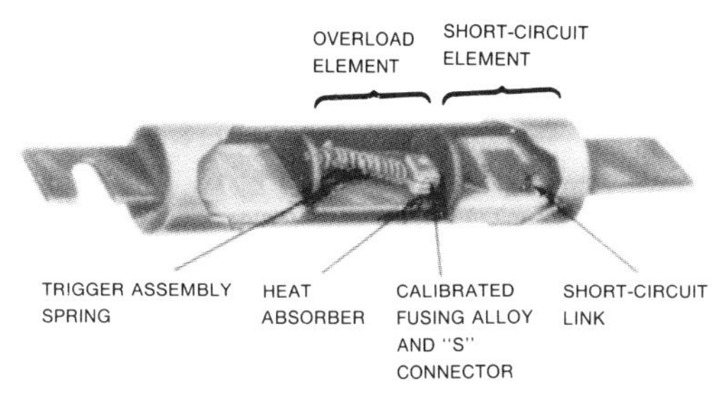

FIGURE 4–8. Typical dual-element fuse.
(*Courtesy Bussman Division, McGraw-Edison Co.*)

FIGURE 4–9. Dual-element fuse opened under overload conditions.
(*Courtesy Bussman Division, McGraw-Edison Co.*)

FIGURE 4–10. Dual-element fuse opened under short-circuit conditions.
(*Courtesy Bussman Division, McGraw-Edison Co.*)

Figure 4–11 illustrates what you might face if the overcurrent devices on the transformer are overrated. Figure 4–12 depicts the explosive nature of short circuit, high-amperage current when the interrupting capacity of the fuse has not been properly selected. Figure 4–13 shows the resulting damage to a panelboard due to the explosion.

FIGURE 4–11. Effects of overload due to the use of too large ampere rating of the protective device.
(*Courtesy Bussman Division, McGraw-Edison Co.*)

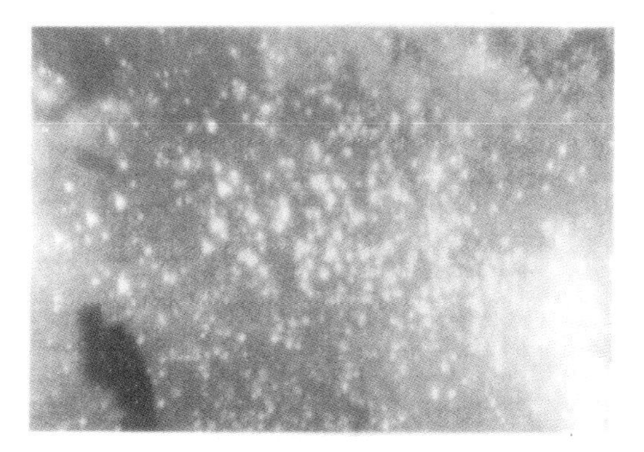

FIGURE 4–12. Explosive nature of short-circuit, high-amperage current.
(*Courtesy Bussman Division, McGraw-Edison Co.*)

These disasters can occur due to inadequate maintenance and carelessness on the part of the electrician who services this equipment. Fuses may have deteriorated over the years and have never been checked as part of the preventative maintenance program. Fuses may have been blown and been replaced without regard to their ampere and voltage ratings or their interrupting capacity. In each case, the causes may be laziness, or a complete lack of knowledge on the part of the mechanic about the possible results of the negligence. Keeping the system in operation at all costs is not applicable in these circumstances.

Loss of life is a possible outcome of these actions. Loss of time, money, and electrical service can also be a major problem. No one except you and your family

FIGURE 4–13. Results of a short-circuit condition on a panelboard.
(Courtesy Bussman Division, McGraw-Edison Co.)

is going to worry about you losing your job and the possibility of a lawsuit. An exception may be your employer if he or she is involved in a lawsuit for hiring you.

Electrical Tests

Test the dielectric strength of the coolant. Take samples from both the top and the bottom of each compartment. If the breakdown voltage is less than that recommended for the coolant, it should be filtered until it passes.

Make an insulation test of the windings of the transformer using a megger. Tests should be run between each winding and ground and between each pair of windings. Low resistance values indicate that the transformer is the problem. Normal readings would not necessarily rule out the transformer. There may have been a fault in the windings which sealed itself with the insulating liquid.

Insulation tests should be run on all conductors entering and leaving the transformer. If ampere readings were possible before the transformer was de-energized, they would provide a definite guide as to which conductors or loads might be the cause of the problem.

If the above tests still prove inconclusive, a dielectric strength test can be run on the windings and the conductors. Someone experienced in this type of testing should perform these tests. Improper procedures may damage the transformer. The local electrical power company will have equipment and trained personnel to provide assistance.

In the event that one of the procedures above results in detecting the problem, and repairs have been made, other tests may be conducted before putting the transformer back in service. A voltage test and a short-circuit test are recommended.

When bringing the transformer back on-line, monitor it carefully. Take temperature readings and look for sudden rises in temperatures. If loads can be connected back one at a time, do so. Listen for any unusual noises. Check the cooling system and temperature to make sure it is operating properly. Follow standard procedures for activating a transformer.

Transformer Fault

Once the fault has been localized to the transformer itself, it will have to be replaced or repaired. In either case, service will be interrupted unless a replacement is already available to install, or there is an alternative way of supplying power to the disconnected loads. How much backup a system has will be dependent on the economics of the particular situation, or how critical the need for continuity of service is. If an alternative power system is available, it should have been activated when the faulty transformer was identified as the cause of the fault.

An internal inspection of the transformer will need to be conducted. If the transformer uses an askarel coolant, the tank will need to be purged prior to removing the oil. Hydrogen chloride gas may have been produced if arcing has taken place inside the transformer. This gas when combined with the moisture in the air forms hydrochloric acid which attacks the internal parts of the transformer. The gas if combined with the moisture in one's nostrils or mouth is harmful.

To purge the gases from the transformer, connect a tank of nitrogen gas to the bottom drain valve and allow it to bubble up though the coolant. It will neutralize the hydrogen chloride gas before it can form an acid. Provide adequate ventilation to take off any gases that may have been formed and are not neutralized by the nitrogen. These procedures may vary with different transformers and types of coolants. The manufacturer's instructions should be consulted for the particular installation.

Once the gases have been purged from the transformer, the coolant should be drained. If the transformer has several compartments, the coolant from each should be stored in separate and clean containers. Each container is to be identified as to which compartment it was taken from.

After the coolant is removed and stored, the inside of the transformer can be inspected. Use protective gear and procedures when working with askarel-filled transformers. Disconnect any conductors and equipment necessary for entering the transformer. Remove the cover. Protect all parts, such as insulators, from being damaged during this operation.

Remove the coils from the transformer. Be careful not to damage them in the process. Examine the coils for evidence of electrical breakdown which is evidenced by burnt insulation where arcing has occurred. The coils may have to be removed from the core in order to see where the damage is. Check for damaged core sheets and core insulation.

The coil assembly will have to be crated and shipped to the local repair shop or back to the manufacturer for repair. Be sure that the coil assembly is properly packaged and protected before it is shipped.

Once the coil assembly has been repaired and returned to you, take the same precautions you used when removing it from the transformer to install it. After the transformer has been assembled, the same test procedures used for a new installation need to be followed before putting the transformer back on-line.

Questions

1. List two reasons why an electrical maintenance program is important.
2. Why does a routine maintenance program make sense when a plant has a repair crew?
3. When should the maintenance program on transformers begin? Explain.
4. Because transformers do not have moving parts, and when properly selected, they operate for years, maintenance needs to be performed infrequently or only when a problem arises. Explain.
5. How should frequency of maintenance be established on a new transformer installation?
6. Where are damaged conductors most often found around a transformer installation?
7. Why is it necessary to clean the contacts of the tap changer? Are there any precautions that should be taken when this task is performed?
8. Dielectric tests on mineral oil cooled transformers should register at least ______ kilovolts.
9. How can oxidation be detected in a coolant?
10. Rubber hoses should not be used to transfer coolants. Explain.
11. If askarel tests below 25 kilovolts, what should be done?
12. A high reading on the pressure gauge in Figure 4–3 would indicate ______.
13. What is the one best indicator that a transformer is in trouble and is about to fail?
14. If it is impossible to obtain proper ventilation for a coolant-filled transformer, what type of device should be on the transformer? Why?
15. How are deposits in the coils of a water-cooled system removed?
16. Are there any precautions you should take when mixing an acid solution?
17. What is the purpose of using nitrogen in a gas-sealed transformer?
18. As a general rule, how often should the fans on an air blast transformer receive maintenance?
19. When can the periods between maintenance on a transformer be safely extended?
20. What three factors are the primary cause of transformer failure?
21. The best way to detect an overload on a transformer that does not have instrumentation is through using your vision. Explain.
22. A well-prepared electrician assigned to transformer maintenance will ______ the maintenance records and become ______ with the distribution system and the loads.
23. What makes troubleshooting high power systems simpler than low-power systems?
24. The primary breaker of the transformer on which you are working was open when you arrived on the scene. All the secondary

breakers were still set. What is the most likely problem?

25. A secondary fuse to a feeder has all of its elements burned open. The most likely condition of the fault is a(an) a. overload. b. short circuit.
26. Besides failure, what may be the effect on a transformer that is operated under a continuous overload condition?
27. When changing fuses, you should never replace a bad fuse with one that has a (higher, lower) interrupting capacity. Explain.
28. What type of instrument would you use to measure the insulation quality of the transformer and conductors leading to and from it?
29. Would a good "hipot" test on the transformer assure good insulation quality of the transformer? Explain.
30. Besides the destructive qualities of arcing and the resulting spikes on the line, is any other danger caused by this phenomenon when it occurs in an askarel-filled transformer?

CHAPTER 5

Distribution Systems

The purpose of any distribution system is to deliver power to the loads as safely and as economically as possible while meeting all requirements of the job specifications, the NEC, and the code of the local jurisdiction.

Design Considerations

The system should meet present needs and take into account any future needs. Some of the basic factors that must be considered are:

- type of structure.
- present utilization and possible future uses.
- projected life of the structure.
- flexibility of the structure.
- location of service entrance equipment.
- load requirements and their locations.
- sources, quality, and continuity of power.
- switchgear, distribution equipment, and panels.
- installation methods.

Distribution Voltages

Distribution voltages, those connected to the primary of the transformers or to distribution panels, and *utilization voltages*, those connected to the loads, determine the voltage rating of the transformer. Adding all of the current loads will provide the needed ampere rating. Multiplying the sum of the load currents times the secondary voltage and dividing by 1000 will provide the figure for the kVA load for a single-phase transformer. For three-phase transformers, the value can be determined by multiplying the voltage times the current times 1.73 and dividing by 1000.

Single-phase $$\text{kVA} = \frac{\text{Volts X Amperes}}{1000}$$

Three-phase $$\text{kVA} = \frac{\text{Volts X Amperes X 1.73}}{1000}$$

To determine transformer size, start with one of the calculations above. This is the absolute minimum rating for the transformer. To this value additional kVA is added to compensate for the various considerations, conditions, and factors listed before.

Distribution voltages are classified as follows.

1. Low voltage—600 volts or less
2. Medium voltage—601 to 15,000 volts
3. High voltage—above 15,000 volts

Until the early 1950s the most common distribution voltage within most commercial buildings was 208/120 V system. With increased loads of central air conditioning, electrical heating, lighting levels, and the addition of numerous appliances and electrical office equipment, this value has been changed to 480/277 V. For industrial plants, 480-volt systems are almost always used.

Grounding

Distribution systems when installed in industrial plants are often ungrounded delta arrangements, resistance grounded delta, or wye systems. Resistance grounded systems may be either high resistance or low resistance. See diagrams in Figure 5–1.

The advantage of using an ungrounded or a high-resistance ground is that a single ground fault will not interrupt service. This is true even in the circuit having the problem. This is the only advantage.

There are several disadvantages of using an ungrounded system. Transient voltages caused by disturbances on the line due to arcing, switching equipment on and off, or lightning strikes have no ground path and may subject the insulation of the wiring and equipment to voltages several times their rated capacity.

This problem is somewhat alleviated by the capacitance between the conductors of the system which provides a capacitive grounding system. A small system might deliver approximately 0.1 ampere to ground, whereas a very large system may have currents as much as 20 amperes.

When a ground fault does occur on an ungrounded system, it may go unnoticed for an extended period of time. Because of the nature of these faults, repeated arcing may occur causing transients and stressing the system. The fault, even without considering transients, will put stress on the insulation at 1.73 times the voltage on a wye connected transformer. This is not important for low-voltage systems where the insulation is rated for 600 volts, but for medium voltage systems, cable insulation is rated at 100% for phase to neutral voltage, and higher voltage ratings of the cable should be considered if operations must continue under ground fault conditions.

For systems having insulation rated at 100%, clearance of the fault is recommended within 1 minute. A ground fault sensing device to warn of shorts should be installed so that the fault can be cleared before the event of a second short, which may cause serious damage to the installation. For the system to continue to operate for an indefinite period, it is recommended that the insulation rating be 173% of the phase to neutral voltage.

On ungrounded systems, it is often difficult to locate the fault even when its presence is detected by a warning system. Unless there is physical evidence that is observable, the loads will have to be taken out of service one at a time. The conductors feeding the loads and the loads would have a "hipot" test run on them.

Most of the difficulties associated with ungrounded systems can be overcome by using high-resistance grounding systems. The system will continue to operate under fault without developing transient overvoltages, and the location of the problem is less difficult to determine.

This can be accomplished on a wye system by placing the resistor between the neutral and ground. On a delta system, the use of a zigzag transformer connected to the three phases of the delta can be used. The resistor should be placed between the neutral of the zigzag transformer and ground.

With high-resistance grounding, fault currents are limited to 0.1% of three-phase fault current to ground which is more than sufficient to reliably operate a fault relay. With the introduction of highly sensitive relays, low-resistance grounding systems are seldom used in modern plants.

The NEC in Article 230 requires ground fault protection only on the service disconnect. If it is necessary to maintain service to other areas of the building, ground fault protection should be extended to feeders and branch circuits with selective tripping of circuits designed into the system. For health care facilities, the NEC in Article 517 requires at least two stages of protection.

Need for Higher Distribution Voltages

Canadian industries have recently begun to install 600/346 volt distribution systems to compensate for the increasing current loads in modern structures. This is in violation of the NEC in this country where the branch circuit voltage to ground shall not exceed 300 volts.

Voltages that were only used for outdoor distribution systems are becoming commonplace in modern interior wiring designs for large industrial plants, high rise commercial buildings, and large shopping centers. In each of these cases, long feeders are needed, and the same economies that the power company realizes by using high voltage can be achieved by the customer.

In the medium voltage range, the most common values over 600 volts nominal are 2400, 4160, 6900, and 13,800 volts. Depending on the requirements, this

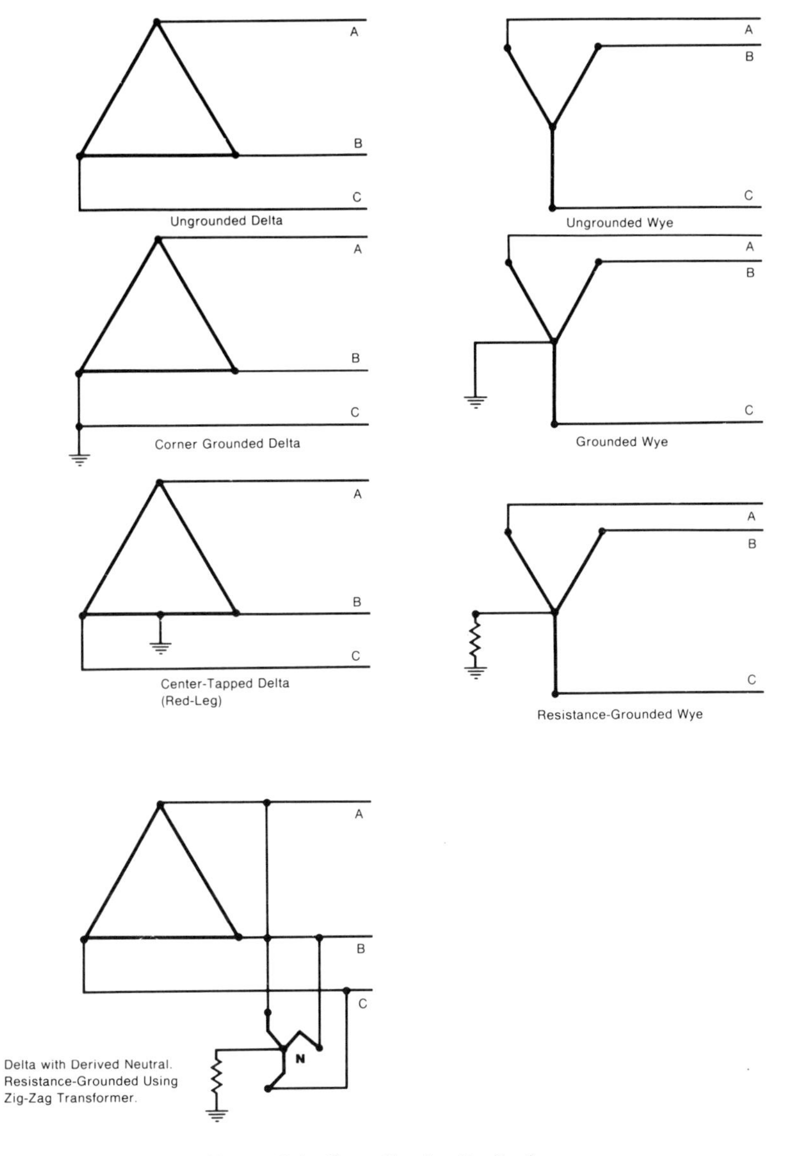

FIGURE 5–1. Grounding for distribution systems.

may go as high as 34,500 volts. Values above 34,500 volts are rarely specified unless for a special purpose. The trend now is toward 13,800 volts because it provides greater growth possibilities, is more flexible, and for loads exceeding 20,000 kVA, is more economical than lower voltages. For loads up to 10,000 kVA and for larger motors, a 4160-volt system is recommended along with 208/120 volts for lighting, appliances, and receptacles.

There is no limit in the NEC as to the maximum voltage that approved conductors may carry when installed in rigid steel conduit. There are requirements as to where such conductors and conduits must terminate. Article 230 of the NEC requires service conductors either to enter metal enclosed switchgear when the voltage exceeds 15,000 volts, or enter a transformer vault where walls and roof have not less than a 3-hour fire-resistance rating. This also meets the requirements of the NEC, Article 450.

Article 250 of the NEC provides for grounding of all exposed metallic parts of the system. This is to assure that potential between devices is zero and at ground potential. This should not be confused with system grounding, which may or may not be operating at ground potential. Further study of Article 250 is a necessity.

Types of Systems

No particular system can be recommended for all installations due to the differences in usage, equipment, safety, structures, requirements of the customer, and a multitude of other variables. The remainder of this section will explore various power distribution systems along with some of the advantages and disadvantages of each.

The following systems will be examined:

1. Simple radial system
2. Modern simple radial system
3. Loop primary radial system
4. Banked secondary radial system
5. Primary selective radial system
6. Secondary selective radial system
7. Simple network system

Simple Radial System

In the majority of cases, power is supplied to the building at the utilization voltage, and the conventional simple radial system is employed for the distribution of the power. Figure 5–2 provides a diagram for this system.

Under these conditions, a single source of power is supplied to a single substation transformer and converted to the utilization voltage which is connected to the main low-voltage bus. From here, it is distributed throughout the building at the utilization level to the various panels. Generally, the utility company has responsibility for the system to the output of the kilowatt-hour meter.

The loss of the power source, transformer fault, or interruption of the primary feeder or the main low-voltage breaker will remove power from the entire building. Service cannot be restored until the problem is corrected.

Economies are recognized due to the simplicity of the system. The utility company has responsibility for the source voltage, transformer, and primary feeder breaker, thereby relieving the customer of these responsibilities.

However, restoration of the system may be delayed due to the power company's involvement. After power is again made available, it may be discovered that the fault is within the structure.

It is recommended that the low-voltage feeder beakers be opened prior to having power reestablished. The load bus units can then be energized one at a time to determine which of the panels contains the fault. If selective tripping has been utilized, and a breaker is not defective in the load bus unit, this step should not be necessary since the defective load has already been identified by the open protective device and is off-line.

Because of the low-voltage distribution system used and voltage drops on the feeders and branch circuits, the conventional simple radial system is not efficient and may require larger breakers and disconnects to offset any savings. When the system requires more than 1000 kVA, other systems should be considered.

Modern Simple Radial System

When power demand is above 1000 kVA, the modern simple radial system may be utilized. See Figure 5–3 for a diagram of this system. The distribution voltage is delivered through the primary feeder breaker to power centers in areas where the loads are located within the structure. These centers are normally factory assembled. They consist of a primary fused switch, a step-down transformer, a transformer breaker, and a series of load circuit breakers.

The use of the higher distribution voltage allows the installation of smaller conductors to carry lower currents to deliver the same amount of power to the power centers. This in turn reduces line losses and improves the voltage regulation of the system. The high current–low voltage feeder breakers are eliminated from the system resulting in additional savings.

Transformers in the power centers may be either air-cooled or liquid-filled types. Each transformer is protected by a fused disconnect switch and a breaker. Because the transformers are located within the load area, they must have the capacity to carry the peak load for the area. If there are great differences in the loads among areas, the modern radial system will require greater transformer capacity than the simple-radial installation. Therefore, the design engineer should take care to balance the loads among centers.

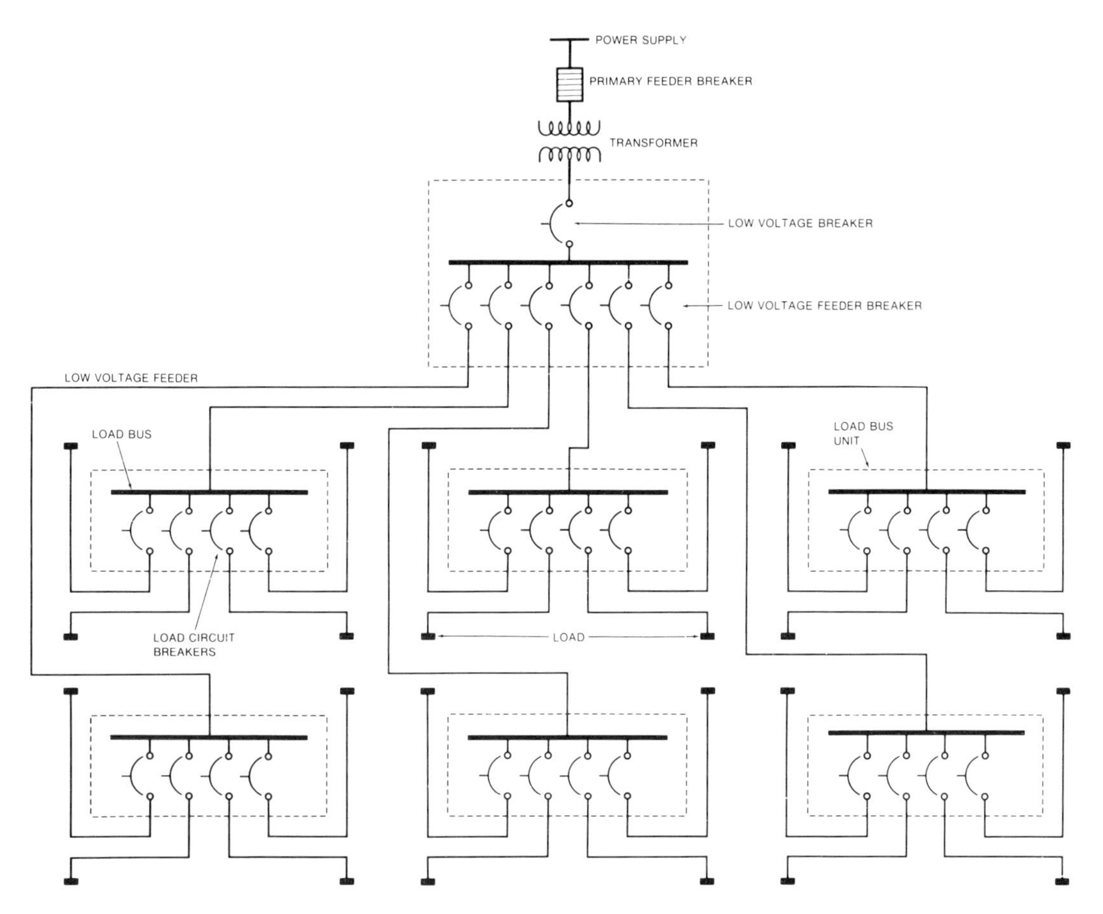

FIGURE 5–2. Simple radial distribution system.

When a fault occurs in one of the power centers, power can be restored to the rest of the building by opening the primary fused switch at the center and resetting the primary feeder breaker. The system, however, still suffers from the lack of flexibility and continuity of service. Disruption in the single power source, opening of the single primary feeder breaker, trouble with the primary feeder, or a fault with any one of the transformers will interrupt power to the entire building.

Where peak loads are above 1000 kVA, the modern simple radial system is the least expensive. Improvements can be made in its flexibility and continuity of service by modifying it as shown in Figure 5–4.

Each transformer in the diagram is supplied by its own primary feeder which is individually protected by a distribution circuit breaker. A fault in a feeder or a power center will disrupt only the loads assigned to that particular station. This arrangement increases the cost of the system but improves the continuity of service to the areas not affected by a fault.

Economies can be realized by supplying more than one transformer from a single primary feeder. Also the automatic feeder circuit breakers can be replaced with load break switches backed up with one automatic circuit breaker. By doing this, however, the number of areas affected by any fault will be increased, and the advantages of using the system are reduced.

Loop Primary Radial System

Figure 5–5 depicts the wiring of the loop primary radial system. A sectional primary loop is used rather than the primary radial. It has all the advantages of the modified modern simple radial system at less cost, and at slightly greater cost than the modern simple radial system.

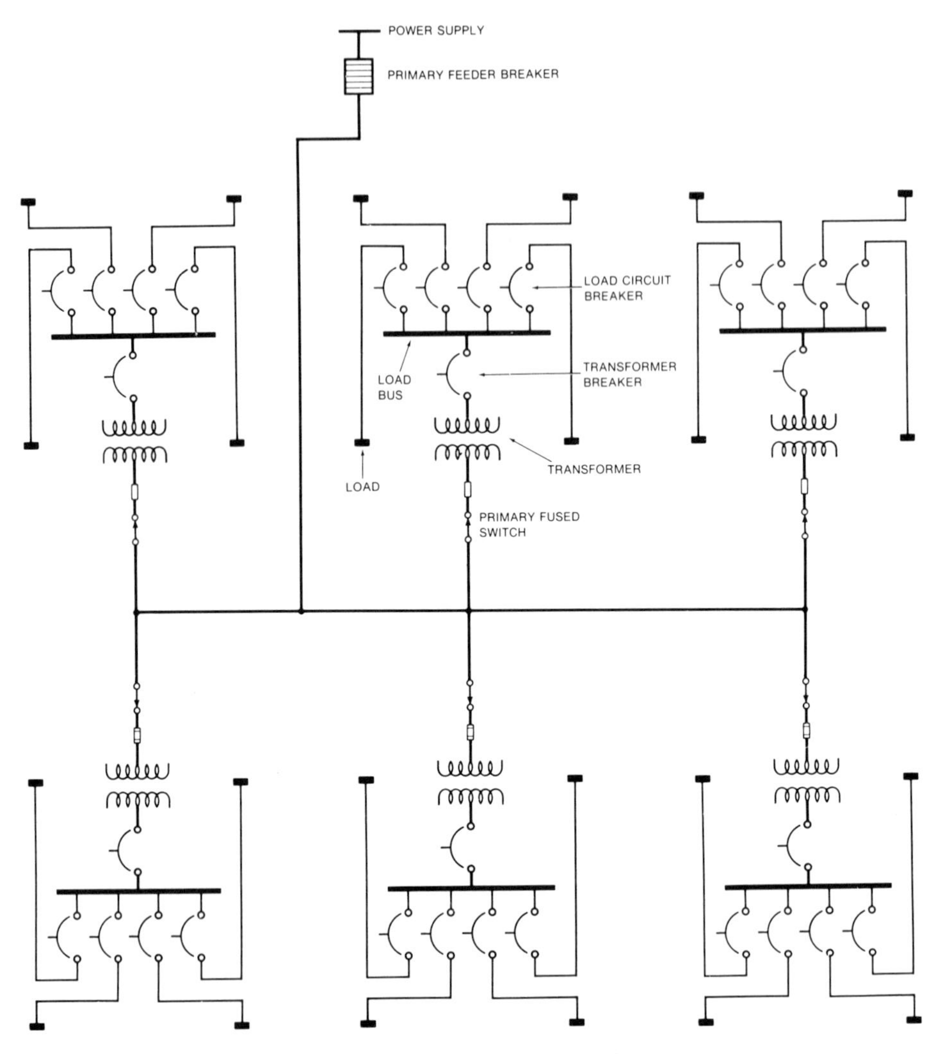

FIGURE 5–3. Modern simple radial distribution system.

Each power center is equipped with a primary load break switch. In normal operation, all these switches except one are closed. The open switch is selected to balance the current load on the loop feeder. Another load break switch is located after the primary feeder breaker.

When a fault occurs in one of the centers or on the primary feeder loop, the primary feeder breaker opens, disrupting power to the entire system. The primary load break switches can be opened to the defective center or feeder line, and power can be restored to the rest of the system. The normally open primary break switch will need to be closed under these conditions unless it is associated with the fault. Power can only be restored to the faulty section after repairs have been made. The loop primary radial system is a good compromise between the modern simple radial and the modified modern simple radial installations.

Banked Secondary Radial System

Figure 5–6 shows the power center for the banked secondary radial system. This arrangement utilizes the loop primary radial distribution system along with a secondary loop. Figure 5–7 illustrates how the power centers are connected.

Power can quickly be restored to this system with malfunctions in the primary feeder or any transformer. The section of bad feeder can be isolated by opening

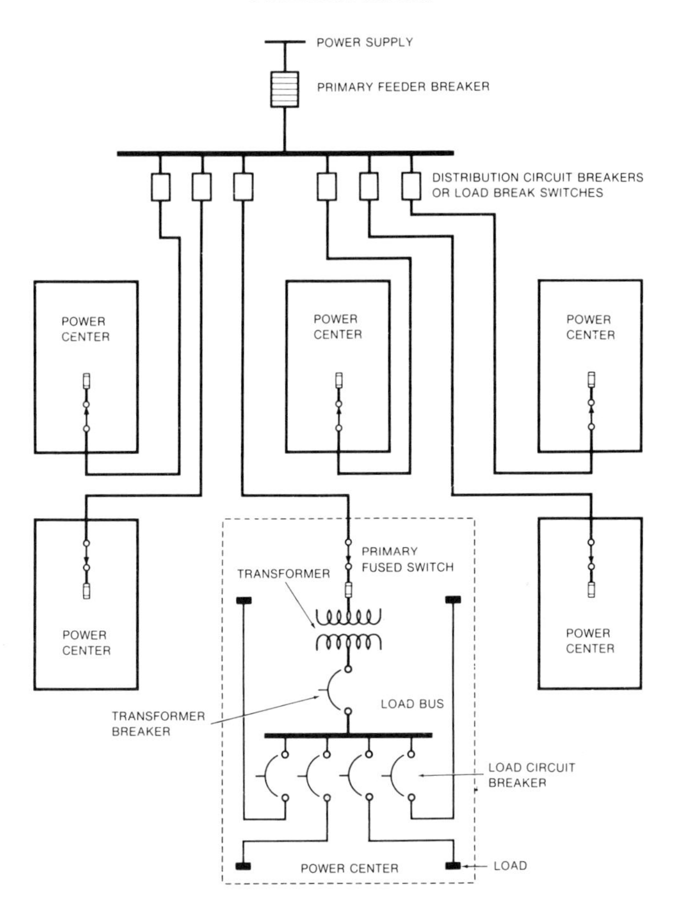

FIGURE 5–4. Modified modern simple radial distribution system.

both of the loop break switches in the primary circuit on both ends of the defective cable while closing the remaining switches to maintain power to all of the transformers.

Under normal operating conditions one of the loop primary switches will normally be opened to balance the load on the feeder. If this switch is not associated with the feeder fault, it will need to be closed.

In the case of a transformer fault, the primary fused load switch can be opened, removing primary feeder power from the defective transformer. It may not be necessary to open the feeder loop switches. At the same time, the transformer's bus breaker must be opened or a back feed will result through the limiter bus circuit. This would result in the transformer having both the secondary voltage present on its windings, and if still operative, a voltage on the primary equal to approximately the distribution voltage on the feeder. This could be dangerous for an electrician working on the transformer.

When a problem occurs in one of the load circuits, the defective circuit can be dropped off the line by opening the load circuit breaker. If the problem were to occur on the load bus itself, the load break switches along with the transformer breaker in the secondary in the power center would be opened to remove power from the load bus until repairs could be completed.

Secondary loops provide a number of additional advantages other than providing emergency power to the loads of a defective circuit.

1. They help to equalize the load on all transformers, therefore only one kVA rating can be established for all the transformers in the system without the need to match different loads to the individual transformer. The transformers, however, need to be matched for

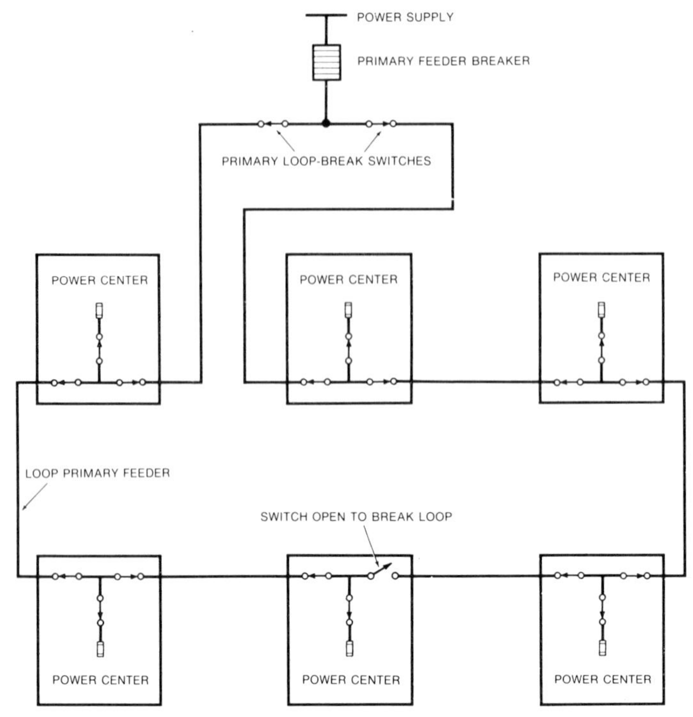

FIGURE 5–5. Loop primary radial distribution system.

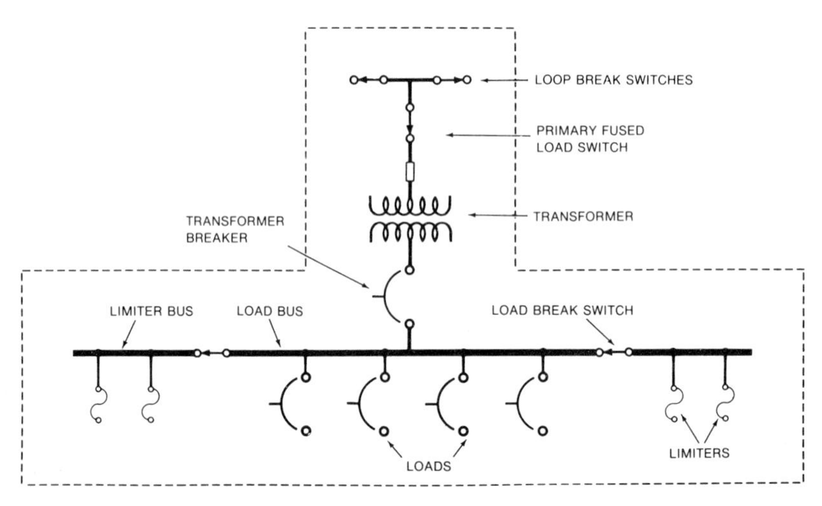

FIGURE 5–6. Power center for banked secondary radial system.

voltage and impedance. Matching loads to the transformers is a real problem in the other systems with the exception of the simple radial installation.

2. Because all the transformers are connected in parallel, there is usually savings in the need for additional transformer capacity.
3. The system is better for across-the-line starting of motors and may provide savings in motor starting equipment.
4. It is the most satisfactory arrangement for combined lighting and power loads.

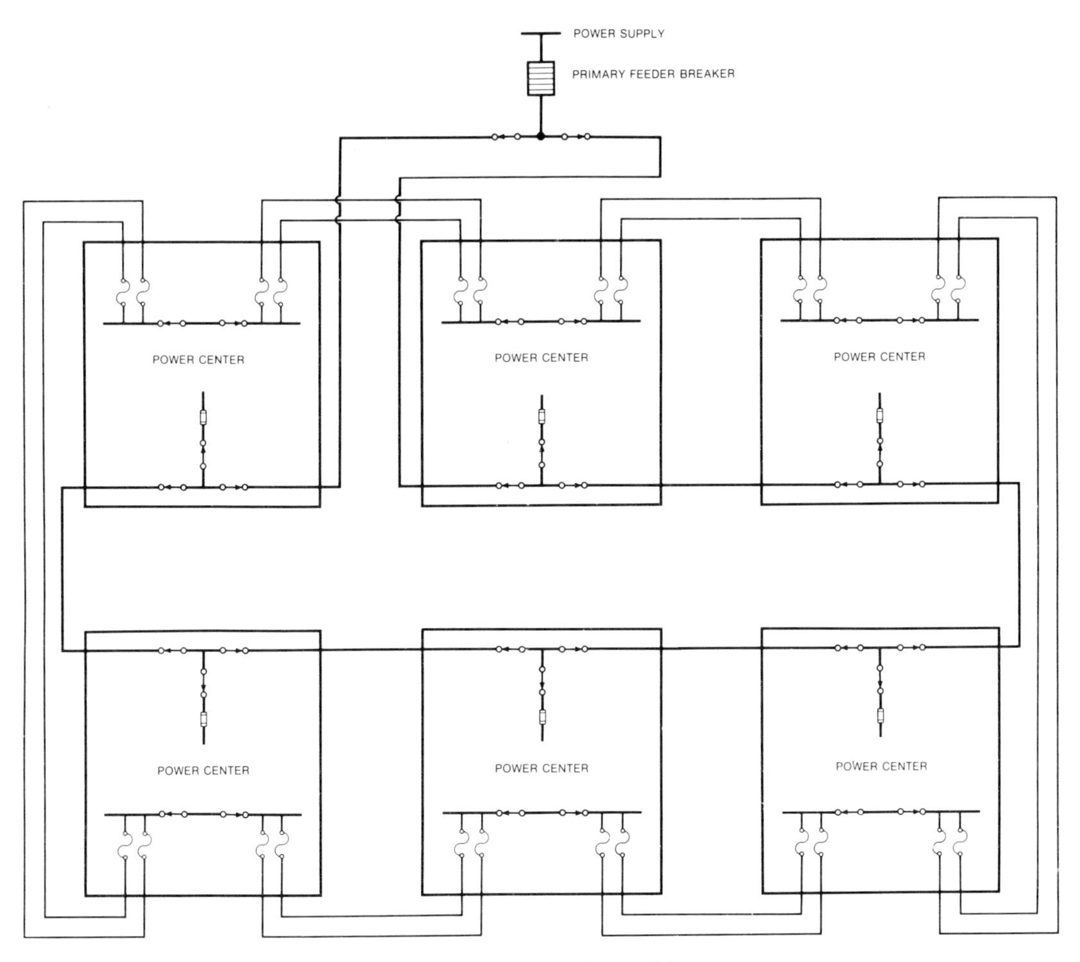

FIGURE 5–7. Banked secondary radial system.

The bus loads of the defective transformer are supplied power by the other transformers in the system through current limiters (see Limiters, below).

Primary Selective Radial System

The primary selective radial system allows quick restoration of power to the system in the event of primary feeder or breaker failure. Figure 5–8 provides an illustration of this system.

In the event of a fault with power being supplied to the primaries of the transformers, only half the transformers will be out of service. The loads are equally divided between the two feeders. Because each feeder has been designed to carry the total load of the building, the transformers can be switched over to the second supply that is available in the power center, and service will be returned to the affected areas.

When the problem exists in one of the power centers, the station can be dropped from the line. This would not be possible using the single-pole–double-throw switch as shown in Figure 5–8. In order to disconnect the power center, the single-pole–double-throw switch is usually connected in series with a single pole disconnect switch as shown in Figure 5–6. This switch is identified as the primary fused load switch. Another arrangement would be to have the single-pole–double-throw switch replaced by two single-pole load break switches with a common terminal of each wired to the primary fuse.

Secondary Selective Radial System

The secondary selective radial system uses the primary-selective radial system with all of its advantages as part of the circuit arrangement. In addition, a

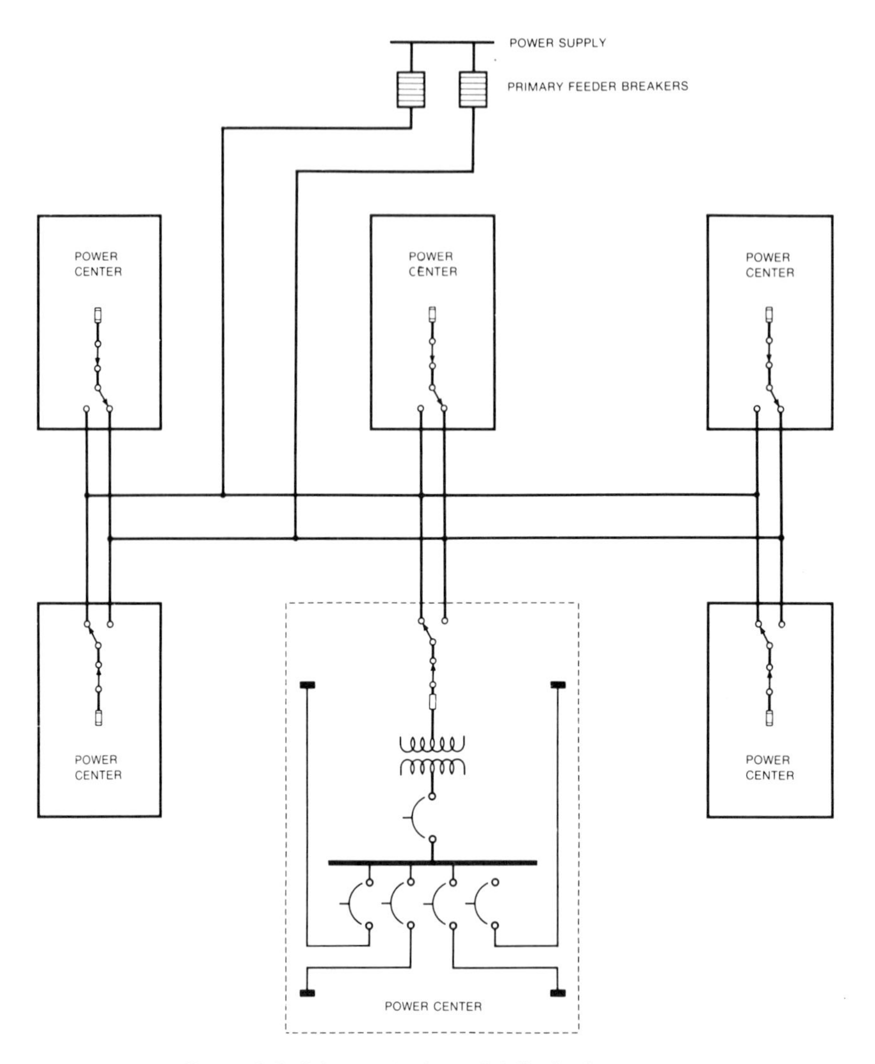

FIGURE 5–8. Primary selective radial distribution system.

redundancy of transformers is incorporated in each power center. Each transformer is rated to carry the total load of the power center. Under normal operating conditions, the bus-tie breaker is open, and the load is divided equally between the two transformers. The bus-tie breaker is mechanically interlocked with the transformer breakers so that it cannot be closed unless one of the transformer's breakers is opened. Figure 5–9 depicts the circuit for this system.

A transformer malfunction in this system will cause the primary feeder breaker to open. Power can be restored to the system by opening the disconnect switch and transformer breaker on the defective transformer and closing the bus-tie breaker. The primary feeder breaker can then be reset, and the second transformer in the power center will pick up the load of the transformer that has been removed from service.

Secondary selective radial systems are very expensive. This is primarily due to the duplication of transformer capacity and power feeders. A modified design can be incorporated at less cost. This would consist of a single transformer at each power station with a secondary tie between each pair of transformers. Each transformer in the pair would have to have the kVA rating required to carry the load for both power centers. This will reduce the number of transformers required

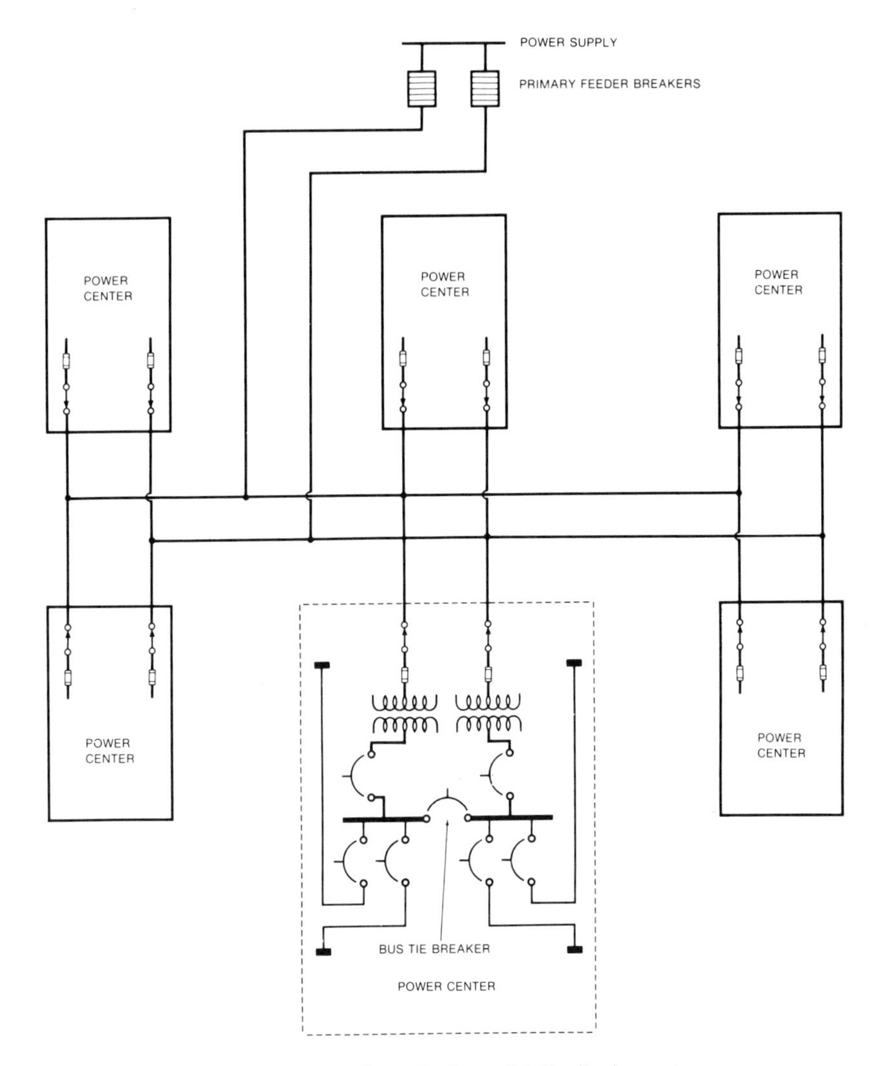

FIGURE 5–9. Secondary selective radial distribution system.

by one-half. Even though the total power capacity is the same, a single 1000 kVA transformer would cost less than two 500 kVA transformers.

Simple Network System

A simplified drawing of the power center for the simple network system is given in Figure 5–10. One significant change was made in the diagram. The transformer breaker is replaced with a network protector in a network system.

The network protector is a special heavy-duty air circuit breaker which is opened and closed automatically by an electrical motor. The motor is controlled by network relays which automatically sense circuit conditions. The breaker will close only when voltage conditions are such that the associated transformer will supply power to the load. It will open at any time there is a backfeed, or when the secondary is feeding power back to the primary. The primary purpose of the network protector is to protect the secondary loop and loads from transformer or primary feeder faults.

Figure 5–10 depicts in simplified form a simple network system. This arrangement has three primary feeders. They are designed so that any two can carry the full load if the third one fails.

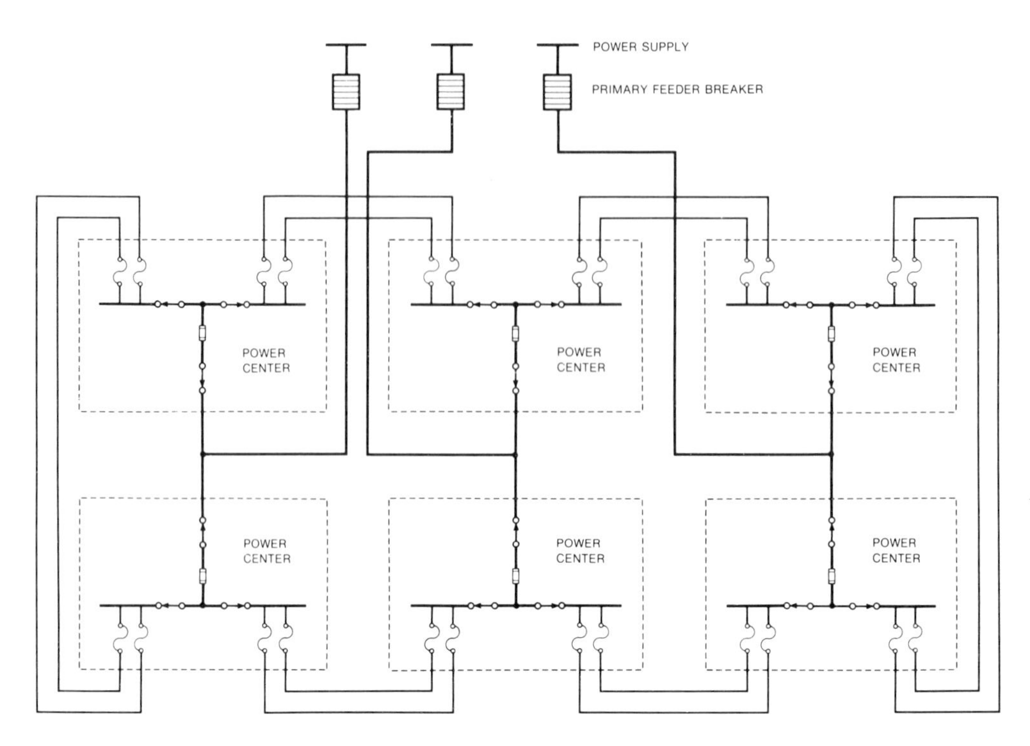

FIGURE 5–10. Simple network distribution system.

The system has a secondary loop from which the loads are supplied over short radials. The secondary loop will supply any of the loads automatically in the event of a transformer failure within the system. When one of the transformers is underloaded, it can supply power to other circuits that have larger loads. This tends to equalize the loads on all of the transformers.

Some of the advantages of the simple network system are

- large motors can be started across the line and at lower voltages, which results in lower cost for motor starting equipment.
- the secondary loop provides greater flexibility.
- additional loads can be supplied from the nearest supply bus without the necessity of recalculating and redistributing loads in order to provide balance to the system.
- power centers can be connected or disconnected from the system depending on the need for power.
- voltage regulation on the system is very good.

Other Distribution Systems

Eight different distribution systems have been examined in the course of this chapter. There are many more that were not taken into account. It does not take much imagination to see that various combinations can be put together to meet the particular loads and needs of the customers. Additional feeders or transformers can be added or removed. Reliability of secondary loops can be improved by having three sets of three phase cables instead of two sets or one set. Primaries can be made either selective or nonselective. The questions to answer by the design engineers as to which system to use would include:

- How much reliability must be built into the system?.
- How much flexibility must the system have?

It is the responsibility of maintenance electricians to know the systems they are responsible to service. Time spent studying systems prior to a failure will pay great dividends when problems arise.

System Protection

Distribution protection is usually designed to protect the equipment in the system as well as the circuit to the loads. The protective devices must be sensitive to abnormal conditions. When a fault occurs, fuses, breakers, and limiters must be capable of isolating the fault from the other circuits served by the distribution system. They must be sensitive to overloads, and at the same time, be capable of clearing a short circuit fault.

Small Plant Distribution System

The distribution system of a small plant has a single power source supplied to it by the utility company. The system consists of a single circuit breaker or fuse between the source and the transformer that steps the distribution voltage down to the utilization voltages. Secondary circuits of the transformer are also protected by circuit breakers or fuses. Both primary and secondary protection is provided. A fault on a secondary circuit should be cleared by its protective device. If it fails to trip, the primary device will open and clear the fault. Problems with the transformer or primary feeder will be cleared by the circuit breaker or fuse between them and the distribution voltage source. Figure 5–11 depicts two such arrangements for a small plant.

Additional protection has been provided in Part B of Figure 5–11. The primaries of the transformers have a primary fuse as well as the circuit breaker. The fuses act as backups for faults not cleared by the secondary breakers.

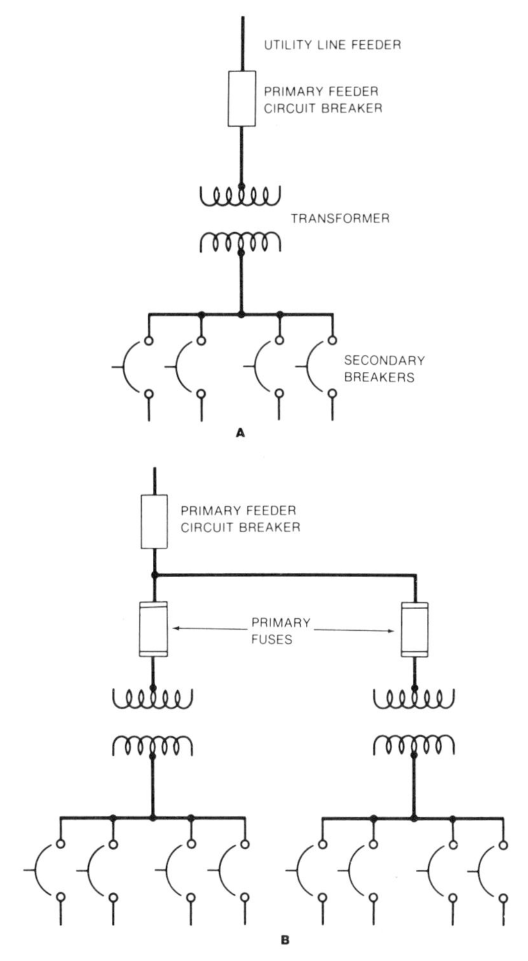

FIGURE 5–11. Small plant distribution systems.

Distribution System for a Commercial Office Building

Figure 5–12 illustrates a typical single-line riser diagram showing electrical protection devices and their arrangements for a 60-story office building. The building is supplied by the utility company with a one-spot network from a vault under the sidewalk and another in a fireproof vault on the 40th floor of the building. The second vault on the 40th floor is used to reduce the length of the runs for the secondary feeders. Primary voltage for the system is 13,800 volts.

Each vault has six 2500 kVA network transformers and each supplies 4000 amperes at 480/277 volts. Fault currents at each service would be close to 100,000 amperes.

The main and feeder circuit breakers must have the interruption capacity to withstand the high short circuit current available. Smaller feeder circuit breakers can be of the current limiting type, integrally fused breakers, or they may have the interrupting capacity necessary for the circuit. Breakers are coordinated so that under fault conditions only the circuit containing the fault is dropped from the line. Electrical service is continued to the other parts of the building.

Fuses may be used instead of breakers because of the lower costs involved. Their use would involve more time off-line when a fault occurs. In addition, the savings may be offset by the need to carry a complete line of spares as a backup for repairs. Also, there is danger if qualified maintenance personnel are not used of having replacement fuses installed that have incorrect values. This could cause a reduction in the selective coordination of the system, and in some cases, reduced safety.

There are many variations in the design of a high rise installation. These would be determined by build-

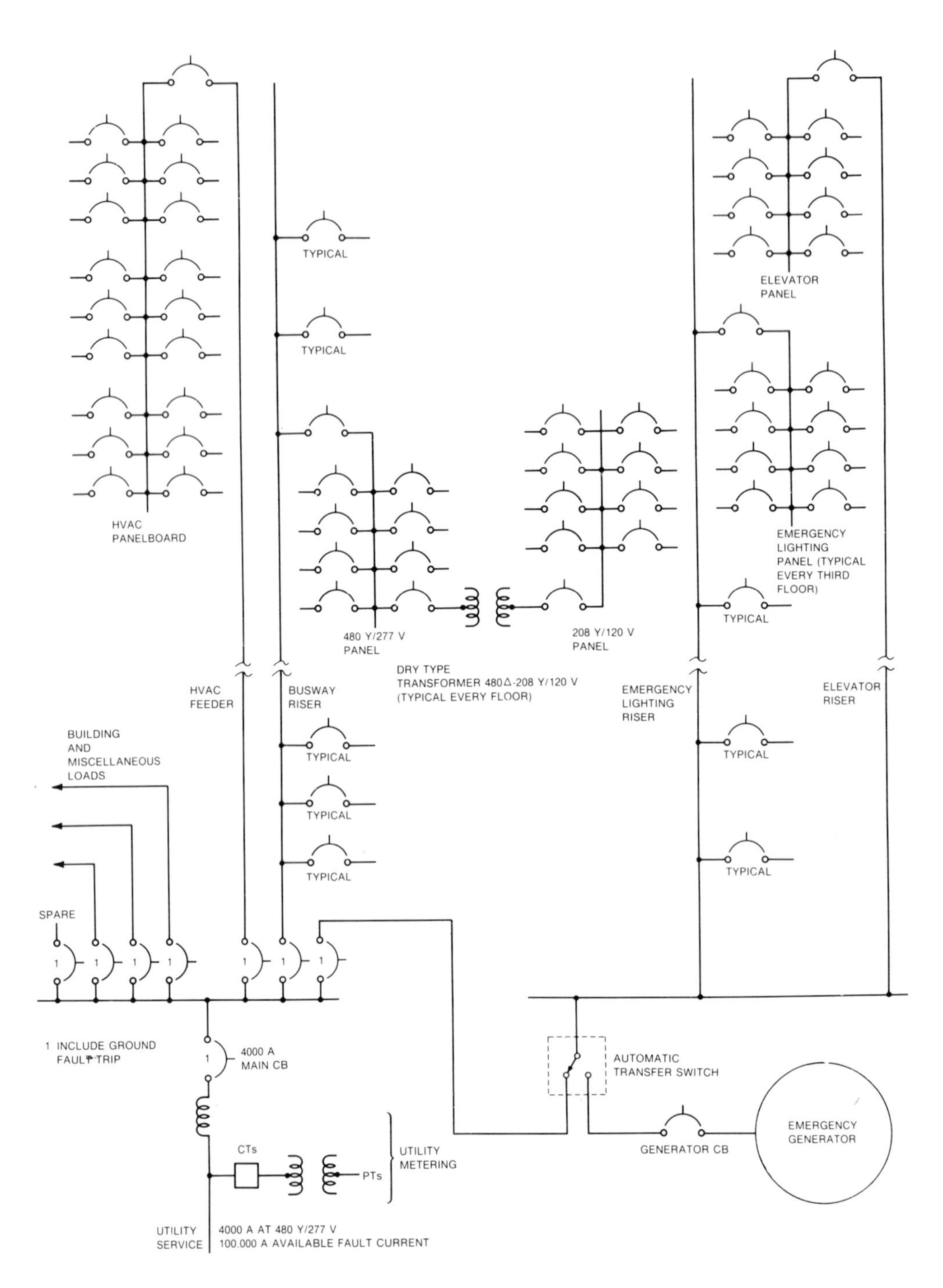

FIGURE 5–12. Typical power distribution and riser diagram for a commercial office building. *(Courtesy Westinghouse Corp.)*

ing size, utilization equipment, special requirements, or any of the considerations presented at the beginning of this chapter. For example, a cable riser might replace the busway riser. Line losses would be reduced using this installation. However, the cost of this arrangement would be higher and some diversity would be sacrificed. Another approach would be to use more than one busway riser in large buildings with more than one electrical closet on each floor.

Selective Coordination

It is not usually desirable to have the service network protector or the service disconnect open when a fault occurs on a feeder or branch circuit. Under these conditions, power would be interrupted to the entire plant or installation.

In order to avoid blackouts, one system used is elective coordination of breakers and fuses. Figure 5–13 shows such a system.

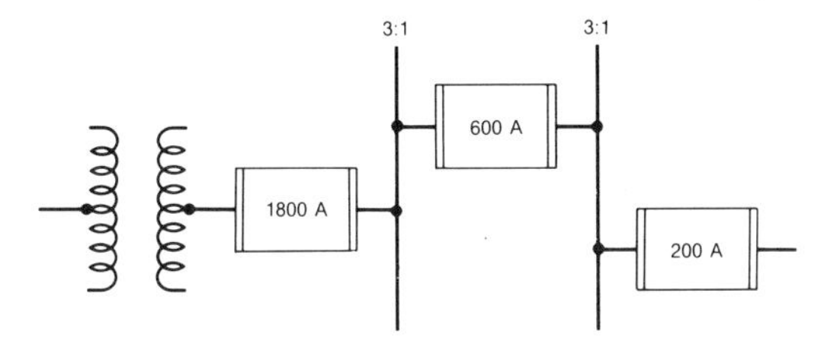

FIGURE 5–13. Selective coordination of protective devices.

Although the figure depicts 3:1 ratios, the ratio may be different for each circuit. The ratio may be as high as 10:1 or low as 1.5:1. The ratio will vary with the type of protective device and the manufacturing design technology of the device. Manufacturer's recommendations must be followed in order to achieve the selective coordination desired.

Circuit Breakers

Many older circuit breakers previously installed may not be current limiting and will not handle the high fault currents available in modern installations. A common practice is to use a fuse with the required interrupting capacity in front of the breaker to protect it from high currents. This was illustrated in Part B of Figure 5—11. The fuse can be coordinated to open on currents that exceed the interrupting capacity of the breaker.

If a system is fully rated, each circuit breaker will have the interrupting capacity for the part of the circuit which it serves as protection. It will be equipped with a long time delay and an instantaneous overcurrent trip device. The main breaker's time device will be long enough to allow a feeder breaker to react to fault. The time delay for the feeder breaker will be long enough to allow a branch circuit breaker to operate.

Usually a maximum of four low-voltage breakers can operate reliably in series with each other. The time required for operation of each in case of fault should not overlap. Each breaker should allow enough time within the safe operation of the system for the breaker downstream from it to trip.

Only recently has it been possible to use breakers with instantaneous trip for selective coordination. This has been achieved through the use of solid state technology.

Limiters

Limiters are special fuse elements which are current limiting in nature and have special characteristics. When heated beyond their capacity by overcurrents, which is approximately three to three and a half times the continuous current rating of the cable, they will blow as any fuse does, but will not emit smoke, flame, or noise.

The higher current rating of the limiters is to insure that they do not blow under overload conditions and disrupt power to other circuits that normally would not be affected by the problem. They are designed to provide short circuit protection.

Limiters are supplied to be used with various gauges of conductors and are rated by cable size. The largest limiter is rated at 1000 MCM. Limiters have an interrupting rating of 200,000 amperes and can be used in circuits up to 600 volts. Cable limiters are also available with ratings of 250 volts or less for light commercial and residential applications. Figure 5–14 shows the use of limiters in the power supply for residential customers. Only one limiter is required in this arrangement, because there is no other voltage being supplied on the secondary loop.

For commercial and light industry, the secondary loop should have a minimum of two sets of three-phase conductors to provide maximum continuity of service. Each loop should be connected to the bus of each transformer through limiters. Limiters must be installed on each end of the secondary loop in order to be effective. If a fault occurs on the cable, current from the source would destroy the limiter between it and the fault. Current in the loop-secondary would blow the

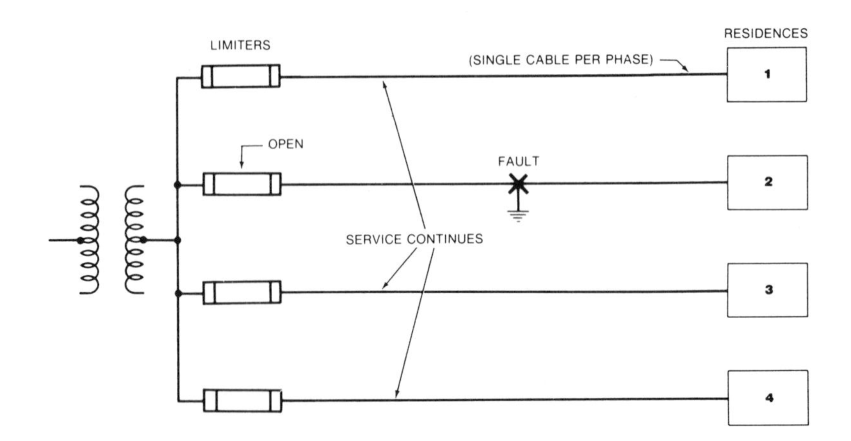

FIGURE 5–14. Residential service entrances protected by limiters.

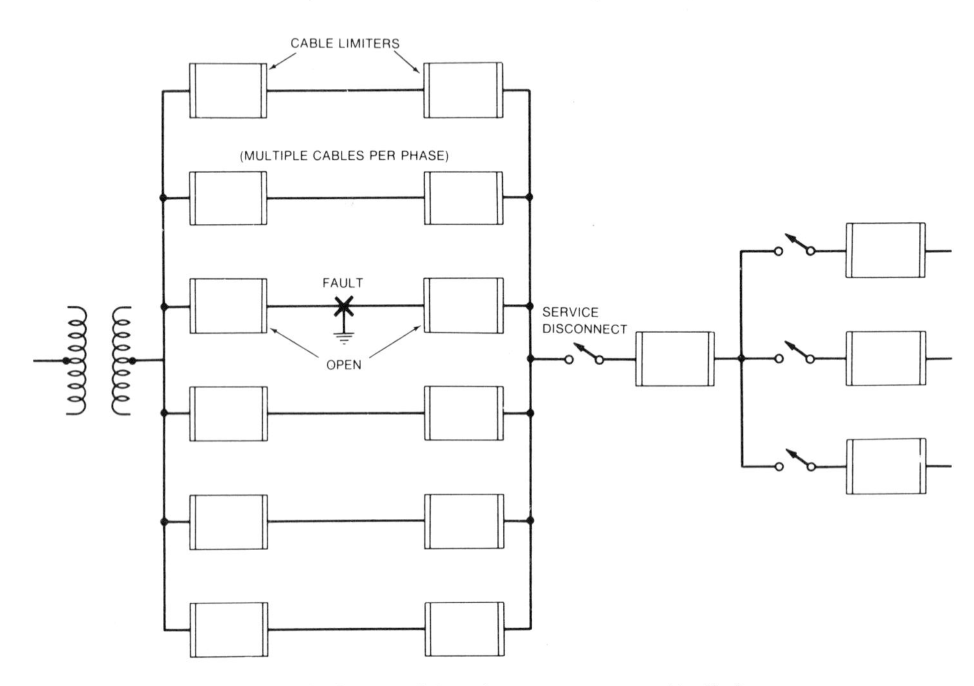

FIGURE 5–15. Commercial service entrance protected by limiters.

limiter between the customer and the fault (Figure 5-15).

Although the NEC prohibits equipment connected to the supply side of service disconnect equipment to be connected to the service disconnecting means, cable limiters are an exception. There are definite advantages provided for by this exception. These include

- isolation of faulted cables without interruption of service on the other cables supplied by the common source.
- convenient scheduling of repairs.
- reduced chances of burnout of the equipment due to a fault on the line side of the main overcurrent protective device. Without cable limiters, a fault between the transformer and the service entrance equipment is given little protection.

• protection against short circuit currents. Often a short circuit fault may exceed the interrupting rate of service circuit breakers.

Other Types of Protection

Depending on the design and the requirements of the loads placed on a distribution system, other types of protection are available. Some of these are listed below.

1. Low-voltage protection. If the source voltage drops to predetermined level, the system can be designed to drop off the line. This is usually accomplished by relays in a control circuit which continually monitors the input voltage. They are usually associated with the circuit breaker or magnetic switch through which power is provided to the equipment. When the voltage falls to the set level of the relay, the relay will drop out, and through mechanical linkage with the breaker or switch, open it, removing power from the circuit. Overvoltage protection can also be provided in the same manner, but is seldom used.

2. Single-phasing. If a three-phase system loses one of its phases, the system can be provided with relay control that will remove it from the line. This type of protection is usually desirable on a distribution system that is providing power to polyphase motors.

3. Ground fault interrupters. Current relays are used to sense any imbalance in the current in the conductors supplying the load. These devices are sensitive to differences in the low-milliampere range and will disconnect the power from the load in milliseconds when a ground fault occurs.

4. Phase reversal. Considerable damage can occur to some loads if the phase of the power system is accidentally reversed. Protection can be provided to remove power to these loads, once again using phase-sensitive relays.

5. Surge protection. Modern electronic devices are often very sensitive to occurrences on the distribution system. Special provisions for them are often provided in the form of various types of surge protectors. With more computer installations being built every day, this type of protection has become necessary.

6. Alarm systems. On supervised systems, alarms can be installed to sound if any critical element begins to malfunction. Alarms can be used for temperature, flow, pressure, current, voltage, or for any function that can be measured or detected.

Questions

1. List some of the factors associated with the design of any distribution system. Be prepared to discuss each of them in class.
2. A three-phase transformer is going to supply power to motor rated at 480 volts/25 horsepower. What is the minimum kVA rating of the transformer? Each horsepower is equal to 746 watts.
3. How much current will each phase supply to the 25-horsepower motor in Question 2?
4. Medium-range voltages are classified as those voltages between ______ and ______.
5. What is the most common voltage used for distribution in a large industrial plant?
6. Explain the advantage of using ungrounded distribution systems.
7. List some disadvantages of ungrounded systems.
8. What is the maximum allowable branch circuit voltage to ground according to the NEC?
9. What is the maximum allowable distribution voltage according to the NEC? Under what conditions?
10. Are equipment grounding and system ground the same? Explain.
11. If one of the load bus units is defective in a simple radial distribution system, how can power be maintained to the other units?
12. How does the modern simple radial system differ from the older simple radial system?
13. Why is the modified modern simple radial system more reliable than the modern simple radial system?
14. Give some possible uses of loop-break switches used with the loop-primary distribution system.
15. What is the primary advantage of a banked secondary radial system over the previous systems studied in this text?
16. Specify the purpose of the limiters in the banked secondary system.
17. Explain how the highest degree of reliability and flexibility can be achieved from a distribution system.

18. In Figure 5–12, how is the 120-volt utilization voltage obtained?
19. How much interrupting capacity must the main circuit breaker have for the building depicted in Figure 5–12?
20. Using the same diagram, name the riser diagrams.
21. What does the term "selective coordination" mean when applied to fuses and circuit breakers?
22. How are current limiters rated?
23. Why must a banked secondary system have two current limiters on each cable?

PART

II

Motors

CHAPTER 6

Fundamental Concepts: Motors

When Alessandro Volta discovered the principles of the electric battery in the early 1800s, he opened the way for expanding experimentation in electricity.

Volta noticed that while cutting up a frog, the frog twitched each time he entered the abdominal cavity with two dissimilar metal instruments. The frog's fluids acted as an electrolyte, and electrons were transferred from one piece of metal to the other.

The battery provided a reliable source of electrical power. Up to this time only static charges were available for making scientific inquiry into the nature of electricity. Appropriately, the unit of electrical potential energy is called the volt.

In 1819, Hans Christian Oersted discovered the principles of electromagnetism. He noticed while lecturing at the university that each time he brought a compass close to a current carrying conductor, the compass aligned itself perpendicular to the conductor. Figure 6–1 depicts this phenomenon. The unit of magnetizing force, ampere turns per unit length (NI/L), is named for him.

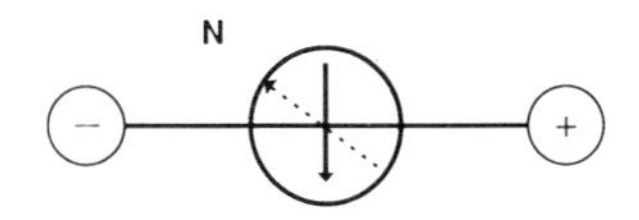

FIGURE 6–1. Effect of the magnetic field created by current flow on a compass.

In 1821, Michael Faraday used Oersted's findings to create motion with the magnetic field. Faraday is credited with inventing the first electric motor. Figure 6–2 illustrates his device.

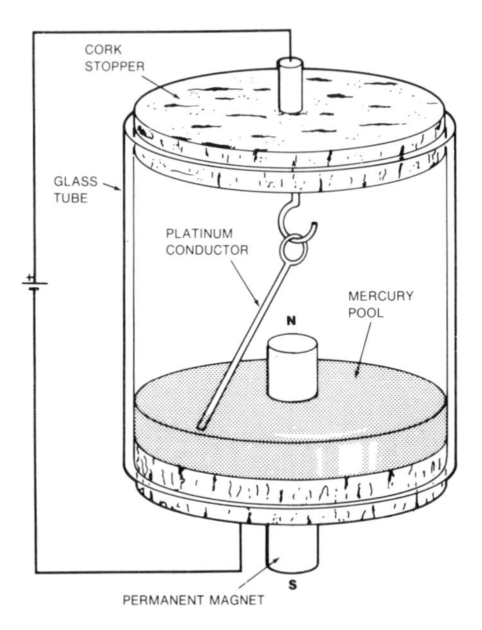

FIGURE 6–2. Faraday's motor.

Faraday achieved continuous motion of the wire which rotated around the magnet. Because the wire was enclosed in the glass tube, there was no way of having the motor perform work. Although the motor had no practical value, it demonstrated the possibility of obtaining a workable machine.

The first U.S. patent for a motor was issued to Thomas Davenport in 1837. Davenport developed his motor with very little money and under varying hardships. He used silk taken from his wife's wedding gown as the insulation for the conductors. His brother Oliver helped finance him in his endeavor. Oliver sold his horse and wagon used in his business for this purpose, resulting in his own bankruptcy.

Because the magnetic generator had yet to be invented, Davenport's motor lacked the needed electrical energy to be commercially feasible. He died penniless without recognizing his giant contribution to the world. It was not until 1879 at the international exhibition in Berlin that the first motor intended to be purchased for commercial applications was shown.

Electric motors play a tremendous role in modern life, much of which we are not even aware until we are affected in some adverse way. Have you counted the number of electric motors used in your home? Take a few minutes and list them. Are you surprised?

My count came to over 20 when I attempted this exercise, and I realized that I had missed at least 2, my electric pencil sharpener and the motor in my printer. There are probably more. Yes, there was my smokeless ashtray.

What about these motors found in the home? Are they all the same? Obviously not. They come in various sizes and configurations. They have different applications. Some must be more powerful than others. Some must have a longer life span and higher reliability. The motor in my smokeless ashtray is not nearly as important to me as the motor in my water pump.

As we leave the home and prepare to drive away in our automobile, we encounter several more electrical motors. We depend on an electric motor to start the gasoline engine so that we can begin our journey. In addition, electric motors are used to roll the windows up and down, blow hot and cool air at our command, pump liquid washer onto the windshield, wipe the windows clean, raise and lower the antenna, and adjust the position of the seat. Electric motors drive fan blades to cool the radiator and the condenser of the air conditioning system. The list of motors used only in the automobile is indeed impressive.

Our journey is either speeded up or delayed according to signals transmitted by devices using electric motors. Signal lights are controlled by timing circuits driven by small electric motors. A drawbridge may be open or closed by a large motor. Toll gates are dependent on these devices.

When we arrive at the end of the journey, the chances are we will again encounter more motors. They are used in elevators and escalators to save us from climbing up and down stairs. They help maintain a comfortable environment in which we work or shop. They move conveyors and drive machinery of all types. No matter what our job is, motors normally play an important role.

With all these motors in our environment, it is important that there are skilled electricians who know how motors operate, how to select and install them, how to maintain them, how to repair them, and how to replace them. The intention of Part II of this book is to provide the necessary information to help train the electricians needed to keep our motors running.

Definitions

A *motor* is defined as an electromagnetic device that converts electrical energy into mechanical energy. Electrical energy is measured in watts, and the mechanical energy is measured in horsepower.

Horsepower

Power is defined in physical terms as the rate of doing work. *Work* is defined in terms of weight in pounds multiplied by distance in feet, or

$$\text{Work} = \text{Pounds} \times \text{Distance (Foot-Pounds)}$$

$$\text{Power} = \frac{\text{Work}}{\text{Time}}$$

(Foot-Pounds/Second)

A *horsepower* is defined as 33,000 foot-pounds per minute or 550 foot-pounds per second.

Watt

The watt is the measure of electrical power and is defined in terms of voltage and current in amperes, or

$$\text{Watts} = \text{Voltage} \times \text{Amperes}$$

This formula is related to the physical units in that the *volt* is defined as 1 joule per coulomb of electrons, and the *ampere* is defined as the flow of 1 coulomb of electrons per second.

The *joule* is the metric unit of physical work and is equal to approximately 0.737 foot-pound. The *coulomb* is a quantity of electrons. According to authorities at the National Bureau of Standards, 1 coulomb is equal to approximately 6.15 X 10^{18} electrons. Substituting these values for the voltage and current we find

$$\text{Watts} = \frac{J}{C} \times \frac{C}{T}$$

Coulombs cancel in the formula, and it becomes

$$\text{Watts} = \frac{J}{T}$$

or 1 watt is equal to approximately 0.737 foot-pounds per second.

From these definitions, it is possible to calculate the relationship between the electrical watt and horsepower. This is found to be

$$1\ \text{HP} = 746\ \text{W}$$

When the current and voltage are not in-phase in an AC circuit, the formula for determining the wattage of the circuit must be modified. This out-of-phase condition occurs when the circuit is either inductive or capacitive.

In an inductive circuit, the current will lag voltage. In the capacitive circuit, current will lead the voltage.

The volt-ampere calculation must be modified in either of these two cases in order to determine the actual power used in watts. This is accomplished by multiplying by the cosine of the angle between the current and voltage.

$$\text{Watts} = \text{Volts} \times \text{Amperes} \times \text{Cosine}\ \theta$$

Purpose

Motors may be manufactured for general purpose, definite purpose, or special purpose applications.

1. *General purpose motors* are designed for continuous 40°C operation with an open enclosure. They have standard ratings, construction, and operating characteristics. They have no restrictions as to the type of application for which they are used under normal operating conditions.
2. *Definite purpose motors* are those designed to operate under conditions other than normal. They also have standard ratings, construction, and operating characteristics. However, their design takes into account operations under conditions other than the usual or for a particular type of application. For example, a submergible water pump would be a definite purpose device.
3. *Special purpose motors* are those with special mechanical construction and/or special operating characteristics. They are normally designed for a particular application. Any motor that does not fall within the classification of general purpose or definite purpose would be a special purpose motor.

DC/AC Motors

Motors may be designed to operate on either DC or AC. Alternating current motors may use a single phase, two phases, or three phases of power and are classified accordingly. Most AC motors over 1 horsepower are polyphase and use three-phase power. Motors may range from small fractional-horsepower devices up to large units rated for hundreds of horsepower.

Speed

Motors may be single speed, dual speed, multiple speed, or variable speed. The latter can be adjusted to any value of speed within the range of the design.

The speed of a motor is given in revolutions per minute (RPM). The speed of DC machines is more easily adjustable than that of the AC motors.

The synchronous speed of AC motors is dependent on the frequency of the power source and the number of poles. This is expressed mathematically as

$$\text{Synchronous Speed} = 120 \times \frac{\text{Frequency(Hz)}}{\text{Number of Poles}}$$

or as

$$\text{Synchronous Speed} = 60 \times \frac{\text{Frequency(Hz)}}{\text{Pairs of Poles}}$$

The multipliers of 120 and 60 are used to convert the frequency from cycles per second to revolutions per minute for the motor. To illustrate these two methods, let's take an induction motor having four poles and operating at 60 hertz. Method one would yield

$$\text{Synchronous Speed} = 120 \times \frac{60}{4} = 1800\ \text{RPM}$$

With method two, the same answer is found using

$$\text{Synchronous Speed} = 60 \times \frac{60}{2} = 1800 \text{ RPM}$$

It appears more logical to use the second formula than to use the first. The reason is that the multiplier 60 relates to the conversion of seconds to minutes more so than using the value of 120 and the total number of poles.

Slip

Slip is strongly associated with synchronous speed. If the motor turned at the same RPM as the magnetic field, there would be no relative motion between the rotor and the field. Therefore, no current would be induced into the rotor, and no magnetic field would be created to cause it to turn.

Slip is the difference between the synchronous speed of a motor and its actual speed. Mathematically it is determined by

$$\text{Slip} = \text{Synchronous Speed} - \text{Actual Speed}$$

Using the example above for synchronous speed of a motor with four poles operating at 60 hertz, the value was calculated to be 1800 RPM. The value most likely given for the RPM of the motor on the nameplate will be 1725 RPM. Using the formula for slip

$$\begin{aligned}\text{Slip} &= 1800 \text{ RPM} - 1725 \text{ RPM}\\ &= 75 \text{ RPM}\end{aligned}$$

Electrical Degrees

In a motor or generator, a conductor must move past a north and a south pole in order to complete one cycle or 360 electrical degrees. For a two-pole device, this means that the conductor must move through a full 360 mechanical degrees, or make a complete circle, to obtain the 360 electrical degrees. If the device has four poles, the conductor needs to move only 180 mechanical degrees. For six poles, only 120 degrees would be required.

Direction of Rotation

Motors may rotate only in one direction, or they may be reversible. Standard manufacturing procedure for motors designed to rotate in one direction is for the shaft to turn in a counterclockwise direction as viewed from the front of the motor. The *front* is the end opposite the shaft (Figure 6–3).

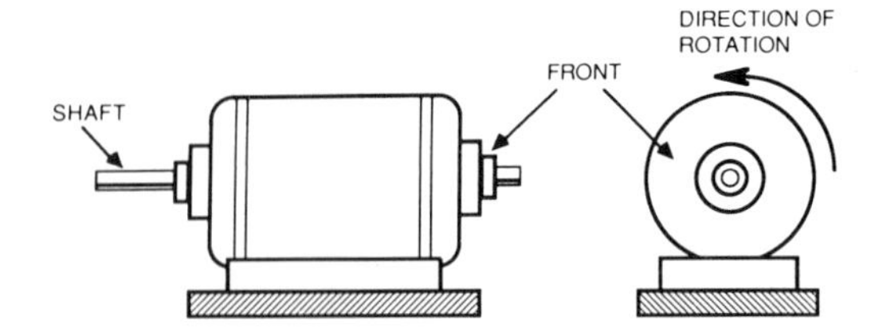

FIGURE 6–3. Standard direction of motor shaft rotation is counterclockwise.

Efficiency

The percent of *efficiency* of a motor can be determined by dividing the output horsepower by the input watts and multiplying the fraction by 100. Horsepower can be converted to watts by multiplying the figure by 746.

$$\%\ \text{Efficiency} = \frac{\text{HP} \times 746}{\text{Watts}} \times 100$$

Efficiency could also be expressed in the following way:

$$\%\ \text{Efficiency} = \frac{\text{Input} - \text{Losses}}{\text{Input}} \times 100$$

Input power is easily measured using electrical instruments that are readily available. Output power, the mechanical energy transferred to the load, is not so easy to measure. Special equipment is required for this purpose. A more detailed discussion is provided in the section Torque, below.

Power Factor

A term closely associated with efficiency is power factor. The *power factor* is the ratio of the watts consumed by the circuit divided by the volt-amperes delivered to the load. This ratio is expressed for a single-phase circuit as

$$\text{Power Factor} = \frac{\text{Watts}}{\text{Volts} \times \text{Amperes}}$$

For the three-phase circuit, the formula is modified to

$$\text{Power Factor} = \frac{\text{Watts}}{\text{Volts} \times \text{Amperes} \times 1.73}$$

Circuits having poor power factor will also operate at a lower efficiency than those with a high power factor. Power factor will be discussed in greater detail in Chapter 9.

Torque

Torque is defined as a turning force. It can exist even if there is no movement. An example would be where a wrench is used to loosen a bolt that has rusted. The bolt refuses to move, but the torque is being applied. Even though the bolt refuses to turn, the torque can be so great that the head of the bolt will separate from the threaded end.

The unit of measure for torque is in "pound-force feet." By increasing the length of the handle of the wrench, more leverage is provided, and the torque will be increased. The horsepower of the motor can be determined using the following formula:

$$HP = \frac{\text{Force} \times \text{Length of Lever} \times 2\pi \times \text{RPM}}{33{,}000}$$

The force is measured in pounds, and the length is in feet. To determine the distance the force is moved (work), 2π is the multiplier used to obtain the circumference of a circle. The RPM is the rate at which the work is being done (power). The conversion factor to horsepower is the 33,000 foot-pounds per minute rate. Figure 6–4 illustrates a laboratory test setup for determining the values to be used in this formula.

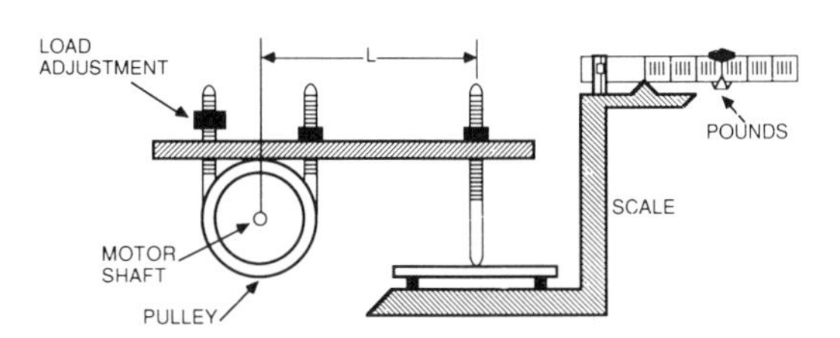

FIGURE 6–4. Using a prony break to determine motor horsepower.

This test setup allows for the adjustment of the load on the motor by the tightening of the band on the pulley. Torque and the horsepower can be determined experimentally for locked-rotor conditions up to the motor's maximum operating speed while under load. The mechanical energy delivered to the load is converted to heat due to friction between the pulley and the band. While performing this test, the heat is removed by a coolant such as water or oil.

Although the field electrician will normally never run this type of test, the formula is useful for determining the motor horsepower required to drive a known load. Also, the illustration is very useful in helping one visualize the changing of electrical energy into mechanical energy.

Using the test setup in Figure 6–4, the length of the lever is set at 9 inches, and the force on the scale reads 8 pounds. The motor's RPM is measured using a precision-type hand tachometer at 1725 RPM. What is the horsepower delivered to the load by the motor? Substituting these values into the formula, we find

$$HP = \frac{8 \times 0.75 \times 6.28 \times 1725}{33{,}000} = 1.97$$

or nearly 2 horsepower.

If the rated RPM at full-load was twice the value in the example, the calculations would show twice the horsepower. On the other hand, if rated RPM has been one-half, the horsepower would have been approximately 1. The fact is that every motor at a given speed must have a definite torque to develop its rated horsepower.

A customer may specify a motor of a given horsepower that will develop the required torque close to or equal to the torque required for a mechanical load. It is important to consider the overall efficiency of the system and how the motor will be operated before filling the order.

If the efficiency is low, and the motor is programmed for continuous operation, it may well overheat even though it develops the required torque to do the job. At the same time, if the system's efficiency is high, and the motor is operated only intermittently and has time to cool, the specified motor may perform the intended functions.

Most pony break apparatuses are obsolete because of instability and lack of accuracy. There was always some question about the accuracy of the results achieved using this type of device. Pony break devices have been replaced by the dynamometer (Figure 6–5).

The *dynamometer* is an electric generator whose base has been modified. On a generator, the housing is normally attached to the base so that it cannot move. The dynamometer has a second set of bearings mounted outside the bearings for the armature shaft. This arrangement allows both the armature and the case to turn freely. The movement of the case, however, is limited by a spring, which is calibrated into pounds force, and can be read from a scale. A counterweight is used to balance the weight of the lever and dynamometer case so that the scale does not need constant calibration.

The dynamometer is driven by the motor under test. Being a generator, the dynamometer converts the mechanical energy of the motor into electrical energy.

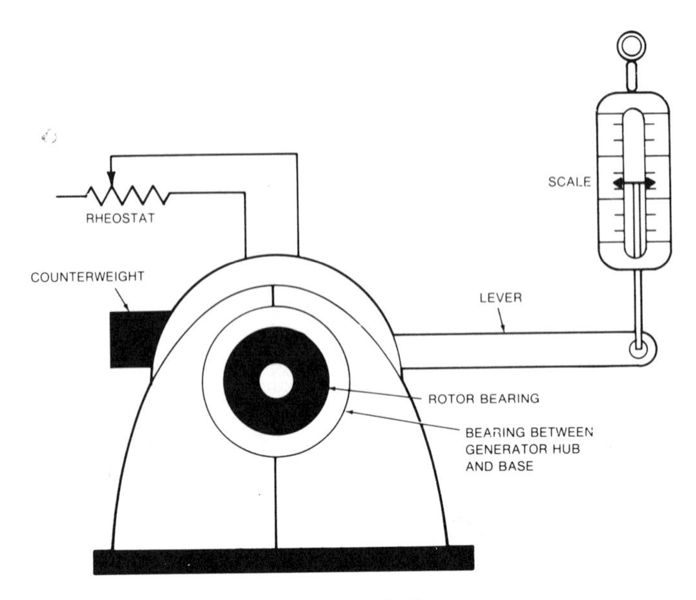

FIGURE 6–5. The electric dynamometer.

The load on the motor can be adjusted by drawing varying amounts of current from the dynamometer by adjusting a rheostat for the load desired.

When current is taken from the dynamometer, a magnetic field is set up that opposes the field which caused it (Lenz's law). This causes the case of the dynamometer to rotate in the opposite direction to the motion of the driven armature. Rotation of the case is opposed by the spring and the torque applied to the spring can be read from the calibrated scale.

Some dynamometers incorporate electronic tachometers. Tachometers measure the RPM. The accuracy of some of these tachometers is very high, and some will read as low as one-half RPM.

Four types of torque are important when working with motors. These are: pull-up torque, full-load torque, breakdown torque, and locked-rotor torque.

1. *Pull-up torque* is measured when full power and frequency are applied to the motor when it is at rest. It is the minimum torque developed during the period from when it is started until it reaches full speed.
2. *Full-load torque* is the amount of turning force the motor produces at its operating speed in order to develop the rated horsepower.
3. *Breakdown torque* is the maximum torque a motor develops when operated at its rated voltage and frequency without a sudden drop in RPM. If the load on a motor is increased enough beyond its horsepower rating, the speed of the motor will drop without any warning and the motor may stop turning.
4. *Locked-rotor torque* is the minimum torque a motor will develop when rated voltage and frequency are applied.

The voltage applied to a motor greatly affects the magnitude of the torque. Torque is proportional to the square of the voltage applied. If the voltage is doubled, the torque will be four times greater. On the other hand, if the voltage is reduced by one-half, the torque will only be one-quarter as great.

Motor Losses

Motor losses are categorized under two conditions: no-load losses, and losses when operating under load. The no-load losses include

- Windage
- Friction
- Core losses

Motor losses under load are

1. Stator I^2R loss

2. Rotor I^2R loss
3. Stray load loss

Windage and Friction

Because the shaft of the motor turns and is often attached to a fan load for cooling purposes, the air around it must be displaced. This movement of air offers resistance to the turning force, thereby consuming energy. This loss is defined as the *windage*.

The efficiency of a motor is not as good as that of the transformer. Unlike the transformer, parts of a motor must rotate in order to convert electrical energy into mechanical energy. This in turn causes friction between the rotating parts and the stationary parts of a motor. Friction is converted to heat rather than mechanical energy, thereby lowering the overall efficiency of the machine.

Core Losses

Motors have the same type of core losses as transformers. These include eddy currents, hysteresis, saturation, and flux linkage losses. Core losses are increased or decreased mainly through the rated frequency of design, flux density, type of steel used, and the insulation quality of core laminates.

Motors have two kinds of eddy current losses. There are the eddy currents within the individual laminate, and eddy currents between laminates where the insulation is not sufficient to prevent them. Damage to the insulation can cause motors to operate at a much lower efficiency than their ratings.

Most motors are manufactured using number 24-gauge steel. The laminations are about 0.025 inch thick and can be purchased with a thin coat of varnish on both sides to provide the insulation quality needed to reduce the circulating currents in the core material.

Another method of providing insulation is also used. Because the core has its slots and contours punched out on a press, this causes stress within the material and reduces the magnetic quality of the steel. To overcome this problem, the core material is heat treated at approximately 1500°F in a controlled atmosphere before it is punched to specifications. The heat treatment provides an additional benefit in that an iron oxide forms on the surface to provide insulation between the laminates.

Unlike the transformer, a motor must have an air gap between the moving and the stationary part. This air gap increases the reluctance of the magnetic circuit which reduces the total number of flux lines. Also the loss is increased due to reduced magnetic coupling between stator and rotor. This is known as the *flux linkage loss*.

I^2R Losses of Windings

Losses due to the resistance of the windings of a motor can normally be ignored under no-load conditions. Under load conditions, however, they become significant. These losses are usually addressed in the manufacturing of the motor by reducing the resistance of the wire to make a more efficient device. How the stator and rotor are wound is also considered. Some winding configurations provide for greater efficiency than others. If an older motor needs to have its windings replaced, the overall loss of the motor can be reduced at that time by using these techniques.

I^2R Losses of Rotor

It is more difficult to determine this loss than that of the stator windings, the reason being that the currents in the rotor bars and end rings are unevenly distributed. It can be proven, however, that this loss is directly related to the slip of the motor. Slip is normally within a 3 to 5% range of the synchronous speed. At a given speed if the slip was in the range of 4%, then the loss would equal approximately 4% of the power induced into the rotor.

Stray Load Losses

Factors that affect the stray load losses involve the engineering design of the motor and the processes used in the manufacturing of the motor. Some of these are

- Geometry of stator and rotor slot
- Number of slots
- Length of the air gap between the rotor and stator
- Rotor slot insulation

Stray load losses vary from one manufacturer to another due to their nature. Because they involve both design and the manufacturing process, these losses can vary widely among companies and will affect the overall efficiency of a given machine.

Efficiency

When we moved into our all-electric home in 1971, the power company put us on the budget plan at $35 per month. We did not take their advice and paid $50 per month due to the number of members in the family. At the end-of-the-year settlement, the average came to $49 per month.

This year, with a reduction of family size from ten to five, the introduction of various energy saving devices, and a significant reduction in the kilowatt-hours used, our budget is set at $203 per month. This represents a fourfold increase in cost while receiving less electricity.

At the same time, the energy crunch caused the author's family to change over to smaller, more efficient automobiles. This dampened our love affair with the big "gas burners" formerly used by our large family. With the price of gasoline increasing daily and with increased income not responding nearly as fast, little choice remained.

The typical energy efficient car driven an average number of miles costs approximately 20% of its purchase price to operate each year. In any case, over the life expectancy of the vehicle, the operating costs will usually exceed the purchase price by one and one quarter to three times the initial cost. The price of energy (gasoline in this case) is a major factor in determining the multiple.

Motor Costs

On the other hand, a 100-horsepower, continuously used, 91% efficient motor will cost nearly 900% of its purchase price to operate yearly with the cost of electrical energy at $0.04 per kilowatt-hour. Compared to a motor's purchase price in the range of $3200, operating costs would be over $28,000. With an expected life of 20 years, this would represent over one-half million dollars.

Economists have predicted that the cost of electrical energy will increase dramatically by the year 2000. These forecasts show rates in the range of $0.30 to $0.40 per kilowatt-hour. This would raise the operating costs to between 7000% and 9000% of the purchase price of the motor. If the cartel formed by the oil producing nations exerts enough influence, these rates may well be realized. Is it any wonder that industrial leaders are looking for more efficient motors?

Manufacturers of electrical machinery have responded to this need. Several companies now produce more energy efficient machines. Although the initial cost is greater than the older, lower-efficiency motors, substantial savings can be realized by the user over the life cycle of the motor. Even with the higher purchase price, the energy efficient motor may prove less costly than a less-efficient motor over the life of the two devices.

Although there are exact ways of determining the long-term real costs of motors, these exercises are not usually performed by the typical electrician. The electrician, however, represents his or her employer and should be energy conscious. This is beneficial for the employer, the customer, and the producer of energy saving machines.

Estimating Savings

A rough approximation of savings can be estimated for the customer. This should be sufficient for the customer to do an actual cost analysis.

For every one-tenth percent increase in efficiency, the customer will save $1 per horsepower, if overall cost is assumed to be $1000 per kilowatt over the life cycle of the machine. If the cost is $2000 per kilowatt, the savings would be $2 per horsepower. With reduced costs of $500 per kilowatt, which are highly unlikely at this time, the savings would be reduced accordingly to $0.50 per horsepower. Table 6–1 illustrates the savings at $1 per horsepower.

TABLE 6–1. Savings Due to Motor Efficiency (Life-Cycle Savings $1/HP @ $1000/kW)

HP	Efficiency (%)		Differential		Life Cycle
	High	Industry Avg.	Points	Tenths	Savings ($)
10	90.2	85.0	5.2	52	520
20	91.7	88.5	3.2	32	840
80	94.1	91.0	3.1	31	1880
100	95.0	93.0	2.0	20	2000

(Courtesy of General Electric Corporation)

It should be noted that the savings are greater as the horsepower goes up. At the same time, smaller horsepower motors should not be neglected. If a 100-horsepower motor were replaced by ten 10-horsepower motors, the savings on the ten motors would be $5200. Likewise, for five 20-horsepower motors, the savings would be $3200. Because smaller motors are less efficient than larger motors, a larger percentage of savings is realized with the higher-efficiency motors.

Operation Near Full-Load

A motor will be more efficient when operated near its horsepower rating. This is due to the fact that, like a transformer, a motor will act more inductive when unloaded. Under these conditions, not all the current flowing in the circuit is being converted to output power. The current, however, is passing through the wires, which have resistance. Heat is generated, resulting in I^2R losses. These losses are due to low-power factors and should be considered when making the installation.

Defining Efficiency

Efficiency is defined in several ways by different persons and manufacturers. A motor purported to be 90% efficient may be more or less efficient than the rating.

The term "efficiency" by itself as a guide when purchasing a motor is much like the Environmental Protection Agency's miles per gallon rating of an automobile. You need to know much more to determine the exact mileage you will obtain from a particular automobile.

Terms such as full-load, 4/4 load, average, nominal, and calculated efficiency are not always sufficient either and have the same limitations as the Environmental Protection Agency ratings for the automobile. Individual motors may perform well below any of ratings described by these terms.

Another term used is *apparent efficiency*. This value is determined by multiplying the motor's power factor by the efficiency rating. However, the power factor varies with loading and other factors, and once again this rating does not give the true efficiency of an operating motor. Specifications for efficiency should not be written in reference to this value.

Using the minimum or expected minimum efficiency is a better way of expressing a specification. Some manufacturers, however, further define this term by stating that a given percentage of their motors will meet the minimum expected efficiency. A manufacturer's guarantee along with the buyer's confidence in the manufacturer is the best way for selecting an efficient motor.

Further assurance of a motor's efficiency can be obtained if the standard used by the manufacturer or testing laboratory is known. Studies contracted by the Federal Energy Administration have shown that the user of electrical motors cannot be assured of the true efficiency of motors due to the various testing procedures used by different manufacturers as well as by those in foreign countries. Table 6–2 makes some comparisons among standards used in four countries.

NEMA has adopted Test Standard MG1-12.53a (IEEE-112, Test Method B). Method B requires a dynamometer to be used when testing polyphase squirrel cage motors rated 1-125 HP. I^2R losses are to be adjusted for temperature rise, and stray load loss data are averaged using linear regression analysis to reduce random errors. As can be deduced from Table 6–2, this standard reports the lowest efficiency rating of all the standards shown. It is strongly recommended that this standard be specified when requesting efficiency ratings for motors.

TABLE 6–2. Comparison of Motor Ratings Using Different Standards

Standard	Full-Load Efficiency (%)	
	7.5 HP	20 HP
International (EC 84—2)	82.3	89.4
British (BS—269)	82.3	89.4
Japanese (JEC—37)	85.0	90.4
US (IEEE—112, Method B)	80.3	86.9

(Courtesy of General Electric Corporation)

Nameplate Nomenclature

Article 430 of the NEC specifies the information that the manufacturer must provide on the nameplate of a motor. The requirements are modified to take into account specific types of motors. For example, field current and voltage for DC excitation of a synchronous motor would need to be included for that particular type of motor.

A typical nameplate that meets the requirements of the NEC for a fractional horsepower AC induction-type motor is shown in Figure 6–6. A brief explanation for each block is provided.

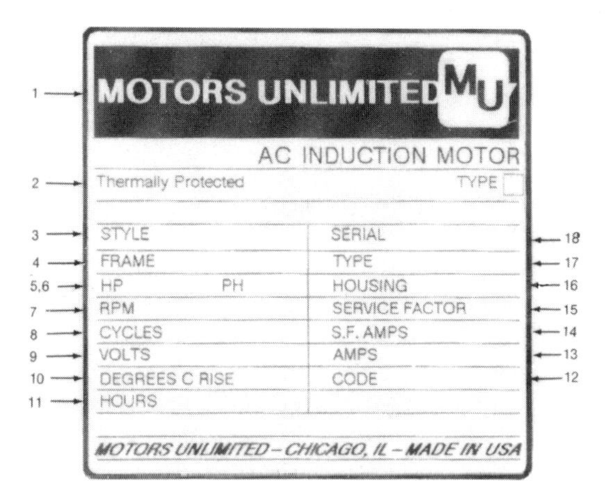

FIGURE 6–6. Nomenclature for a typical motor.

1. The manufacturer's name must be shown.
2. If a motor is provided with thermal protection it must be marked. The abbreviation T.P. may be used for motors rated 100 watts or less that comply with the NEC, Article 430. Standard codes for thermal protected motors are

    - A—automatic reset, Underwriters Laboratories (UL) approved
    - M—manual reset, UL approved
    - D—automatic reset, time delay, UL approved for oil burner service
    - X—automatic or manual reset, not UL approved.

3. The style or model number generally are derived from the manufacturer's specifications and drawings necessary to produce the particular motor and others exactly like it. Appendix A provides a description of how the model number is determined by General Electric.
4. Standards for frame size have been established by the NEMA and numbers are assigned to correspond to specific dimensions. Occasionally, motors will be manufactured with non-standard frames due to the special needs of a user. When motors are manufactured according to NEMA standards, motors with the same frame and type can be replaced with a motor of the same type from another manufacturer. Mechanical construction and dimensions will be the same as will be the ratings and operating characteristics. Figure 6–7 gives the dimensions for a motor manufactured by General Electric using NEMA-defined frames 2512, 2513, 2812, and 2813.
5. This value provides the horsepower, or part thereof, for the motor at the rated RPM.
6. Motors are designed to operate on one, two, or three phases. The abbreviation "DC" will appear in this block for DC motors.
7. This is the speed at which the motor will run to develop the rated horsepower when the proper frequency and voltage is applied.
8. The required frequency of the power source is given in this block. This value will almost always be 60 hertz in this country, although 50-hertz motors may be encountered. Motors may be designed to operate on a wide range of frequencies. If a 60-hertz motor is to operate on 50 hertz it must have its voltage and horsepower reduced. Motors designed to operate at either frequency are non-NEMA-defined motors.
9. Rated voltage in this block is normally specified at a lower value than the nominal voltage supplied by the power company. This design feature was incorporated at an earlier time in the electrical industry to compensate for voltage drop on the line. Modern power distribution systems which include utilization transformers closer to the loads, power factor correction, and shorter runs of wire have greatly overcome this problem. Nominal voltages of 120/240 volts are used with motors rated 115/230 volts, 208 volts are used for motors rated at 200 volts, and this continues through the standard voltages up to 4160 volts for a 4000-volt rated motor. The voltage applied to the motor must be within 10% of that given on the nameplate. Tables 6–3 and 6–4 provide the nominal operating voltages of power systems as well as the nameplate voltages of the motors.

Table 6–3. Nominal Power System Voltage Versus Motor Nameplate Voltage at Standard 60 Hertz

Nominal Power System Voltage	Motor Nameplate Voltage	
	Below 125 HP	125 Hp and Up
Polyphase 60 Hz		
208	200	
240	230	
480	460	460
600	575	575
2400	2300	2300
4160	4000	4000
Single Phase 60 Hz		
120	115	
208	200	
240	230	

(Courtesy of General Electric Corporation)

10. *Temperature* rise is the difference between the operating temperature of a motor and the ambient temperature. Most motors are designed to operate at 40°C above the ambient temperature. Some motors are designed for a 50°C rise above the ambient temperature. The heat generated by a motor

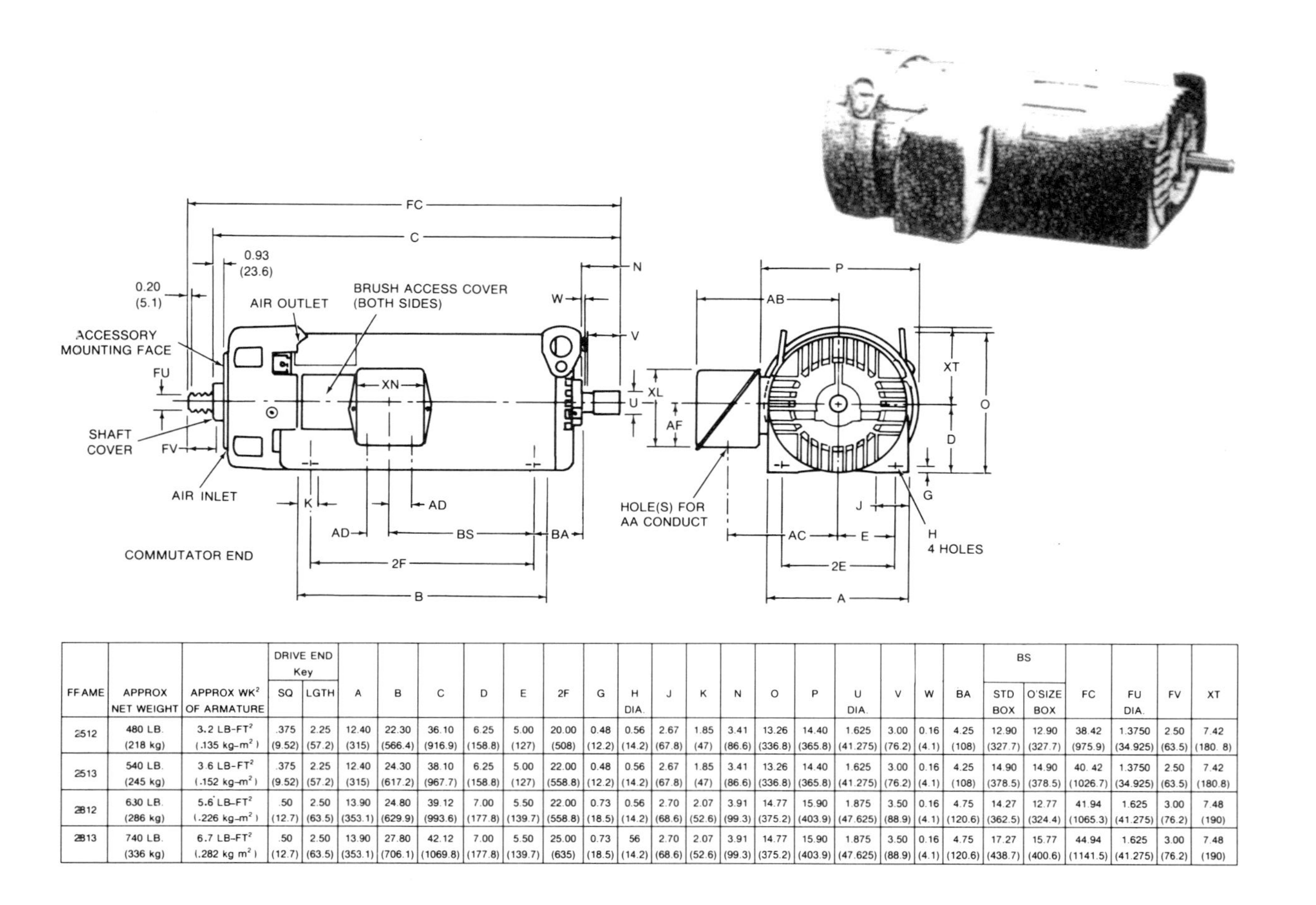

FFAME	APPROX NET WEIGHT	APPROX WK^2 OF ARMATURE	DRIVE END Key SQ	DRIVE END Key LGTH	A	B	C	D	E	2F	G	H DIA.	J	K	N	O	P	U DIA.	V	W	BA	BS STD BOX	BS O'SIZE BOX	FC	FU DIA.	FV	XT
2512	480 LB. (218 kg)	3.2 LB-FT^2 (.135 kg-m^2)	.375 (9.52)	2.25 (57.2)	12.40 (315)	22.30 (566.4)	36.10 (916.9)	6.25 (158.8)	5.00 (127)	20.00 (508)	0.48 (12.2)	0.56 (14.2)	2.67 (67.8)	1.85 (47)	3.41 (86.6)	13.26 (336.8)	14.40 (365.8)	1.625 (41.275)	3.00 (76.2)	0.16 (4.1)	4.25 (108)	12.90 (327.7)	12.90 (327.7)	38.42 (975.9)	1.3750 (34.925)	2.50 (63.5)	7.42 (180. 8)
2513	540 LB. (245 kg)	3.6 LB-FT^2 (.152 kg-m^2)	.375 (9.52)	2.25 (57.2)	12.40 (315)	24.30 (617.2)	38.10 (967.7)	6.25 (158.8)	5.00 (127)	22.00 (558.8)	0.48 (12.2)	0.56 (14.2)	2.67 (67.8)	1.85 (47)	3.41 (86.6)	13.26 (336.8)	14.40 (365.8)	1.625 (41.275)	3.00 (76.2)	0.16 (4.1)	4.25 (108)	14.90 (378.5)	14.90 (378.5)	40. 42 (1026.7)	1.3750 (34.925)	2.50 (63.5)	7.42 (180.8)
2812	630 LB. (286 kg)	5.6 LB-FT^2 (.226 kg-m^2)	.50 (12.7)	2.50 (63.5)	13.90 (353.1)	24.80 (629.9)	39.12 (993.6)	7.00 (177.8)	5.50 (139.7)	22.00 (558.8)	0.73 (18.5)	0.56 (14.2)	2.70 (68.6)	2.07 (52.6)	3.91 (99.3)	14.77 (375.2)	15.90 (403.9)	1.875 (47.625)	3.50 (88.9)	0.16 (4.1)	4.75 (120.6)	14.27 (362.5)	12.77 (324.4)	41.94 (1065.3)	1.625 (41.275)	3.00 (76.2)	7.48 (190)
2813	740 LB. (336 kg)	6.7 LB-FT^2 (.282 kg m^2)	.50 (12.7)	2.50 (63.5)	13.90 (353.1)	27.80 (706.1)	42.12 (1069.8)	7.00 (177.8)	5.50 (139.7)	25.00 (635)	0.73 (18.5)	56 (14.2)	2.70 (68.6)	2.07 (52.6)	3.91 (99.3)	14.77 (375.2)	15.90 (403.9)	1.875 (47.625)	3.50 (88.9)	0.16 (4.1)	4.75 (120.6)	17.27 (438.7)	15.77 (400.6)	44.94 (1141.5)	1.625 (41.275)	3.00 (76.2)	7.48 (190)

FIGURE 6–7. NEMA standards for frames 2512, 2513, 2812, and 2813. (*Courtesy General Electric Corp.*)

TABLE 6–4. Nominal Power System Voltage Versus Motor Nameplate Voltage for 50 Hertz

Nominal Power System Voltage	Motor Nameplate Voltage	
	Below 125 HP	125 HP and Up
Polyphase 50 Hz		
Nominal Power (System Voltage Varies from Country To Country)	200	
	200	
	380	380
Motor Voltage (Should Be Selected For the Country in Which It Will Be Used)	415	415
	440	440
	550	550
	3000	3000
Single Phase 50 Hz		
	110	
	200	
	220	

(Courtesy of General Electric Corporation)

during operation is sometimes referred to as *watt losses*, or the current squared times the resistance. Contributing to the heat are eddy currents, and friction in the bearings. Heat is a major factor in motor failure. Motor applications requiring frequent starts, and those having frequent overloads may require special motors. The class of the insulation system and rated ambient temperature may be substituted for the temperature rise.

Class	Maximum Operating Temperature
A	105°C
B	130°C
F	155°C
H	180°C

11. Normally the time rating on a motor is "continuous." When a value is given it shall be 5, 15, 30, or 60 minutes. This is the amount of time a motor may be operated without overheating.
12. The kilovolt-amperes per horsepower with locked rotor is provided by the NEMA code. The value corresponding to the alphabetical letter provided in this block can be found in the NEC, Article 430. This code is to be used to determine branch circuit short circuit and ground fault protection.
13. This is the full-load current for the motor when operated at the rated voltage and frequency. This value must be given for each speed of multispeed motors with the exception of shaded-pole and permanent-split capacitor motors where the amperage is required only for the maximum RPM.
14. *Service factor amperes* is the amount of current the motor will draw when operated at its maximum service factor value.
15. *Service factor* is the permissible overload that the motor has been designed to withstand. The value can be determined by multiplying the service factor times the horsepower of the motor. Rated voltage and frequency must be maintained in order for this value to be applicable. If a motor is operated at this acceptable level, both the current and temperature rise values shown on the nameplate will be exceeded. Common values for the service factor range from 1 to 1.25.
16. There are two general types of enclosures. They are totally open (TO) and totally closed (TC) motors. Totally open motors are modified so that the up-end is enclosed and ventilation holes are at the bottom of the motor, or so that they are drip proof. Totally closed motors are completely sealed but may have forced fan air cooling.
17. Manufacturers generally use this space to indicate the type of motor such as split-phase, capacitor-start, and so on. Other codes may be included to help the producer identify the motor.
18. This is the serial number of the motor. It may be coded to indicate the date of manufacture, special order, particular customer, or any number of other purposes.

Parts of Motors

The essential parts of a motor vary among the different types. All motors, however, must have a stator, rotor, and a means of supporting the rotor so that it does not come in contact with the stator poles as it rotates. The stator and rotor will be examined first. Afterward, other parts used on motors will be explored.

Stator

The stator, which is the stationary part of the motor, consists of the core and the windings of copper wire on the core. Cores are constructed using laminated electrical grade steel to reduce losses due to eddy currents and to insure a good magnetic path. The purpose of the stator is to form a strong electromagnet in which the rotor turns. In some small electric motors, the core and windings are replaced by a permanent magnet, but in most cases the stator is an electromagnet. Figure 6–8 shows a core of a stator for a DC motor.

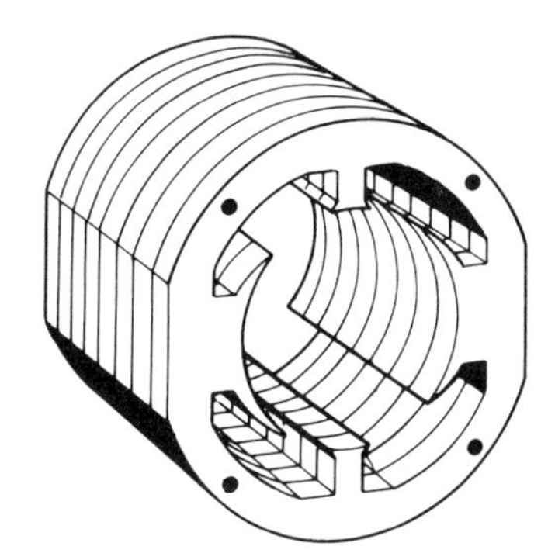

FIGURE 6–8. Laminated stator.
(Courtesy General Electric Co.)

This is a two-pole motor. The number of poles on a DC motor are usually easily distinguishable. The small projection poles at the top and bottom of the core are for the interpole windings. Their purpose will be discussed in Chapter 7.

Pole pieces are shaped to fit the contour of the rotor. This reduces the air gap and lowers the reluctance (the opposition to the passage of magnetic flux lines) of the magnetic field between the stator and rotor. Air has a much higher reluctance than does the special steel used in the core. For AC induction motors, the air gap as a rule is from 0.015 to 0.040 inch. On the synchronous motor, the gap is usually greater.

Pole pieces may be individually formed and bolted to the frame of the motor. Individual poles are known as salient poles. Figure 6–9 illustrates a four-pole DC motor using salient poles.

On the AC induction motor, the number of field poles is not as easy to distinguish. This is because there are pole pieces for the starting windings, and in some cases, the compensating windings. Poles for these windings are spaced in with the field poles. Figure 6–10 depicts the stators of two different types of AC motors. Both of these are four-pole motors.

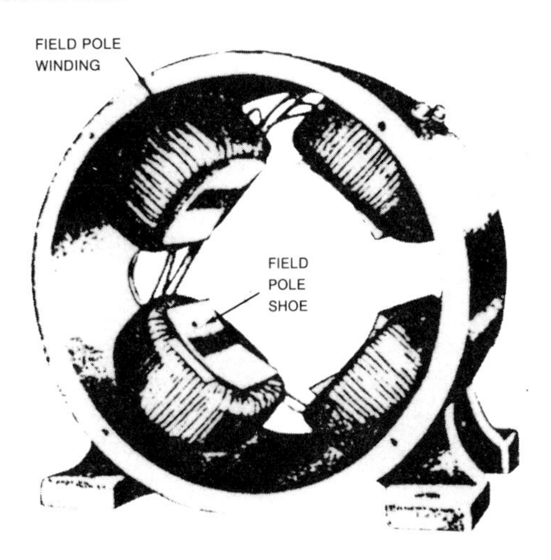

FIGURE 6–9. Four-pole DC motor using salient poles.

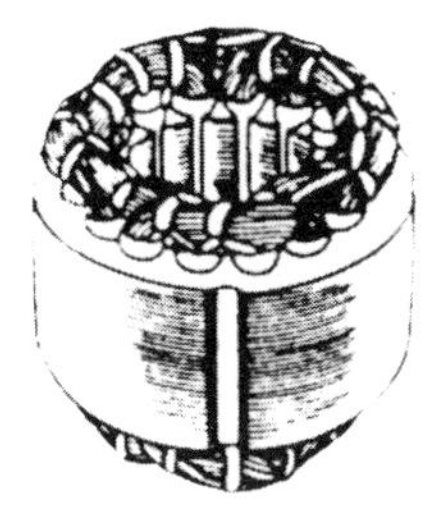

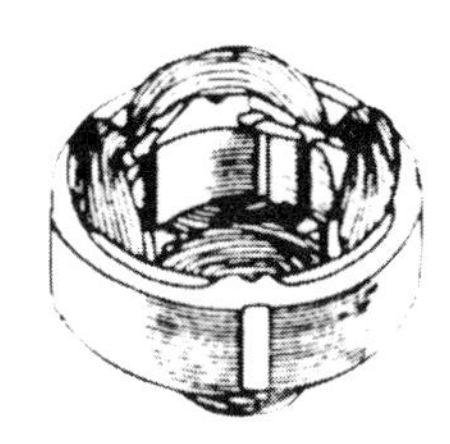

FIGURE 6–10. Stators of two types of four-pole induction motors. *(Courtesy Fasco Distributing Co.)*

Stator Coils

Most electricians will never rewind a motor unless assigned to a motor shop where the skills can be acquired. However, they should be aware that the coils are merely turns of insulated wire which are formed to be placed into the slots of the stator. The insulation may be just a thin coat of varnish on each individual wire, or the entire coil containing several turns may be wrapped with additional insulation. These coils are interconnected to each other to meet the electrical requirements of the device.

It is more important that the electrician be aware of how to test the coil windings, and be able to isolate problems involving them. They also need to know how to make the proper connections to meet the electrical requirements of the motor.

There are several wiring configurations and methods used to wrap and form the coils so that they can be properly located in the appropriate insulated slots. Each of the configurations has certain advantages that contribute to the efficiency of the motor.

Rotor

The *rotor* is the part of the motor that is free to turn. It is located inside the stator and should turn easily. It is also called an *armature*, and the terms are used interchangeably when referring to a motor. This is not true in the case of a generator. On the generator, the field windings may be rotated (rotor), and the armature is held stationary (stator). This is done so that the high current being generated in the armature can be taken off with direct fixed connections without the use of brushes.

Rotors have three parts. These are the core made of magnetic steel, windings of copper wire, and the shaft. The core is pressure-fitted onto the shaft. The shaft transmits the mechanical energy converted from electrical energy by the motor to the load.

There are two types of motor rotors. They are the wound rotor and the squirrel cage rotor. The wound rotor has coils of wires wound in the slots of the rotor. Whereas the squirrel cage rotor consists of bars of copper or aluminum the length of the rotor which are electrically connected at each end with shorting rings. The rings are also constructed with either copper or aluminum. This arrangement provides for a closed electrical circuit in which current can flow.

The name "squirrel cage rotor" comes from the similarity of appearance between this type of rotor and an exercise device used for squirrels when they are kept in captivity. Figure 6–11 depicts the wound rotor, and Figure 6–12 shows a representation of the squirrel cage rotor.

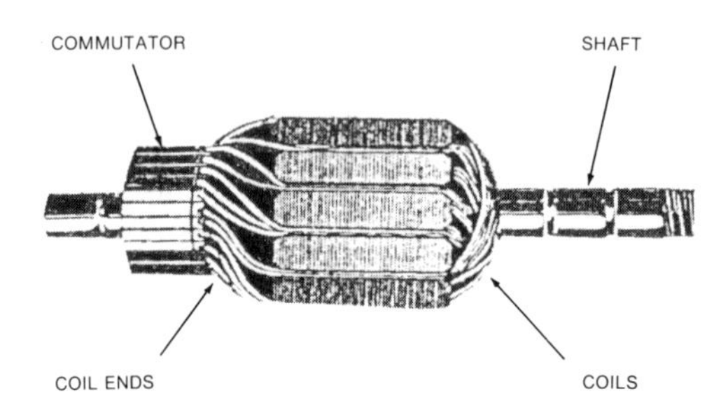

FIGURE 6–11. Wound rotor.

Direct current motors will always have a wound rotor. In addition, the DC armature will always have a commutator.

Alternating current motors may have either wound or squirrel cage rotors. The operating characteristics of the two types of rotors differ in AC motors. Certain types of AC motors also have commutators.

FIGURE 6–12. Squirrel cage rotor.

The core of the rotor is laminated to eliminate eddy currents. Laminations, like those in the transformer and the motor stator, are electrically insulated from each other.

Rotor Support

Some method must be used to mount the rotor in the stator field. The rotor must be free to turn and not come in contact with the stator poles. Figure 6–13 shows a support system for the rotor. The bearing housing and the bearings keep the rotor centered in the stator field.

In this case ball bearings are pressed into the motor's in-plates, or the bearing housing. The outer race of bearings is held firmly by the housing so that it will not turn. The shaft of the motor is run through the inner race of the bearings. The inner race should turn easily and is pressed fitted to the shaft so that it will turn as the shaft turns. The inner and outer races are designed and constructed to hold the individual bearings. Because there is no friction between the races and the turning shaft, the races and ball bearings can be made of high-grade steel to ensure a long life.

The bearing housing helps protect the bearings from dust, dirt, other abrasives, and contaminants that might attack the bearings. End caps and other protective components are usually used on the motor to prevent entry of these destructive materials which will greatly reduce the life of the bearings.

To enhance the cooling process of the bearings, the end plates in this case have fins. This increases the surface area exposed to the air surrounding the motor to effectively remove heat from the bearings and other electrical parts of the motor.

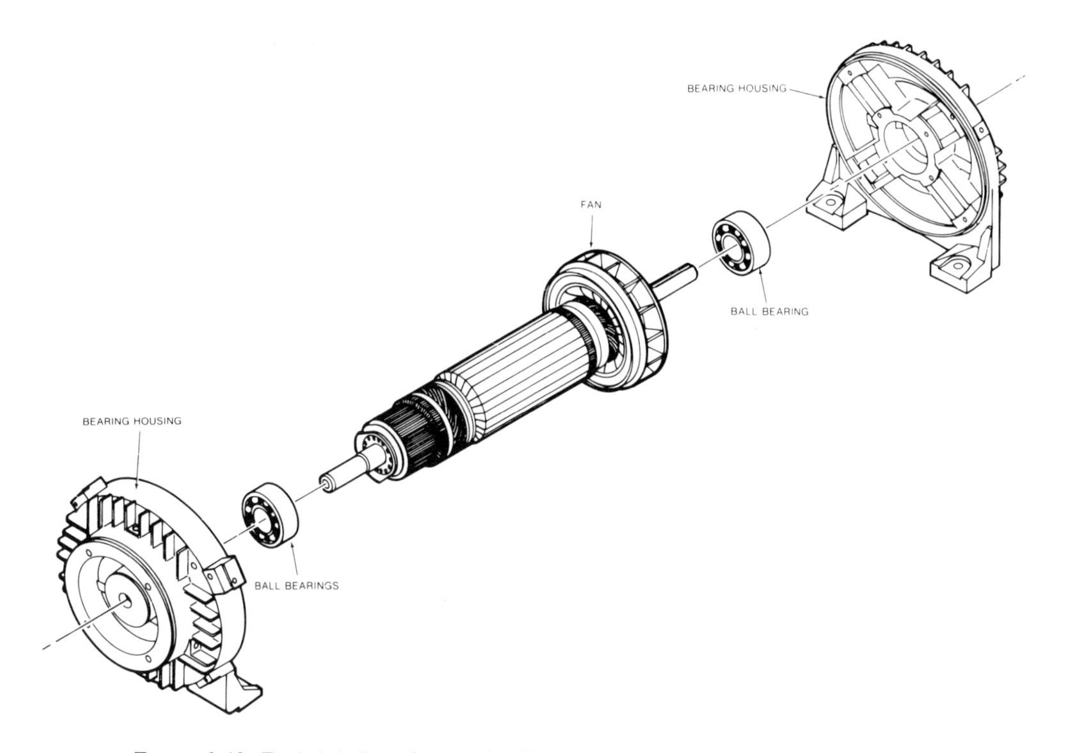

FIGURE 6–13. Exploded view of motor showing rotor support. (*Courtesy General Electric Corp.*)

A fan is mounted on the rotor shaft to help cool the total machine. The housing may be entirely closed or it may have ventilation holes to allow the free passage of air through the interior of the motor. In some cases, the ventilation holes may be in both ends of the motor, or they may be in only one end. This depends on the motor's application and the type of enclosure selected. Like the transformer, high temperatures cause failures in motor parts, and some method must be used to reduce the heat to acceptable levels.

Bearings

Motors use two types of bearings. These are *ball bearings* and *sleeve bearings*, (sometimes called a *bushing*). Sleeve bearings are normally, but not always, used on small fractional-horsepower motors up to one-half horsepower. Ball bearings are used on larger motors, but they may also be used on smaller motors.

Where there is end thrust on the motor, ball bearings are recommended. End thrust is present any time the motor and load are connected together with a belt, chain, or gears. End thrust causes abnormal wear on sleeve bearings, resulting in the interior circle becoming egg-shaped.

Sleeve bearings are less expensive than ball bearings, and they have a lower noise output. They can be used more effectively when the mechanical load is connected directly to the shaft. This includes such applications as fan, blower, and pump operations.

Ball bearings are of three basic types. They are the open, shielded, and sealed bearings. The environment in which they are going to operate determines which type is used. Sealed bearings can be used in all cases, whereas open bearings could not be utilized in areas where the bearing is exposed to dirt and corrosive materials. Under these conditions, shield or sealed bearings would be needed.

When the motor comes from the manufacturer, the bearings are packed in grease. Because of the relative high temperatures involved, the grease is usually rated at 150°C or above. Special grease is used by some manufacturers, and substitutes cannot be used without completely cleaning out the grease already in the housing. The different types of grease are not always compatible and will not mix well together. In some cases they will form a gummy mixture which will completely destroy the bearings. Manufacturer's specifications should be carefully followed when replacing the grease.

This problem is overcome if sealed bearings are used. These bearings come prelubricated and need no maintenance. The sealed bearing is designed for the life of the machine. They do fail, however, and should be checked during regular maintenance procedures. Sealed bearings are more expensive than other types.

The type of bearings used in motors should be selected carefully based on how the motor is going to be used. Consideration is needed in terms of the motor speed and if continuous or intermittent operation is proposed for the motor. How the motor is to be mounted is important. Horizontal, vertical, or some angular displacements all have different requirements in terms of the stress on the bearings.

Vertical mounted motors used as turbine pumps often have several sets of bearings along the shaft to compensate for the different forces placed on them. Figure 6–14 depicts the typical upper bearing construction for a vertically mounted pump. Three sets of thrust bearings are used on the shaft.

The thrust bearings are angular-contact ball bearings. They are easily manufactured and have an adequate life expectancy at reasonable cost. Usually, no special cooling considerations need be given to them (Figure 6–15).

These bearings have an extended or heavy shoulder on one end of the outer ring. A counterbored shoulder is used on the other end. With the application of heat, the counterbored shoulder allows more bearings to be inserted into the race than is possible with the standard radial deep-groove bearing. The extended shoulder allows angular contact between the shaft and bearing during times of high thrust.

Angular-contact bearings may be used singularly or may be stacked as shown in Figure 6–14. The heavy or extended shoulder on one end, and the counterbored shoulder on the other, provide for a high thrust in only one direction. The bearings may be mounted opposite to each other in a stack to provide protection in both directions.

When replacing bearings on one of these motors, the electrician should note the particular arrangement. It would be easy to make an improper application.

Angular-contact bearings used singularly will handle thrust loads of 8000 pounds at 1760 RPM. Used in tandem or stacked, thrusts of approximately 15,000 pounds can be applied without failure of the bearings.

For loads greater than 15,000 pounds, spherical-roller thrust bearings are used to carry the down-thrust. The lower guide bearing will carry the upthrust. Radial stability is insured by keeping the bearings in contact with the shaft under these conditions by a set of springs

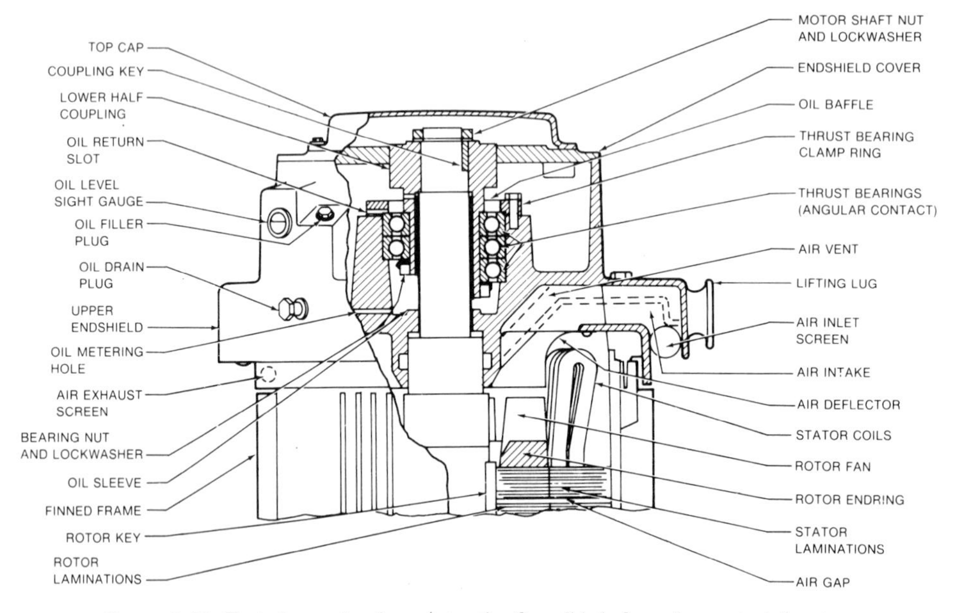

FIGURE 6–14. Typical upper bearing construction for solid-shaft weather-protected motor. *(Courtesy General Electric Corp.)*

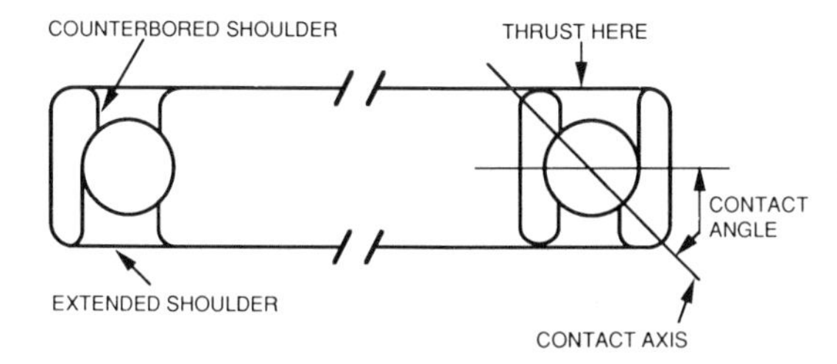

FIGURE 6–15. Cross-section view of angular-contact ball bearing. (*Courtesy General Electric Corp.*)

which push up against the lower race of the thrust bearings. Because spring pressure will be several thousand pounds, depending on design, considerable pressure is imposed on the guide bearing if the external down-pressure is not sufficient to overcome the spring pressure. It is important that consideration be given to the minimum pump total down-thrust requirement to insure that the springs are not loaded under normal operating conditions.

Spherical-roller thrust bearings are often water cooled. Cooling coils with water flowing through them are installed in the oil reservoir.

Special conditions can also influence the type of bearing to be specified for a motor. There may be end play limitations imposed by the pump seal or the possibility that the motor may be off-line for considerable periods of time. Other possibilities are that on-off cycles may be frequent, or that the motor must be reversed for each stop. There are many factors pertinent to any given installation that could influence the type of bearing. These special conditions should be discussed with the supplier.

Specifications requiring the motor manufacturer to meet unreasonable thrust requirements and extended life for the bearings should be avoided if at all possible. These specifications can be met with larger bearings and special cooling features. However, larger bearings cause a loss in efficiency, and oil foaming, vaporization, and leakage become more difficult to prevent.

Bearings eventually fail due to metal fatigue, even when operated under normal conditions. When specifying bearing life expectancy, the hours of operation and conditions under which the motor is going to perform must be considered. The lower cost of smaller bearings, and the necessity for replacing them more frequently, need to be balanced against the increased cost for larger bearings and their inherent inefficiency losses.

The minimum life expectancy of bearings is statistically predictable based on the operating conditions. A B-10 rating for vertical motors means that under continuous operation for 1 year (8760 hours) at rated speed, only 10% of the bearings are expected to fail. For in-line published thrusts, this rating is good for 2 years.

When rated average life is given, 50% of bearings will provide satisfactory service under the conditions specified. This value is usually about five times the minimum life of the bearing.

Ball bearing failures vary inversely with the third power of the imposed load. If the average life of a bearing at a given load is 3 years, then under half load, it will last eight times longer, or 24 years. It is not reasonable to assume this long a life for any bearing with maintenance procedures being what they are in most cases. A 15- to 20-year life expectancy would probably be closer to the truth provided routine maintenance is performed.

There is no guarantee that any particular bearing will achieve the life expectancy predicted. The manufacturer's warranty, usually 1 year, is the only assurance available.

Sleeve bearings, which are often used on fractional horsepower motors, are made from babbitt, brass, or bronze. Graphite may be impregnated into the metal to enhance the lubrication. Because the metal in these bearings is much softer than the steel shaft of the motor, the bearing will wear rather than the motor shaft.

Sleeve bearings are merely a modified tube or pipe with the inner surface machined to the proper diameter for the shaft. The shaft of the motor is separated from the bearing by a thin film of oil. Slots are cut into the bearing to allow oil to be picked up by a felt wick from the oil reservoir and distributed between shaft and bearing. Figure 6–16 depicts a sleeve bearing, and Figure 6–17 shows the bearing installed in the bell housing.

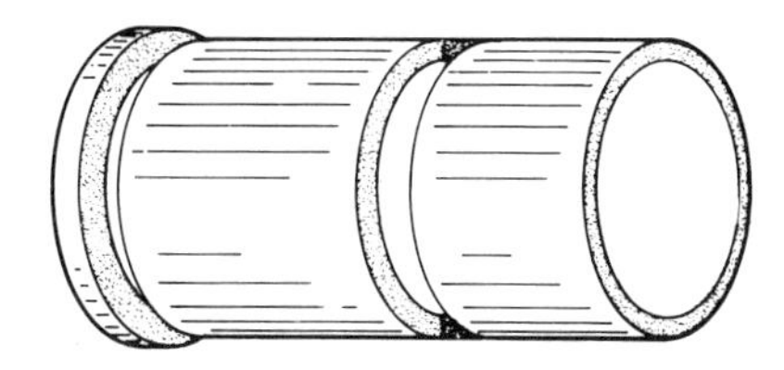

FIGURE 6–16. Sleeve bearing.

Motor Enclosure

The motor enclosure consists of the two bell housings (bearing housing) on the ends of the motor and the coil

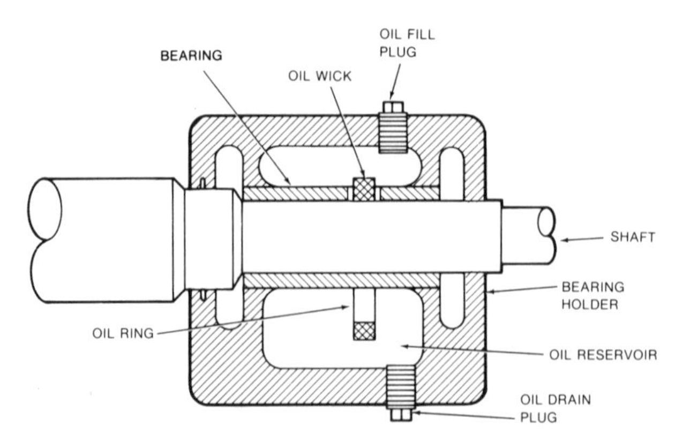

Figure 6–17. Cross-sectional view of installed sleeve bearing.

frame bolted between them. Mounting brackets may be part of the bell housing, welded to the coil frame, or there may be a specially designed frame to hold the motor.

There are two basic defined enclosures. They are the totally enclosed (TE) motor and the totally open motor (TO). Open motors depend on the free flow of air through the motor to ventilate the heat. Totally enclosed motors depend on the surface area of the motor for the same purpose. Either type of motor may have auxiliary methods for cooling such as forced air, fans, piped ventilation systems, addition of fins to increase the surface area, or even water jackets to provide cooling.

Totally enclosed motors have no ventilation holes either in the peripheral or the end plates. Cooling is completely dependent on the transfer of heat from the case to the surrounding atmosphere. There is no free flow of air through the windings. If the motor is driving a fan that causes air to move across the case, the motor is designated as having a TE enclosure. If the motor is not to be located in a moving airstream and is TE, the designation would be totally enclosed nonventilated (TENV).

Figure 6–18 illustrates both types. The motor at the top with the single shaft would be the nonventilating type if it does not drive a fan as a load which pulls air across the frame. The TE motor with the shaft extending from both ends of the motor has the provision for the fan regardless of the type of load. The fan is mounted on the shaft opposite the drive end to direct the air currents over the motor's case. A shroud may be used to force the air in contact with the motor. This type of cooling is also known as the "blast cooled" method.

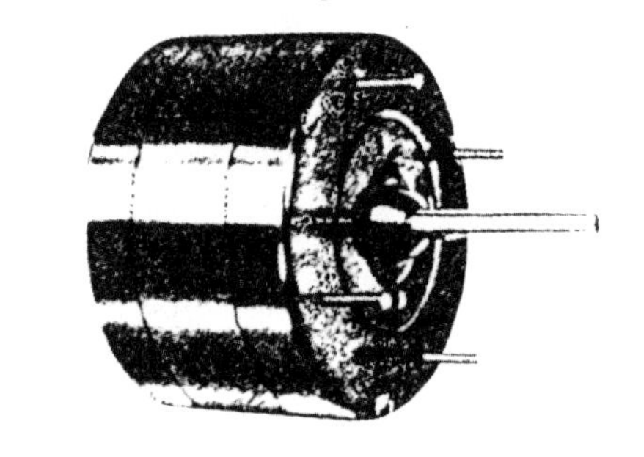

Figure 6–18. Totally enclosed motors. *(Courtesy Fasco Distributing Co.)*

The TE fan-cooled enclosure is designated TEFC. A fan is mounted on the rotor shaft inside the motor. Air going into the motor may be filtered, and the forced air coming out prevents the entry of contaminants. Figure 6–19 depicts a General Electric motor of this type.

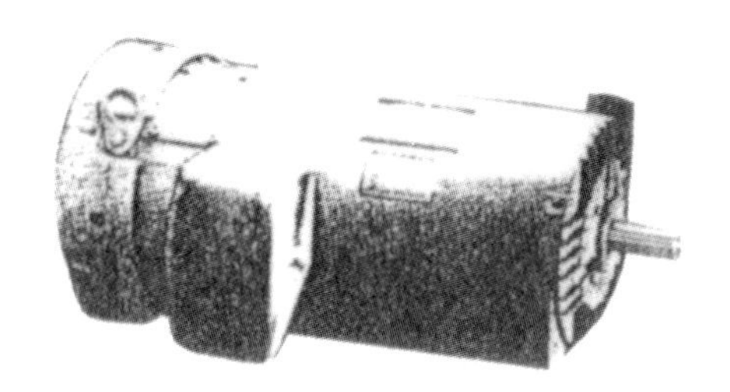

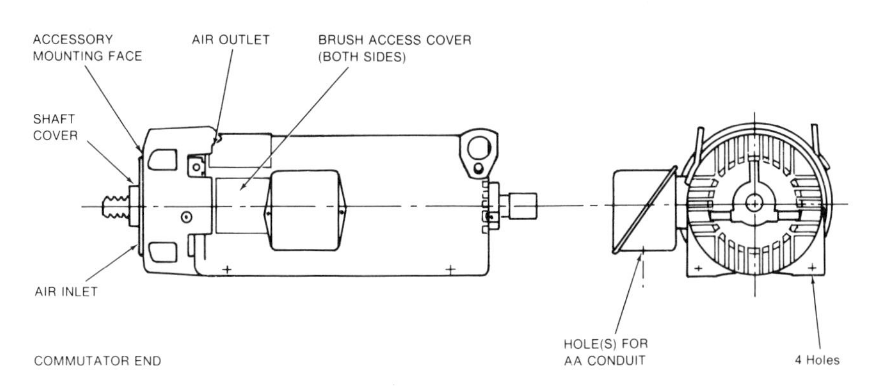

FIGURE 6–19. Totally enclosed fan-cooled motor. *(Courtesy General Electric Corp.)*

A totally open (TO) enclosure provides the maximum ventilation. Figure 6–20 shows this type of enclosure. There are holes in the bell housing as well as all the way around the coil housing. Air is allowed to circulate freely through the motor windings. Usually a fan is mounted on the shaft to enhance the flow of air through the motor.

FIGURE 6–20. Totally open enclosure. *(Courtesy Fasco Distributing Co.)*

The TO enclosure may be modified so that the ventilation holes are located only at the bottom of the motor. This type of arrangement is often used in environments where moisture may enter the motor through the top openings. Typical applications for this type of enclosure are for window air conditioners or for condenser fan motors.

There are other modifications made on the TO to meet environmental needs and still insure adequate ventilation. Only the up end is enclosed on some motors. This design is used on motors that operate in the vertical position. The shaft may be either up or down. The end plate which faces up is enclosed to prevent the entrance of rain or other foreign matter.

Another modification which is frequently used is the open dripproof enclosure. Ventilation holes are located so that moisture cannot enter the motor easily. Usually the end plates will be closed as is the top of the motor. Peripheral holes will only be located in the lower half of the enclosure. Figure 6–21 illustrates this design.

The enclosures are sometimes modified to permit a blower attachment to the housing. Figure 6–22 depicts a blower mounted on a motor. This design allows the blower to be rotated in 90° increments. The blower may be mounted on either side or on the top as shown. The air being drawn into the motor can be filtered if necessary.

Commutator and Brushes

Commutators and brushes are used on all DC generators and DC motors. They are also used on some AC

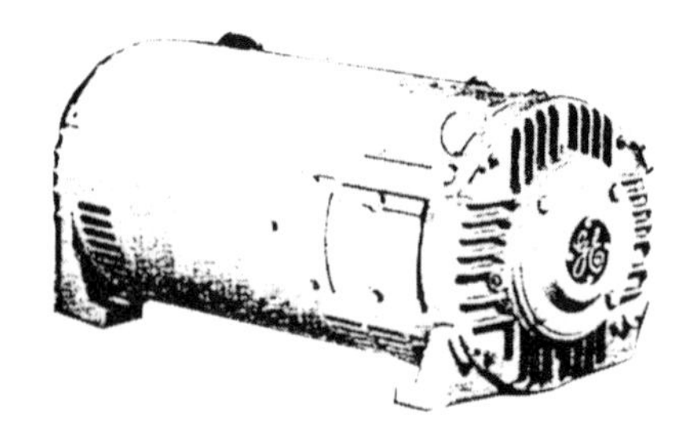

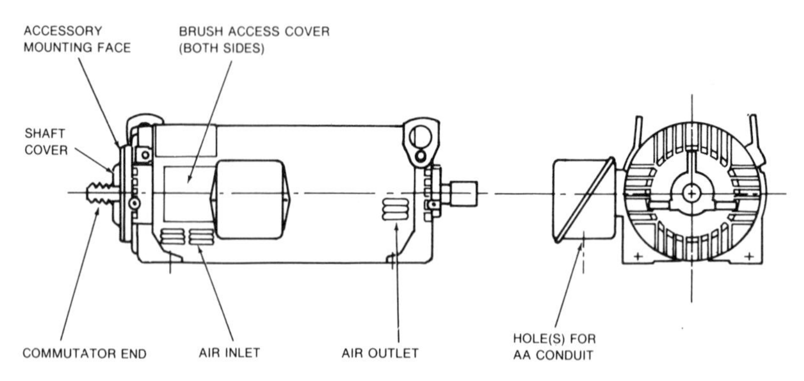

FIGURE 6–21. Dripproof fully guarded motor. *(Courtesy General Electric Corp.)*

motors such as the repulsion, synchronous, and universal motors.

All generators produce a sine wave, or AC currents when the rotor turns in the magnetic field. The commutator on the DC generator converts the AC into pulsating DC. The commutator assures that the current from the generator always flows in one direction. The brushes ride on the commutator and make good electrical connections between the generator and the load.

On DC and most AC motors the purpose of the commutator is to insure that the current flowing through the rotor windings is always in the same direction, and the proper coil on the rotor is energized in respect to the field coils. By mechanically positioning the brushes on the commutator, an angle of displacement can be set up between the magnetic force of the field windings and the magnetic force of the rotor windings. The theory of operation will be discussed in principles of operation, below, and in Chapter 7 when commutators are used for a particular type of motor.

Figures 6–11 and 6–13 show the commutator mounted on the shaft with the armature. Figure 6–23 identifies the parts of the commutator.

The segments of the commutator are usually made of copper and are separated from each other by mica insulation. The mica is cut so that it lies below the copper segments. Slots are cut in the riser on the commutator to facilitate the soldering of the ends of the coils. There are twice the number of segments on the commutator as there are slots in the laminated core for the coils.

Pressed against the commutator and held by spring tension are the brushes. Brushes are usually made of a graphite substance which is softer than the copper segments. Brush wear will, therefore, be greater than that of the commutator, and they will need to be replaced more often. Figure 6–24 depicts the exploded view of the brush holder, bush, spring clip, and cover of a DC motor manufactured by General Electric Corporation.

Centrifugal Switches

Centrifugal switches are mounted inside the motor. These switches have two parts, the stationary part which is the switching mechanism, and the rotating part which is a set of spring loaded weights. The rotating spring-loaded weights are sometimes referred to as a *governor*. The governor will be activated when the speed of the motor reaches a predetermined RPM to create a centrifugal force necessary to overcome the force of the springs holding the weights. Contacts on the switching mechanism can either be opened or

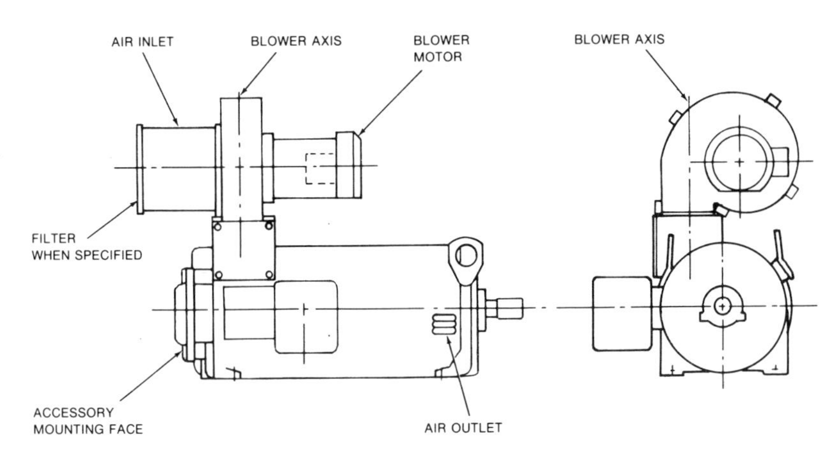

FIGURE 6–22. Motor with blower attachment. (*Courtesy General Electric Corp.*)

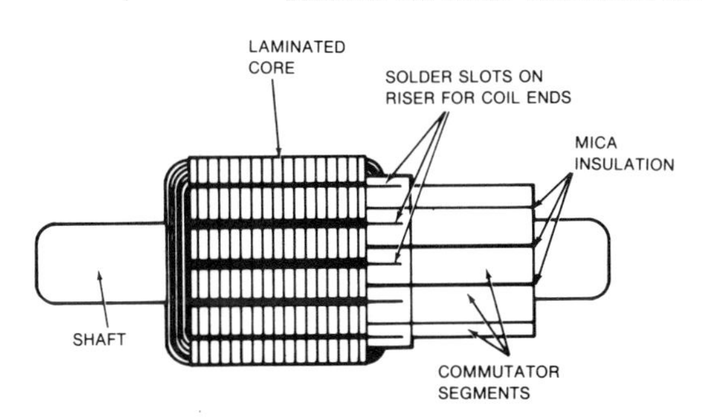

FIGURE 6–23. Parts of the commutator.

closed depending on the requirements of the particular type of motor.

There are many different types of design for centrifugal switches. Figure 6–25 illustrates one of these switching arrangements. Figure 6–26 depicts the action of the switch when the rotor is not turning, and after the rotor is up to speed.

As shown in Figure 6–26, the force of the governor's spring action overcomes the force of the spring on the centrifugal switch when the rotor is stopped. The spring on the centrifugal switch is compressed by this action, wanting to push the switch open. When the motor comes up to speed, the governor opens allowing the compressed spring on the switch to force the contacts open.

These switches are commonly used to open the starting windings on motors when they reach 80% to 90% of their synchronous speed or to switch components such as capacitors on capacitor-start-run motors. They are also a trouble point on motors and require maintenance.

Capacitor

A capacitor is used in some motors to provide greater starting torque or to improve running conditions. Capacitors are constructed of two conductive plates separated by an insulator. They store electrical energy and provide for a leading current. All capacitors have these qualities, but they differ in construction and in the type of dielectric used.

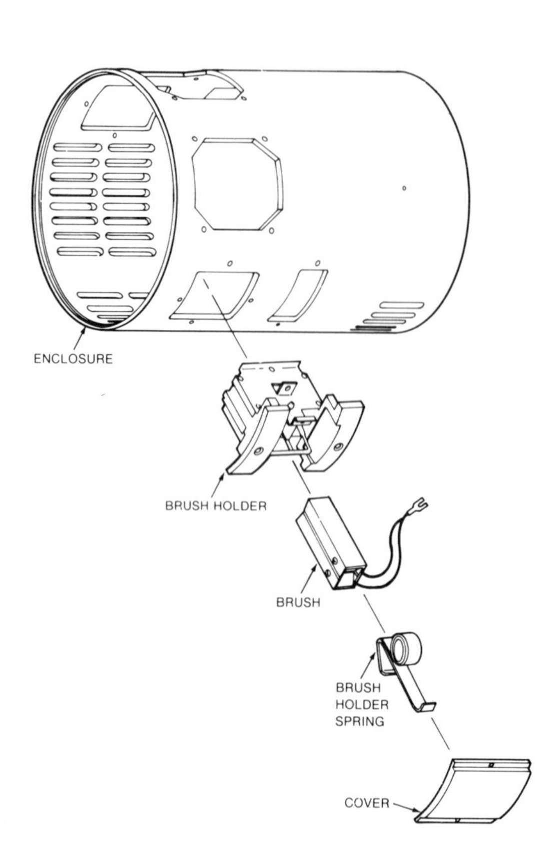

FIGURE 6–24. Exploded view of a brush assembly. (*Courtesy General Electric Corp.*)

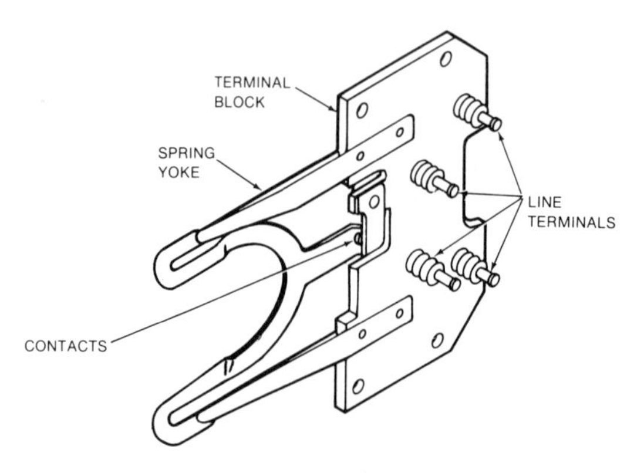

FIGURE 6–25. Centrifugal switch.

Motors utilize two basic types of capacitors in their operation. These are the oil-filled capacitor and the electrolytic type.

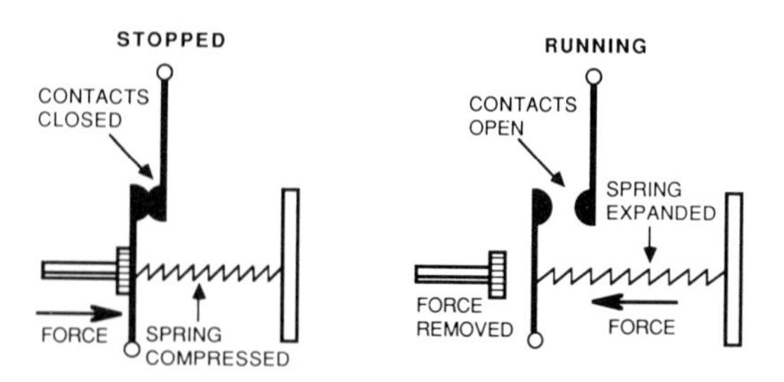

FIGURE 6–26. Operation of the centrifugal switch.

Oil-filled capacitors are used mostly on permanent-split capacitor motors. Electrolytic capacitors, which can have very high capacitances for their size, are used for motor starting. Electrolytic capacitors can only be kept in the circuit for a short period of time because alternating current breaks down their thin film of insulation. If the centrifugal switch fails to open, there is a good chance the electrolytic capacitor will be destroyed.

The capacitors are rated in microfarads and volts. Capacitors for motor starting may range from 2 to 800 microfarads depending on the use, type, and size. The voltage rating will normally be given in volts alternating current (VAC).

Capacitors are often mounted on the top of the motor. Figure 6–18 shows this arrangement on the motor at the bottom of the illustration.

Capacitor problems can cause a motor not to start or to run improperly. The capacitor may open, short, or change in value to cause these problems. Under these circumstances, the capacitor will have to be replaced. Care should be taken to replace it with the original value of capacitance and voltage rating. Improper application of the farad rating can cause voltage rises in excess of the voltage ratings of the capacitor and the windings. A capacitor rated at a higher voltage can be used, but a smaller-value capacitor must never be installed.

Thermal Protectors

Although thermal protectors are sometimes called thermal relays, these devices do not have a coil. They consist of a bimetallic strip, and when heated beyond a given temperature, the movable contacts will curl upward breaking the electrical circuit. This is due to the difference in expansion rates of the two dissimilar metals.

These devices are not found on all motors, but they are installed on many as added protection. They are commonly used on hermetically sealed units and on motors that have loads that are likely to cause a lock rotor condition. A good example of the latter is the use on a garbage disposal where a piece of silverware or bone will inadvertently become lodged in the shredder at some point during the life of the device.

Some thermal relays have heaters as shown in Figure 6–27. Others depend solely on the heating of the bimetallic strip by any overcurrent flowing through it to deactivate the motor.

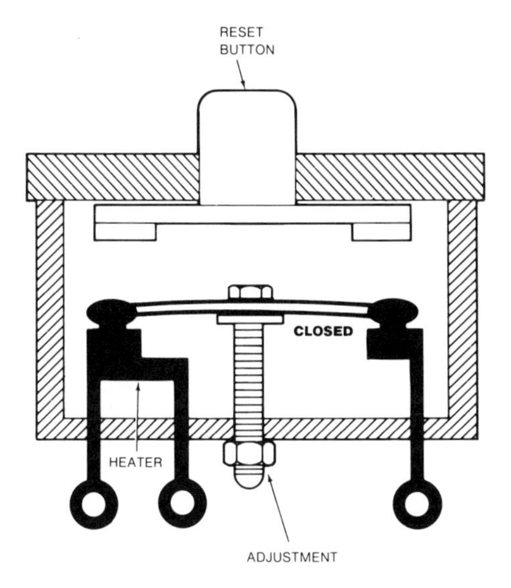

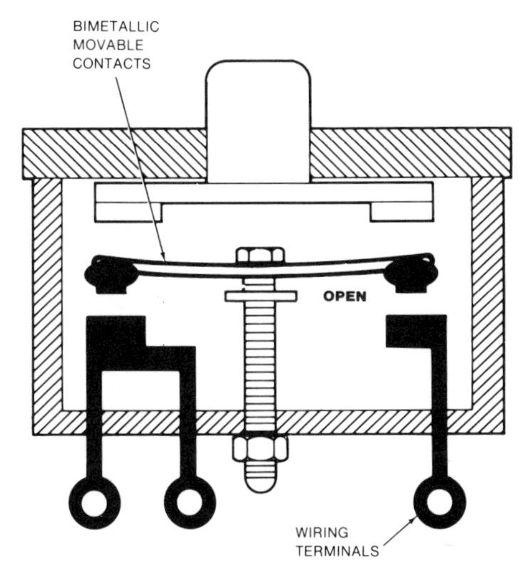

FIGURE 6–27. Thermal protector for motor.

The adjustment screw at the bottom of the device provides for setting a higher or lower temperature at which the relay will deactivate the motor. This is normally set at the factory. The value at which it is set is based on the class of insulation used.

A manual reset button allows the thermal breaker to be reset after the bimetallic strip has cooled. Some thermal breakers do not have this feature and must be replaced once they have operated. These are seldom used for modern motor applications.

Others automatically reset when the strip has cooled. When a motor has the automatic reset feature, care should be taken when working on the motor. It can suddenly restart without warning unless the control circuit is designed to prevent this.

The contacts to complete the circuit to the coils are normally 1 and 2, and the heater is connected between terminals 2 and 3. When thermal protectors are used on single-voltage motors and the higher voltage of dual-voltage motors, terminal 2 is not used. Figure 6–28 illustrates this arrangement.

If the same wiring were used for the lower voltage on a dual-voltage motor, twice the current would flow through the heater with the motor windings in parallel with each other. This would result in four times the power being dissipated by the heater to the bimetallic strip. The result would be that the motor would be disconnected from the line even when operating under normal conditions.

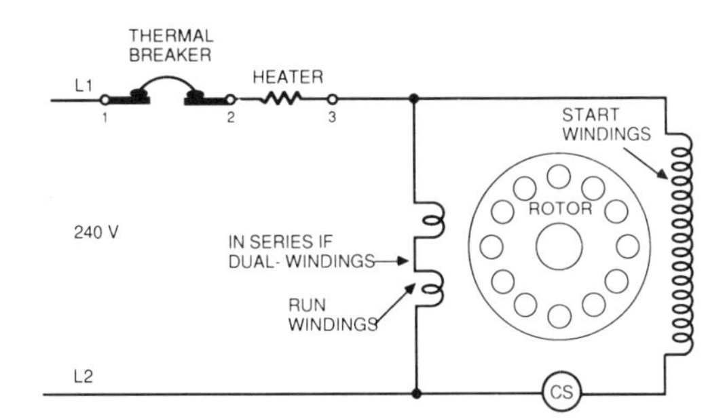

FIGURE 6–28. Thermal protection wiring for single-voltage motors and the higher voltage on a dual-voltage motor.

Figure 6–29 provides the wiring diagram for the thermal protector when the lower voltage is used for a dual-voltage motor. In this case, the heater is in series with only one of the running windings. This allows the same amount of current to pass through the heater as the higher voltage configuration in Figure 6–28. Note that terminal 2 is used under these conditions.

Thermal breakers are sometimes mounted on the terminal board of the motor or on the enclosure. At

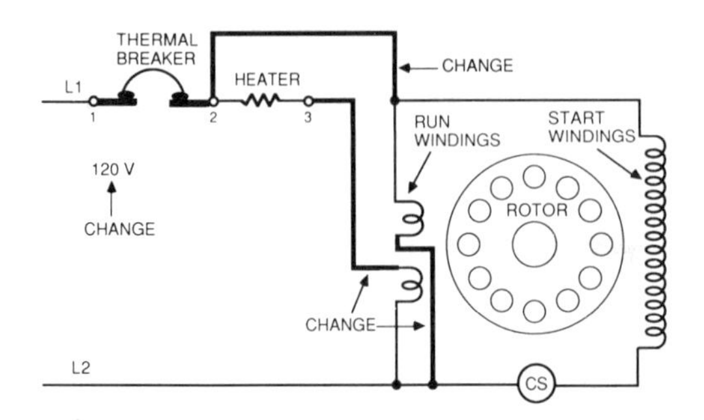

FIGURE 6–29. Thermal protection wiring for the lower voltage on a dual-voltage motor.

times they are embedded into the windings of the motor to better sense the temperature on the insulation itself. When this arrangement is used, care should be taken in the event the breaker must be replaced. It is easy to damage the wiring insulation which could result in motor failure. When the breakers are embedded, they are normally of the automatic reset type.

A motor may have thermal protection provided under three different sets of circumstances. These are

- "OVER TEMP PROT 1." Thermal protection is provided for the windings under locked-rotor conditions. Table 6–5 gives NEMA standards for the maximum and average temperatures in degrees Centigrade for the four classes of insulation.

TABLE 6–5. NEMA Standards for Maximum Winding Temperature under Locked-Rotor Conditions*

Type of Protector	Maximum Temperature (°C)			
	Insulation System Class			
	A	B	F	H
Automatic Reset				
During First Hour	200	225	250	275
After First Hour	175	200	225	250
Manual Reset				
During First Hour	200	225	250	275
After First Hour	175	200	225	250

*Assumed ambient temperature of 25°C.

- "OVER TEMP PROT 2." Under these circumstance the motor is running continuously under full load. The maximum allowable temperatures are shown in Table 6–6. When a motor is marked with this notation, it does not have locked-rotor protection.
- "OVER TEMP PROT 3." When this marking appears on the motor, the manufacturer needs to be consulted for the type of protection and the maximum allowable temperature on the windings. Manufacturer's specifications should be followed.

TABLE 6–6. NEMA Standards for Maximum Winding Temperature for Maximum Continuous Load*

Insulation System Class	Maximum Temperature (°C)
A	140
B	165
F	190
H	215

*Assumed ambient temperature of 29°C.

Principles of Operation

Motor operation is dependent on the interaction of magnetic fields. To understand how a motor operates, the rules of magnetism need to be defined along with the relationship that exists between current flow and magnetic fields.

Magnetism

The first magnetic material was discovered by the ancient Greeks in an area around Magnesia in Asia Minor over 2000 years ago. The Greeks called this material lodestone, or leading stone, because when the rock was suspended and was free to turn, it always pointed in the same direction.

This brings us to our first definition. The end of the magnet which points toward the north geographic pole of the earth is defined as the north pole of the magnet. Actually, the magnet points to the north magnetic pole of the earth, which is slightly displaced from the earth's north geographic pole. The earth can be seen as an immense bar magnet. Figure 6–30 illustrates this definition.

The lines of magnetic force which cause the compass to align itself in Figure 6–30 are called *flux lines*. They are invisible to the eye, but their presence can be seen if iron filings are sprinkled on a piece of plastic placed on top of a magnet. Figure 6–31 shows the results. Although this is a two-dimensional drawing, the flux lines surround the bar magnet in all directions.

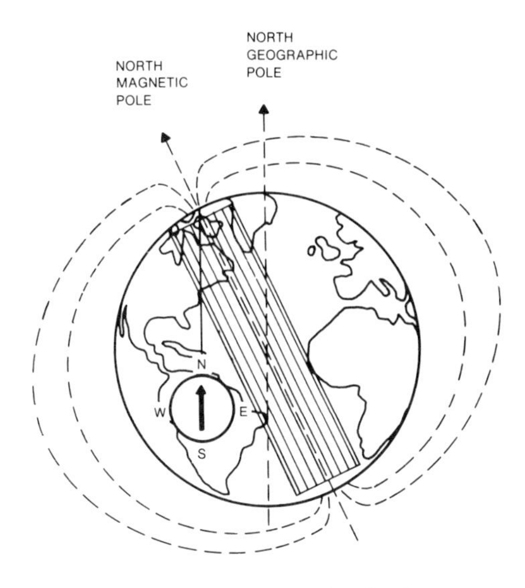

FIGURE 6–30. Defining the north pole of a magnet.

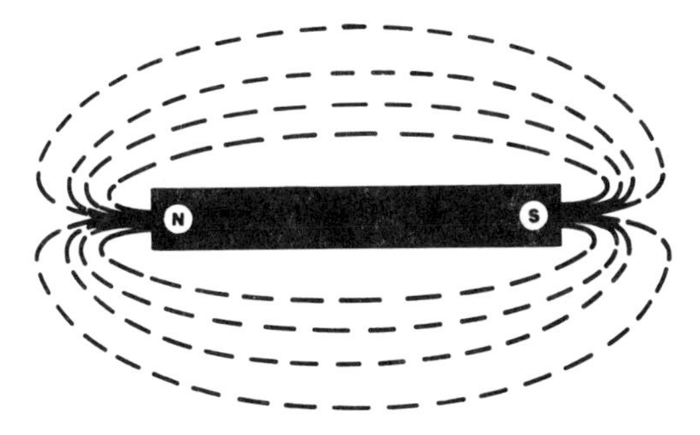

FIGURE 6–31. Effects of a magnetic field on iron filings.

An examination of the pattern produced by the magnet in the iron filings provides no clue as to the direction of the flux lines. In order to explain the effects of magnetism in common terms that could be understood by the scientific community, it became necessary to define the direction of the field. By definition, the flux lines are specified as leaving the north pole of a magnet and entering the south pole. Figure 6–32 depicts this definition.

There are two noticeable effects that can be felt if two bar magnets are brought together. In one case, the magnets will be attracted to each other, and in the other case, they will push each other apart. Because the poles of the magnet have been defined, it can be concluded that *like poles* of a magnet *repel*, and *unlike poles* of magnets *attract* each other.

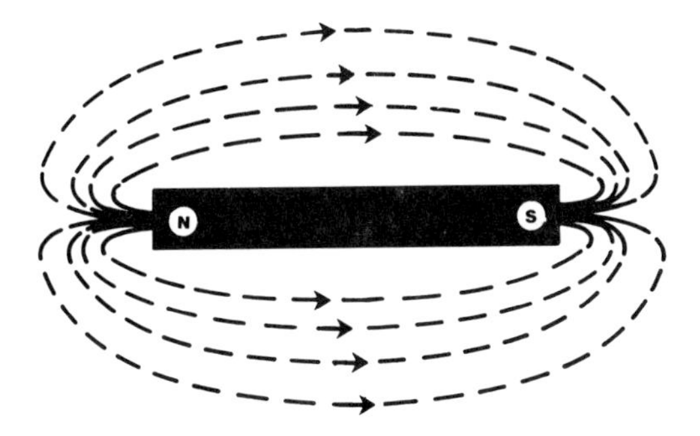

FIGURE 6–32. Direction of flux lines.

Figure 6–33 illustrates the effect on the iron filings when two like poles are brought together. Figure 6–34 provides the pattern when unlike poles are in close proximity of each other. It is the attracting and repulsing forces of the magnetic field that make possible the operation of a motor.

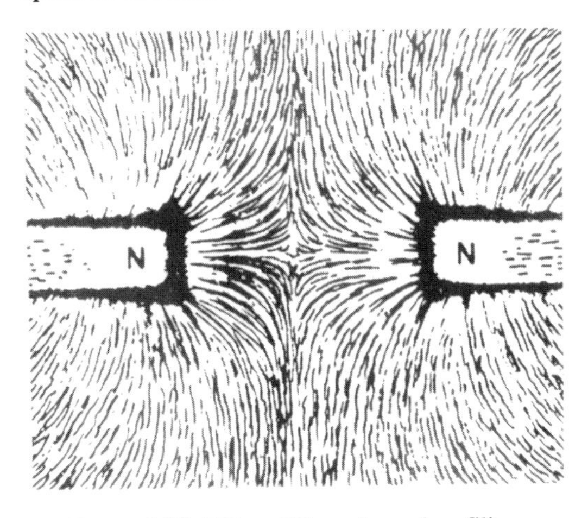

FIGURE 6–33. Effect of like poles on iron filings.

If a magnetic material is placed in the flux field, the flux lines will concentrate themselves through the magnetic material. The flux lines tend to take the path of lowest reluctance. Figure 6–35 depicts this phenomenon. The stator core of a motor is contoured to fit the rotor's configuration to enhance the passage of the flux field between stator and rotor, thereby reducing the amount of flux leakage which would increase the motor's losses and reduce its efficiency.

Using Figures 6–30 through 6–35, some of the characteristics of flux lines can be discerned. The basic rules of flux lines are that they

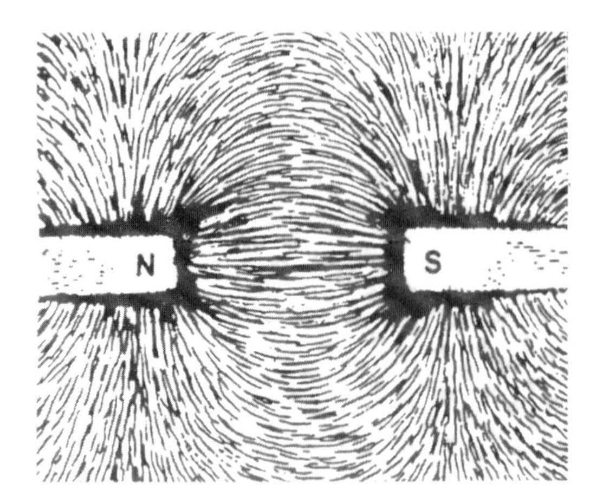

Figure 6–34. Effect of unlike poles on iron filings.

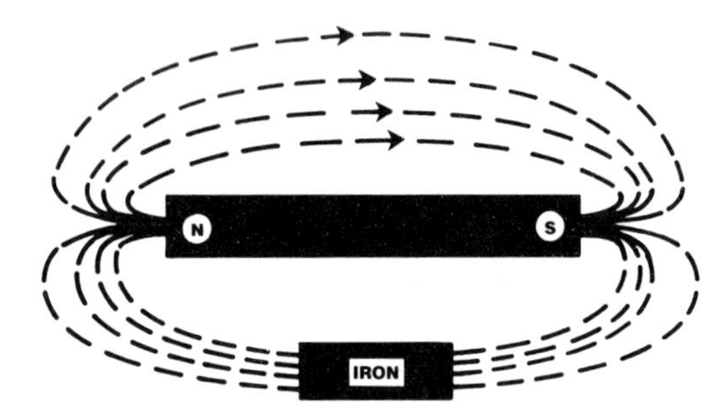

Figure 6–35. Effect of placing magnetic material in the flux field.

1. Always form closed loops. The number of flux lines that leave the north pole of a magnet must enter the south pole. This can be compared to current flow where the number of electrons leaving the negative pole of a source must return to the positive pole.
2. Never cross each other.
3. Repel each other when traveling in the same direction.
4. Magnetically cancel each other's effect when going in the opposite direction.
5. Take the path of least reluctance.
6. Pass through all materials.

The principles of how a motor turns can be illustrated using a compass and a bar magnet. If a compass is used to represent the rotor of a motor, and a bar magnet is used as the field, the motor effect can be shown.

As the bar magnet is rotated around the compass, the compass's pointer rotates with the magnet (Figure 6–36). When the bar magnet is located in one position, the compass needle aligns itself with the field and remains in one position. For the rotor of a motor to turn, it is necessary to have a rotating magnetic field.

Electromagnetism

Although Figure 6–36 illustrates the ability of permanent magnets to interact and cause rotation of an armature, in practice, this is done with electricity. There are a few basic rules that need to be studied in order to understand how current flow and the resulting magnetic fields interact to cause motor action.

Figure 6–1 depicted Oersted's discovery of electromagnetism. It was not by chance that the needle of the compass is pointing down. If the compass is put on top of the conductor, it will swing 180 degrees and point up on the page. Figure 6–37 shows this effect.

Reversing the direction of current through the conductor will once again cause the pointer to swing to the down position. Figure 6–38 shows this change. Which way the needle of the compass will point can be predicted using the "left-hand rule."

If the conductor is gripped with the left hand with the thumb pointing toward the positive pole of the battery, or in the direction of current flow, then the fingers around the conductor indicate the direction of the magnetic field around the conductor. Because the flux field enters the south pole and exits the north pole of a magnet, the pointer on the compass will align itself accordingly. Figure 6–39 illustrates the "left-hand rule" for conductors.

The magnetic field around the conductor forms concentric circles and has no magnetic polarity. That is, the field does not have a north and south pole. If the conductor is formed into a loop, the individual fields will combine to form magnetic poles.

Figure 6–40 shows a conductor formed into a coil around an iron core. The direction of current flow in the coil is indicated by the arrows inside the conductor. The individual magnetic fields around the conductor are represented by the dotted circles with the arrows to indicate their individual direction. Notice that the arrows combine in the same direction in the iron to form magnetic poles.

The polarity of the magnetic field formed in this way can be found using the "left-hand rule" for coils. The fingers of the left hand encircle the coil in the direction of current flow. The thumb will point to the

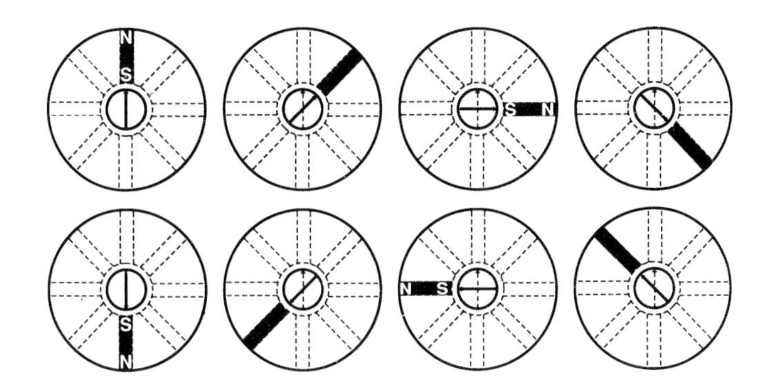

FIGURE 6–36. Effect of rotating a bar magnet around a compass.

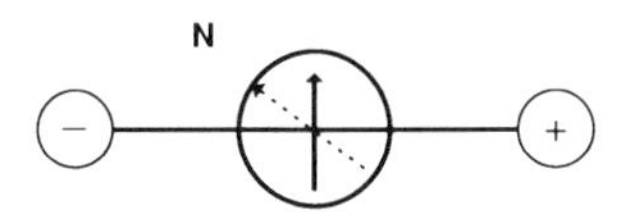

FIGURE 6–37. Needle of a compass reverses direction when compass is placed on top of the conductor.

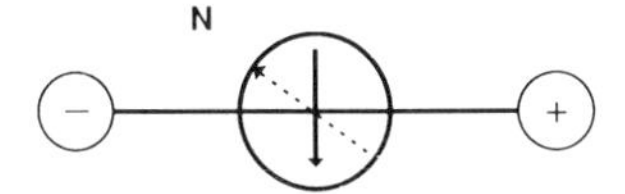

FIGURE 6–38. Direction of pointer reverses with change in direction of current flow.

north pole of the resulting magnetic field. The concentrated flux lines leave the north pole of the electromagnet and enter the south pole.

Reversing the direction of current in the coil will result in the north and south poles switching ends. To predict this, the left hand must be reversed to reflect the change in direction of the current, with the thumb now pointing to the opposite end.

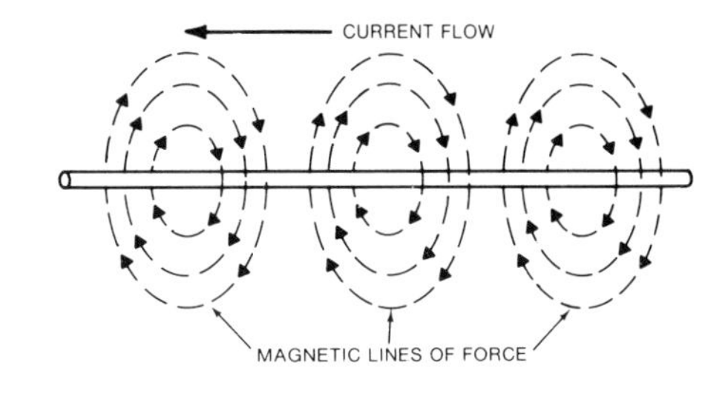

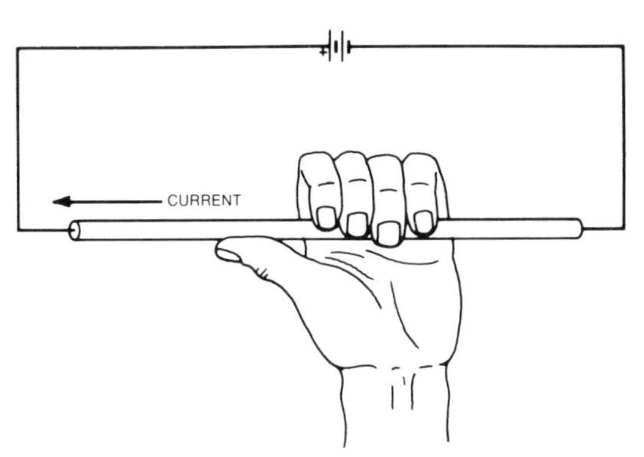

FIGURE 6–39. Left-hand rule for conductors.

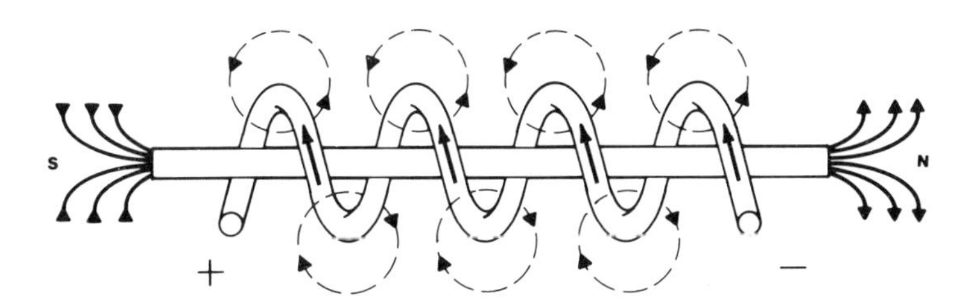

FIGURE 6–40. Left-hand rule for coils.

Generator Principles

A *generator* is an electromagnetic device that converts mechanical energy into electric energy. To better understand the principles of motor operation, it is helpful to know how a generator performs this task.

When the flux field of a magnet cuts through a conductor, a voltage is induced into the conductor, and a current will flow if the conductor forms a closed loop. The secret to the existence of this phenomenon lies in the word "cut." There must be relative motion between the conductor and the flux field. If a magnet is merely setting in the proximity of a coil, no voltage will be induced. If either the coil or the magnet is moved, a voltage will be induced. If both are moved at the same time in the same direction, and at the same rate of speed, no voltage will result in the coil. There must be relative motion between the two.

Maximum voltage will be induced if the field passes through the conductor perpendicularly or at a right angle. Lesser voltages will be induced at angles less than 90 degrees, with no voltage induced when the flux lines are moving parallel to the conductor.

The voltage induced at any given angle is based on the sine wave function of the trigonometry tables. Because of the circular motion of the rotor, all power generators produce AC according to the sine function.

Figure 6–41 illustrates a simple AC generator using a single loop of wire which is rotated clockwise in the magnetic field. The flux field can be provided either by a permanent magnet or an electromagnet. Permanent magnets have limited use for this purpose. They are normally used only on small generators where small currents are required. Electromagnets are normally used because they can produce stronger fields, and the amount of flux can be adjusted by regulating the amount of current in the field windings to compensate for changes in the speed of rotation and voltage drop at the output due to changes in the load.

For each volt of output, a single loop must cut 100,000,000 flux lines in 1 second. To obtain commercial power, multiple windings in series are used to form a coil. Voltages induced in the single loops become additive and provide higher voltages out for the same flux density.

In position 1, the conductors are moving parallel to the flux lines and no voltage is induced. At position 2, the loop is cutting the maximum number of flux lines and the maximum positive voltage is present at the brushes. As the loop passes through this point and progresses to position 3, the voltage is falling until it reaches zero when the loop is once again moving parallel to the flux field. As the loop passes though position three the voltage swings negative reaching its maximum value in that direction at position 4. The loop then returns to its starting point, where, once again no voltage is induced.

Half the maximum voltage will be induced when the conductor is at 30 degrees. Approximately seventy percent will be present at 45 degrees, and the output will be 86.6% at 60 degrees. The amount of voltage can be changed by either changing the rate of speed of the rotor or the strength of the magnetic field.

Each end of the loop is attached to a slip ring which rotates with the loop. A single-phase generator will have one set of slip rings. Two-phase generators will have two sets, and three-phase generators will have three sets. Brushes ride on the slip rings to remove the voltage from the generator.

If either the direction of rotation of the loop or that of the magnetic field is reversed, the polarity of the output will be also reversed. For example, in Figure 6–41, reversing rotation would cause the output to go negative first. The output sine wave would then start at position 3 instead of position 1.

Polarity, which is defined as the direction of current flow, can be determined for a generator by the "left-hand rule" for that device. Figure 6–42 shows how to apply this rule.

The index finger points in the direction of the flux field. This has been defined as being from the north pole to the south pole of the magnet. The thumb is then pointed in the direction of rotation of the conductor through the field, and the middle finger will provide the direction of the current flow in the conductor. Note that within the generator, electrons flow from the positive pole to the negative pole. Externally, electron flow is from the negative pole back to the positive pole.

To convert an AC generator into a DC generator, the slip rings would be replaced by a commutator. Each end of the loop would then be connected to a separate segment of the commutator. The segments are electrically insulated from each other.

Figure 6–43 shows this arrangement with the resulting output. The AC is converted to DC because the brushes always ride on the segment attached to the loop that is passing through the magnetic field in the same direction.

The brushes change from one segment to the other at the point where the loop is moving parallel to the magnetic flux lines. At this instant, no voltage is being induced into the loop of wire. This point is defined as the *neutral plane* of the generator. If the commutation

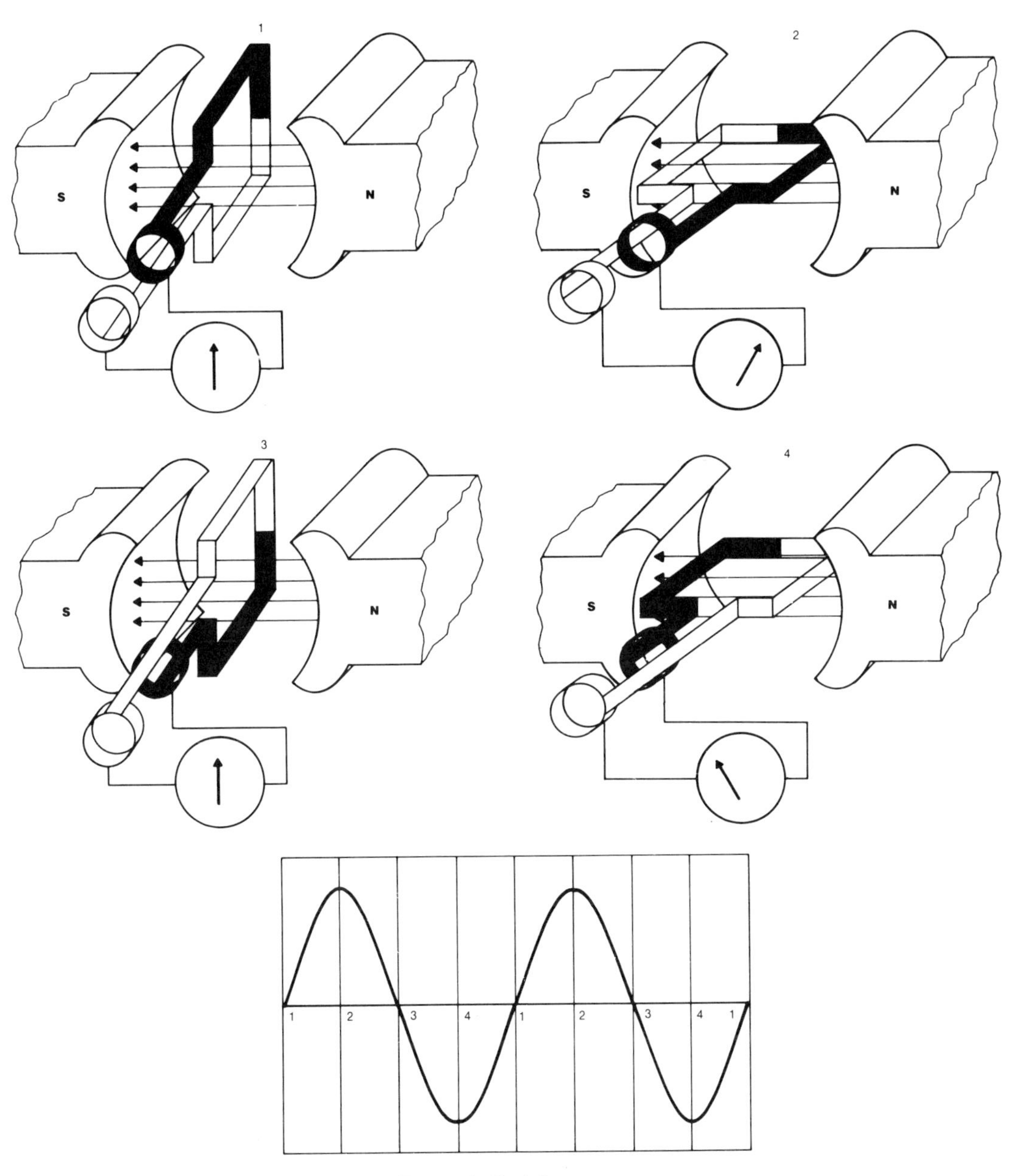

FIGURE 6–41. AC generator.

occurs at any point other than at the neutral plane, high currents will result due to the brushes shorting out the segments when voltage is present. This condition causes considerable arcing. Arcing can destroy the brushes and cause possible damage to the commutator.

Polarity of the output voltage is dependent on the direction of rotation of the generator. The polarity shown in Figure 6–43 is determined by assuming the loop on the right is moving down through the field, while the loop on the left is moving up. Check this assumption using the left-hand rule.

When the generator is loaded, the current supplying

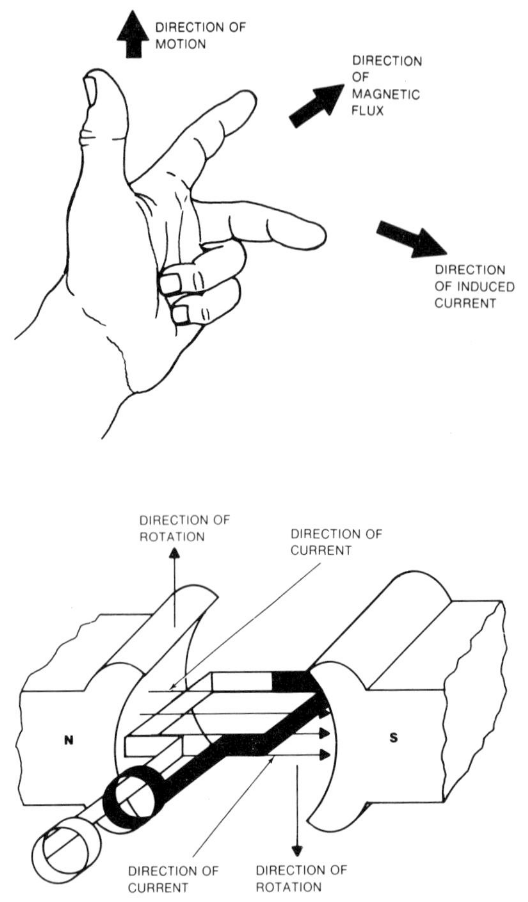

FIGURE 6–42. Left-hand rule for generators.

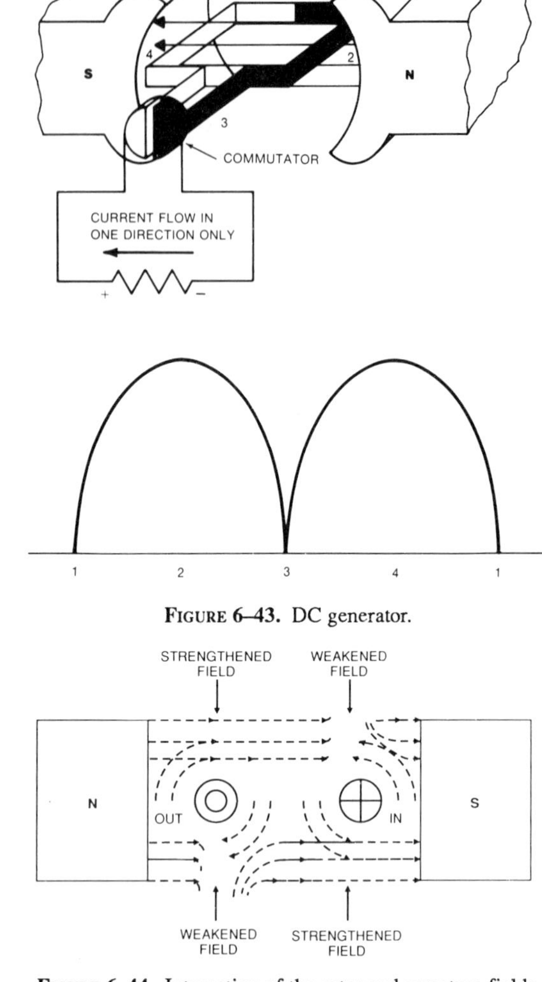

FIGURE 6–43. DC generator.

FIGURE 6–44. Interaction of the rotor and armature fields of a simple DC motor.

this load creates its own magnetic field. This field will be in such a direction as to oppose the field that caused it.

This opposing force can be readily demonstrated using a megger. Crank a megger with its leads disconnected and with them shorted. In the first case, the crank will turn freely. When the output is shorted, however, the crank will be very difficult to turn. Lenz's law is applicable.

Direct Current Motors

Our simple DC motor looks just like the DC generator. In fact, if voltage is applied to the loop of wire and current flows, a magnetic field will be created that will interact with the field of the magnet. The repulsion and attraction of the two fields will cause the loop to turn.

Figure 6–44 helps explain why this is true. Looking at the two ends of the loop, current is flowing into the loop on the right side and out of the loop on the left. Applying the left-hand rule for conductors, the direction of the magnetic field around the conductor on the right is counterclockwise, and the direction of the field around the conductor on the left is clockwise.

Under these conditions, the magnetic field of the conductor on the right opposes the field of the magnet at the top. This weakens the total field at this point. The fields at the bottom of the conductor reinforce each other.

Just the opposite is true of the conductor on the left because of the change in direction of current flow. The field at the top of the conductor is strengthened, and the field at the bottom is weakened.

When the fields oppose each other, the total strength of the field in that area is weakened. On the other hand, when the fields are in the same direction, the total field now becomes stronger. This causes the loop to move away from the strong field toward the weak field. Under these circumstances the loop will turn in a counterclockwise direction. Reversing the direction of current would cause the loop to turn in a clockwise direction.

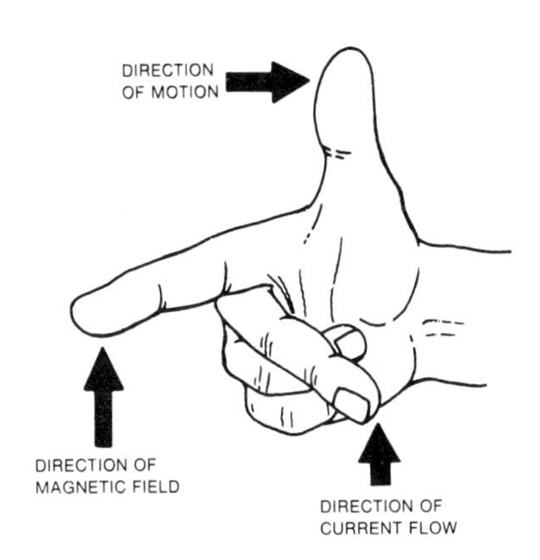

FIGURE 6–45. Right-hand rule for motors.

The direction of rotation can be determined by using the "right-hand rule" for motors. Figure 6–45 illustrates this rule. Apply this rule to Figure 6–44 to verify the direction of rotation of the loop.

Another approach to the theory of operation of a DC motor is to use an iron core inductor as the rotor with each end of the coil connected to a segment of a commutator. By increasing the number of turns in the loop of the armature and adding an iron core, the magnetic field is increased, and increased torque is obtained. Figure 6–46 shows this arrangement.

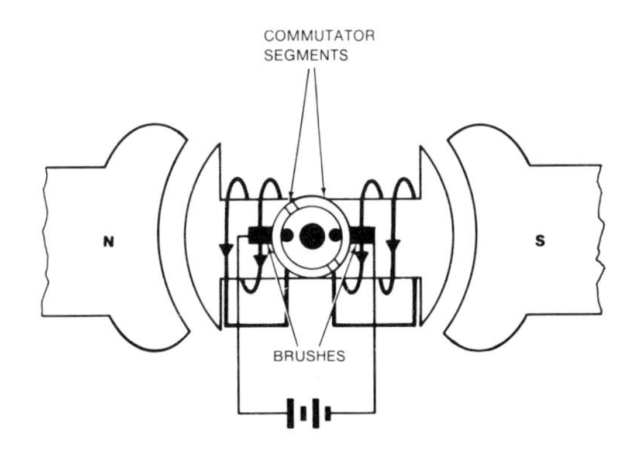

FIGURE 6–46. Basic DC motor.

When the left-hand rule for coils is applied to the drawing in Figure 6–46, it is found that like poles of the field and armature are opposite each other. This is shown in Figure 6–47 as Position 1.

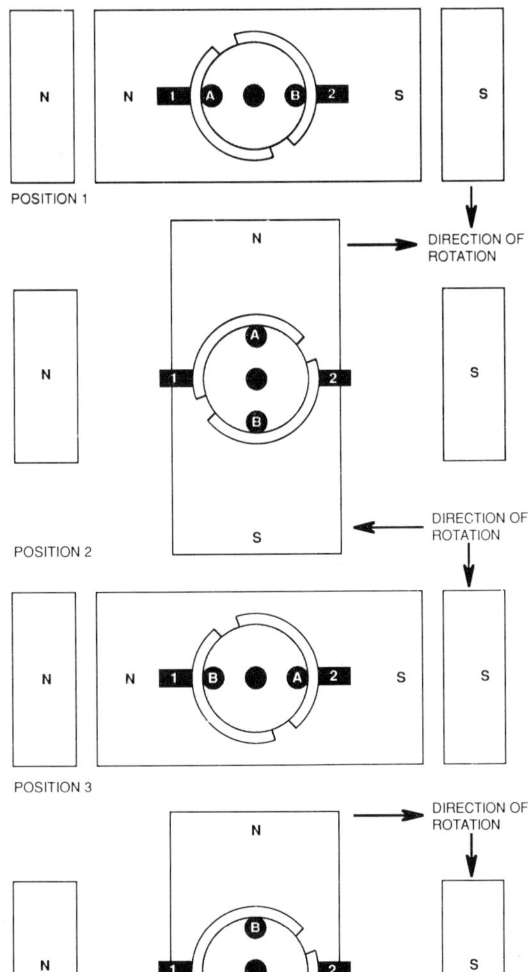

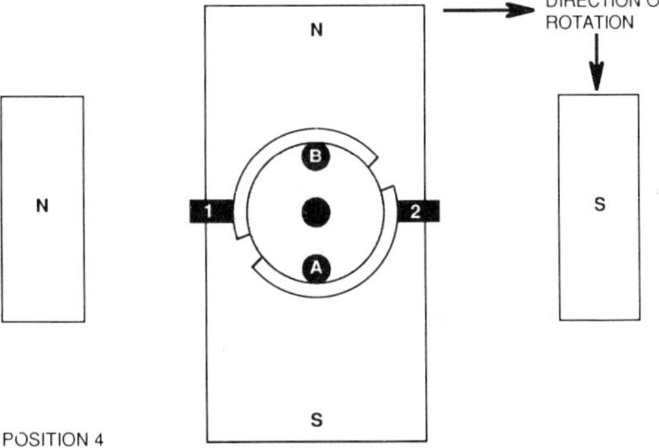

FIGURE 6–47. DC motor action.

Because like poles repel, and the armature is free to turn, it will begin to rotate. The direction of rotation will be in a clockwise direction. The armature will move toward its neutral plane. This is shown as position 2.

The gap in the commutator has been positioned forward of the neutral plane in this case. The reason for this is that when commutation takes place, current in the windings does not reverse immediately due to the self-inductance of the collapsing field in the windings. This in turn causes arcing at the brushes during commutation. The arcing can be reduced by moving the brushes against the direction of rotation and taking advantage of the counterelectromagnetic force (counter-emf) of the previous pole. This counter-emf comes from generator action when the conductors in the armature are turning and cutting through the magnetic flux of the field.

When the rotor is at Position 2, the commutator segment marked "A" is still in contact with Brush 1, and segment "B" is connected to Brush 2. Therefore, the magnetic poles of the rotor have not reversed. Inertia will cause the armature to move through the neutral field, where its north poles will be attracted by the south poles of the field magnet.

Commutation takes place as the armature is moving from Position 2 to Position 3. The two segments will switch brushes with which they were in contact. At this point, the magnetic field of the armature will reverse due to a reversal of the direction of current through the armature's coil.

Even though the pole reversal occurs slightly before the armature arrives at Position 3, and like poles are approaching each other at this point, the inertia of the rotor will carry it to a point where the like poles once again will repel each other and continue the clockwise motion.

Polarity remains the same as the armature moves through Position 4 and the neutral plane. The unlike poles of the two fields will attract each other until commutation takes place as the rotor moves back to Position 1 to once again be repelled toward Position 2.

There are two drawbacks with the motor used to demonstrate DC motor principles. These are

1. This type of motor may not start of its own initiative.
2. Once started, the motor does not run smoothly and may stop when loaded.

The motor does not always start because the loop may be in the neutral plane and the rotor and stator fields do not have any effect on each other. Another condition is when the brushes are located at the gap of the commutator segments. At this point they may short the segments and no current would be supplied to the rotor windings. In each case, the rotor would have to be given a turn to start it. This is not very practical.

Direct current motors can be made self-starting by adding additional loops in parallel with each other on the armature. This type of arrangement requires two additional segments on the commutator for each loop. Figure 6–48 illustrates this arrangement.

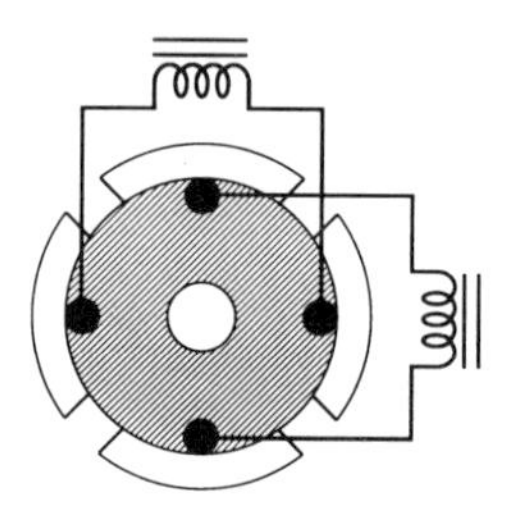

FIGURE 6–48. Adding additional loop to self-start a DC motor.

As shown in Figure 6–48, one loop would always be outside the neutral plane and connected to the battery. This would insure the starting of the motor. It would not help, however, the erratic operation of the motor or its efficiency. One loop is always taking a free ride and not contributing to the motor's torque. How these various problems are resolved is discussed in Chapter 7.

Alternating Current Motors

Unlike DC, 60-hertz current is constantly changing in amplitude and reverses its direction of flow 120 times each second. These reversals are called *alternations*.

Because the amplitude of the current is changing, the strength of the magnetic field is also constantly changing. The alternations cause the polarity of the flux field to reverse with the change of direction of current through the field windings.

In order for an AC motor to operate it must have a rotating magnetic field. This requirement is automatically met with polyphase motors. Figure 6–49 illustrates the rotating magnetic field when three-phase power is applied to the stator windings of a motor.

In three-phase systems each of the phase voltages is

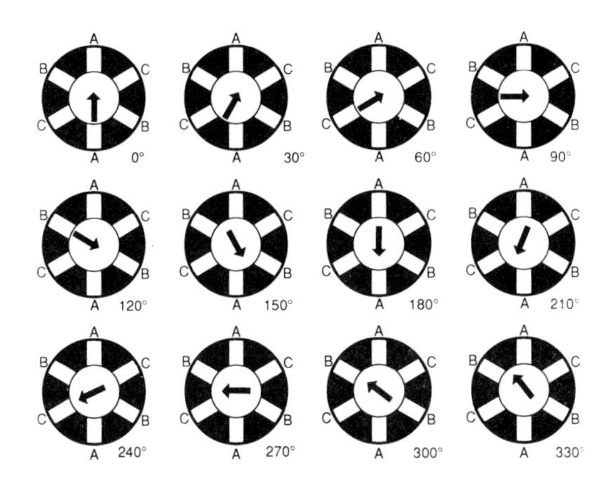

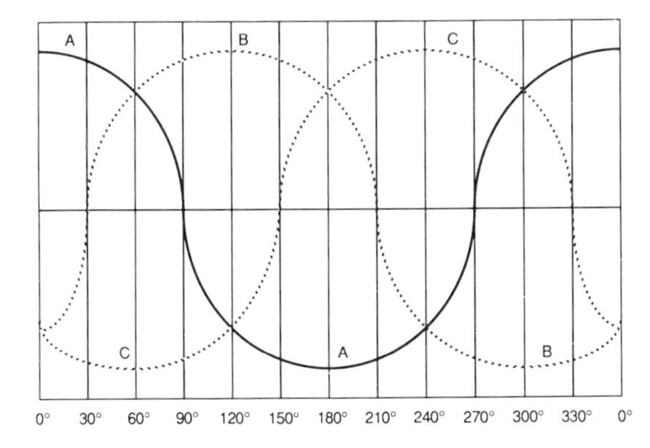

FIGURE 6–49. Rotating magnetic field developed by three-phase power.

displaced 120 degrees from each other. This displacement causes the magnetic field to rise and fall at each pole in sequence.

Because of transformer action, the current flowing through the stator will set up a constant changing magnetic field which will cut the windings of the rotor. The current induced in the rotor causes a magnetic field to form that is opposite that of the pole which caused it. Unlike poles attract.

At zero degrees the current entering Phase A is equal to the two currents of Phases B and C which are leaving the motor. The rotor is attracted to the strong field of Phase A.

When Phase A has moved to 30 degrees, Phase B has no current. A current equal to that of Phase A leaves the motor on Phase C. This causes the field to rotate clockwise with the rotor being attracted in that direction.

At 60 degrees, the positive currents in Phases A and B are equal with the maximum current leaving the stator in Phase C. Voltage on Phase A is falling and the voltage on Phase B is rising. The rotor continues to turn clockwise.

These sequential events continue for the complete cycle until Phase A again rises to its maximum positive potential. The cycle of rotation is again repeated 60 times per second.

Two-phase motors may also be used to produce a rotating magnetic field. They differ from three-phase motors in that they only require two sets of poles rather than three, and the phase displacement is 90 degrees instead of 120 degrees. The process of rotation is identical.

For single-phase devices this is not true. When power is applied to the stator of the single-phase motor, the magnetic pole of the stator attracts the magnetic pole of the rotor. As the cycle progresses, this attraction continues with the rotor remaining stationary. Some method must be incorporated into the motor's design to cause a rotating magnetic field in order to have the rotor begin to turn. How the rotating magnetic field is obtained is what distinguishes one fractional-horsepower, single-phase motor from another. This topic will be discussed in Chapter 7.

Questions

1. Who provided the first reliable source of electrical power? What was this source?
2. Who first noticed the relationship between magnetic electric current and magnetic fields?
3. Who received the first patent for an electrical motor in the United States?
4. Explain why electric motors are so important in modern society.
5. Define the term "electric motor."
6. How many watts are there in 1 horsepower?
7. How many classifications are there for the purpose of a motor?

8. Generally, motors greater than 1 horsepower are ______-phase motors.
9. The speed of ______ motors is easier to adjust than that of ______ motors.
10. What two factors determine the synchronous speed of an AC motor?
11. What would be the synchronous speed of a 240 volt/60 hertz, six-pole motor?
12. Define the *slip* of a motor.
13. When are the electrical and mechanical degrees of a motor equal to each other?
14. Standard operating procedure is to have a motor rotate in the clockwise direction when viewed from the shaft end. True/False.
15. What is the approximate efficiency of a motor that has 1120 watts of input and develops 1.25 horsepower output?
16. Define *torque*.
17. What type of device is recommended for measuring the torque of a motor?
18. What is the purpose of a tachometer when used with motors?
19. If a motor's output is rated at 1 horsepower, what would be its output if the voltage was reduced 10%?
20. List the six classifications of motor losses.
21. List some of the reasons why a motor has greater losses than a transformer?
22. How is the copper loss in the rotor related to slip?
23. Why is the efficiency of a motor so important?
24. The thermal protection on a motor is coded "X." What does this mean?
25. What class of insulation has the highest operating temperature?
26. Why is the time rating on a motor important?
27. If a motor has a service factor of 1.15, what would this mean?
28. What are the essential parts of all motors?
29. There are two types of rotors. Name them.
30. There are two basic types of bearings used with motors. Name them.
31. The input of a motor is measured in ______, and the output is rated in ______.
32. What is the principle of operation for a thermal relay?
33. How is the *north pole* of a magnet defined?
34. Like poles of a magnet ______, and unlike poles ______ each other.
35. Magnetic flux lines will take the path of least ______.
36. In order for the rotor of a motor to turn, a ______ must be present.
37. The north pole of an electromagnet can be determined using the ______.
38. Direction of current flow in the armature of a generator can be determined using the ______.
39. Direction of the rotation of the armature of a motor can be determined using the ______.
40. When a conductor is moving parallel to flux lines, ______ volts will be induced.
41. What is the purpose of the commutator on a DC generator?
42. On what principle does a DC motor operate?
43. Commutation should take place at the ______ plane to reduce arcing.
44. How can a single loop DC motor be modified to assure it will be self-starting?
45. The phase displacement for two-phase and three-phase motors is ______ and ______ degrees, respectively.

CHAPTER 7

Fractional-Horsepower Motors

There are many types of motors on the market today that the electrician is required to install, maintain, repair, or replace. Some of these motors are very small and are classified as fractional-horsepower (FHP) motors. These motors are under 1 horsepower. The vast majority of motors in use today will be fractional horsepower motors.

Power Requirements

Motors are designed to operate on either AC, DC, or on AC or DC (universal motors). Voltage requirements can be as low as 1.5 volts for FHP motors and as high as 6600 volts for multiple-horsepower motors. Alternating current motors are single phase or polyphase.

The standard frequency for AC motors in this country is 60 hertz or 60 cycles per second. Other frequencies may be encountered due to special applications.

Motors under 1 horsepower are normally single-phase AC or DC motors. Those over 1 horsepower usually operate on three-phase AC or DC power. Motor designs used primarily for FHP ratings may exceed the 1 horsepower classification. This is particularly true when we talk about DC motors.

Most three-phase motors are designed with multiple-horsepower ratings. They are, however, also rated below 1 horsepower. In either case, the basic design of the motor and its operating theory will be the same. Three-phase motors are investigated in Chapter 8.

Some motors are of the dual-voltage type. The most common rating for FHP AC motors would be 115/230 volts. These motors have two windings for each stator pole. The windings are wired in series for the higher voltage and in parallel for the lower voltage.

Speed of Operation

Motors operate at various designed speeds. There are five classifications of motor speed. These are constant, adjustable, multispeed, varying, or adjustable-varying speed.

1. Constant speed motors have approximately the same RPM from no-load to full-load.
2. An adjustable-speed motor can have its speed varied over a wide range. Once adjusted, however, the speed remains relatively constant provided the load change is within the limits of the rating of the motor. There is a minimum speed at rated voltage and temperature rise at which these motors may be operated. This is defined as the *base speed*.
3. Multispeed motors operate at two or more definite or fixed speeds. Each of these speeds remain relatively constant with change in load. Most multispeed motors manufactured today are not in fact truly multispeed. They are really multihorsepower. The windings are tapped to produce a weaker magnetic field. The weaker field allows the motor under load

to have a greater slip, and thereby run at a lower speed.

For example, if you were to replace a half-horsepower multispeed motor with one of 1 horsepower, the chances are the motor will run at the same speed for all settings of the taps. The load would not be great enough to cause sufficient slip to change the RPM. Speed of a motor can be checked with a tachometer.

Some motor manufacturers still provide multispeed motors by having two or three sets of field windings. This method, however, is more complicated and expensive.

4. Varying speed motors change speed with load. Usually the speed decreases with an increase in loading.
5. Adjustable varying speed motors have the capability of gradual adjustment of speed. Once the speed is set, the motor will run at the set speed unless the load changes.

Synchronous Speed

The *synchronous speed* of an AC motor is determined by the frequency of the source and the number of poles. The RPM is calculated by multiplying the frequency times 60 and dividing by the number of pairs of poles. This was explained in Chapter 6. Some motors are designed to operate at synchronous speed.

Actual speed of the induction motor will be less than the synchronous speed. The reason for this is if the armature turns exactly at synchronous speed, then the magnetic field of the stator windings no longer cuts the rotor windings. Under this condition, no current would be induced to the rotor windings. With no current flow in the armature, no magnetic field will be developed.

When this happens, there is no longer any torque on the rotor, and its speed drops below the synchronous speed. When the speed is reduced, magnetic flux lines produced by the stator windings again cut the rotor's windings, inducing a current and the resulting magnetic field. Torque is again established.

Actual RPM is approximately 3 to 5% less than synchronous speed. Table 7–1 provides some typical comparative values between the synchronous speed and the actual speed of an induction motor.

TABLE 7–1. RPM of Induction Motors

Pairs of Poles	Synchronous RPM	Actual RPM	Slip (RPM)
1	3600	3492–3420	108–180
2	1800	1746–1710	54– 90
3	1200	1164–1140	36– 60
4	900	873– 855	27– 45

Reversible Motors

Another feature of motors is the ability to reverse their direction of rotation. Some motors are reversible and others are not. How the direction of rotation of a motor can be accomplished is described for each of the motors investigated in this text.

Torque

Torque varies among different types of motors. The load will to a great extent determine the type of motor to use to meet torque requirements. The various definitions for torque are explained in Chapter 6.

Summary

In order for an electrician to be an effective and efficient motor mechanic, he or she must know the characteristics of the different motors, their applications, wiring connections, and how the motor operates. These factors are examined for several types of FHP induction motors in this chapter. After these topics are explored in some detail, the FHP DC motor and the universal motor are described.

Split-Phase Motor

The split-phase motor was one of the first types of single-phase motors used industrially. It is still popular. These motors are used for machine tools, blowers, pumps, washers, dryers, and for a variety of other applications. The majority of these motors are manufactured in the 1/20 to 1/3 horsepower range.

Principles of Operation

All single-phase motors have the common problem of how to start them. Figure 7–1 depicts what happens when voltage is applied to the run windings of an induction motor.

When an AC voltage is applied to the run windings of an induction motor, current passes through the windings. This causes a changing magnetic field to expand, contract, and change polarity in phase with the current. The run windings act as a primary of a transformer. Its magnetic field induces a current into the secondary, the armature of the motor.

The current in the armature creates a magnetic field that is opposite the magnetic field of the run windings. These unlike poles attract each other and prevent the

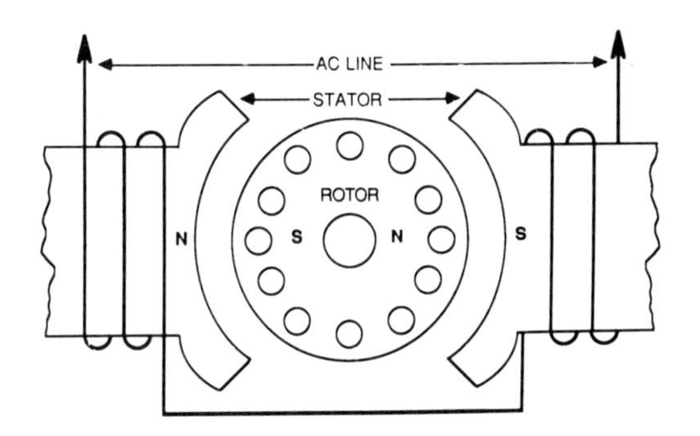

FIGURE 7–1. Magnetic attraction between the running windings, magnetic field, and the rotor's magnetic field.

rotor from turning. Under these conditions, a force exists only in the horizontal plane, with no vertical force to cause the rotor to move.

When the polarity of the magnetic field changes due to the change in polarity of the current, the polarity of the rotor field will change in unison. The two fields are constantly attracting each other. Therefore the rotor is locked in position and will not turn.

If the rotor is given a spin by hand, it will continue to turn and gain speed in the direction in which it is turned. The motor will run at an RPM slightly less than synchronous speed.

The reason the armature continues to turn once it is started is that the moving rotor is now cutting flux lines from the stator. This induces a voltage with a resulting current. The impedance of the rotor is almost totally reactive. The resulting current lags the voltage by almost 90 degrees. This current causes a magnetic field that is 90 electrical degrees out of phase with the run windings' field. Because of this generator action in the armature, a rotating magnetic field is created which is a necessary ingredient to keep it turning.

It is not very practical to start a split-phase motor each time by giving it a turn. Therefore, some way of automatically starting the rotation needs to be incorporated into the motor's design to do this.

Stator Windings

The practical split-phase motor has two sets of windings on the stator. The first set is the main windings or the run windings. The second set is the auxiliary or start windings. These two windings are connected in parallel with each other. The line voltage is applied across both when the motor is energized. Figure 7–2 shows this new arrangement, while Figure 7–3 depicts the electrical circuit for the pictorial.

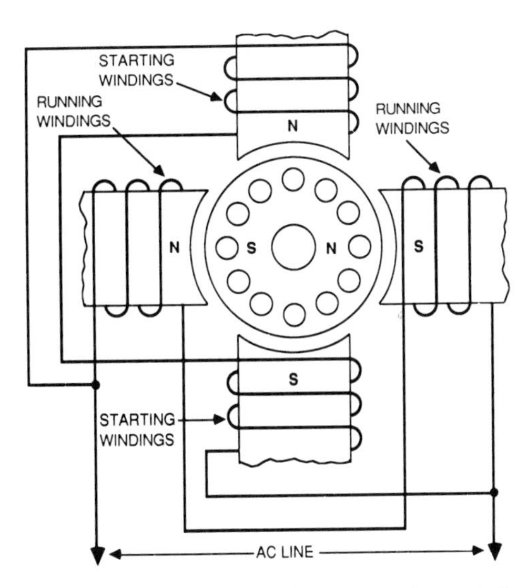

FIGURE 7–2. Displacement of the start and run windings for a two pole, split-phase motor.

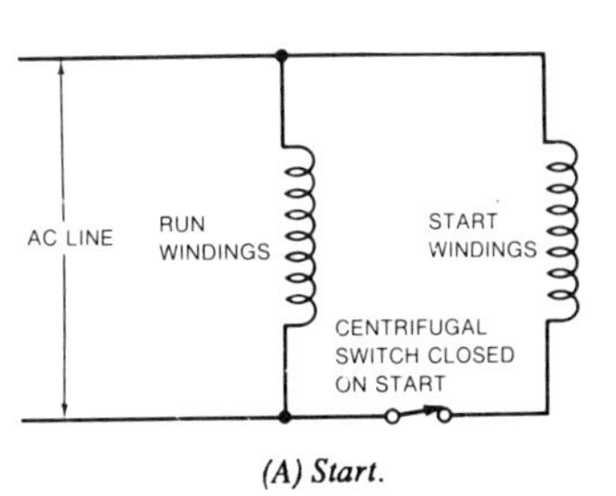

(A) Start.

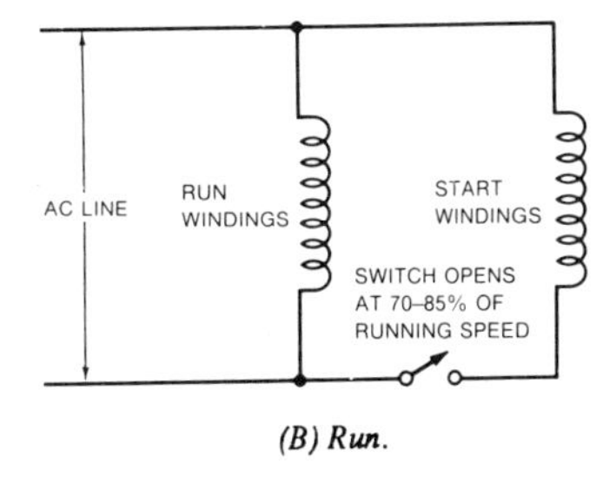

(B) Run.

FIGURE 7–3. Circuit for the split-phase motor.

These two windings differ from each other both physically and electrically. The run windings are formed from larger wire and have more turns than the start windings. They are located in the bottom of the slots on the stator with the start windings wound on top. The start windings have fewer turns formed with a smaller size of wire.

For the two-pole motor, the poles of the run and start windings are 90 mechanical degrees apart. If their windings were electrically the same, the two flux fields would be in phase with each other and combine to form a single field midway between the run and start poles. The rotor would lock in this position with unlike poles attracting each other. This was the case when only the run windings were connected as shown in Figure 7–1.

The two sets of windings do differ electrically. The run windings are more inductive than the start windings. This is due to the fact that the start windings have more resistance than the run windings due to the smaller size of wire used to form its coil.

Figure 7–4 shows the phasor diagrams and sine curves for various conditions when an AC voltage is applied to different loads. The voltage is the reference point at zero degrees. This is because the two windings are connected in parallel with each other. The voltage is the same across a parallel circuit. Therefore, the voltage across the run and start windings will always be in phase.

Parts A and B of Figure 7–4 show two extreme conditions for when the circuit is only resistive or purely inductive. In a resistive circuit, current and voltage are in-phase with each other. Resistance merely opposes current flow. On the other hand, if the circuit were purely inductive, the current would lag the voltage by 90°. The self-induced voltage across the inductor will always be in such a direction to oppose the change in current.

Under normal operating conditions, however, a motor is neither purely resistive or purely inductive. All wire used to form coils has resistance. In the case of the split-phase motor, the run winding has less resistance than the start windings. Therefore, the current through the run windings will lag the current through the start windings by a greater angle. This difference in angular displacement is shown in Parts C and D of Figure 7–4.

Part E combines Parts C and D to show the phase relationship between the currents through the two windings. The resulting phase displacement of current between the two windings is approximately 30°. This is less than the 90° displacement for a two-phase motor or the 120° shift of three-phase power. It is, however, sufficient to create an artificial two-phase rotating magnetic field to cause the motor to start. These two currents when combined are equal to the current supplied by the line.

Regardless of the number of poles a motor has, the displacement in electrical degrees between the main and auxiliary poles will always be 90 electrical degrees. The mechanical degrees will vary according to the number of poles.

For a two-pole motor, the poles of the start windings is displaced from those of the run windings by 90 mechanical degrees. This value would be the same as the electrical degrees. For a four-pole motor, however, the spacing between the two sets of poles will be 45 mechanical degrees. For a six-pole motor this value is reduced to 30 degrees. In each case, the poles of the start windings is halfway between the poles of the run windings, or 90 electrical degrees apart.

Centrifugal Switch

The centrifugal switch is connected in series with the start winding (see Figure 7–3). This switch is normally closed when the motor is at rest. When starting the motor, it remains closed until the motor reaches approximately 70% to 80% of its operating speed. At this point, the centrifugal force of the rotor's mechanism is sufficient to overcome the spring tension which holds the switch closed, and it opens. The exact speed at which this action takes place depends on the weights on the actuator and the tension of the spring(s).

This action removes the start windings from the circuit. The motor will continue to run due to the induced current into the rotor's windings due to its motion.

If the centrifugal switch is not closed when the motor is energized, the motor will not start. It will hum loudly and draw excessive current. Because the rotor does not turn, its impedance is approximately equal to the resistance of the rotor bars, and it acts as a shorted secondary of a transformer. This in turn causes the primary circuit of the transformer, the run windings, to draw too much current. Leaving the motor in this condition for too long a time can cause the run windings to overheat.

Failure of the centrifugal switch to open in a very short time after the motor is started also has dire consequences. The start windings would remain in the circuit, and the motor will run irregularly at a lower RPM and torque than for which it was designed. This will result in an excessive current being continually delivered to the motor.

Because the start windings have a higher resistance, they will immediately begin to overheat and destroy

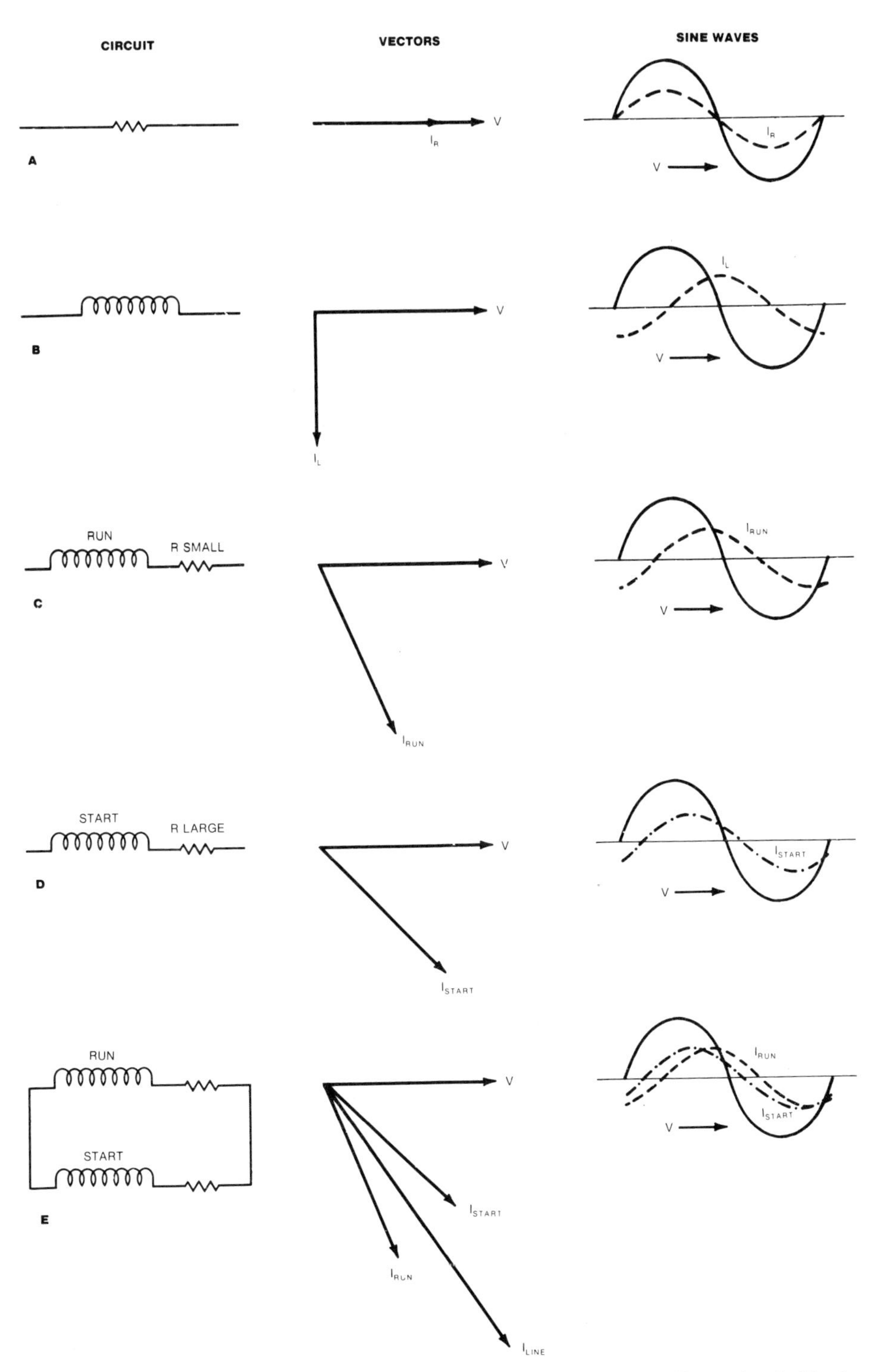

FIGURE 7–4. Phasor diagrams and sine curves of voltage and current with different electrical loads.

their own insulation qualities. Ground faults or shorted turns may occur in as little as 5 seconds under this condition. If the condition continues even for a longer time, the run windings may also be damaged. The net result is replacing or rewinding the motor.

The centrifugal switch plays a very important role in the proper operation of the split-phase motor. When it malfunctions, damage can result almost immediately. This device must be checked and serviced during routine maintenance procedures.

Current Relay

To provide added protection if the centrifugal switch does not open on the split-phase motor, a current relay may also be incorporated in the design in addition to or in place of the centrifugal switch. A *relay* is an electromagnetic device with the coil in series with the run windings. A set of normally open contacts on this relay is placed in series with the start windings. Figure 7–5 illustrates this arrangement.

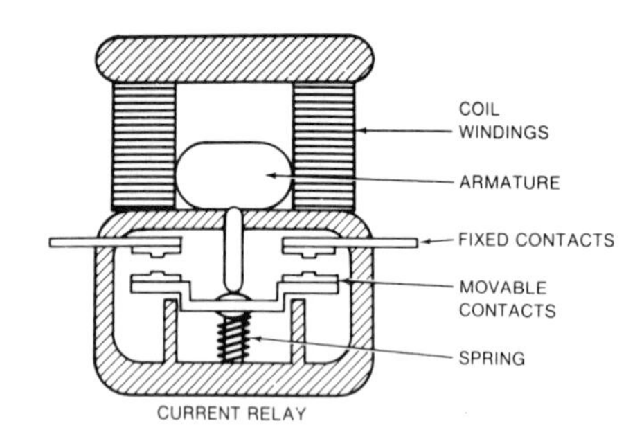

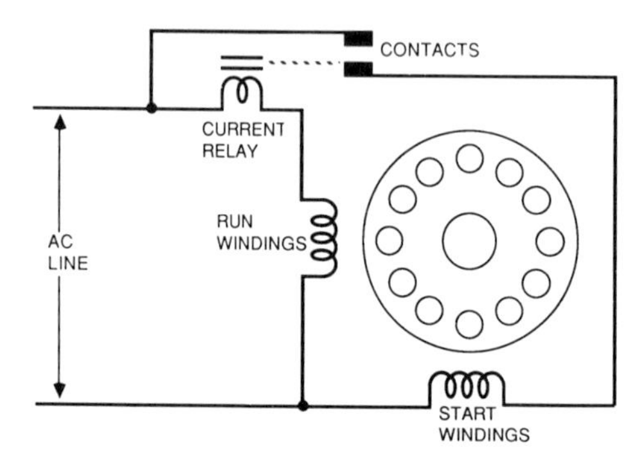

FIGURE 7–5. Current relay used for motor starting.

What makes this operation effective is the starting current delivered to the motor before it comes up to its operating speed. Starting current is in the range of 300% to 600% of the normal running current.

This high current flows through both the run windings and the coil of the current relay. The relay coil activates. This action closes the circuit to the start windings. As the motor comes up to speed, the current through the coil drops. The coil de-energizes. The contacts open and disconnect the start windings.

The pictorial in Figure 7–5 shows the construction of this relay. The higher starting current draws the armature of the relay up into the coil due to magnetic attraction, allowing the contacts to close. As line current drops with the increasing RPM of the motor, there is not sufficient magnetizing force to hold the armature. The armature releases, allowing power to be disconnected from the start windings.

Current relays are extensively used on hermetically sealed motors such as those used in air conditioning and refrigeration. These motors are built so that they are exposed to the refrigerant gas. Exposing a centrifugal switch to the refrigerants is not feasible, and the switch would not be accessible for maintenance. The current relay, however, can be mounted externally from the sealed motor. This removes the device from contact with the gas and provides accessibility.

Current relays are selected very carefully by the designer of the motor. This is to assure that the start windings do not drop out of the circuit at too low of a speed. At the same time, the start windings cannot remain in the circuit for too long a time. This would result in the windings overheating. The temperature rise can cause damage to the motor.

When replacing the current relay, an exact replacement number is recommended. If this is not possible, the specifications of the relay need to be matched as closely as possible with the new relay.

Split-Phase Motor Connections

NEMA has established a standardized numbering and color coding system to assist in the identification of the leads on motors. The number of each lead is prefixed with a "T." Figure 7–6 depicts the code for the split-phase motor having one main winding and one auxiliary winding.

Figure 7–7 shows the same motor when thermal protection is provided. The leads connected to the thermal protector are designated with the prefix "P."

Figure 7–8 provides the NEMA standard for the split-phase motor with dual-run windings and thermal protection. This design allows for dual-voltage connections.

When the leads are all one color, usually black, they will be tagged using the number designations shown in

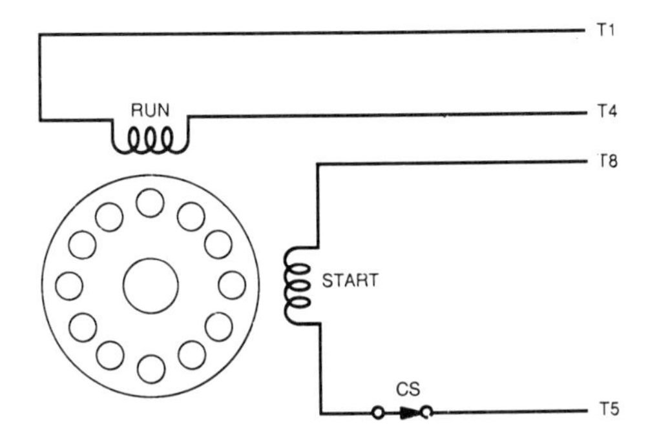

FIGURE 7–6. NEMA standardized numbering system for a split-phase motor having one main and one auxiliary winding.

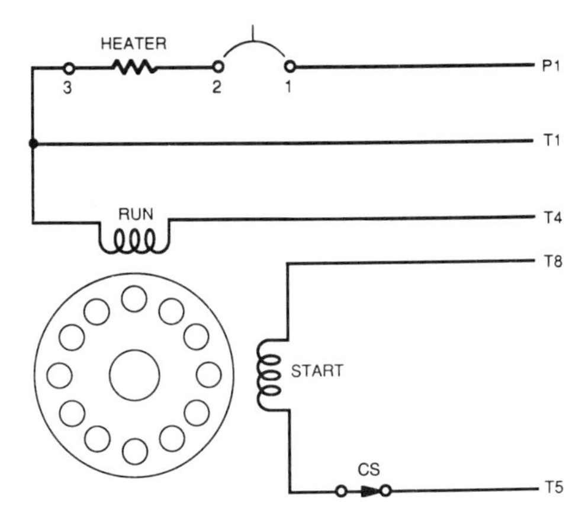

FIGURE 7–7. NEMA standardized numbering system for a single-voltage split-phase motor with thermal protection.

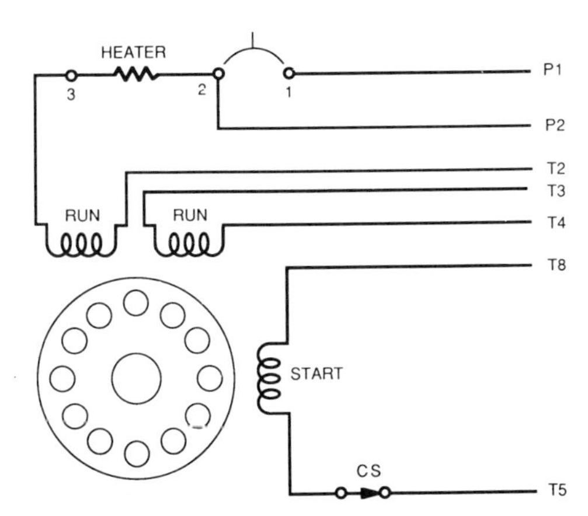

FIGURE 7–8. NEMA standardized numbering system for a dual-voltage, split-phase motor with thermal protection.

Figures 7–6, through 7–8. If the leads are color coded instead, NEMA has designated the following combinations.

T1—Blue	T5—Black
T2—White	T8—Red
T3—Orange	P1—No color assigned
T4—Yellow	P2—Brown

It is not unusual to find the markings on the leads or their colors faded on older motors. In this case, the run and start windings can still be identified using an ohmmeter. Pairs of leads can easily be identified this way. The same instrument may be used to differentiate the run windings from the start windings. The run windings will normally have a lower resistance than the start windings.

On larger motors, an ohmmeter may not be sufficiently sensitive on the low ohm scale, and a Wheatstone bridge instrument or a shunt-type ohmmeter would be needed.

If the leads cannot be identified using test instruments, a less convenient method of visual inspection can be used. In order to do this, the end plates of the motor will have to be removed.

The run windings will normally be of a larger size of wire and in the bottom of the slot with the start windings on top of them. One side of the start windings will be connected to the centrifugal switch. One side of the run windings will be connected to the thermal breaker if the motor has one. A continuity check can then be made using an ohmmeter to find the matching leads.

It should be noted that standards for marking the leads have not been developed for some definite-purpose and multispeed motors. It is not practical to do so for the definite-purpose motors, and in the case of the multispeed devices, there is a great variety of methods employed to obtain the speed changes.

Reversibility

A split-phase motor is normally wired when it comes from the supplier to run in the counterclockwise direction as viewed from the front, or bell housing end. Figure 7–9A shows this wiring arrangement.

The motor can be made to operate in the clockwise direction by a wiring change. This is generally achieved by reversing the direction of current flow through the start windings. To reverse current flow in these windings, merely interchange the two connections of the start windings to the power line. Figure 7–9B shows leads T5 and T8 interchanged between L1 and L2.

The same results can be obtained by reversing the direction of current flow through the run windings. Figure 7–9C illustrates this change. The same procedure is used as with the start windings, only in this case, T1 and T4 are switched between L1 and L2.

In summary, the direction of rotation of a split-phase motor can be reversed by *either* reversing the run windings' connections *or* the start windings' connections, but not *both*. The latter procedure is the same as merely reversing the plug in the receptacle. Because both currents are reversed at the same time through the two windings, the direction of rotation remains the same (see Figure 7–9D).

Dual Voltage

Some split-phase motors are designed to operate on either 115 or 230 volts. These motors will generally have two main (run) windings and a single auxiliary (start) coil. Figure 7–10 depicts this configuration.

Figure 7–10A illustrates the wiring when the motor is operated at 115 volts. The coils of the run windings are connected in parallel with each other for the lower voltage. Note that only the current of one of the run windings is passing through the heater of the thermal breaker for the low-voltage arrangement.

Figure 7–10B depicts the motor wired for 230 volts. The run windings are now connected in series with each other. In this case, the heater of the circuit breaker is in series with the run windings, and the same amount of current is available to it as for the low-voltage connection.

Regardless of which voltage is used, the same amount of current flows through the run windings. The net result is that the current ratings of the windings is

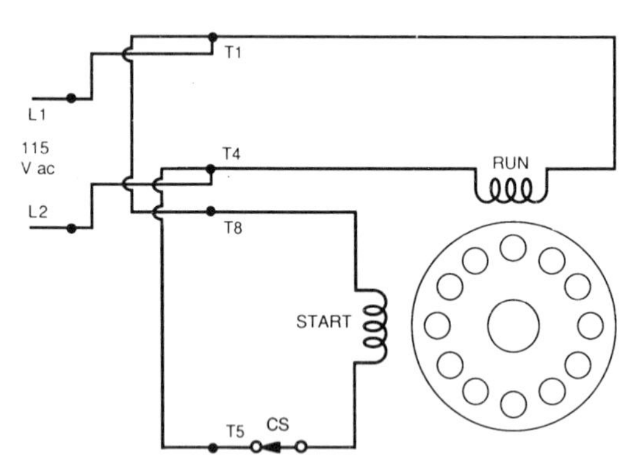

(A) Factory wired; counterclockwise rotation; L1 to T1, T8; L2 to T4, T5.

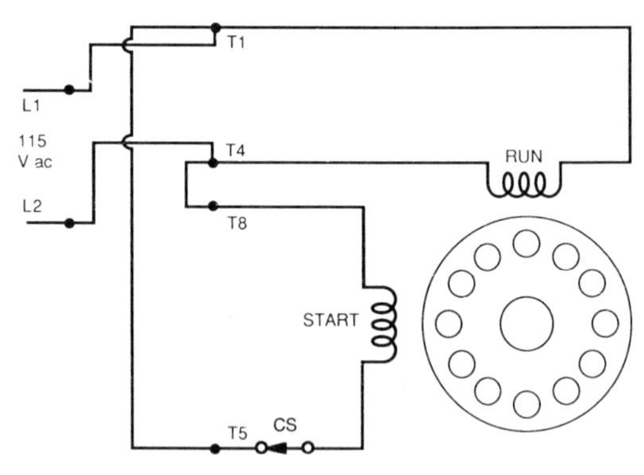

(B) Clockwise rotation; reverse current through start windings; L1 to T1, T5; L2 to T4, T8.

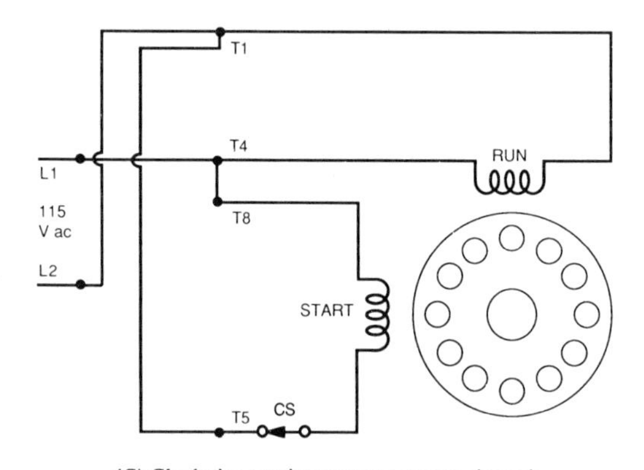

(C) Clockwise rotation; reverse current through run windings; L1 to T4, T8; L2 to T1, T5.

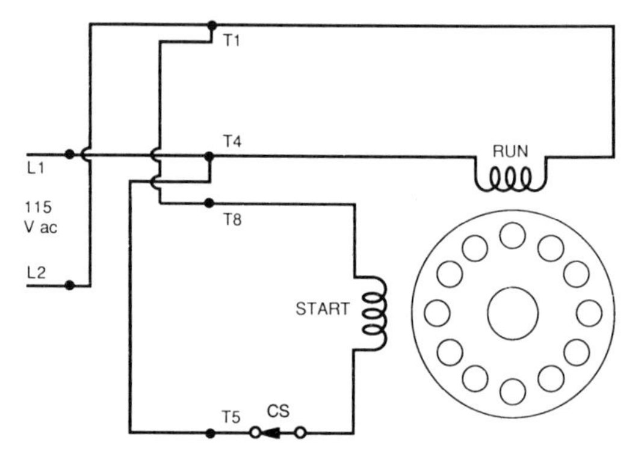

(D) Counterclockwise rotation; reverse direction through run and start windings; L1 to T4, T5; L2 to T1, T8.

FIGURE 7–9. Reversing the direction of rotation of a split-phase motor.

not exceeded, and the voltage drop across each winding is always the same. Each of the two methods results in the same flux density, and the same horsepower is developed. The advantage of the higher voltage is that the source will need to deliver only one-half the current as compared to the low-voltage selection.

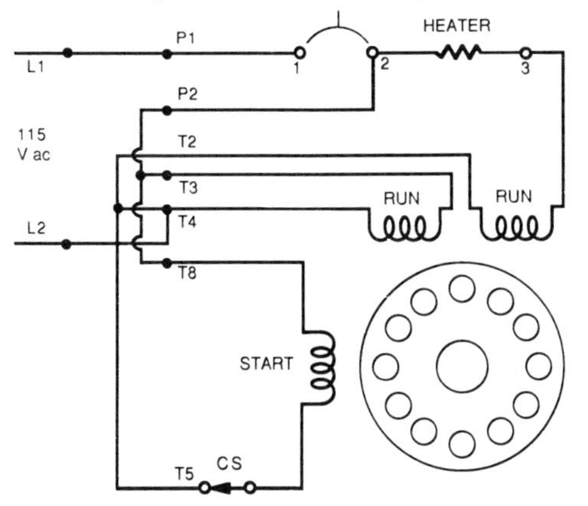

(A) Split-phase motor wired for lower voltage; L1 to P1; P2 to T3, T8; L2 to T2, T4, T5.

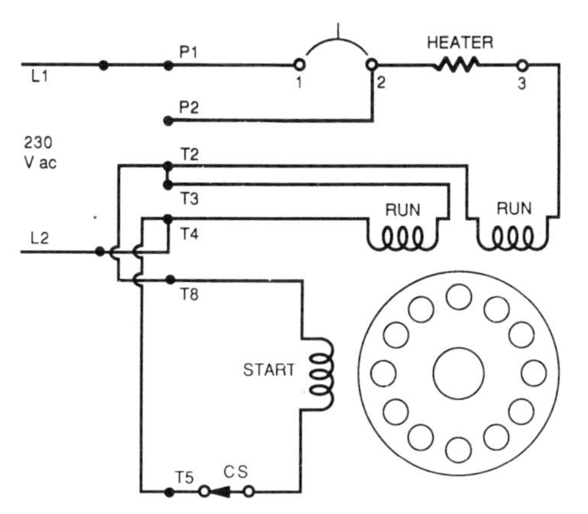

(B) Split-phase motor wired for higher voltage; L1 to P1; L2 to T4, T5; T2 to T3, T8.

FIGURE **7–10.** Dual-voltage split phase motor.

Two-Speed, Split-Phase Motor

The split-phase motor by its nature runs at the synchronous speed minus the slip. The synchronous speed is determined by either the frequency of the power source or the number of poles incorporated in the motor's design.

Changing the frequency is less feasible than changing the number of poles. The standard frequency of the power source in the United States is 60 hertz, and it seldom varies more than a fraction of 1 cycle per second.

Even if we were to vary the frequency to the motor, it would present some complicated design problems. Lowering the frequency to make the motor run slower would also lower the inductive reactance ($2\pi fL$) and the resultant impedance of the windings. This would cause the motor to draw more current and provide less torque than that developed at the higher speed. The higher current would increase the copper loss, even if the windings were of sufficient gauge to prevent them from overheating.

Increasing the frequency would increase the inductive reactance and increase the impedance. This would lower the current through the windings, thereby lowering the flux density and the horsepower of the motor. This change would normally be unacceptable.

For these reasons, frequency change to obtain a different speed is not normally considered. Instead, the number of poles are changed. There are three commonly used methods to achieve this. These are

1. Add an additional run winding but not an additional start winding.
2. Add a run and a start winding to the design.
3. Use the consequent-pole concept without the addition of any other windings.

Three windings are used on the stator with Method 1. There are two run windings and one start winding. These motors are usually wound for six and eight poles with speeds of 1150 and 875 RPM. The most common use of these motors is to drive electric fans. Figure 7–11 shows a schematic diagram for this method.

A single-pole switch is used to select one of the two speeds or to turn the motor off. This motor always starts at its high speed. When the low speed is selected when starting the motor, the centrifugal switch is connected to the line through the LOW position of the selector switch. This applies power to the starting windings and the six-pole run windings (high speed). As the motor approaches running speed, the centrifugal switch changes position. This change disconnects the start and six-pole windings and connects the eight-pole run windings to the power source.

If the motor is started in the HIGH position, the power source is connected through the selector switch directly to the six-pole windings. The six-pole contact on the centrifugal switch also receives power which places voltage across the start windings. As the motor

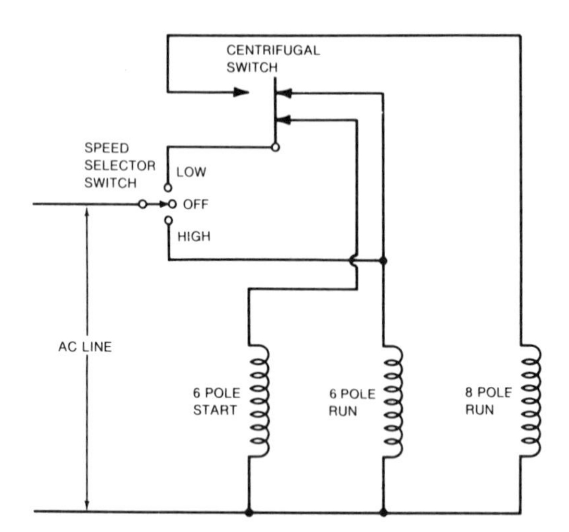

FIGURE 7–11. Two-speed, split-phase motor with two run windings and one start winding.

comes up to speed the centrifugal switch opens, removing power from the start windings.

The second method of obtaining two speeds with the split-phase motor is to use two run and two start windings (Figure 7–12). This motor has the advantage over that used in method one. It can start on either low or high speed because of the two start windings.

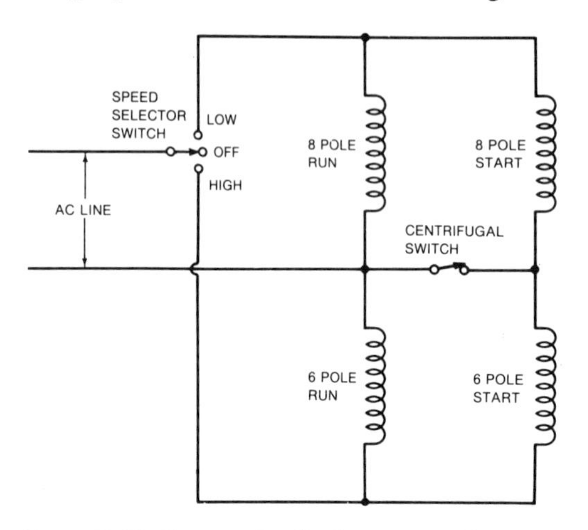

FIGURE 7–12. Two-speed, split-phase motor with two run windings and two start windings.

In this case power is supplied to either the high speed windings or the low-speed windings through the selector switch. Voltage is also provided to the respective start winding through the centrifugal switch. As the motor once again approaches the selected run speed, the centrifugal switch opens and disconnects the start windings.

Consequent-Pole Windings

Consequent-pole motors are more economical to manufacture than the other two methods used to obtain a two-speed motor. The reason is consequent-pole motors require less copper.

Consequent poles are formed when the stator has run windings only on one-half the poles. Figure 7–13 depicts this arrangement using a single pole.

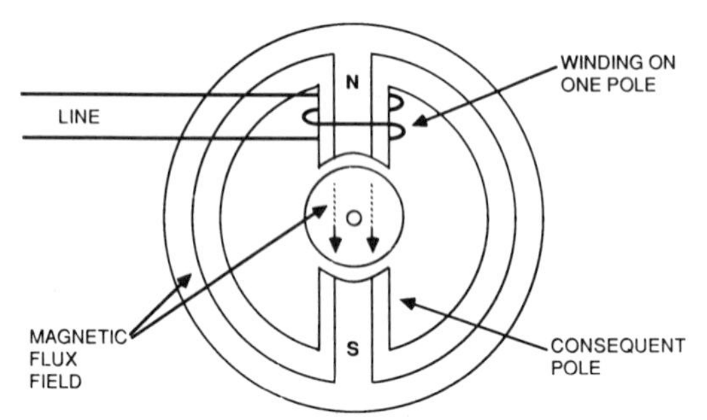

FIGURE 7–13. Consequent-pole motor circuit.

When the coil around the pole is energized with AC, the polarity of the pole will continually reverse with the change in direction of current flow through it. Because it is impossible to have a magnetic circuit with only one magnetic pole, the pole opposite the energized pole will become the alternate pole of the field.

The magnetic field will pass through the rotor circuit inducing currents into the rotor. The frame of the motor provides a low reluctance path for the magnetic field to complete the magnetic circuit.

In Figure 7–14A, the motor is wound as a standard two-pole motor. The stator poles are wound so a north and a south pole are formed when current passes through the windings.

When the double-throw, double-pole switch is reversed in Figure 7–14B, the current through the windings of the bottom pole is reversed. This causes both poles of the motor to have the same polarity at the same time. An equal number of consequent poles will be formed. This changes the motor to a four-pole motor which will run at approximately one-half the speed of the two pole motor.

Summary of Characteristics

Starting current for split-phase motors is usually three to five times the normal full-load current. This

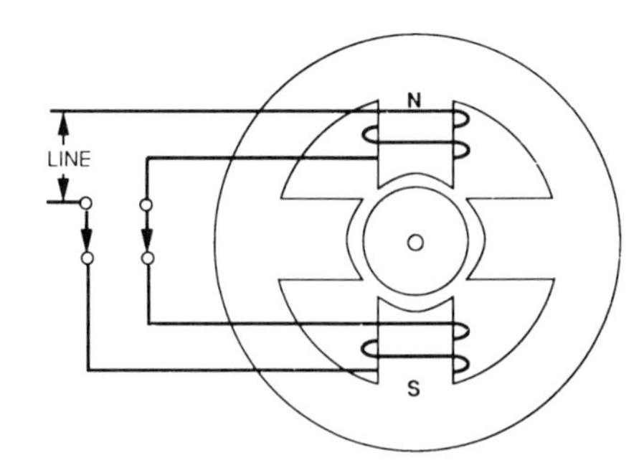

(A) Coils wound for two-pole motor.

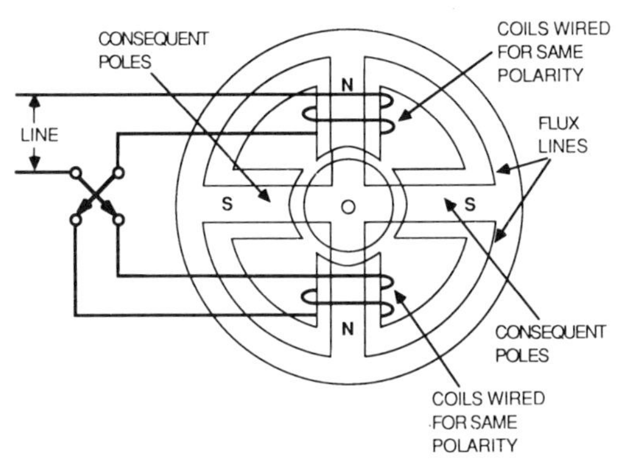

(B) Two-pole motor changed to four-pole motor.

FIGURE 7–14. Changing speed of induction motor using consequent poles.

causes the motor to heat rapidly if it fails to start and may cause the line voltage to drop. A fall in line voltage reduces the motor's starting torque.

The no-load current of these motors ranges from 60% to 80% of the full-load current. This is high when compared to three-phase motors.

Most of the no-load current is used to create the magnetic field. Very little current is used to overcome the motor's losses. Because of the need for a large current to create the magnetic field, these motors have a poor power factor. The power factor is in the range of 60% even at full load.

Split-phase motors are noisier than three-phase motors. This is due to the mechanical vibration due to the 120-cycle alternations as current and magnetic fields reverse.

The speed of these motors is dependent on the number of poles and frequency. Even with reduced voltage under a no-load condition, the RPM will remain relatively constant. Speed change ranges from 3% to 5% from no load to full load at normal operating voltage.

Capacitor Motors

There are three basic types of capacitor motors. These are the capacitor-start motor, permanent-split capacitor motor, and the two-capacitor motor.

Capacitor motors are single-phase machines designed to produce a high starting torque to mechanical loads like power tools and compressors of all types. They are manufactured in sizes ranging from 1/8 to 10 horsepower.

The capacitor motor is very similar to the single-phase induction motor. They both have a start and run windings displaced 90 electrical degrees apart.

A capacitor motor differs from the split-phase motor in that it has a capacitor placed in series with its start winding. This additional component provides for a higher starting torque at a lower current.

Capacitors

Capacitance is defined as the ability of a component to store an electrical charge. The capacitor is the only known electrical device that can perform this function. A battery, for example, stores chemical energy, not electrical energy. In earlier times, the capacitor was referred to as a *condenser*.

Capacitors are rated in farads and volts. A 1-farad capacitor will store a charge of 1 coulomb of electrons for each volt applied to it. This is expressed mathematically as $C = Q/V$ where Q is the charge in coulombs and V is the voltage in volts. Only recently has the technology been available to manufacture capacitors of one farad rated at several volts. Prior to this the physical size of these devices prohibited their practical application. Almost all capacitors used in motor circuits are rated in microfarads and volts AC.

The physical construction of a capacitor consists of two conductors separated by an insulator. This means that all electrical components have some capacity. Electrical wires, for example, consist of a piece of copper wire surrounded by an insulating material to prevent them from shorting. Wires run parallel to each other form capacitors. Capacitance also exists in respect to ground, which is also a good conductor. It is almost impossible to escape from the effects of capacitance in any electrical circuit.

Capacitors are classified according to the type of dielectric used as the insulator. Some of the common types of dielectrics used in the manufacturing of capacitors are air, paper, ceramic, oil, glass, and mica. Each of these materials differ in their dielectric constants and dielectric strength.

Air is given a dielectric constant of unity, or 1. Titanium dioxide compounds when used as insulators may be as high as 170.

Dielectric strength is rated in volts per 0.001 inch. Air has a dielectric strength of 80, and mica has the highest rating of 2000. This determines the maximum voltage rating that a capacitor will withstand based on the material used and its thickness. When this voltage is exceeded, the molecules of the dielectric breakdown, and the capacitor begins to conduct electrons through the dielectric. This process will destroy the capacitor, and it will have to be replaced.

Given the physical parameters of a capacitor, the farad value, can be determined using the following formula.

$$C = \frac{KA(N-1)}{T}$$

where,

C is in farads,
K is the dielectric constant of the material,
N is the number of plates used in the capacitor,
A is the area of the plates,
T is the distance between plates.

As can be determined from the formula, the larger the area of the two conductors, and the closer they are spaced together, the greater the capacity of the device.

Increasing the area of the conductors will result in a larger physical size of the finished product. Decreasing the spacing between the conductors raises the value of the farad rating, but at the same time reduces the voltage rating. Therefore, there is always a compromise to be made among farad rating, voltage rating, and the physical size of the component.

The ratings for capacitors used in capacitor motor circuits are usually calculated for 60-Hz operation. These capacitors are oil-filled and electrolytic types. The electrolytic capacitors are used in the start cycle, and the oil-filled capacitors are used during the run cycle.

Electrolytic capacitors are specially designed devices which have relatively high capacitance for their size. This is achieved by having a very thin dielectric. The electrolyte oxidizes the sides of the aluminum foil which come into contact with each other in the capacitor. The oxidation forms a high resistive dielectric which allows the two conductors to be very close to each other, thereby creating a high capacitance. They must be used within their voltage and design ratings.

Unlike most other capacitors, electrolytic capacitors are polarized. When a reverse polarity is applied, they will conduct heavily and quickly overheat.

This action is similar to that of a solid state diode. The diode conducts heavily with one polarity applied. It has a high resistance when under reverse polarity.

Unlike the diode, the capacitor will overheat and explode if the polarity which causes the high leakage is continued for a very short period of time. The explosion is like that of a shotgun shell. The interior parts will blow out of the crimped end of the capacitor. This action can cause injury to any person working on the device. The eyes are particularly in danger due to the hot electrolytic solution. Polarity of the electrolytic capacitor must be observed when used in DC circuits.

Electrolytic capacitors used in motor starting circuits are subjected to reverse polarity due to the alternating current. These devices are specially designed to withstand the reverse currents during the start cycle and do not have their polarity marked. They are referred to as "dry-type" capacitors. A very thin sheet of gauze absorbs the electrolytic. The gauze is placed between the aluminum foil of the conductors. For the capacitor to remain functional, the gauze must remain moist. The electrolytic is not in the liquid form under this condition.

Electrolytic capacitors used in motor circuits have a vent hole. This allows any pressure created by the heating of the capacitor to escape without exploding the capacitor.

Direct current electrolytic capacitors must *never* be used in AC circuits. Do not use a capacitor rated in volts DC. Always use the capacitor designed for the AC motor on which you are making repairs. The device will have an AC voltage rating. For 115-V motors the rating will be 130 to 150 V AC, 60 Hz. For 230-V motors, the rating will be 250–300 V AC, 60 Hz.

The farad values will range from 20 to 550 microfarads. This is the minimum values of capacitance. The actual values may be 20% to 50% higher.

Motor start capacitors are also rated for a duty cycle. The common value is twenty, 3-second periods per hour. Equivalent duty cycles may be given, such as sixty 1-second cycles.

When capacitance is added to a circuit, it opposes changes in voltage. The current in this circuit leads the voltage across the capacitor by 90°. That is, electrons must first flow to and collect on one of the plates of the capacitor. This sets up an electrostatic field which forces electrons on the other plate to return to the source. This action is based on the laws of electrostatics which states that like charges repel each other.

The electrons do not flow through the dielectric of the capacitor. The plate that has an excess of electrons is the negative plate, and the plate with a deficiency of electrons becomes the positive plate. This process creates a polarized potential across the capacitor. Figure 7–15 illustrates the current flow in a capacitive circuit.

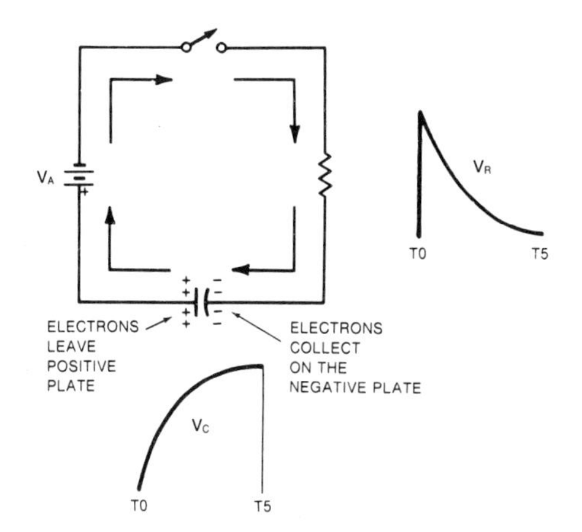

FIGURE 7–15. Current flow in a capacitive circuit.

At the instant the switch is closed, t_0, the current will be maximum and equal to V/R. The IR drop across the resistor will be equal to the source voltage, and the voltage across the capacitor will be zero.

As time passes, the voltage charging it. This causes the current to fall as shown by the waveshape of the voltage drop across the resistor. The value of current at any point in time will be equal to $(V_a - V_c)/R$, and $V_c + V_r = V_a$. When the capacitor has fully charged to the value of the source voltage, no further current flow will take place. The voltage drop across the resistor will be zero. This happens after approximately five time constants.

A time constant is equal to

$$T = RC$$

The current will decay and the voltage across the capacitor will increase at an exponential rate. After one time constant, the voltage across the capacitor will equal 63.2% of the source voltage, and the current will have decreased to 36.8% of its maximum value. This process continues until the current falls to zero, and the voltage across the capacitor is equal to the applied voltage.

The same process takes place in an AC circuit. The charge across the capacitor that opposes the change in the source voltage is what causes the capacitor to have reactance and oppose current flow. This AC reactance can be expressed mathematically as

$$X_c = 1/(2\pi fC)$$

Figure 7–16 shows the sine wave relationships in an AC series RC circuit between the current and the voltages across the resistor and capacitor. Current is the same at all points in a series circuit. It is used as the reference at zero degrees. The voltage developed across the resistor is in-phase with the current, and the voltage across the capacitor lags the current by 90 degrees. The vectors for this circuit are also shown.

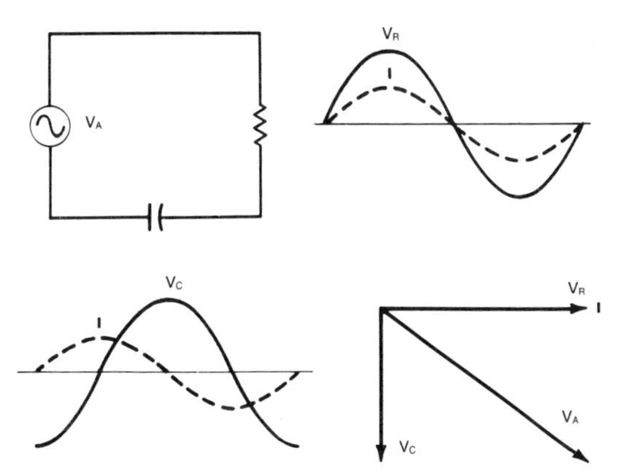

FIGURE 7–16. Phase relationships between current and voltages in an AC series RC circuit.

Capacitor-Start Motors

A split-phase motor can be modified to operate as a capacitor-start motor by placing a capacitor in series with the start windings. The schematic shows this arrangement in Figure 7–17.

Principles of Operation

The introduction of the capacitor in the start windings form a series LRC circuit. This circuit can act as a pure resistance if the inductive reactance is equal to the capacitive reactance. The circuit is said to be at resonance when this occurs. This point is usually carefully avoided for a motor circuit. Under resonance conditions, the voltages across the capacitor and the start

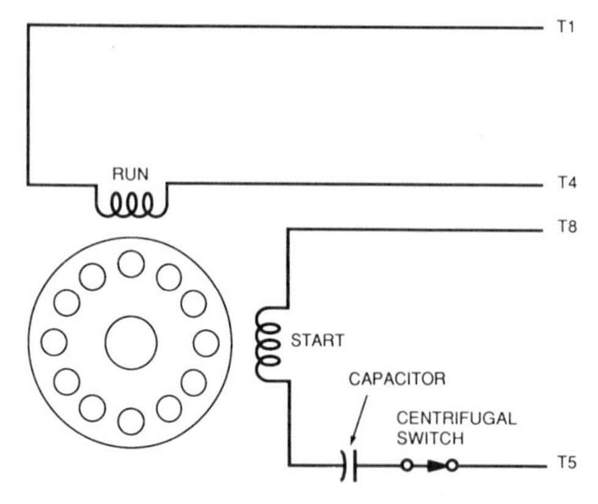

Figure 7–17. Schematic for a capacitor-start motor.

windings can greatly exceed the source voltage. This could result in the dielectric of the capacitor failing, or the insulation on the start windings breaking down under voltage stress.

If the inductive reactance is greater than the capacitive reactance, the circuit will act as a series RL configuration. This was the case for the split-phase motor. However, the intent of the design is to have the circuit be sufficiently capacitive so that the current through the start windings will lead the current through the run windings by 90 degrees. This is accomplished by selecting a capacitor of specific value which will provide a capacitive reactance great enough to overcome the inductive reactance of the start windings. This makes the circuit look electrically as an RC circuit. Table 7–2 provides typical values of electrolytic capacitors used for starting circuits of capacitor-start motors at different horsepower ratings.

Table 7–2. Typical Values of Electrolytic Capacitors for Capacitor-Start Motors

Horsepower	Capacitance (μF)	Voltage	Frequency (Hz)
1/8	72/88	115	60
1/6	88/108	115	60
1/4	108/145	115	60
1/3	161/193	115	60
1/2	216/259	115	60
3/4	378/440	115	60
1	378/440	115	60

Figure 7–18 compares the current vectors for the split-phase motor and the capacitor start motor.

The greater displacement of the currents in the capacitor-start motor provides for the necessary rotating magnetic field to start the motor. The advantage is

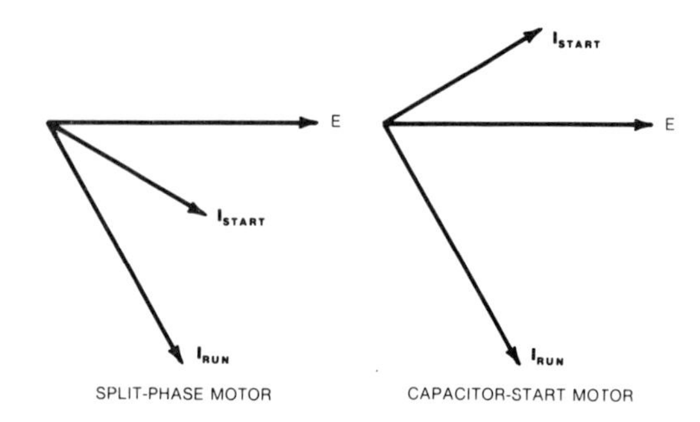

Figure 7–18. Comparison of current vectors between split-phase and capacitor-start motors.

that it will have a higher starting torque than that of the split-phase motor. Once the capacitor-start motor has achieved approximately 70% of its running speed, and the centrifugal switch disconnects the start windings, it is identical in all characteristics to the split-phase motor.

Commercially manufactured capacitor-start motors are not merely split-phase motors with a capacitor added to the start winding. They are carefully designed so the start windings and the capacitor complement each other for optimum results. Like the split-phase motor, the capacitor-start motor has run and start windings with the run windings having larger wire. The starting windings of the capacitor-start motor has more turns than the comparable split-phase motor.

Dual-Voltage Motors

Capacitor-start motors may be designed for dual voltages. When this feature is available, they normally have two run windings and one start windings like the split-phase motor. The run windings are connected in parallel for the lower voltage, and they are connected in series for the higher voltage. The numbering sequence and color coding of leads is the same as that for the split-phase motor. Motors less than ¼ horsepower are usually single-voltage types, and those over ¼ horsepower will usually have the dual-voltage feature.

Thermal Protection

Thermal protection may be provided for the capacitor-start motor. Special purpose motors often have this feature. This will disconnect the motor from the line in the case of overload or overheating. It does not provide for short circuit protection which is designed into the line fusing circuit. When thermal protection is available, the leads have the same numbering sequence as the split-phase motor.

Reversing Direction of Rotation

The direction of rotation can be altered on some capacitor-start motors. Others do not have this feature depending on the application. When the motor can be reversed, both the run windings' and the start windings' leads are brought to the terminal block. Either the run windings' or start windings' leads are reversed to the line leads in order to reverse the motor. This is the same principle used for the split-phase motor.

Capacitor-start motors have some different wiring configurations for reversing their direction of rotation. One of these is the three-lead reversible motor. Figure 7–19 depicts this arrangement.

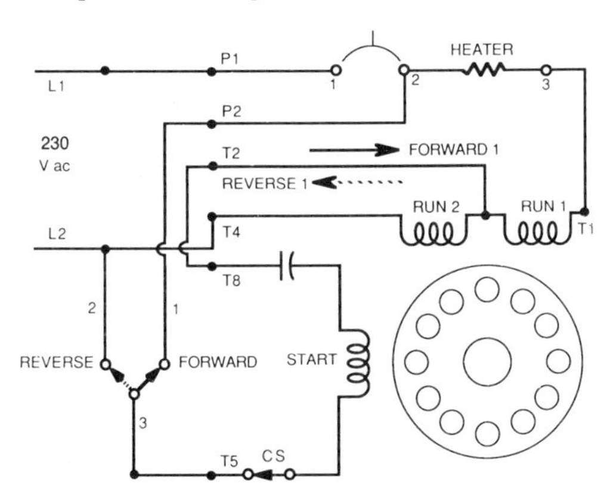

FIGURE 7–19. Three-lead reversible capacitor-start motor.

When only three leads are used, the motor must have two run windings as shown in Figure 7–19. The current flow through the run windings is in the same direction regardless of the direction of rotation. A single-pole, double-throw switch reverses the current flow through the start windings.

If the switch's position is changed while the motor is running, it will have no effect. The motor must be disconnected from the line long enough for the centrifugal switch to close so that power can once again be applied to the start windings. A distinguishable click can usually be heard as the motor slows. This indicates that the switch has closed.

An instant reverse feature is also possible on a capacitor-start motor. This is accomplished using a relay, a drum controller, and by modifying the centrifugal switch so it has a double contact (Figure 7–20).

When the motor in Figure 7–20 is started in the forward direction, the run windings are connected to L1 and L2 through the drum switch. L1 is connected to the top of the start windings through the drum switch, the capacitor, and the double-contact centrifugal switch. L2 is connected through the drum switch. During this period of starting, the voltage across the capacitor is also across the relay which is in parallel with it. This opens the relays contact.

After the motor has achieved approximately 70% of its running speed, the centrifugal switch will change positions. This will remove the capacitor from the circuit. Because the relay contact is open, the relay is now in series with the start windings. The relay's coil has a high resistance. Just enough current flows through it and the start windings to keep the relay energized and the contacts open.

To reverse the direction of rotation of the motor, the handle of the drum switch is thrown from the right-hand position to the left-hand position as shown in Figure 7–20. All the movable contacts are ganged together as indicated by the dotted line among them.

As these contacts are moved, the line connections to the motor are broken instantaneously. This removes power from the relay, and its set of contacts close.

When the drum switch has seated in its new position, L2 is now applied to the top of the start windings through the drum switch and the normally closed contact of the relay. L1 is connected to the bottom of the start windings. This arrangement has reversed the current through the start windings in respect to the run windings. Current flow through the run windings is still flowing in the same direction as before.

This action applies reverse torque on the motor, bringing it to a rapid stop. The centrifugal switch returns to the position as shown in Figure 7–20. This puts the capacitor in series with the start windings. The motor is once again started using the capacitor, but now its rotation is in the reverse direction. When voltage is again applied across the relay, the cycle is repeated.

Drill presses and small lathes use capacitor motors with these features. They must be special motors that are designed to withstand the stress of the fast reversals.

Motor Speed

Two-speed capacitor-start motors are also available. The different speeds are obtained by changing the number of poles. This is achieved by having two separate running windings. These motors are usually

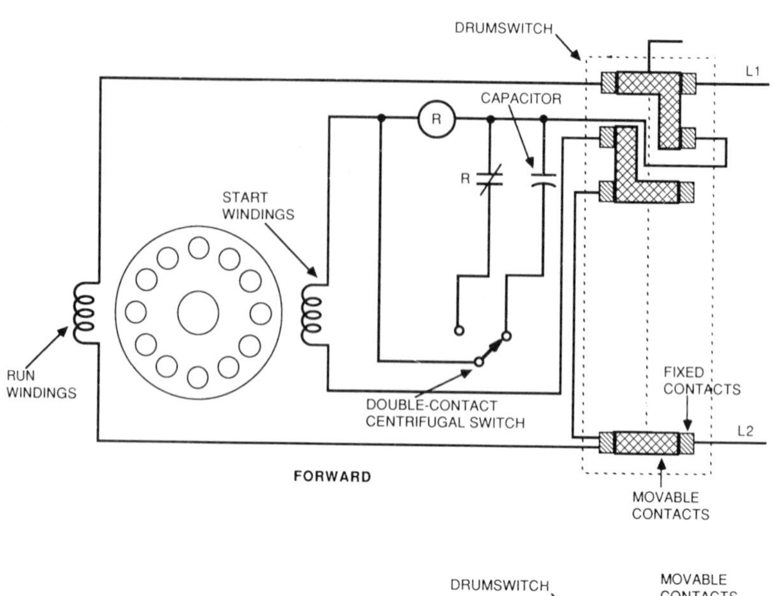

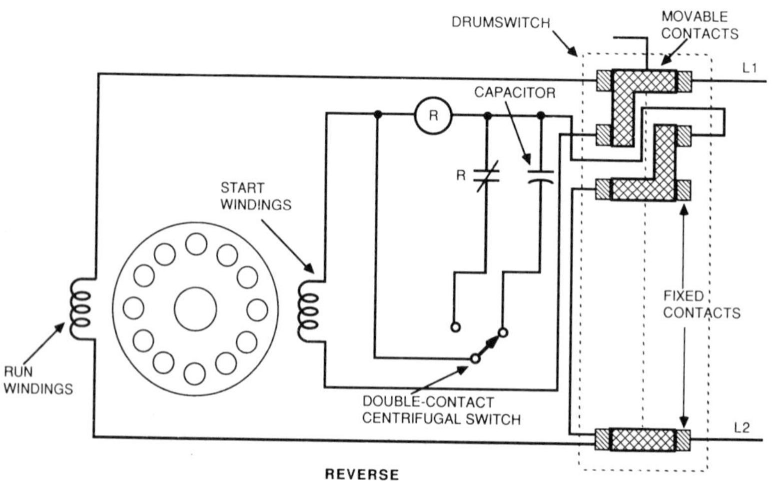

FIGURE 7–20. Instant reverse capacitor-start motor.

six or eight pole types. Figure 7–21 shows one arrangement using a single capacitor and a single start winding.

This motor always starts at its high speed regardless of the position of speed selector switch. If the switch is set for low speed when power is applied, current will flow through the starting-winding circuit and the high-speed run winding. When the motor approaches running speed, the centrifugal switch will change position. This action will disconnect the high-speed winding and the start circuit and connect the low-speed run winding.

When the motor is started in the high-speed position, a direct connection exists between the source voltage and the high-speed run winding. The centrifugal switch has no effect on this arrangement. It will still disconnect the start windings when the motor has reached the appropriate design speed for the switch to activate.

Another design of the two-speed capacitor-start motor does allow the motor to start on either the high or the low speed. This motor has two start windings, two capacitors, and two separate sets of run windings

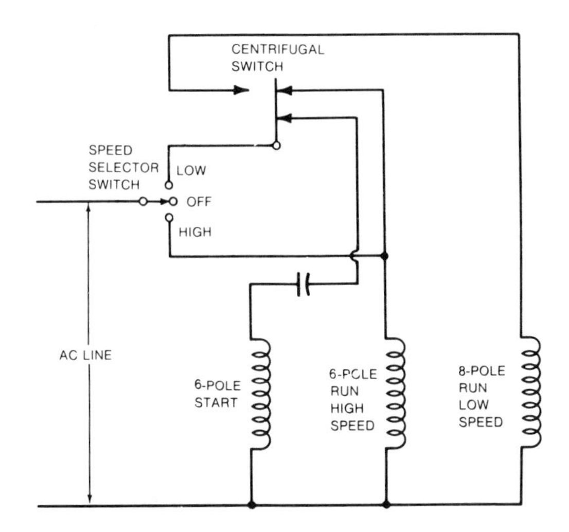

FIGURE 7–21. Two-speed capacitor-start motor using a single capacitor.

which have a different number of poles. Figure 7–22 provides a schematic diagram of this type of motor.

The capacitors in this circuit have different values for proper operation of this type of motor. The centrifugal switch is a double-pole type that disconnects the start windings at the proper speed.

Capacitor-Start, Capacitor-Run Motor

Capacitor-start, capacitor-run motors are very similar to capacitor-start motors. The difference is that the start windings in series with a capacitor remain in the circuit while the motor is running at normal speed. Because of this, the start windings must use larger wire than that used for the split-phase or capacitor-start motors.

The capacitor used during the run cycle may be the same one used to start the motor, or it may be a different, smaller capacitor. When two capacitors are used, the start capacitor is usually an electrolytic type, and the run capacitor is oil filled. If the same capacitor is used for both the start and run cycle, it will be oil filled.

These motors run quietly and smoothly, similar to a two-phase motor, thereby reducing the vibration and noise caused by the 120-hertz vibrations. They are sometimes called two-value capacitor motors.

Figure 7–23 depicts the circuit for the two-value capacitor-type motor. The centrifugal switch is eliminated if the motor runs on the same capacitance used to start it.

This motor starts using both the electrolytic and the oil-filled capacitor. When the centrifugal switch opens, the electrolytic capacitor is dropped from the circuit, and the motor runs using the oil-filled capacitor.

Although the overall capacitance has been reduced, there is still higher torque than if no capacitor was in the circuit. The start windings and capacitor are designed to complement each other so the magnetizing force is approximately equal to that of the run windings.

The result of having the capacitance is almost a 90° phase shift between the currents through the run windings and start windings. This reduces the magnetizing currents in the rotor, reducing copper losses, and improving the overall efficiency of the motor. Full-load torque is relatively constant. The power factor of the motor is better than that of split-phase or capacitor-start motors.

These motors may be reversible and of the dual-voltage type. The principles of obtaining these features are the same as for the motors previously discussed.

Permanent-Split Capacitor Motors

The permanent-split capacitor motor is similar to the capacitor-start motor, but it does not have a centrifugal switch. An oil-filled capacitor of 3 to 25 microfarads is connected in series with the start windings and remains in the circuit during the run cycle. Because the phase shift of the currents in the run and start windings is less than ninety degrees, this motor has a medium starting torque as compared to the capacitor-start motor. The start and run windings are identical in this motor. These motors are sometimes called single-value capacitor motors.

Reversibility

A typical method of reversing the direction of rotation of these motors is to switch the capacitor from the start windings to the run windings. Figure 7–24 illustrates this circuit.

Voltages

Permanent-split capacitor motors are either single or dual voltage rated. When dual voltage designs are used, the windings are connected in parallel for the lower voltage, and in series for the higher voltage. Usually, only single voltage motors have the forward-reverse feature or speed control.

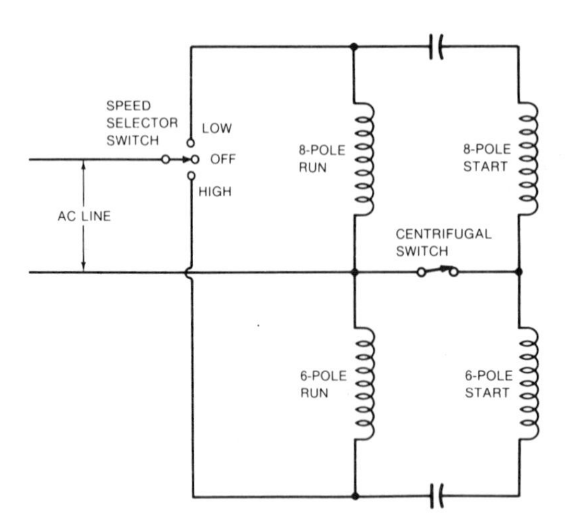

FIGURE 7–22. Two-speed capacitor-start motor using two capacitors and two start windings.

Speed

Permanent-split capacitor motors are available as single-, two-, or three-speed machines. Figure 7–25 shows the schematic for the two-speed motor, and Figure 7–26 depicts the three-speed device.

The permanent-split capacitor motor uses slip as the method of obtaining dual or three-speeds. A single auxiliary coil is wound on top of the run windings for the two-speed motor. This does not add any additional poles to the motor to change its speed.

When the speed selector switch is in the high position, the full line voltage is impressed across the run windings. The auxiliary coil is placed in series with the start windings when the switch is in this position. The motor will run at its synchronous speed minus the normal slip.

When the speed selector is set for low speed, the auxiliary coil is placed in series with the run windings. This reduces the voltage across the run windings with the resulting lower current. Because of the lower current, the magnetizing force of the run windings is reduced. Under load, the slip will increase, and the motor will run at a lower RPM. This method of speed control can also be used with the split-phase motor.

The same principle is applied for the three-speed motor (Figure 7–26). Two auxiliary coils are used to obtain the additional speed.

At high speed, the full voltage is applied across the run windings. Both auxiliary windings are in series with the start windings.

For medium speed, auxiliary winding 1 is placed in series with the run windings, and auxiliary winding 2 is connected in series with the start windings. Current is reduced through the run windings, and the slip is increased due to the weaker magnetic field. The RPM will be lowered due to the slip when the motor is under load.

To obtain the low speed for this motor, both auxiliary windings are placed in series with the run windings. This further reduces the current through them, reduces the strength of the magnetic field and causes the slip to increase further. When the selector switch is in this position, full line voltage is across the capacitor and the start windings which are in series with each other.

Characteristics

This motor runs with greater noise level when under a no-load condition. It is only under full-load that it acts as a true two-phase motor. With the proper selection of capacitance, this motor comes close to operating at 100% power factor. It has a low starting torque, however, and is not recommended for starting heavy loads. The starting current for the single-phase motor is approximately three times the normal running current.

Shaded-Pole Motors

Shaded-pole motors are used for applications such as fans and blowers where there is a low horsepower requirement. Their values range from as low as 0.0007 HP to 1/4 HP. Most are manufactured in the range of 1/100 to 1/20 HP.

These motors have an advantage in that they are very simple to construct, reliable, rugged, and have a low cost. They do not require auxiliary parts such as capacitors, brushes, commutators, or movable switches. Maintenance on them is minimal and simple. Under locked-rotor conditions these motors do not draw much more line current than when operating normally. This provides automatic circuit protection, and the motor will not overheat and destroy itself.

A motor with all these advantages also has some disadvantages. The shaded-pole motor has a very low starting torque and very low efficiency. Maximum efficiency for the 1/20 horsepower motor is only about 35%. For the smaller motors it is as low as 5%. Power factor is also poor. Because these motors use such small power, however, the advantages far outweigh the disadvantages when the application is appropriate.

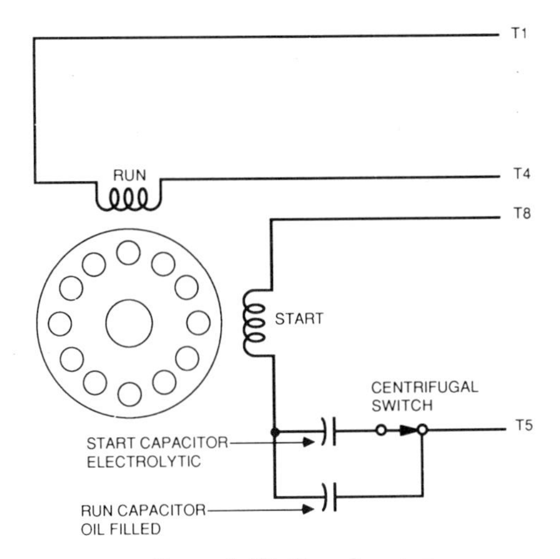

FIGURE 7–23. Capacitor-start, capacitor-run motor.

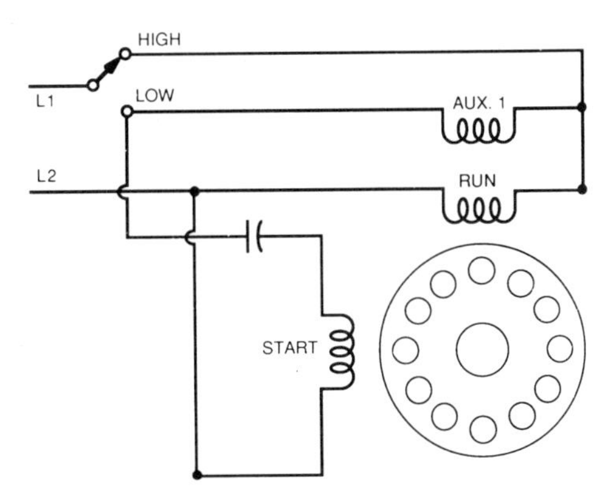

FIGURE 7–25. Two-speed permanent-split capacitor motor.

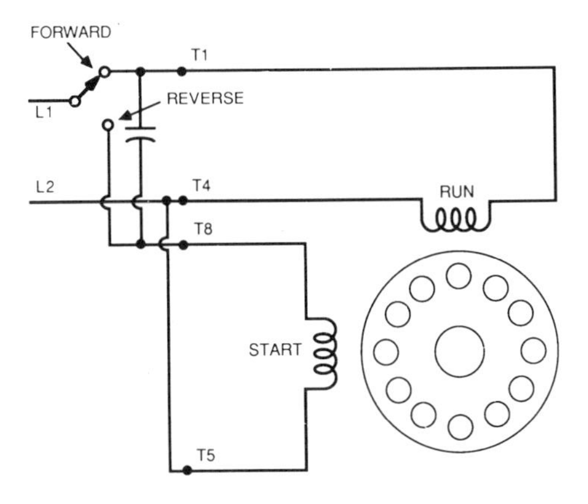

FIGURE 7–24. Reversing the direction of rotation of the permanent-split capacitor motor.

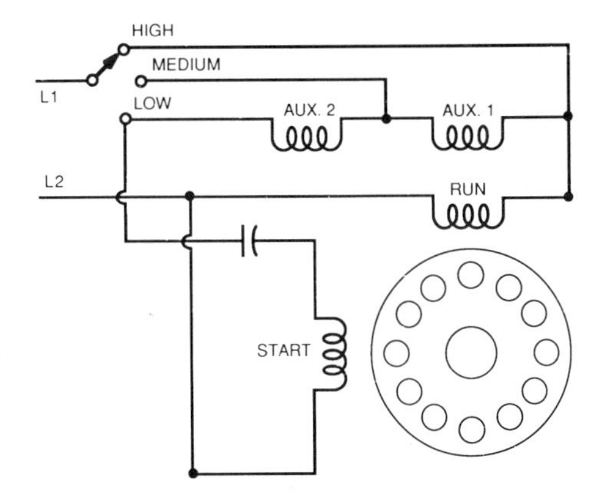

FIGURE 7–26. Three-speed permanent-split capacitor motor.

Principles of Operation

Shaded-pole motors operate on the induction principle as the previously discussed single-phase motors. They obtain their name from the method used to produce a rotating magnetic field to provide starting torque. The pole pieces have a slot cut in them on which is wound the shaded pole coil. The shading coils are normally a short-circuit single turn of heavy gauge copper wire. The wire is formed fitted on the shaded-pole which is a notch cut into the main stator pole. The shaded-pole coil is held in place with pins or wedges. All shaded-pole motors use the squirrel cage rotor. Figure 7–27 illustrates this construction.

Although there are exceptions, the shaded-pole motor usually uses salient pole construction for the main field windings. The salient pole, or definite-pole as it is sometimes referred to, is shown in Figure 7–28.

The shaded-pole motor is actually a split-phase motor. It differs in that it obtains its starting torque by induction in the stator poles.

Power is not applied to the winding of the shaded-pole, but rather only to the main run windings. The magnetic field created by current from the power source cuts the winding of the shaded pole. This induces a voltage due to transformer action. Because the secondary is shorted, a high current will flow in the shaded pole coil. The magnetic field developed by the shaded-pole opposes the magnetic field that caused it. Its magnetic field is nearly 90 electrical degrees behind the magnetic field of the main pole. A rotating magnetic field is developed which gives the motor its starting torque. Direction of rotation is toward the shaded pole. Figures 7–29 and 7–30 illustrate this action.

At the beginning of the current cycle in the main windings, or at zero degrees as shown in Figure 7–29, no current is flowing through the main windings. Therefore, no magnetic field is formed in the main pole pieces. Because of the nature of the sine wave, the most rapid change is occurring in this current as it passes through the base line. From zero to 30 degrees the current reaches 50% of maximum current. This means that the flux field is building most rapidly during this period.

The faster the flux changes, the greater the current induced into the shaded-pole winding. This current will be maximum when the main windings' current is zero. The current in the shaded-pole windings sets up a magnetic field which will oppose the magnetic field which caused it. This opposition of fields causes the magnetic field to be concentrated in the main poles.

Current through the shaded-pole windings will decrease as current in the main windings is increasing from zero to 90 degrees. This is due to the decreasing rate of change of the flux field.

It takes the current in the main windings twice as long, or 60 degrees, to rise the next 50%. Current in the shaded coil will decrease to zero at 90 degrees when current in the main windings has reached its maximum value. At that instant the flux is neither expanding or contracting, and no current is induced into the shaded-pole winding. The magnetic field will now be distributed uniformly across both poles. The axis of the magnetic field has shifted toward the shaded-pole, providing the magnetic rotating field. Figure 7–30 shows the magnetic fields of the two windings expanding and contracting during this period and at precisely 90°.

From 90° to 180° the current in the main windings is decreasing causing its flux lines to collapse. This action induces current into the shaded-pole windings. The current in the shaded-pole windings will reach its maximum value at 180° when the current in the main windings is zero and alternates to go negative. The reverse magnetic pole is produced in the shaded pole during this period due to the change in the direction of the flux lines cutting it.

When the main windings current reaches its maximum negative current at 270°, its flux field will be maximum once again. The polarity of the field is

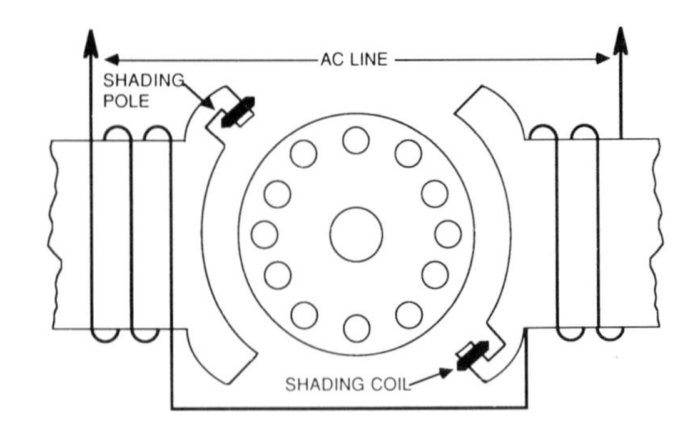

FIGURE 7–27. Two-pole shaded-pole motor.

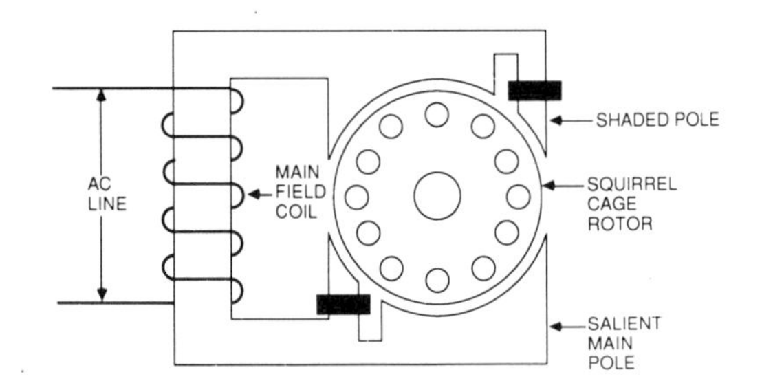

FIGURE 7–28. Salient poles used with shaded-pole motors.

opposite to the field created at 90°. The shaded-pole field will once again be zero at this point. The cycle continues back to 360° where the process is repeated.

This arrangement produces an uneven magnetic field. The torque, therefore, varies during each cycle. This makes the shaded-pole motor noisier than the split-phase motor or the capacitor type motor of the same horsepower.

Reversibility

Although some shaded-pole motors can be reversed with switching, most of those manufactured cannot. To reverse the direction of rotation of the latter, the motor must be disassembled. Usually only one end plate is removable, with the other being cast as part of the frame.

The stator is removed and reversed by turning it end to end. This puts the shaded pole on the opposite side of the main pole. Rotation will once again be toward the shaded pole, and the motor will run in the opposite direction. Figure 7–31 depicts the two different positions for the shaded pole.

There are two common methods of reversing the shaded-pole motor externally. One method uses two shaded poles and one main winding. The other uses a single shaded pole and two main windings.

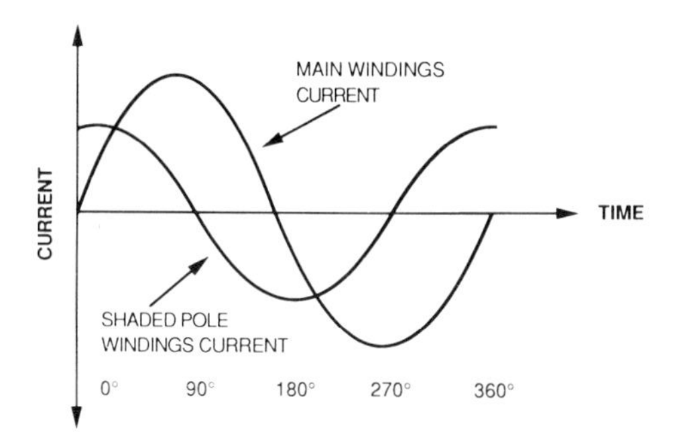

FIGURE 7–29. Comparison of the current flow in the main and the shaded-pole winding.

Figure 7–32 shows a motor having two shaded poles for each main winding. Only one of the shaded poles is active at a given time. An external switch is used to complete the circuit for the shaded-pole winding that is in use. The other winding is open, and no current flows through it. To reverse the direction of rotation, the switch is moved to the other position, connecting the alternate windings, and disconnecting the one which was being used. This places the shaded pole on the other side of the main windings. This arrangement causes the rotor to move in the opposite direction.

When two main coils are used to reverse the direction of rotation of the shaded-pole motor, they are wound in the slots so that the shaded pole is positioned on the opposite sides of each coil. Because the rotor will move toward the shaded pole, this arrangement effectively causes the rotor to reverse when the alternate coil is used.

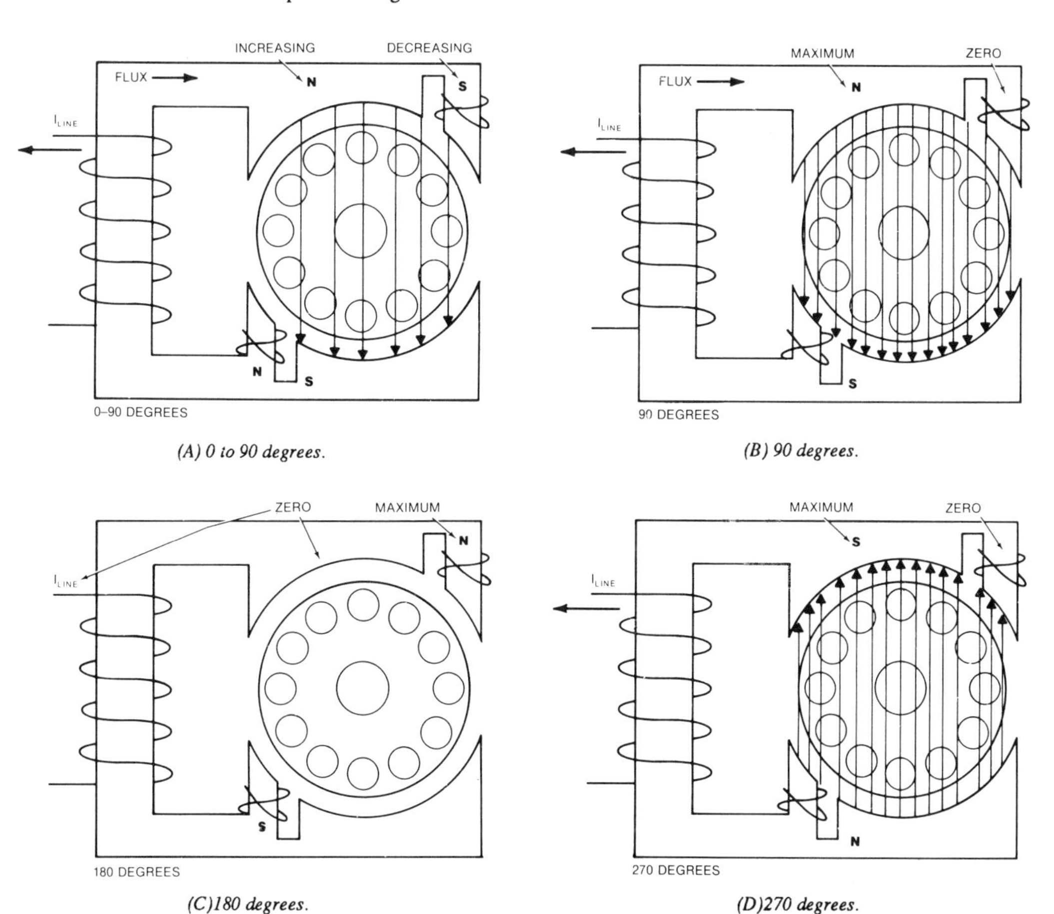

(A) 0 to 90 degrees.

(B) 90 degrees.

(C) 180 degrees.

(D) 270 degrees.

FIGURE 7–30. Comparison of the magnetic flux fields for the main and shaded poles.

Only one of the main coils is connected to the circuit at any one time. This is accomplished by an external switch. Figure 7–33 illustrates the use of two main windings.

Voltages

These motors may be rated as either 115 or 230 V AC. They are single-voltage motors and must be operated at their designed voltage and frequency ratings.

Speed

Shaded-pole motors are manufactured as two-, four-, six-, and eight-pole motors. This provides a variety of single-speed motors.

The speed of the shaded-pole motor may also be of the varying type. This is accomplished by using the slip characteristics of the induction motor. Speed varies with load due to greater slip.

One method of varying the speed of the motor is to use two sets of windings. The windings for each set use the same slots, so the number of poles is not changed.

When only one set of windings is used, the motor will operate at its higher speed. Switching the second set of windings into the circuit will cause the motor to operate at its lower speed.

Connecting the second set of windings in series lowers the voltage across each winding to one-half the value for the higher speed. Because the windings of each set are identical and wired in series, only one-half the current will flow through them on the low speed arrangement. This, in effect, reduces the strength of the magnetic field and causes the slip to increase. Figure 7–34 illustrates a four-pole, two-speed, shaded-pole motor. A two-, six-, or eight-pole motor could have as easily been used to depict this type of speed control.

Another method provides for two speed operation without using multiple windings. This motor is known as the "consequential-pole" motor. Figure 7–35 depicts the schematic of this machine. Voltages may be either 115 or 230 V AC.

When the motor operates at its high speed, only five of the windings are connected to the line voltage. The windings are identical, and an equal amount of voltage is dropped across each of them. The motor in Figure 7–35 is shown in its high-speed position.

To obtain the lower speed, a sixth coil is switched in series with the other five windings. This winding has a smaller size of wire and more turns. The voltage drop across it will be greater than the individual voltage drop across each of the other five windings. The other five windings divide the remaining voltage equally. The motors slip increases due to the lower current flow through the windings.

Repulsion Motor Types

Before the general acceptance of the capacitor-start motor, the repulsion-type motor was used extensively to drive high-inertia loads where only single-phase power was available. This motor develops more starting torque per ampere than any other type. It requires about 35% of the starting current as that of the capacitor-start motor to deliver the same amount of torque. Repulsion motors have horsepower ratings from 1/8 to 15.

There are three general classifications of repulsion type motors. These are

1. Repulsion-start, induction-run
2. Repulsion
3. Repulsion-induction

Some confusion is caused by the similarities in the names of the different types of repulsion motors, and it

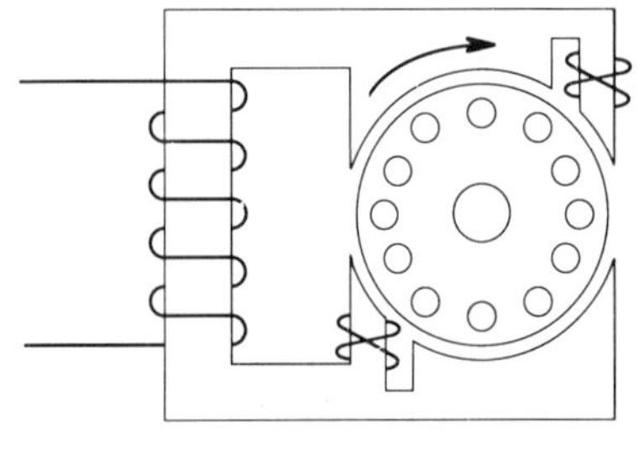

(A) Clockwise.

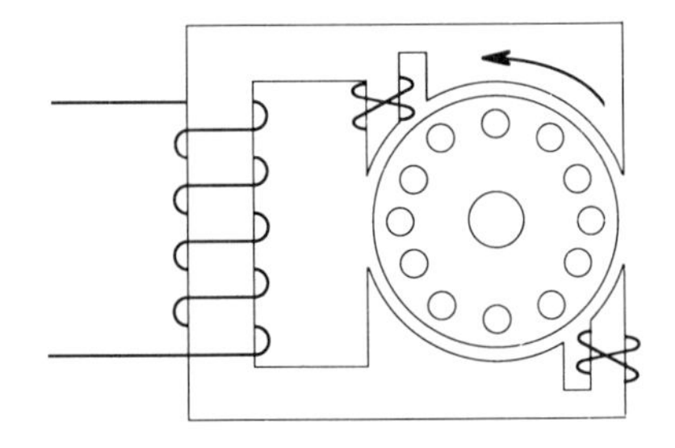

(B) Counterclockwise.

FIGURE 7–31. Reversing the direction of rotation of the shaded-pole motor by turning the stator.

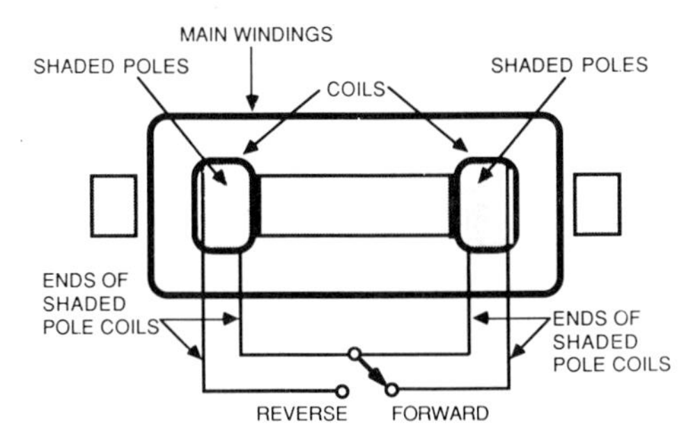

FIGURE 7–32. Reversing the direction of rotation of a shaded-pole motor with two shaded poles.

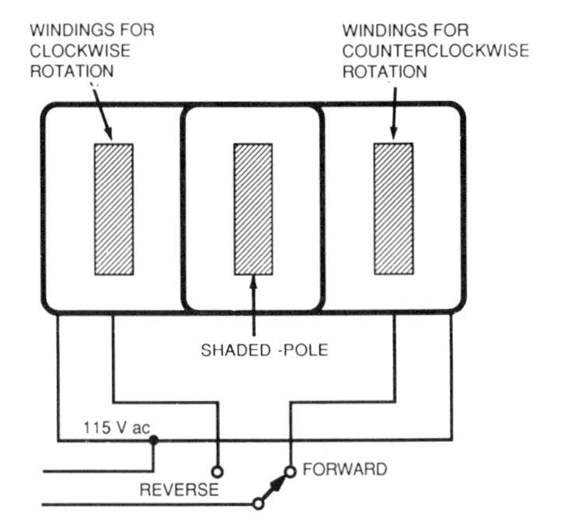

FIGURE 7–33. Reversing the direction of rotation of the shaded-pole motor using two main windings.

is difficult to distinguish one from the other. Each of these motors, however, has its own operating characteristics, and there are differences in the physical construction of these different types of repulsion motors.

All types of repulsion motors have a stator similar to that of the split-phase motor. The stator has a single-run winding. In addition all the types of repulsion motors have a wound rotor and a commutator with brushes.

The rotor has a slotted core in which the rotor windings are placed. The slots are usually skewed. They are not parallel to the motor's orientation and are set at an angle. With this design the same starting torque is produced regardless of the position of the armature when the motor is started. Skewed slots also help to reduce magnetic hum.

Commutators for these motors may have their segments arranged in either an axial or radial position. Axial segments are arranged parallel to the shaft, and radial segments are perpendicular.

Brushes ride on the commutator and provide a conductive path of current flow through the selected armature coils. Pairs of brushes are shorted together by a flexible conductor or through the conductivity of the brush holder which may be mounted on the armature shaft or the front end plate of the motor. The arrangement depends on the particular type of repulsion motor.

Repulsion motors are more complicated than capacitor-start motors, and they require more parts. Repulsion motors have stator windings similar to that of the capacitor-start motor and may have two, four, six, or eight poles. The rotor, or armature of the repulsion motor, however, is more complex than the simple squirrel cage rotor used with the split-phase and capacitor types of motors.

The number of brushes used with these motors would normally correspond to the number of poles. This is not always the case, however, because only two brushes may be required, depending on how the armature is wound.

Two of these motors, the repulsion motor and the repulsion-induction motor, can have their directions of rotation reversed while operating at full speed. When this is done for the purpose of bringing the motor to a rapid stop, the operation is called *plugging*.

By electrically switching the leads on the stator, the armature's direction of rotation will be reversed. This will bring the motor to a quick stop. The direction of the motor's rotation can then be reversed almost instantly.

Most single-phase motors cannot be plugged because of the centrifugal device used in them. These motors must have power removed first, and the speed of the motor must be reduced until the centrifugal switch closes. This is true for the repulsion-start, induction-run motor. The only other single phase motor that can be plugged is the instantly reversible capacitor motor. In general, single-phase motors are not designed for this function.

Because repulsion motors have more parts than capacitor-type motors, this makes their manufacturing more difficult and costly. The use of brushes, commutator, and centrifugal switches increases the amount of maintenance required to keep these motors operating. Due to their complexity and lower overall economy, they have mostly been replaced in industry by capacitor-type motors.

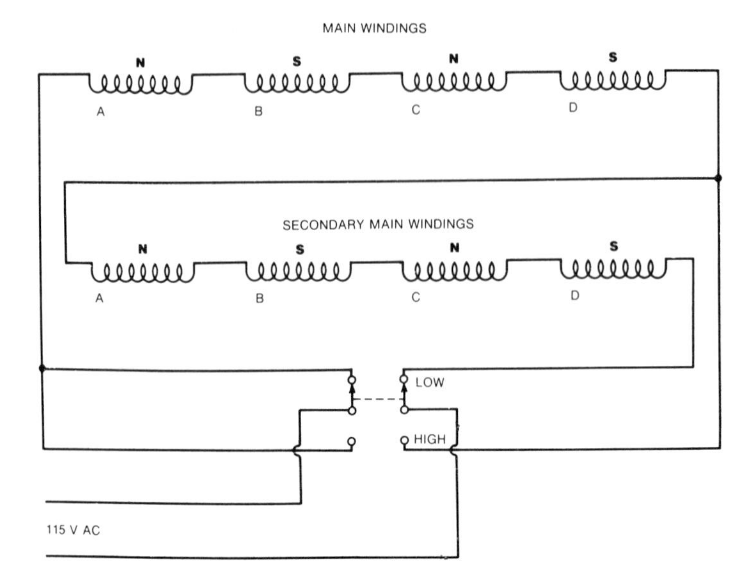

FIGURE 7–34. Four-pole, two-speed, shaded-pole motor.

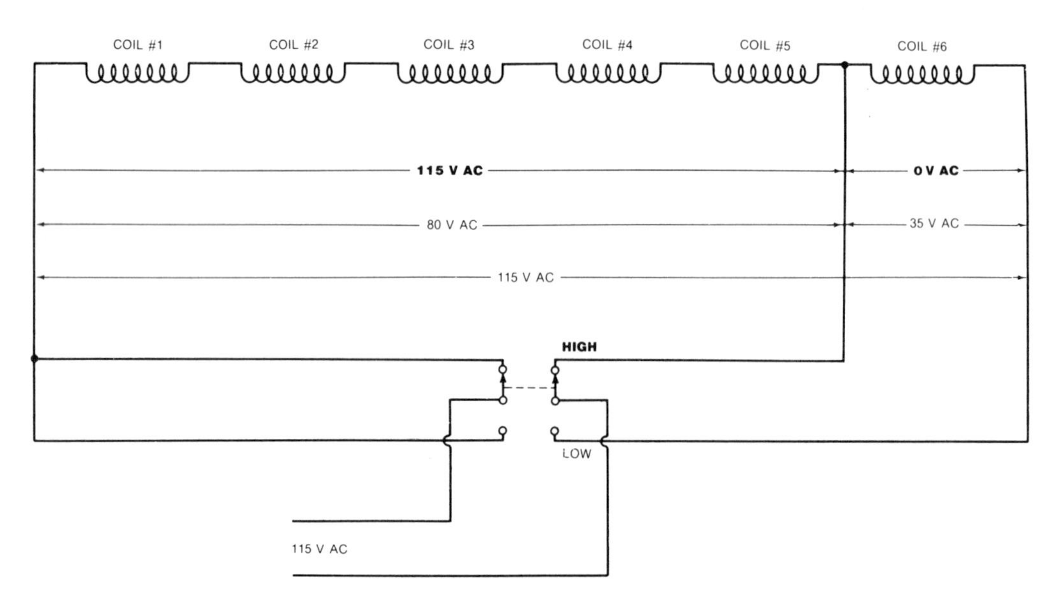

FIGURE 7–35. Consequential-pole, shaded-pole motor.

Repulsion-Start, Induction-Run Motors

The rotor of a repulsion-start, induction-run motor is shown in Figure 7–36. This armature is similar to a DC motor's armature in that it is a wound rotor and has a commutator and brushes. It also has a centrifugal device that activates at approximately 70% to 80% of its full RPM. At that point the centrifugal device moves forward pushing a shorting ring across the commutator's segments. The motor then runs as a squirrel cage induction motor similar to the split-phase motor.

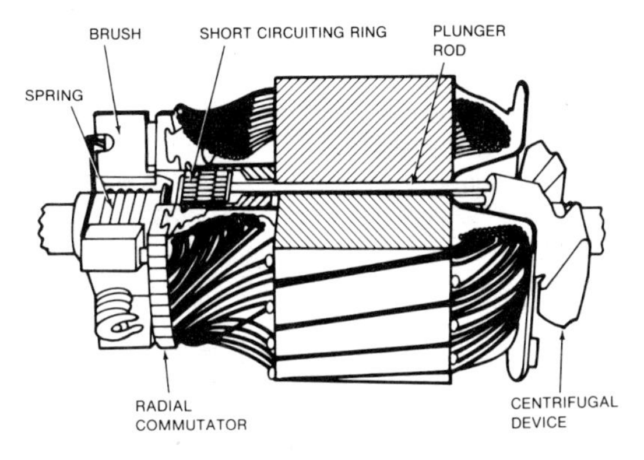

FIGURE 7–36. Armature of a repulsion-start, induction-run motor.

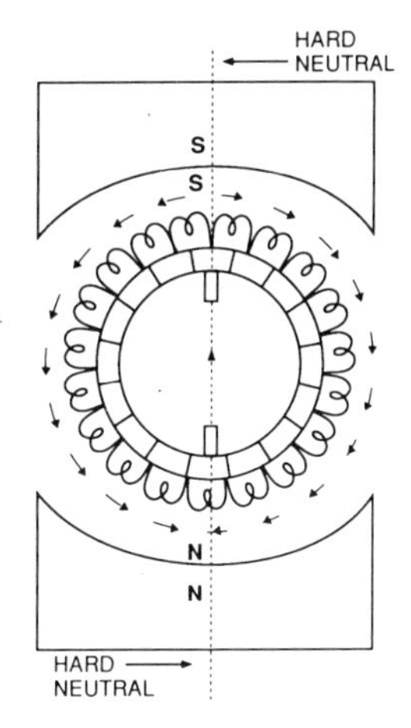

FIGURE 7–37. Current flow in the windings of the armature of a repulsion motor at hard neutral.

On some armatures the brushes are in contact with the commutator during the start and run cycles. These are classified as a brush-riding type. The brush-riding-type motor usually has an axial commutator.

Another type provides a mechanism to lift the brushes from the commutator when they are no longer performing a function. The brush-lifting type has a radial commutator. Figure 7–36 illustrates the design for the brush-lifting mechanism.

When the motor comes up to speed, the plunger connected to the centrifugal device continues forward after shorting the commutator segments. This action pushes the brushes away from the radial commutator. The life of the brushes and commutator is increased due to decreased wear.

When the motor is running, the spring on the armature is put under tension. When the motor is brought to a stop, this tension returns the brushes to the commutator and forces the short-circuiter away from the commutator segments. The motor is ready to be started once again.

Principles of Operation

On a DC generator the brushes remove power from the armature to deliver it to some load. On a DC motor, the brushes provide a current path to the armature. The brushes of a repulsion motor perform neither of these functions. Instead, the brushes are shorted together by a conductor. Their function is to connect the proper segments of the commutator together to set up a current flow through the windings of the armature so that a magnetic polarity is established. Figure 7–37 shows this arrangement.

As can be seen in Figure 7–37, each segment of the commutator has two windings connected to it. The brushes short a pair of segments, providing a current path between segments. When the brushes are in the position shown, an equal amount of current will flow in each of the two sets of windings, establishing like poles in the armature opposite the stator poles. The poles of the stator and rotor always change together as the current alternates.

Like poles of magnetic fields repel each other. The rotor should move away from the stator. In this case, however, only a vertical force is being impressed on the rotor. The stator and rotor fields remain aligned with each other. This is defined as the *hard neutral position* for this motor. A horizontal force is also needed in order for the rotor to begin to turn.

The horizontal as well as vertical force is accomplished by shifting the brushes away from the hard neutral position (Figure 7–38). This shifts the axis of the magnetic field created by the armature current. The rotor will begin to turn.

If the brushes are shifted a full 90°, the rotor will again be in a neutral position. It will not turn of its own accord in this position. This is called the *soft neutral position*. Under these conditions, no current flows through the brushes. An equal potential exists across them.

There is a weak magnetic force field developed between the stator and armature. The tendency is for the stator to turn in either direction, if at all. Shifting the brushes from this position develops a directional

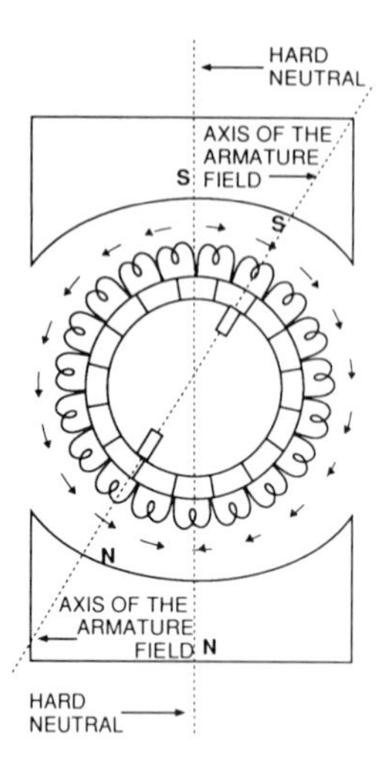

Figure 7–38. Shift in axis of armature's magnetic field due to the change in position of the brushes.

rotational tendency, and the motor may start if the load is not too great (Figure 7–39).

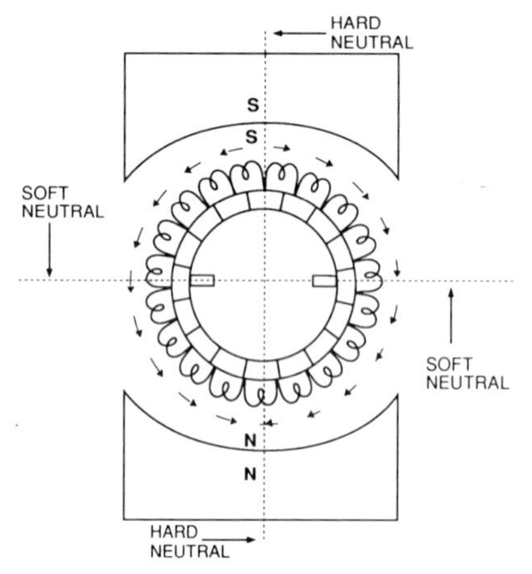

Figure 7–39. Current flow in the armature when brushes are set to soft neutral.

Because of the weak field when operating from the soft neutral, repulsion motors are always set to operate from the hard neutral position. The hard neutral position can be determined by moving the brushes in either direction. If the rotor turns in the direction in which the brushes are shifted, the brushes are set to hard neutral. Shifting the brushes from soft neutral results in the armature turning in the opposite direction of the brush shift. Maximum torque is obtained when the brushes are set at about 17° from hard neutral.

The motor, therefore, starts on the repulsion principle of like poles repelling each other. Because of the transformer effect of the stator windings magnetic field inducing currents into the armature windings, these two windings are effectively in series with each other. This provides the speed and torque characteristics of a DC series motor which will be discussed. The starting torque of the repulsion-start, induction-run motor is at least twice as great as that of the split-phase motor.

When the motor achieves 70% to 80% of its operating speed, the short circuiter activates and converts the motor to a simple induction motor. Under these conditions it has the same operating characteristics of the split-phase motor. Its overall efficiency and power factor is slightly lower.

Reversing Direction of Rotation

Repulsion-start, induction-run motors are designed to be reversed. The brushes are mounted on an adjustable ring. This device is called the *brush yoke*. There are either two or three marks on the brush yoke. The center mark is hard neutral. The other two marks are the settings for forward or reverse rotation of the motor. If there are only two marks, they are for the forward and reverse operation. The hard neutral position is not marked.

The motor will have the same characteristics if set to run in either direction. The motor is designed to turn in the direction the brushes are positioned.

If the motor begins to rotate when the brushes are set to the neutral mark, then neutral has shifted in the motor. When this happens, a brush shift using the marks established by the manufacturer will result in too great a shift in the brushes for rotation in one direction, and too little shift for the opposite direction. In either case, optimum starting torque will not be available.

To correct this problem, lock the brush yoke so the pointer is set to the hard neutral mark. Loosen the end plate bolts on the motor. Turn the end plate slightly in the opposite direction to which the motor turns. Apply approximately one-half the voltage to determine if the motor will turn. The neutral is correctly set if the motor does not start.

If difficulty is experienced in setting the neutral, replace the regular brushes with wedged shape brushes. Brushes designed for the motor can be filed down until the tip is about 1/16 inch in width. This will reduce the number of commutator segments the brush may be touching at any one time. Neutral can be established more precisely using this modification. Once neutral is set, replace the wedge-shaped brushes with those designed to be used with the motor.

Many of these motors, especially those with brush riding features, have brush holders that cannot be moved. Some of these motors are designed so the field poles are off center. To reverse them, the motor must be disassembled and the pole frame reversed.

All three types of repulsion motors may also be designed to be electrically reversed. When this feature is available, the stator has two sets of run windings which are wound 90 electrical degrees apart. Both windings are connected in series. To reverse the motor, the leads of one of the windings are reversed.

Another method is to wind the reversing winding in two sections. For forward direction, the run winding and one section are connected in series. To reverse the motor, the other section is used in series. The sections are wired so they provide opposite polarities when used individually. Figure 7–40 illustrates these two methods.

Voltage

Repulsion-start, induction-run motors are manufactured as single- and dual-voltage motors. When wiring a dual-voltage motor for 115 V AC, connect T1 and T3 together. Wire this junction to one side of the power line. Then connect T2 and T4 and wire this junction to the other side of the power line.

To wire the motor for its higher voltage, 230 V AC, connect T2 and T3 together. Connect the power line to T1 and T4, respectively. Figure 7–10 depicts these connections for the run windings of a split-phase motor.

Speed

The repulsion-start, induction-run motor is a constant speed motor. To change the speed, the number of poles on the stator is changed.

Repulsion Motor

The repulsion motor starts and runs on the repulsion theory. The commutator is active both in the start cycle and the run cycle. This motor does not have a centrifugal device or a shorting ring to convert the motor to a simple induction motor. Brushes are of the short-circuit type. They are continuously in contact with an axial commutator.

These motors have a high starting torque and variable speed characteristics similar to the DC series motor. Repulsion motors are sometimes called inductive-series motors.

The speed of this type of motor is adjusted by shifting the brushes further from the hard neutral position. This reduces the torque of the motor and the slip increases. Base speed is established by the number of poles the motor has. These motors are usually wound for four, six, or eight poles.

Repulsion motors are reversible. This is accomplished by shifting the brushes from hard neutral in either direction. Direction of rotation will be in the direction of the brush shift. Electrically reversible motors are also available.

Voltage

Generally, these motors are dual voltage. Four leads are available from the stator to wire the motor for 115/230 V AC. Wiring codes are the same as those previously discussed.

Compensating Windings

A second set of windings is sometimes added to the repulsion motor to cancel the cross-magnetization in the armature. Cross-magnetization causes the neutral plane of the motor to shift against the direction of the rotation of the motor. The amount of shift depends on the load on the motor. The greater the current being drawn, the greater the shift. The shift in the neutral plane causes poor commutation and a lower power factor.

Compensating windings are smaller than the main windings. They are wound on the inner slots of the main poles, and they are displaced 90 electrical degrees from the main windings. The compensating windings are connected to the armature through a set of brushes.

The magnetic field of the main windings induces a voltage into the compensating windings by transformer action. The size of the voltage will be in proportion to the current being drawn by the motor. Therefore, the current drawn by the armature from the compensating windings will be proportional to the load. The effect is to cancel the cross-magnetization field. Figure 7–41 provides a schematic of the compensated repulsion motor.

Compensating windings are sometimes inductively coupled to the armature. When this method is used, the brushes are not necessary. Instead, the ends of the compensating windings are merely shorted together. This provides a current path to establish the magnetic field which will cancel the cross-magnetization effect.

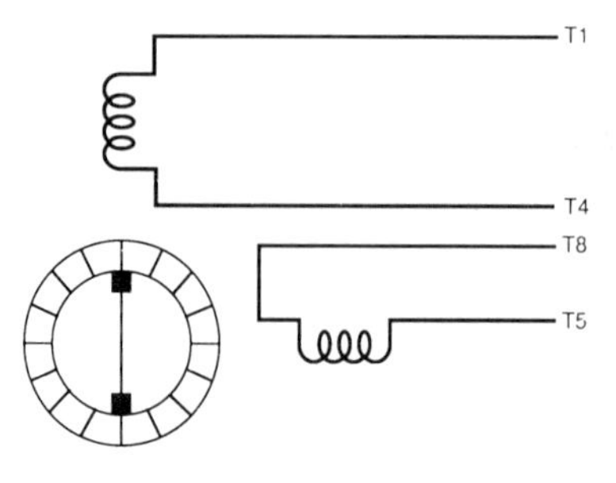

	L1	L2	CONNECT
COUNTER-CLOCKWISE ROTATION	T1	T5	T4, T8
CLOCKWISE ROTATION	T1	T8	T4, T8

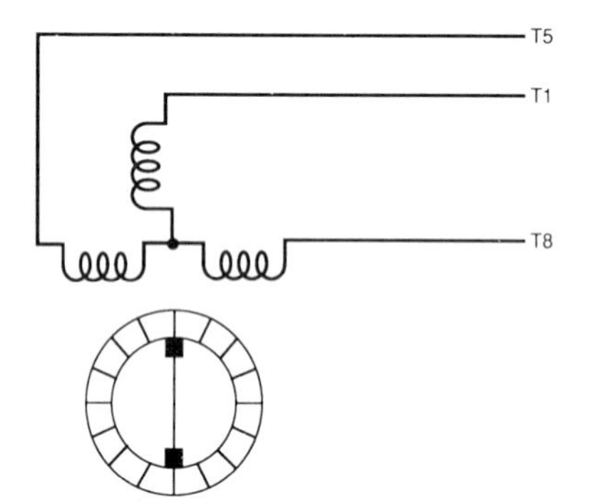

	L1	L2	INSULATE
COUNTER-CLOCKWISE ROTATION	T1	T5	T8
CLOCKWISE ROTATION	T1	T8	T5

Figure 7–40. Reversing direction of rotation electrically on a repulsion-type motor.

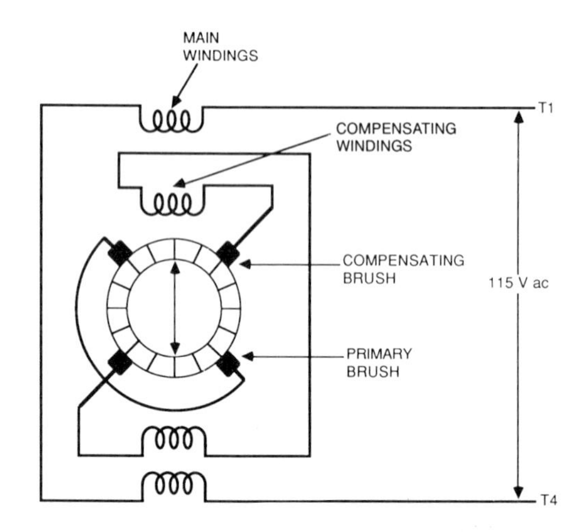

Figure 7–41. Compensated repulsion motor.

Repulsion-Induction Motor

Although it is difficult to distinguish this motor from the other two types of repulsion motors, its electrical characteristics are different. Electrically, it is comparable to a DC compound motor which will be described (see Compound DC Motor).

The repulsion-induction motor has a squirrel cage rotor embedded into the windings of the armature. It does not have the centrifugal short-circuit device of the repulsion-start, induction-run motor. The brushes remain in contact with the commutator at all times to provide torque. Figure 7–42 provides an illustration of this motor.

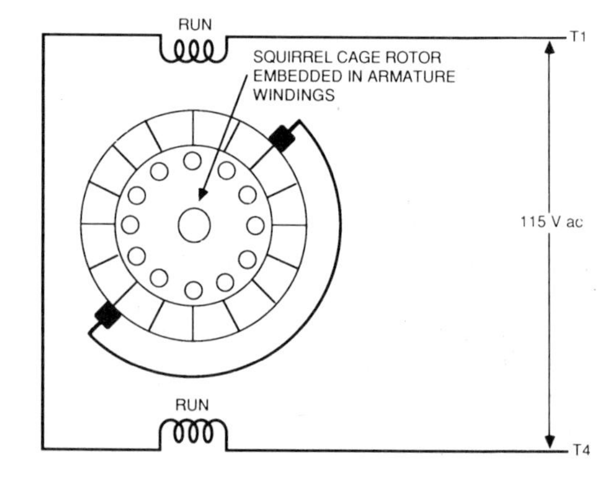

Figure 7–42. Repulsion-induction motor.

This motor has high starting torque and good speed regulation. To differentiate between repulsion-induction motor and the repulsion motor, connect it to the power line and allow it to come up to full speed while under load. Lift the brushes off the commutator. If its speed remains constant, it is a repulsion-induction motor.

Repulsion-induction motors are manufactured in sizes up to 10 HP. They are normally of the dual-voltage type and are used for general purpose application. They may have compensating windings to improve their power factor and commutation. They are also reversible. If compensation windings are used on the motor, the leads to them must also be reversed when reversing the motor.

Direct Current Motors

Direct current motors have become increasingly more popular in modern industry due to their varying speed characteristics, high starting torques, and the ability to reverse their direction of rotation. They are ideal for such applications for use on elevators, electric railroads, cranes, hoists, steel rolling mills, and hundreds of other purposes. The control of these motors is so good, they are often used as the drive for delicate laboratory instruments. Direct current motors are manufactured from the millihorsepower range to several hundred horsepower.

Application and Control

The elevator is as good an example as any to illustrate why these motors are selected as the best possible choice to drive this specific type of load. The motor must start under a load consisting of the weight of several people plus that of the elevator. This load will seldom be exactly the same. The number of people and their weights will vary.

The motor must accelerate the load smoothly without sudden changes which would cause the passengers to lose their balance. It must raise or lower the load rapidly, often making several stops on the way. Deceleration must be as smooth as the acceleration so sudden stops are avoided. Finally, the motor must stop the elevator at precisely a given level every time. For a single machine to meet all these requirements makes it a truly remarkable device.

Direct current motors stop and reverse direction of rotation more smoothly than any other type of motor. This is accomplished by opening the armature circuit and by placing a resistor across the armature. This causes the motor to act as a generator. Current flow will be from the armature through the resistor. The armature's magnetic flux field will oppose the field that is causing the current. The motor is brought to a rapid stop. This action is called *dynamic braking* of the motor. The motor can now be smoothly accelerated in the reverse direction.

Another type of braking is *regenerative braking.* When this principle is used, the energy produced by the generator action is returned to the source. Electric cars make use of this principle. The current from the generator is used to recharge the car's batteries each time the car decelerates or stops. This reduces the wear on the brakes, and at the same time, helps keep the batteries in a charged condition.

Voltage Ratings

Industrial motors are supplied with a variety of voltage requirements. They may be 6, 12, 24, or 48 volts for battery operated equipment. FHP motors supplied by a motor generator set or rectified AC voltages, are usually voltage rated at 115/230 volts. Typical voltage ratings for multihorsepower motors are 240 and 500 volts. They may, however, be supplied by a much lower or higher voltage depending on the application.

Motors for transit operations are provided in typical ranges from 250 to 750 volts. These values also vary, and may be much higher. Small millihorsepower motors usually have voltage ratings of single cells, 1½ volts, or multiples of this value.

DC Armature

The armature of a DC motor has a shaft, core, commutator, and windings. Figure 6–11 provides an illustration of the DC armature.

Direct current motors are often required to develop considerable torque. In order to meet this requirement, they must have a large number of conductors wound as coils on the armature. These windings are placed in the slots designed for them. The slots provide the necessary strength needed by the coils to remain in place. These motors operate at high speeds, and very high centrifugal forces are exerted on them. The force is on the iron of the armature instead of the windings (Figure 7–43).

Bands are placed around the armature to hold the coils in place against the centrifugal forces. Waxed cord may be used for this purpose on small motors. Larger motors require steel bands.

This design allows for a reduced air gap between the armature and the field poles. The small gap reduces the

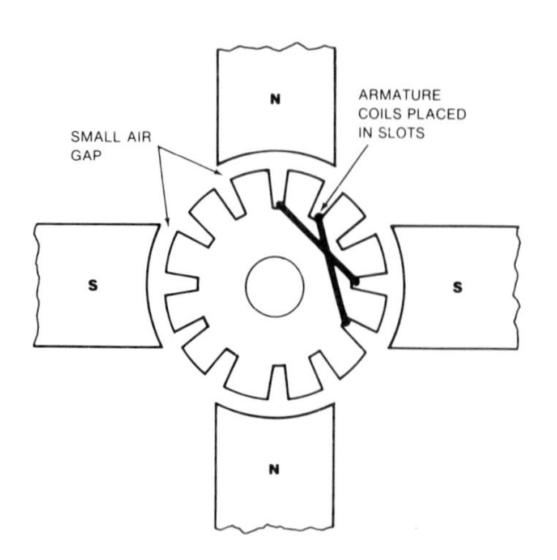

FIGURE 7–43. Placement of coils in slots on a DC armature.

flux linkage loss and provides a maximum field strength between the armature and the field coil.

Losses

Every DC armature has losses due to the resistance of the coils, or *copper loss*; induced currents into the core, or *eddy current loss*; and friction of the revolving magnetic particles, or *hysteresis loss*. Each of these is taken in consideration in the design of the armature.

Copper losses are directly proportional to the length of the wire used in the core windings and the square of the current flowing through them. It is inversely proportional to the circular mil area of the wire.

Temperature rise increases this loss. The resistance of the core increases about 1% for each 2.5°C rise in temperature. Mechanical loading of the motor increases current in the armature and increases loss.

Proper design and sufficient ventilation to limit the temperature rise controls copper loss. Fans mounted on the armature shaft are common. Mechanical loads must be kept within the limits of the machine.

Eddy currents are caused in DC machines due to the armature rotating in the magnetic flux created by the stator's field. The higher the flux density of the field, and the faster the armature turns, the greater this loss.

Eddy current loss is limited by using laminated cores. It varies as the square of the thickness of the core laminates. Each time the core is subdivided, the loss is reduced by the square of the number of laminates. For example, subdividing the core into two parts will reduce eddy current loss to one-fourth its previous value. Subdividing into four parts will cut loss to one-sixteenth. This loss becomes negligible when the core is divided into many sections.

Hysteresis losses will also increase with flux density and the speed of the machine. In addition, the more iron used in the armature, the greater the loss. The use of heat treated silicon steel for the core reduces this loss. The heat treatment process is called *annealing*. It consists of heating the laminates until they are a dull red and then allowing them to cool slowly.

Counterelectromotive Force

A 240-volt, 10-horsepower motor turning at its rated RPM will draw approximately 38 amperes at full load. At the same time, the DC resistance of the armature is only 0.06 ohm. Based on Ohm's law, 240 volts/0.06, 4000 amperes is calculated to be the current flow in this circuit. The difference between the calculated value and the actual current is explained by the presence of the counterelectromotive force (CEMF).

As the armature of the motor rotates in the field of the stator, a voltage is induced into it (generator action). This voltage opposes the applied voltage, and reduces the amount of current required by the motor to develop its full horsepower.

The CEMF can be calculated by subtracting the actual voltage drop across the motor from the applied voltage

$$\begin{aligned} \text{CEMF} &= E_a - IR_a \\ &= 240 - 2.28 \\ &= 237.72 \end{aligned}$$

The CEMF can never be equal to the applied voltage. There must be armature current to produce the torque to turn the armature in the stator field. This motion is necessary to induce the CEMF. The value, however, is relatively close to the value of the applied voltage. Figure 7–44 typifies the conditions which exist in a DC motor. It should be noted that the polarity of the CEMF is opposing the line voltage. This reduces the voltage across the armature.

Using these new values, we find that Ohm's law does apply. The voltage across the armature is 2.28 volts instead of 240 volts. 2.28 volts divided by 0.06 ohm, the armature's resistance, gives the 38 amperes the motor draws under full-load conditions.

Principles of Operation

The basic principles of how torque is developed in the DC motor and how it continues to rotate were explained in Chapter 6. This section should be reviewed

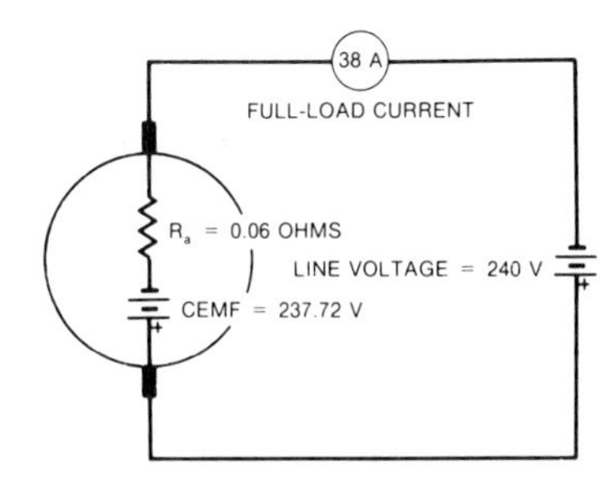

FIGURE 7–44. CEMF of the armature of a DC motor.

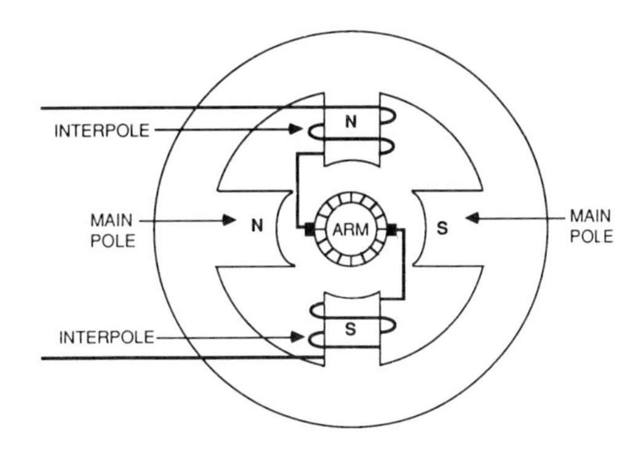

FIGURE 7–45. Interpole wiring for a two-pole DC motor.

in the event that a firm understanding of this theory was not achieved.

For the DC motor to perform according to its specifications and operate over its predicted life expectancy, good commutation is necessary. Poor commutation results in a lower efficiency for the motor and in excessive wear on the brushes and the commutator.

Commutation changes the direction of current flow through the armature windings. The point of commutation should occur when the potential is zero between the coils being shorted. This reduces the arc between brushes and commutator.

When the load changes, however, the current through the armature changes. The inductive nature of the armature opposes this change in current. This is the *armature reactance*. Armature reactance shifts the neutral plane causing poor commutation. To maintain the neutral plane, the brushes would have to be shifted in a direction opposite of that of rotation of the armature. This is not practical in most cases. This is particularly true under conditions of varying loads.

At the point of commutation, the self-induction of the armature coils opposes changes in current. This causes excessive short-circuit current to flow though the coils. This problem along with armature reaction must be addressed in the DC motor.

Interpoles

Interpoles are designed in DC motors to overcome the effects of the armature reactance and the self-induction of the machine. Most shunt and compound DC motors over one-half horsepower have interpoles located 90 electrical degrees from the main poles. Some motor designs use only one interpole with satisfactory results. Figure 7–45 shows this arrangement for a two-pole motor with two interpoles.

Interpoles are wired in series with the armature. Any changes in the armature current due to armature reactance, loading, or self-inductance pass through the interpole windings. This creates a changing magnetic field equal to and opposite that of the armature, thereby cancelling the effect.

Polarity of an interpole depends on the direction of rotation. The interpole's polarity will always be the same as that of the main pole which precedes it. In Figure 7–45, the direction of rotation is clockwise as viewed from the commutator end of the motor. When direction of rotation of the motor is changed, the direction of current flow through the interpoles must be reversed at the same time.

Figure 7–46 illustrates the effects by the interpoles on the magnetic field. Current flow in the windings of the main poles and the interpoles are also shown. Armature rotation is clockwise as viewed from the commutator end of the motor.

Torque

Factors that affect the amount of torque a DC motor develops are

1. Amount of armature current.
2. Strength of magnetic field in which the armature must turn.
3. The number of armature conductors and their active length.
4. Radius of the armature in feet.

Torque is expressed mathematically as

$$T = Fr$$

where,

F is the force exerted in pounds,
r is the radius of the armature in feet.

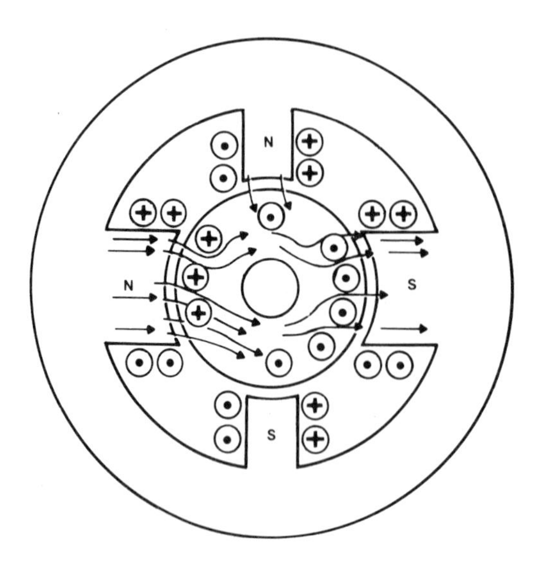

FIGURE 7–46. Effect of interpoles on the magnetic flux field of a DC motor.

The unit pounds-feet is used to distinguish it from work which is given in foot-pounds.

On a given motor, the torque can be controlled by varying the current to the armature or by changing the strength of the magnetic field. The other factors, number of conductors, length of the conductors, and the radius of the armature are fixed and become constants for a given motor.

Work

The amount of work the motor is capable of performing would equal the force times the distance. In the case of a motor, the distance is equal to the circumference of a circle. Assuming a force of 100 pounds and a radius of 1 foot, the amount of work performed would equal

Work = Force X Distance

Work = 100 X 2 X 3.14 X 1

Work = 628 Foot-Pounds

Horsepower

Power is the rate of doing work. The faster the motor turns, the more power it will develop. Assuming 1200 RPM for the example above, the amount of power developed by the motor will equal

Power = Foot-Pounds X RPM

Power = 628 X 1200

Power = 753,600 ft-lb/min

Dividing this result by 33,000 foot-pounds per minute will provide the horsepower the motor is developing.

Horsepower = 753,600/33,000

Horsepower = 22.83

If the speed is reduced by one-half, the horsepower is equal to 11.4. Doubling the speed will double the horsepower to 45.66.

Motor horsepower, therefore, is a function of torque and speed. A small motor operating at a high speed can develop the same horsepower as a large motor operating at a low speed.

Reversing Direction of Rotation

The starter motor in your automobile is a DC motor. If you were to accidentally reverse the battery polarity, the DC motor would still rotate in the same direction. Reversing polarity of the battery will not cause the motor to rotate in the opposite direction.

To reverse the direction of rotation of this type of motor, either the current through the stator winding or the current through the armature must be reversed. Reversing both of them will result in the same magnetic polarities between the armature and the stator poles. This results in the same direction of rotation.

The industry's standard for reversing the direction of rotation of a DC motor is to reverse the direction of the current through the armature. When a DC motor has more than one set of windings, shunt and series, as well as interpoles, the currents through all the stator windings would need to be reversed in order to change direction of rotation. This is far more complicated than merely reversing armature current.

Speed Regulation

Motor speeds vary as they are loaded. This is an inherent characteristic of most motors. *Speed regulation* is the ability of a motor to maintain its speed from a no-load condition to a full-load condition. Speed regulation of motors is usually given as a percentage and is expressed mathematically as

$$\% \text{ Speed Regulation} = \frac{\text{No-Load Speed} - \text{Full-Load Speed}}{\text{Full-Load Speed}} \times 100$$

If a motor operates at 1200 RPM at no load, and its speed drops to 1150 RPM under load, then its speed regulation would equal

$$\% \text{ Speed Regulation} = [(1200 - 1150)/1150] \times 100 = 4.35\%$$

A motor with a small value of % Speed Regulation will provide less change in the RPM of a motor under varying load conditions. The motor with a higher value will change speed by a greater RPM from no load to full load.

Types of DC Motors

Direct current motors are classified and named according to how they are connected for field excitation with respect to the armature. There are three basic types of DC motors.

When the field is connected in parallel with the armature, it is called a *shunt-type motor*. If the field is connected in series with the armature, it is classified a *series motor*. The third type, *compound motor*, has two field windings. One is connected in series with the armature, and the other is wired in parallel.

The compound motor can have the magnetic field of its series windings either aid or oppose the magnetic field of the shunt windings. When the fields aid each other, the motor is called a *cumulative compound* motor. If the fields oppose each other, the motor is classified as a *differential compound motor*.

DC Shunt Motor

Figure 7–47 provides a diagram of a DC shunt motor. The shunt windings are constructed of many turns of wire. There may be thousands of turns for this coil. It is connected in parallel with the armature of the motor.

Direction of Rotation

For normal rotation, which is counterclockwise, A1 is wired to F1, and A2 to F2. To reverse rotation, switch leads on the armature so A1 is connected to F2, and A2 to F1.

Speed Control

Speed of the shunt motor varies directly with the CEMF and inversely with the magnetic strength of the field windings. The larger the CEMF, the faster the motor will run. The stronger the shunt field, the lower the RPM. To keep the speed constant, it is necessary to keep these two factors as independent of the load as possible.

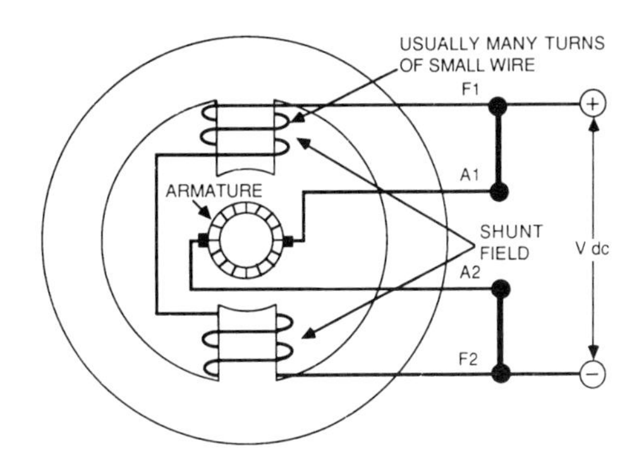

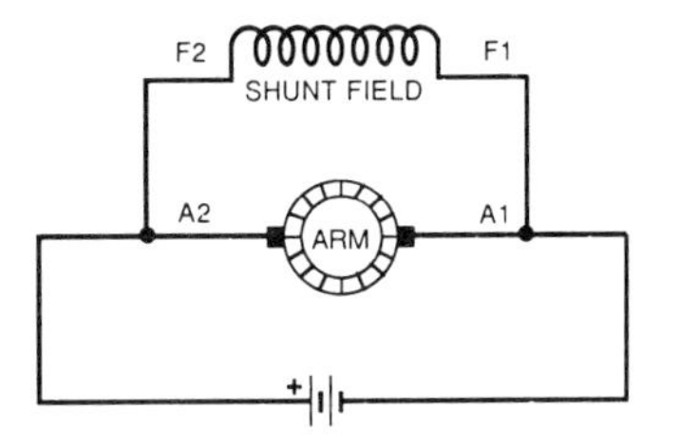

FIGURE 7–47. DC shunt motor.

Field strength is kept constant by connecting the windings directly across the power source. Unless voltage varies, the magnetic flux will remain constant.

Two factors effect the CEMF. These are the armature current and the armature resistance. The use of interpoles will minimize the effects of demagnetization on shunt field flux and help maintain a constant speed. Some shunt motors have a short series winding added to the motor to help stabilize the speed. Because the series field has only a few turns and is weak, this motor is still basically a shunt type and no name change is necessary.

The shunt DC motor is a constant speed motor. Its RPM changes very little from no load to full load. Figure 7–48 shows a graph of these two variables.

The starting torque for this motor is about 1.5 times the full-load torque (Figure 7–49). This is sufficient to start this type of motor while under full load. The shunt motor, however, has the lowest starting torque of any of the other DC motors. The amount of torque depends on how quickly the motor can come up to its rated speed when started under load.

It is fairly simple and relatively inexpensive to convert the shunt motor into an adjustable speed mo-

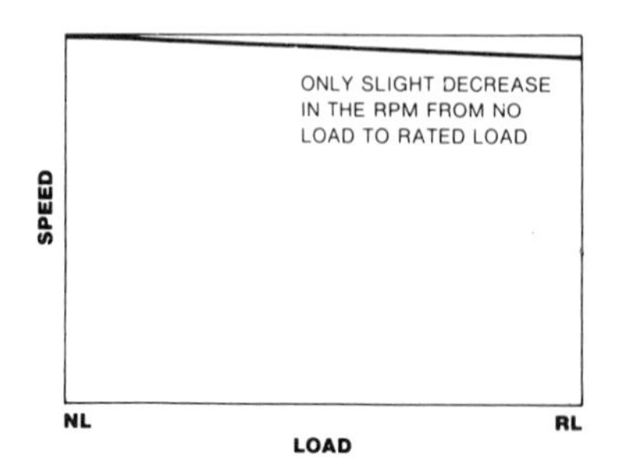

FIGURE 7–48. Speed vs. load for a DC shunt motor.

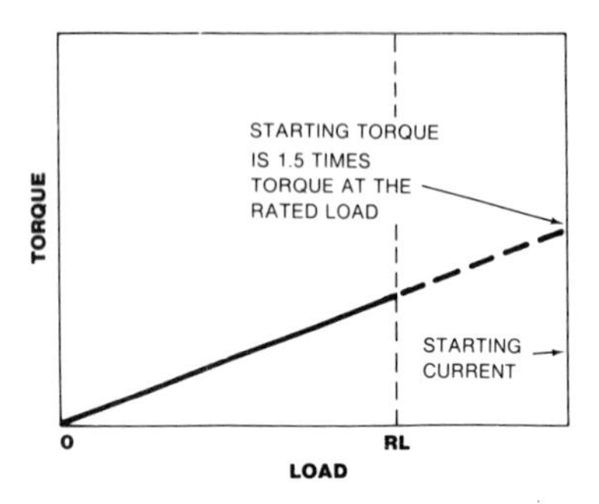

FIGURE 7–49. Torque vs. load for a DC shunt motor.

tor. This can be accomplished by placing an adjustable resistor in series with the field windings (Figure 7–50).

The addition of resistance in series with the windings will reduce the current through the coil. Because the field current is relatively low, the wattage of this resistor need not be very large.

A lower current results in a weaker magnetic field, and the motor will run above base speed. Motor torque is reduced by this process, and the maximum permissible load must also be reduced for the motor to continue to run. See the graph in Figure 7–50.

The motor can be made to run below the base speed by adding resistance in series with the armature. Because the high armature current will flow through this resistor, considerable power is consumed, and the overall efficiency of the motor is reduced accordingly. This arrangement is also shown in Figure 7–50 along with the graph of percent of full-load speed versus percent of full-load torque.

With the resistance in series with the armature, light loads will cause a small voltage to be dropped across the series resistance. The higher voltage is across the armature.

Heavy loads result in just the opposite effect. Because the voltage drop across the armature varies, this results in the speed of the motor changing. The motor would now be classified as a varying speed rather than a constant speed motor. More precisely, it remains an adjustable, varying speed motor. The operator can adjust the speed by changing armature resistance.

Open Shunt Field

The shunt field must never be opened while the motor is running. If the shunt field suddenly opens while the motor is being operated under full-load conditions, only a small voltage due to the residual magnetism of the pole pieces will be induced into the armature. Current will increase considerably. The magnitude of this current may be sufficient to burn out the motor.

If the shunt field opens under a no-load condition, the armature current will once again increase. Because very little torque is needed to overcome the windage and friction, the motor would continually increase in speed. The centrifugal forces acting on the armature at these high speeds may be sufficient to cause it to fly apart. Under these conditions, the motor will act as a series DC motor.

Series DC Motor

Figure 7–51 illustrates the wiring diagram for a typical series DC motor. The field windings consist of a few turns of heavy gauged wire which are connected in series with the armature. The ends of the series coil are designated S1 and S2. Unlike the shunt motor, the field flux of this motor varies with the armature current.

Torque

Series motors have the characteristic of developing very high starting torques. The reason for this is the same current that passes through the armature also passes through the field windings. For example, if the start current through the armature is two times the current at normal operating speed, then the current through the field windings will also be two times as great. These currents develop a torque on the armature that is 4 (2 X 2) times that of the normal operating speed. Figure 7–52 depicts these conditions.

When the load increases on this motor, the armature slows down. This reduces the CEMF and allows the current through the armature and field windings to increase. The increase in current will increase the torque.

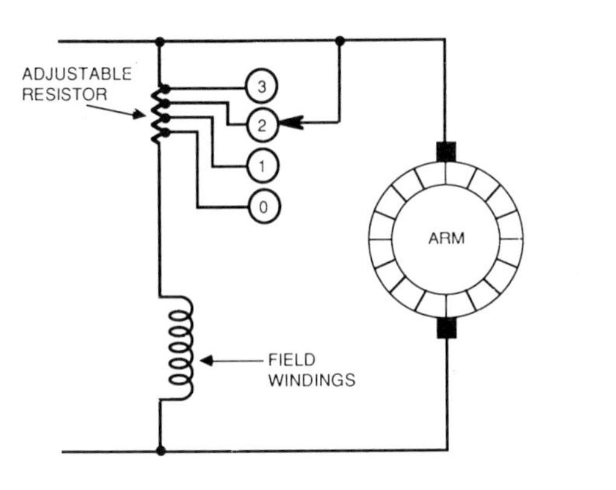

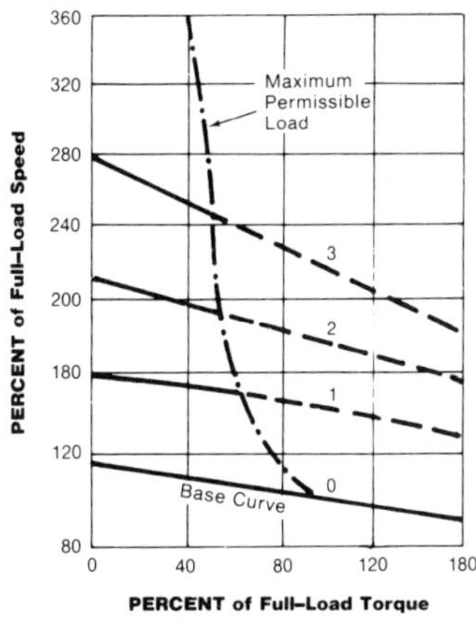

(A) Shunt field control for speeds above base speed.

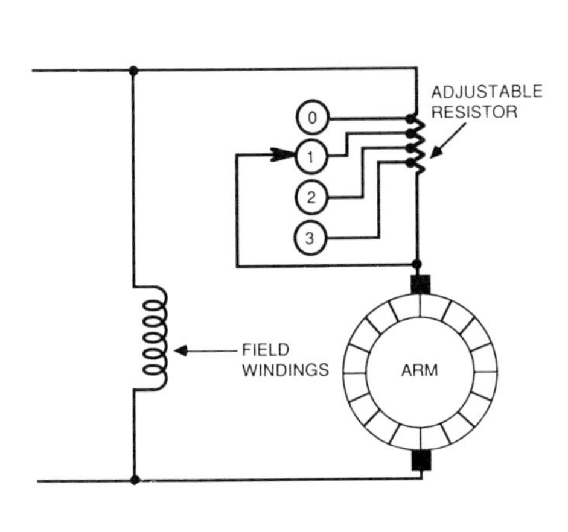

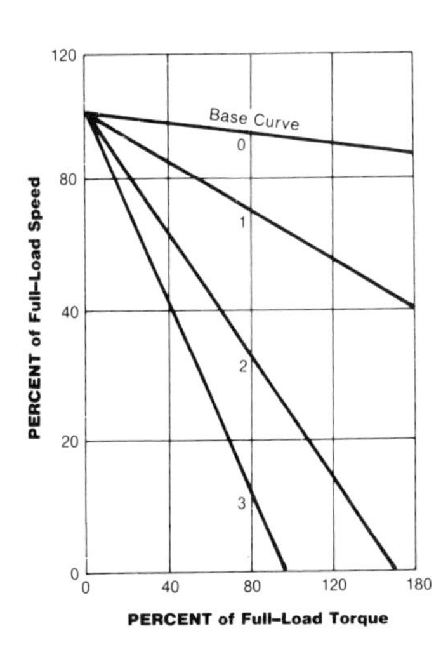

(B) Resistor in series with the armature for speeds below base speed.

Figure 7–50. Adjustable speed DC shunt motor.

Speed

The speed of the series DC motor varies greatly depending on the load. This type of motor runs at very high speeds when unloaded. Under the no-load condition, the current through the armature and the series field decreases. This reduces the flux density and therefore the CEMF. The armature must then turn faster to develop the CEMF to limit the current. The RPM of the motor can be six times the normal operating speed. This is high enough for some of the larger motors to destroy themselves. The brush friction, bearing losses, and windage are usually high enough on small motors to limit the speed to a safe level.

If the load is increased, the armature current will increase. This creates a stronger magnetic field, and the armature will turn slower to develop the CEMF necessary to limit the current flowing through it. This wide range of speed due to loading is called the *series characteristic*. When an induction motor is designed to have a high slip with increased load, it has the series characteristic.

Because of the high-speed series DC motors can develop, these machines should be either direct-coupled

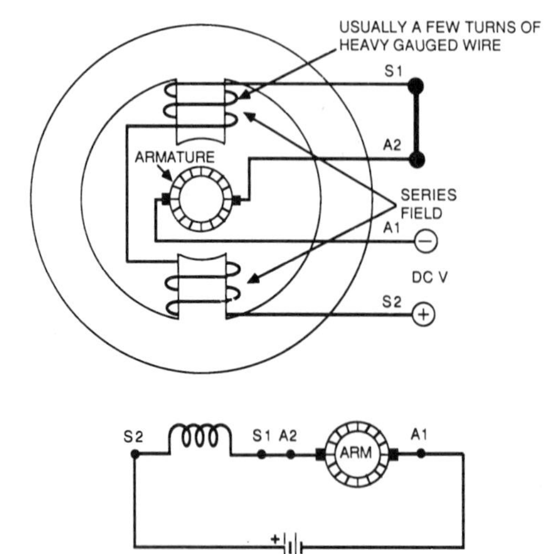

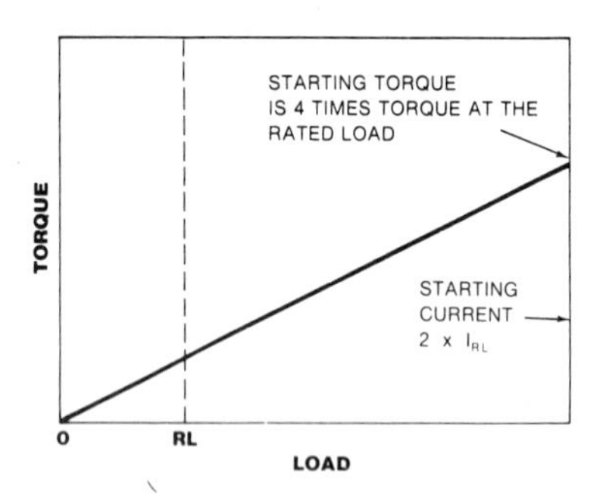

FIGURE 7–51. Series DC motor.

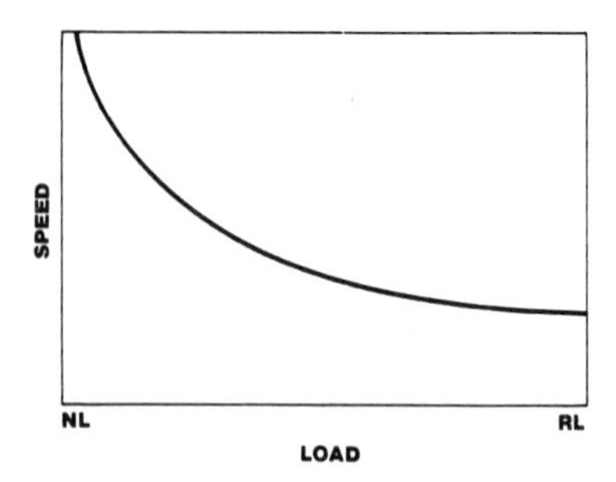

FIGURE 7–53. Speed vs. load of a DC series motor.

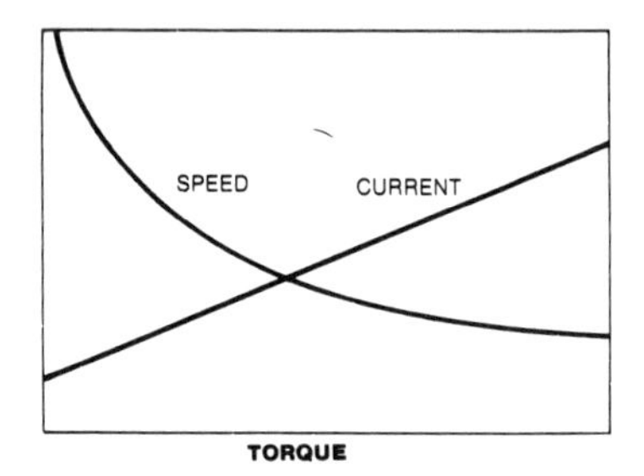

FIGURE 7–54. Speed and current vs. torque of a DC series motor.

STARTING TORQUE IS 4 TIMES TORQUE AT THE RATED LOAD

STARTING CURRENT $2 \times I_{RL}$

TORQUE

0 RL LOAD

FIGURE 7–52. Torque vs. load of a DC series motor.

or geared to the load. Belts are never be used for this purpose. If the belt was to break, the load would be removed, and the motor's RPM could rise to a dangerous level.

Figure 7–53 graphs the speed versus the load of this motor. Figure 7–54 shows the combined graphs of the speed and current versus torque.

The speed of a DC series motor can be controlled by reducing the voltage across the motor. This is illustrated in Figure 7–55A. A series-adjustable resistor is wired in series with the motor. The resistor drops some of the voltage being applied to the motor. This causes the motor to run at a lower speed. Maximum speed is obtained when the adjustable resistor is set at "0," and minimum speed occurs when the resistor is set at "3."

Figure 7–55B depicts another method of controlling the motor's speed. In this case a resistor is also placed in parallel with the armature. This resistor reduces the armature current but leaves the current through the series field windings relatively constant. With reduced armature current the motor will run slower.

Compound DC Motors

Compound DC motors combine the characteristics of the shunt and series types of motors into a single machine. This motor has both the shunt and series coils. Figure 7–56 illustrates the connections for the compound motor.

This motor does not have the disadvantage of a runaway speed as does the series motor. The shunt field has a constant current, thereby limiting the motor's maximum speed when the load is suddenly removed.

At the same time, this motor has good torque characteristics due to the series windings. When the load is increased, and the armature turns slower, the increased current through the armature provides for a greater torque.

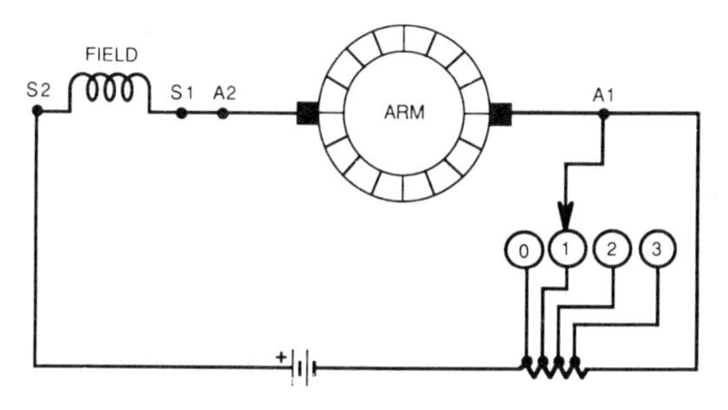

(A) Series adjustable resistor.

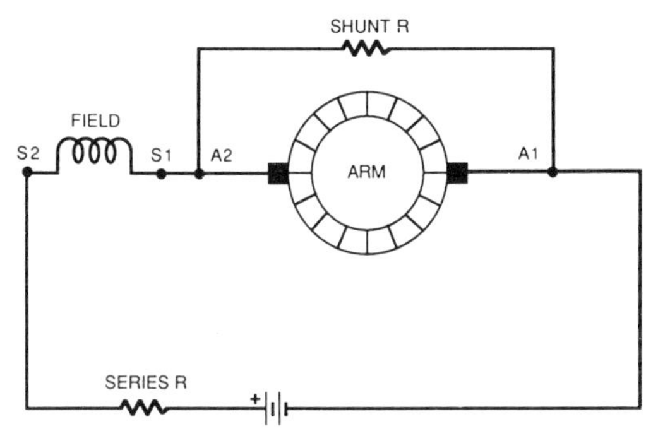

(B) Shunt resistance.

FIGURE 7–55. Methods of controlling the speed of a DC series motor.

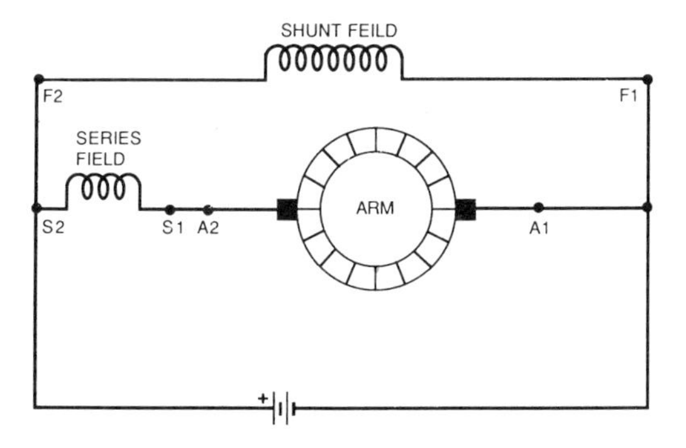

FIGURE 7–56. Compound DC motor.

The DC compound motor is a compromise between the series and shunt types. It limits the maximum speed the motor can achieve to a safe level, but its speed regulation is not as good as the shunt motor. At the same time it provides for relatively good torque which is somewhat lower than that of the series motor.

Applications

These motors are used to operate conveyors, elevators, air compressors, hoisting equipment, paper cutters, printing presses, centrifugal pumps, and for a variety of purposes where the load fluctuates and a constant speed is not essential. They are often used in conjunction with a flywheel which stores energy under light loads and puts mechanical energy back into the system under increasing load. Flywheels help smooth out the fluctuation of line current and line voltage. Direct current compound motors are usually selected as the motor of choice when the line voltage at the plant changes suddenly or fluctuates greatly.

Cumulative Compound Motor

When the shunt and series coils are connected so that an increase in armature current in the motor causes an increase in the field flux, the motor is called a *cumulative compound* motor. This arrangement causes the motor to slow down faster than the shunt motor when the armature current increases.

Some cumulative compound motors have controls that provide for starting as a series motor in order to develop a high torque. Once the motor is up to speed, the series windings are shorted, and the motor is run as a shunt motor for the better speed regulation characteristics.

Differential Compound Motor

The *differential compound motor* has its series and shunt coils wired so that an increase in armature current will cause a decrease in the flux field. This allows the motor to increase in speed as the motor's load is increased. Because the field flux varies at a slower rate than the armature current in this motor, the torque will exceed the demand of the load. This is the only type of motor that provides for negative regulation.

This type motor is sometimes referred to as a "suicide motor." While the motor is operating, if the shunt field suddenly was to open, the series field would then take control. This would cause the polarity of the field poles to reverse. The motor will come to a sudden stop and begin rotating in the opposite direction as a series motor. Under certain conditions, this can cause extreme damage to the driven parts. Under other conditions of coupling, the load could be removed with reverse rotation. The motor would accelerate in speed and destroy itself.

These motors also have a tendency to start in the reverse direction if under a heavy load. Once again, damage may result. For these reasons, the differential compound motor is seldom used in industry.

Universal Motors

Universal motors get their name because they will operate either on DC or AC up to 60 Hz. Their performance will be essentially the same when operated on DC or at 60 Hz.

Applications

One does not merely go out and buy a universal motor. They are designed for special applications. They are commonly used in common electrical hand tools such as power saws, portable drills, routers, and joiners. They are also used for vacuum cleaners, mixers, and a host of household appliances that require an electric motor.

Theory

The theory of operation of the universal motor is the same as that of the DC motor. Direct current motors do not operate well on AC because of poor efficiency, low power factor, and arcing at the brushes. The design of the universal motor reduces these deficiencies to an acceptable level.

Shunt-type designs are not used for universal motors because of the high reactance present when the motor is operated on AC. This would reduce the current flow through the shunt field and weaken the magnetic effect below the required torque to start the motor.

Universal motors are designed as special types of AC series motors and have the series characteristic of that of DC motors. The series field consists of few turns of heavy gauge wire.

To develop the required torque, the armature is wound with more coils, each having fewer turns than the DC motor. The commutator of the universal motor will, therefore, have more segments than that of the DC motor. Inductive effect is reduced and the magnetic field is approximately the same if operated on DC or AC.

Because the armature and field are in series, the current is the same through them, and there is no time lag or phase displacement in the magnetic fields. The attraction and repulsion forces will nearly be the same if operated on DC or AC. Runaway speeds are avoided because the motor is either directly coupled or geared to the particular appliance in which it is being utilized.

Losses

Universal motors have copper, hysteresis, and eddy current losses as do other motors. Copper losses are held to the minimum by using larger-diameter wire. Hysteresis losses are limited by the annealing of special silicon steel. Laminated cores reduce the effects of eddy current loss. All of these methods were detailed before.

Efficiency of universal motors is fairly low. They range from about 30% for small motors to 70% or 75% for larger types.

Compensating Windings

The armature reaction is greater in the universal than a DC motor of the same ratings. This is due to the larger number of armature conductors and armature magnetomotive force (mmf) of this motor. Compensating windings are almost always used to reduce these effects. The compensating windings may be wired conductive (Figure 7–57A) or inductive (Figure 7–57B). These windings are set at 90 electrical degrees from the main field axis. The magnetic flux created by these windings will be proportional to the armature current.

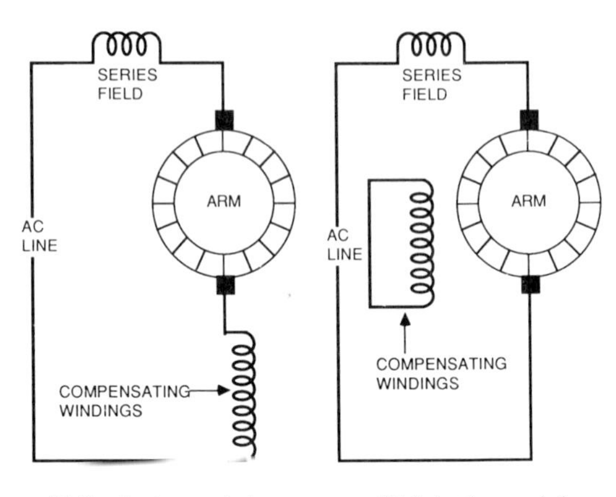

(A) Conductive coupled. *(B) Inductive coupled.*

FIGURE 7–57. Compensation windings for the universal motor.

Speed

Full-load operating speeds are usually in the range of 4000 to 16,000 RPM and may be over 20,000 RPM. Because of the high speeds available with these motors, they develop more horsepower per unit pound than any other AC motor. This makes them particularly applicable for portable tools.

For example, a typical one-half horsepower induction motor with a 5/8-inch shaft operating at 1750 RPM will weigh approximately 30 pounds. A 2½-pound universal motor develops the same horsepower with a 5/16-inch shaft at 19,000 RPM.

Most universal motors are coupled to their loads through a speed reducer. Spur and worm gears are common. Belts may be used with pulleys for this purpose. The friction and windage of the smaller universal machine is sufficient to limit the RPM below the destructive level in the event a belt breaks. Larger motors may present a danger, and belt coupling is not recommended.

Speed controls are available on many devices that use the universal motor. This is accomplished by putting resistance in series with the series field and armature (Figure 7–58A), or reducing the voltage to the motor using a variable or tapped transformer (Figures 7–58B and C). Silicon control rectifiers (SCR) are often also used (Figure 7–58D). These devices provide almost full torque at reduced speeds, whereas the other devices do not have as good torque characteristics at the lower speeds. Figure 7–58 illustrates these methods.

Reversing Direction of Rotation

Universal motors may or may not be reversible. Non-reversible devices normally have three leads. These include the two leads for power and one lead for grounding the frame of the motor. Grounding is required on portable tools due to the high incidence of electrocution caused by them (Figure 7–59).

Reversible motors have four or five leads available for connection. One of these is the ground.

The four lead motor only requires a single-pole, double-throw switch to reverse the direction of rotation. This arrangement requires two field coils that are wound in opposite directions to reverse their magnetic fields. These are usually concentrated-pole motors.

Five lead motors require a double-pole, double-throw switch. This switch will reverse the direction of current either through either the armature or the series field, but not both. Figure 7–60 illustrates the four and five lead connections.

Summary

The split-phase motor is available as a general purpose or special service motor. General purpose motors may be loaded up to their service factor. They are used for medium-duty applications where starts and stops are frequent. The special service split-phase motor usually

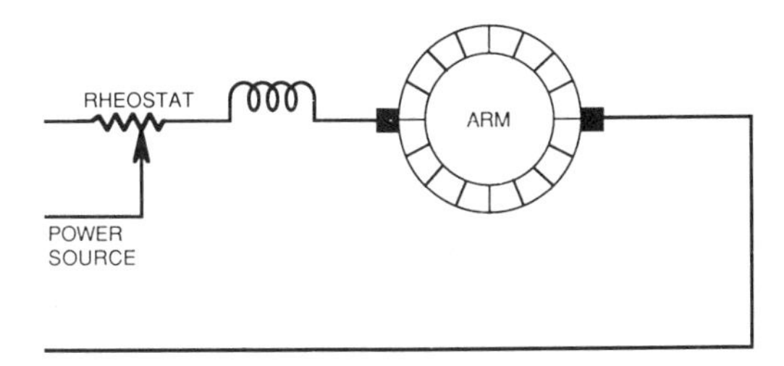

(A) Variable resistance.

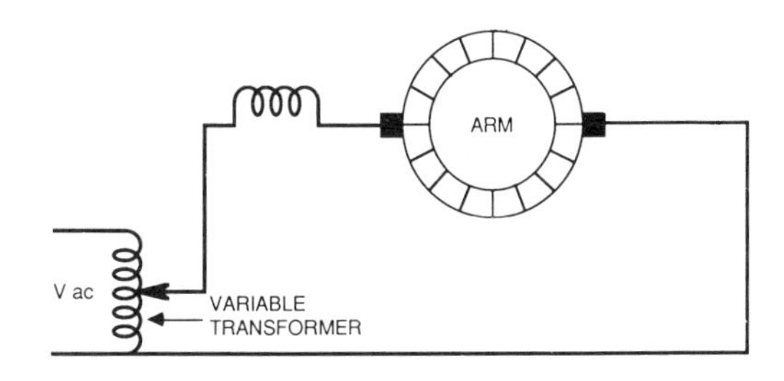

(B) Variable transformer.

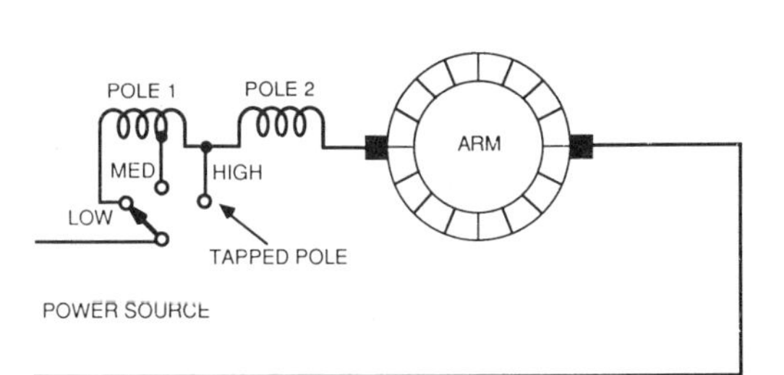

(C) Tapped transformer.

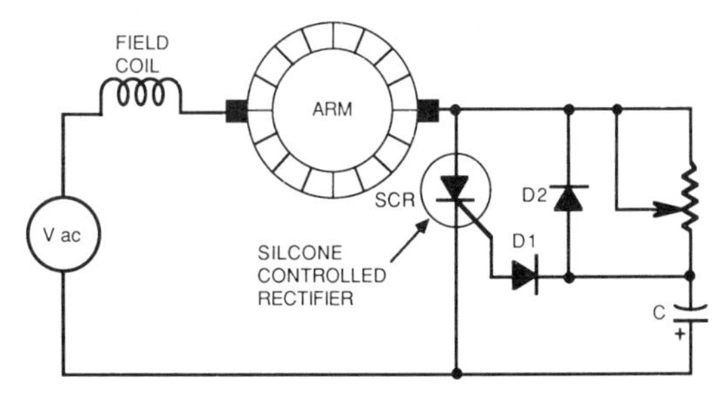

(D) Silicon controlled rectifier.

FIGURE 7–58. Speed controls for universal motor.

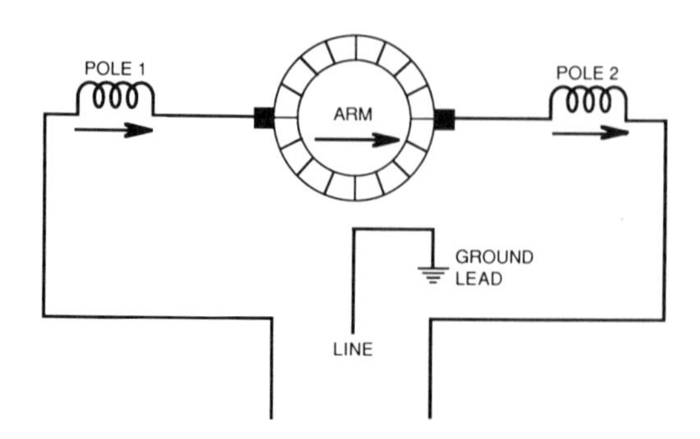

Figure 7–59. Nonreversible universal motor.

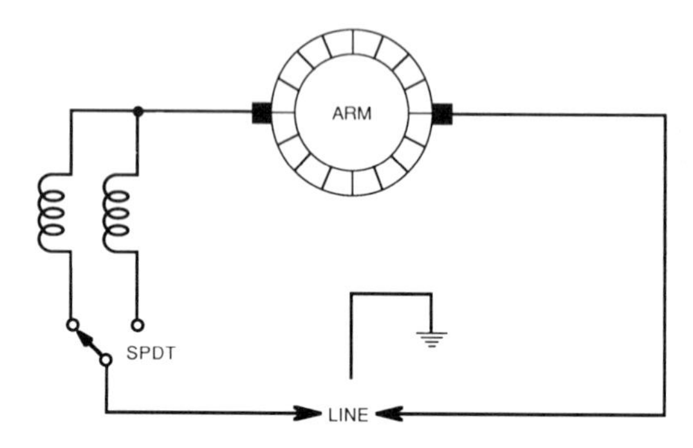

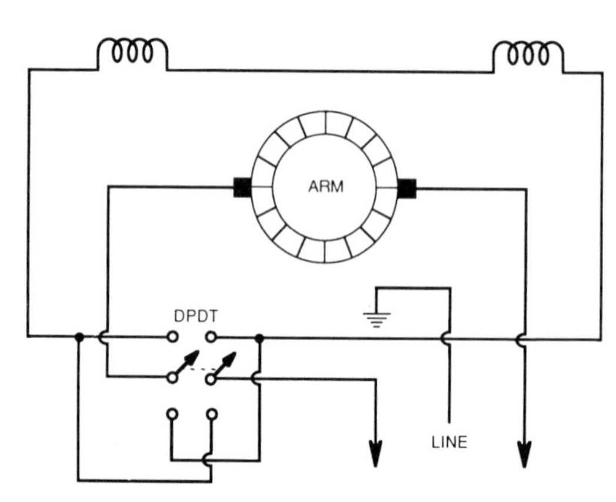

Figure 7–60. Methods of reversing the universal motor.

has a service factor of 1. It should be used only for the load for which it was designed. If used for other applications, the load should never exceed the horsepower rating of the motor. Starts and stops must not be frequent.

Capacitor-start motors are used where the starting load is heavy. They operate as split-phase motors when up to operating speed. Their high starting torque requires only a normal starting current. They can carry small overloads continuously and large overloads on an intermittent basis.

Capacitor-start, capacitor-run motors have the advantage of the high starting torque of the capacitor-start motor. The capacitor used to start the motor is normally an electrolytic. The run capacitor is an oil type. A centrifugal switch is used to change the capacitors from the start to run condition. The torque is higher than the split-phase motor in the run condition.

Permanent split-capacitor motors are designed so that the capacitor delivers equal values in the start or run cycle. They have lower starting torque than the other two types of capacitor motors. They do not require a starting switch and are easy to maintain. They have a high efficiency and a high power factor.

Shaded-pole motors have a low starting torque and their applications are limited to light-duty loads. They are relatively inexpensive and easy to maintain. They do not have a service factor and must be operated at their rated load.

Repulsion motors have largely been replaced by capacitor-start motors. They are more expensive than other types of motors, but they have the highest starting torque per ampere than any other single phase motor. Maintenance is frequently required on these motors with resulting higher cost of operation.

Direct current motors have become increasingly popular in industry. This is due to their characteristics of being highly controllable in speed, torque, and horsepower. Their direction of rotation can be easily reversed. Precision movements can be accomplished with them.

Universal motors are primarily used in small tools and appliances. Their primary advantage is that they can develop high torque and horsepower. This is accomplished with a small, lightweight machine. These machines have a limited life expectancy.

Questions

1. The voltage ratings for most FHP, single-phase induction motors is ____________.
2. What are the general classifications used to designate speed control for motors?
3. Will a split-phase induction motor run at the synchronous speed? Explain.
4. How is a rotating magnetic field obtained in a split-phase motor?

5. What is the displacement in electrical degrees between the main and auxiliary poles of the split-phase motor?
6. The start windings of a split-phase motor is removed from the circuit at 70% to 80% of the operating speed by a ______.
7. What letter of the alphabet is used to designate the leads of a motor?
8. A motor lead has the prefix "P." What does this tell you?
9. If a color code is used instead of a prefix with a number, what would be the color of the leads connected to the start windings of a motor?
10. Can the start and run windings of a split-phase motor be determined without dismantling the motor if there is no distinguishable markings on the leads? Explain.
11. Is it possible to reverse the direction of rotation of a split-phase motor? Explain.
12. How would you connect the run windings of a dual-voltage motor to operate at the higher voltage? Is this true for the start windings also?
13. Under which condition on a dual-voltage motor would a thermal protector be wired in series with both run windings?
14. How is the speed of a split-phase motor usually changed?
15. How are consequent poles formed?
16. What is the advantage of the capacitor-start motor over the split-phase motor?
17. How is the capacitor wired in a capacitor-start motor?
18. Capacitors used with motors are rated in ______ and ______.
19. Does the value of capacitance needed for a capacitor-start motor increase or decrease as the horsepower is increased?
20. How does the capacitor-start motor differ from the split-phase motor when both are in their run cycles?
21. Can all capacitor-start motors be reversed?
22. Is it possible to instantly reverse the direction of rotation of a capacitor-start motor while it is running?
23. Will a two-speed capacitor start motor that uses a single capacitor start on its low speed?
24. What type of capacitor is used on a capacitor-start, capacitor-run motor when the same capacitor is used for both functions? Why?
25. What is the advantage of using a capacitor in the circuit during the run cycle?
26. How is the direction of rotation of a permanent-split capacitor motor reversed?
27. How does the starting torque of a shaded-pole motor compare to that of a capacitor-start motor?
28. In which direction does a shaded-pole motor rotate?
29. Which type of FHP single-phase motor has the highest starting per ampere?
30. The brushes of a repulsion-start, induction-run motor are ______ together.
31. The purpose of the centrifugal device on a repulsion-start, induction-run motor is to ______ the commutator segments. It may also be used to ______ the brushes from the commutator when the motor is running at normal speed.
32. Will a repulsion-type motor start when the brushes are set at either the soft or the hard neutral?
33. When is maximum starting torque developed in the repulsion-type motor?
34. Which type of repulsion motor uses a centrifugal switch?
35. Which type of repulsion motor has a squirrel cage rotor embedded in the windings?
36. Which type of repulsion motor starts and runs on the repulsion theory?
37. Why are direct current motors used more frequently for industrial applications?
38. What is the primary factor that limits current flow through the armature of a DC motor?
39. Which type of DC motor tends to develop high speeds when it is not under load?
40. Interpoles reduce the effects of ______ ______ and of ______.
41. Motor horsepower is a function of ______ and ______.

42. How can the direction of rotation of a DC motor be reversed?
43. Increasing the current flow through the shunt field of a DC motor will cause the RPM of the motor to ______.
44. Which type of DC motor has the highest starting torque?
45. Which type of DC motor has both shunt and series field windings?
46. A differential compound motor is sometimes referred to as a ______ motor.
47. The ______ motor operates well on either AC or DC power.
48. How are runaway speeds prevented for the universal motor?
49. What special arrangement is used in the universal motor to overcome the effects of armature reaction?
50. How is the size and weight of the universal motor reduced while the motor is capable of developing comparably large horsepower?

CHAPTER 8

Polyphase Motors

Polyphase motors are either two- or three-phase motors. Their construction is similar, but the wiring of the coils is different. Two-phase motors are seldom used and will not be discussed in any detail. Suffice it to say that the phase relationship of the voltages in this type of system is 90° apart. Voltage between phases will be equal to 1.414 (square root of 2) times the single-phase voltage. This value is less than the 1.73 multiplier (square root of 3) for three-phase systems. The theory of operation for two-phase motors is the same as for three-phase motors.

General Considerations

Three-phase motors are common in industry. They vary in size from fractional horsepower to several thousand horsepower. Usually they are multiple-horsepower motors.

Maintenance and Repair

Maintenance and repairs on three-phase motors are generally less than that needed for single-phase motors. Three-phase motors do not require the addition of a start winding, centrifugal switches, or capacitors to operate. Maintenance and repair procedures for two- and three-phase motors are essentially the same.

Power Requirements

These motors are manufactured for almost every standard voltage and frequency. Any value of distribution voltage can be used as the basic supply. Transformers are available to step the voltage up or down to meet the requirements of the motor. Often three-phase motors are dual-voltage motors. In most cases they operate at 60 Hz.

Three-phase power systems supplying these motors are three, single-phase systems. The difference is that the voltage of each phase is displaced from the other two phases by 120°.

Operating characteristics of three-phase motors are superior to that of single-phase machines. The reason for this is the single-phase system supplies zero power to the motor as it alternates. Alternations occur two times each cycle. If the power factor is less than unity for a single-phase system, the power going to the load is actually negative during part of the cycle. The power to a three-phase motor is nearly constant for the entire cycle.

Stator

These motors are simple to manufacture. Like the single-phase motors, three-phase motors consist of a stationary part called the *stator*, and a revolving part, the *rotor*.

Power is applied to the stator windings of a three-phase motor. The magnetic field created is due to the interaction of three groups of coils. Each phase will have at least one group of coils. A two-pole motor will have six groups (3 X 2); a four-pole motor will have 12 groups (3 X 4); a six-pole motor will have 18 groups (3 X 6); and an eight-pole motor will have 24 groups (3 X 8). These coils overlap each other as they are positioned in the slots of the stator.

The stator's flux field is a rotating magnetic field. An explanation of how this field is created is given in Chapter 6. Review that section of the text if you do not have a firm understanding of this principle of operation.

Transformer Action

For induction types of three-phase motors, the rotor receives its power through transformer action. The magnetic field created by the current flowing in the stator cuts the conductors in the rotor and causes current to flow. Magnetic flux produced by the stator's current is almost constant. This is due to the stator being directly across the power source. The CEMF of the rotor is nearly equal to the EMF that produces current flow in it.

When current increases in the rotor circuit, its magnetic flux will increase. This flux will oppose the stator's flux and cancel it. Because the voltage across the stator is constant and cannot change, the stator's current will increase due to the reduced inductive reactance to compensate for the effect of the rotor's field.

Rotor Current

The induced current in the rotor creates its own magnetic field. Rotor current varies directly with the load and the induced voltage. It varies indirectly with the rotor impedance. The rotor impedance (Z_R) is equal to the vector sum of its resistance (R_R) and the reactance (X_R).

$$Z_R = \sqrt{R_R^2 + X_R^2}$$

The rotor for three-phase motors is either a wound or a squirrel cage type. Wound rotors have a higher reactance than the squirrel cage. Squirrel cage rotors are designed to have different impedances. The only means of changing the impedance after they are manufactured is to operate the machine at a different frequency.

Rotor reactance will be affected by the frequency of the current passing through it. The frequency of the induced current in the rotor is a function of the frequency of the stator current and the percent of slip. The rotor frequency is 60 Hz, or the line frequency if other than 60 Hz, for a locked rotor, and zero hertz if turning at synchronous speed. Rotor frequency is calculated by means of the following formula.

$$F_R = F_S \times \% \text{ Slip}$$

where,

F_R is the rotor frequency,
F_S is the stator frequency.

Applying this formula to a 60-Hz, three-phase motor having a 5% slip, we find the frequency of the current in the rotor to be 3 Hz. Determine the frequency of the rotor current for an eight-pole, 60-Hz motor that turns at 825 RPM. If your answer is 5 Hz, you are right.

For this particular problem, the % slip must be determined first. In this case, an eight-pole motor would have a synchronous speed of 900. To determine % slip:

$$\% \text{ Slip} = [(900 - 825)/900] \times 100$$
$$= 8.333\%$$

and

$$F_R = 60 \text{ Hz} \times 0.08333$$
$$= 5 \text{ Hz}$$

Torque

Starting currents may be relatively low or very high. Starting torque on these motors may be high, or they can be designed to produce a low torque. The motor may be designed for a high or low torque at its rated RPM.

The torque developed by the induction motor depends mostly on three factors. These are the

1. Rotor current
2. Strength of the magnetic flux field
3. Phase angle between the rotating magnetic flux and the rotor currents

The formula for determining the torque is

$$T = K\varphi\, I_R \times PF_R$$

Torque is equal to a constant for any given motor (K), times the flux in the stator (φ), times rotor current (I_R), times the rotor's power factor (PF_R). The power factor is equal to the cosine of the angle between rotor current and the magnetic flux.

Maximum torque is always produced when the phase relationship between the rotor current and the flux field is 45 electrical degrees apart. The power factor is 70.7% at this point (cos 45°). This condition occurs when the rotor's resistance is equal to its reactance.

Wiring Diagrams

All three-phase motors are either star (wye) or delta wired. Most of these motors have dual windings in each phase to provide for dual-voltage operation. Nine leads are generally available for wiring the motor to the power source. The numbering of these leads has been designated by NEMA standards. Figure 8–1 illustrates the identification of the leads by the numbers assigned to a wye wired motor. Figure 8–2 depicts the delta wired three-phase motor.

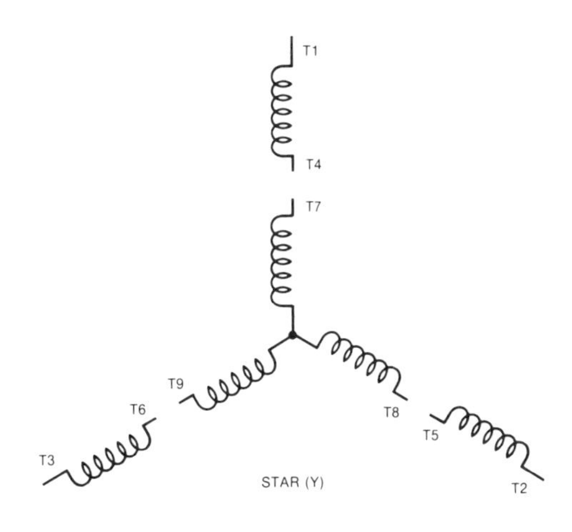

FIGURE 8–1. Standard numbering system for the leads of a three-phase star (Wye) wired motor.

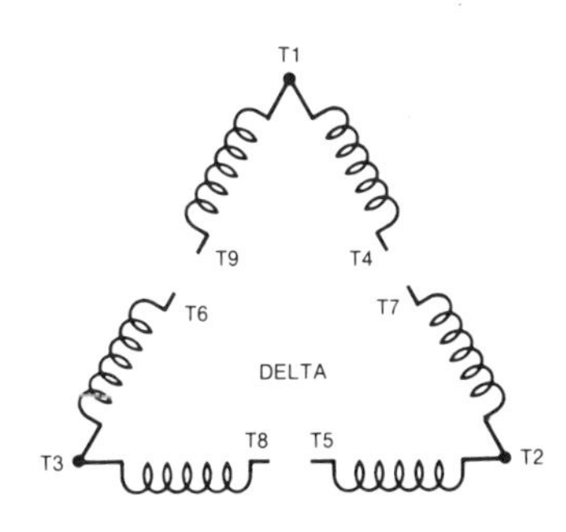

FIGURE 8–2. Standard numbering system for the leads of a three-phase delta wired motor.

In wye connected motors the ends of each of the individual phases join at a common point. In delta wired motors the ends of each phase are connected to the beginning of the next phase. An electrician needs only an ohmmeter or a battery and bell to determine a motor's wiring scheme.

There is continuity between three pairs of wires and one set of three wires for the wye connected, dual-winding motor. The meter will deflect or the bell will ring in the following cases: T1–T4, T2–T5, and T3–T6 for the three pairs; and T7–T8–T9 for the set of three wires.

The continuity check on a delta system will identify three separate sets of three wires. These combinations are: T1–T4–T9, T2–T5–T7, and T3–T6–T8.

Competent electricians are able to visualize the wye and delta connections. They must also know the numbering system to wire the motor correctly. Check the wiring diagram on the motor nameplate.

In the majority of cases, L1, L2, and L3 of the power lines connect to T1, T2, and T3 of the motor leads, respectively, for both the wye and the delta motor. This will provide counterclockwise rotation of the motor as viewed from the bell housing end of the motor. T1, T2, and T3 also corresponds to the start of each winding for each phase. The number of the opposite end of each of these coils is determined by adding three. The opposite end of T1 will be T4, T2 will be T5, and T3 will be T6.

The three remaining leads are T7, T8, and T9. These numbers are the start points for the second winding for each phase. Adding three to the end number of each winding connected to the power source will provide the number for the start point of the second coil which belongs to that phase. T7 belongs with T4, T8 with T5, and T9 with T6.

Dual-Voltage Connections for the Wye Wired Motor

Figure 8–3 illustrates the wye connected motor wired for its higher voltage. The three phases of the power line go to T1, T2, and T3, respectively. The two windings on each phase are in series with each other. This is accomplished by wiring T4 to T7, T5 to T8, and T6 to T9.

Figure 8–4 shows the lower voltage configuration for the wye-wired motor. Leads T1–T7, T2–T8, and T3–T9 are connected together and wired to L1, L2, and L3, respectively. To complete the star connection, T4–T5–T6 are joined. This arrangement places the windings in parallel with each other with two star points. If the common star point at the ends of T7, T8,

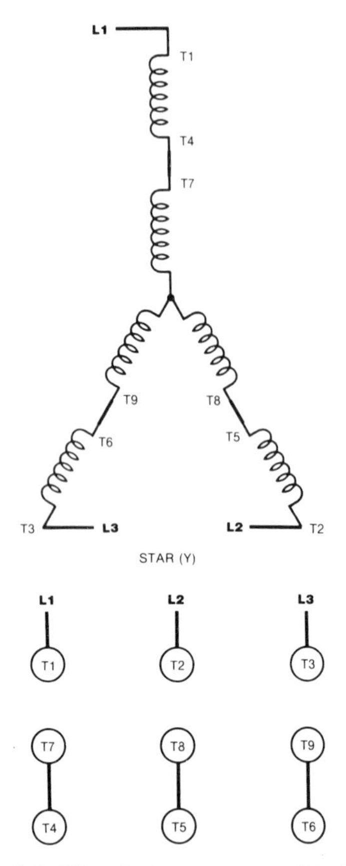

FIGURE 8–3. Wye-wired motor connected for higher voltage.

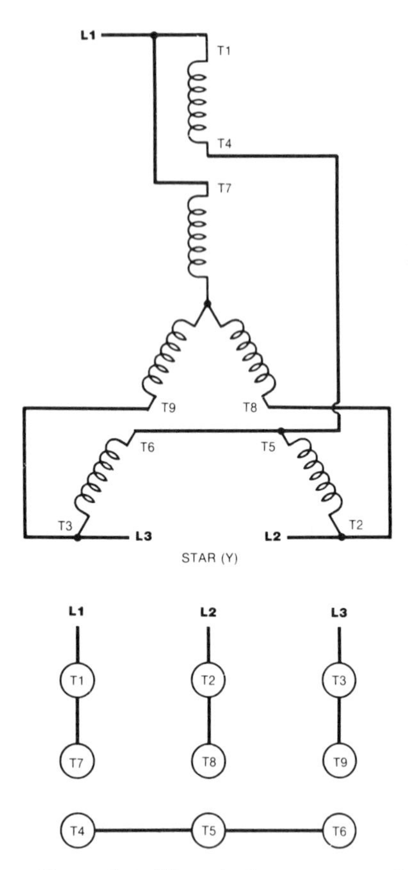

FIGURE 8–4. Wye-wired motor connected for lower voltage.

and T9 are available, the common points of the two stars can be wired together. This provides a common star point for both sets of windings.

Dual-Voltage Connections for the Delta-Wired Motor

Figure 8–5 illustrates the delta-wired motor connected for its higher voltage. In this case, T4–T7, T5–T8, and T6–T9 are connected. The windings of each phase are in series with each other. The three phases of the power line, L1, L2, and L3 connect to T1, T2, and T3, respectively.

Figure 8–6 depicts the delta-wired motor connected for the lower voltage configuration. In this case, T1–T6–T7, T2–T4–T8, and T3–T5–T9 are joined and wired to L1, L2, and L3, respectively. This arrangement puts the two windings of each phase in parallel with each other. Full line voltage will be across each winding.

Another arrangement sometimes provides for the motor to be wired delta for the lower voltage and wye for the higher voltage. The wye-wired motor requires a voltage 1.73 times higher than that of the delta-wired motor. This motor has a single winding for each phase. Six leads are brought out of the motor. Two for each phase. Figure 8–7 shows the numbering sequence of the leads.

For the wye connection, T4–T5–T6 are wired together. L1, L2, and L3 connect to T1, T2, and T3, respectively. Delta wiring for the lower voltage requires T1–T6, T2–T4, T3–T5 to be joined and connected to L1, L2, and L3, respectively. Motors are sometimes wired wye when started. This arrangement provides for reduced voltage during this cycle. As the motor comes up to speed, the windings are switched to the delta wiring scheme for the run cycle.

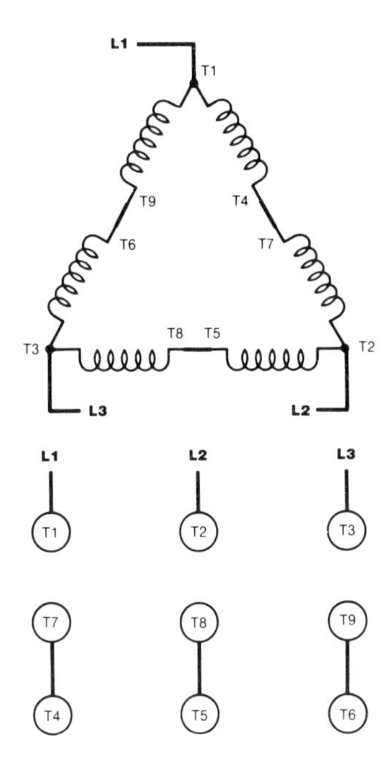

FIGURE 8–5. Delta-wired motor connected for higher voltage.

Motor Speed

RPM of three-phase induction motors is dependent upon the synchronous field speed minus the slip. The synchronous speed changes with the number of poles and the frequency of the power source. The synchronous speed at 60 Hz can be determined by the following formula.

Synchronous Speed = 120 X Hz/Number of Poles

A two-pole induction motor will operate at 3600 RPM minus the slip; a four-pole motor will revolve at 1800 RPM minus the slip; and the RPM of an eight-pole motor would equal 900 minus the slip.

Multispeed motors, depending on the wiring scheme, provide different operating characteristics. They are known as constant horsepower, constant torque, and variable torque motors.

Torque varies inversely with the speed in a constant horsepower motor. More current will flow at the low speed, thereby increasing the torque. The torque decreases in the same ratio as the speed increases. When torque is given in pound-feet, and the speed is in RPM, the constant horsepower can be expressed by the following mathematical formula:

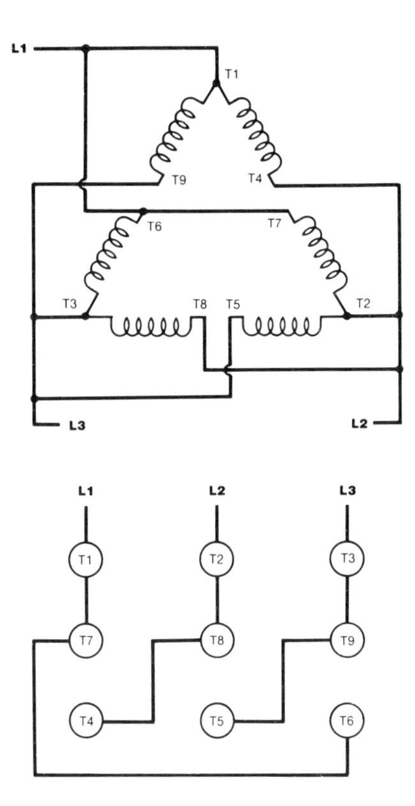

FIGURE 8–6. Delta-wired motor connected for lower voltages.

Horsepower = (Torque X RPM)/5252

Horsepower changes in proportion to the speed in the constant torque motor. The horsepower and the line current increase in the same ratio as the motor speed to provide constant torque. Connections for constant torque motors are usually opposite those of the constant horsepower type.

Torque in variable torque motors varies inversely with the speed. The horsepower varies with the square of the speed change. Torque and horsepower both increase at the higher speed.

There are several methods of controlling the speed of three-phase motors. Some of these are

1. Changing the frequency of the power source.
2. Changing the number of poles.

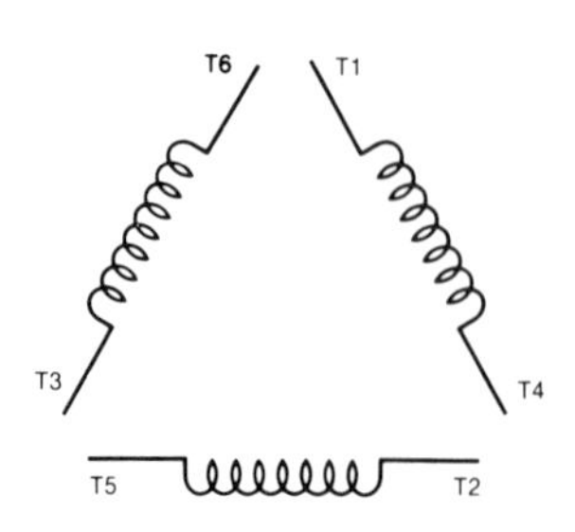

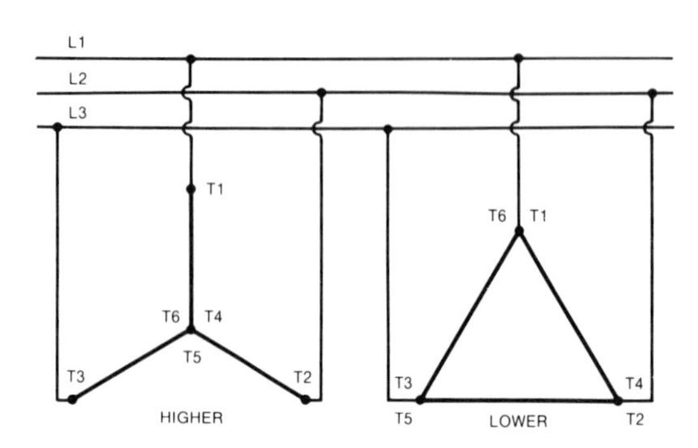

FIGURE 8–7. Wye-delta, dual-voltage motor.

3. Changing the percent of slip.
4. Using a foreign voltage in the secondary windings.

Changing the frequency of the power source is a special application. Each motor requires its own generator or frequency converter. This method is sometimes employed for single large motors that require precision speed control. Broad usage of this method is not feasible for ordinary usage of motors. Electricians working on these systems need specialized instructions and careful study of the equipment for which they are responsible.

Speed of the three-phase motor varies inversely with the number of poles. The number of poles is changed by using two or more sets of windings and poles. Only definite speeds are obtainable by this method. The maximum number of speeds available by changing the number of poles is usually limited to four.

The number of poles can also be changed using the consequent-pole method. This is accomplished by changing the direction of current though the windings so that pairs of poles for each phase have the same polarity. The consequent-pole method effectively doubles the number of poles and reduces the motor's RPM to one-half its value. Using the consequent-pole method and two sets of windings for each phase will provide the necessary conditions to obtain four definite speeds on a given motor.

When motors operate at different definite speeds, the number of leads brought out from the windings on the stator may vary from the standard nine-lead arrangement. Figure 8–8 illustrates a two-speed, constant torque motor where only six leads are available for wiring the motor.

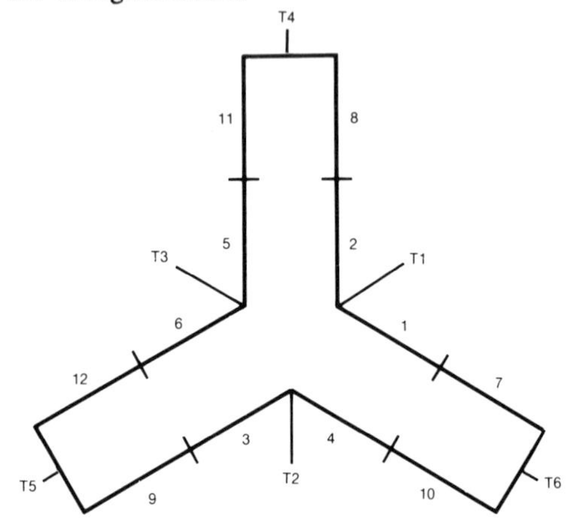

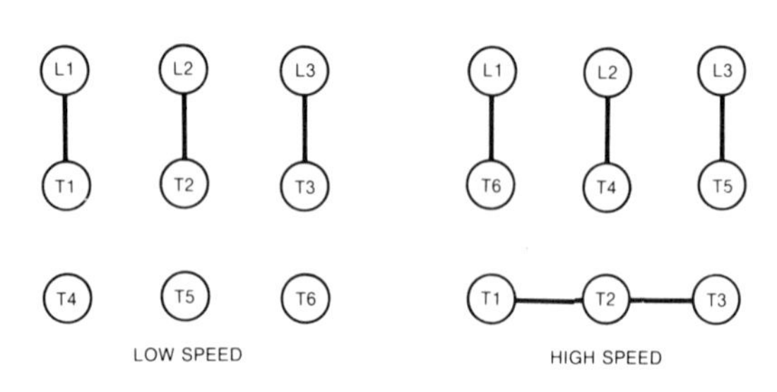

FIGURE 8–8. Two-speed, constant-torque motor using six leads.

This motor has either four or eight poles. Twelve windings are necessary to obtain the four poles for high speed. This corresponds to the three phases times the number of poles. Operation at the low speed requires eight poles. The additional poles are obtained using the consequent-pole method. The number of poles differs due to the wiring changes explained below.

For low-speed operation, T1, T2, and T3 connect to the respective three-phase power lines. T4, T5, and T6 are taped separately, insulated, and are not used. The windings of each phase are connected series-delta.

High-speed operation requires T6, T4, and T5 to be connected to the corresponding power lines. T1, T2, and T3 are joined together and insulated. This provides two-wye circuits in parallel with each other.

Figure 8–9 illustrates another six-lead motor which also provides two speeds with constant horsepower. Two parallel star circuits provide the low speed. In this case, T4, T5, and T6 are joined. The power lines connect to T1, T2, and T3, respectively.

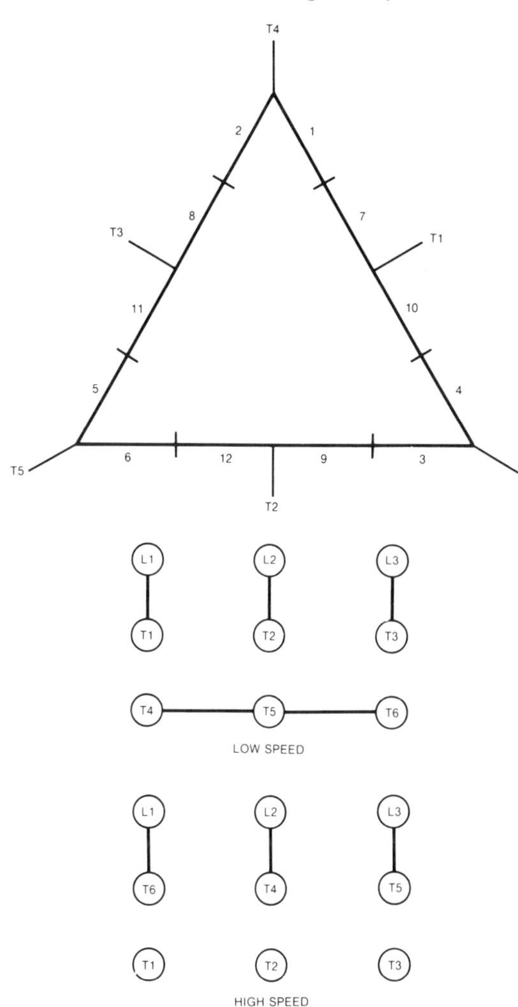

FIGURE 8–9. Two-speed, constant-horsepower motor using six leads.

For high speed, T1, T2, and T3 are taped separately and insulated from the rest of the circuit. T6, T4, and T5 connect to L1, L2, and L3, respectively. This provides a series-delta wiring configuration.

Two-speed, single-winding motors may also have seven leads brought out. The coil to which the seventh lead is attached can be used for normal operation when connected to T3. It can also operate with another winding in the same motor. On three- or four-speed motors the open winding is necessary to prevent unwanted circulating currents. Figure 8–10 depicts the seven-lead motor.

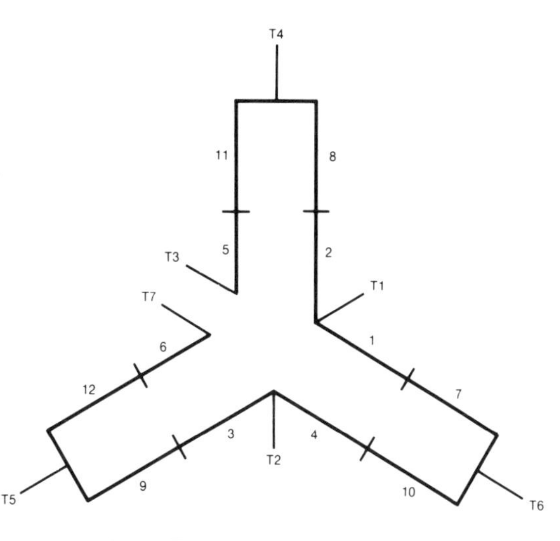

FIGURE 8–10. Two-speed, constant-torque motor using seven leads.

Slip varies proportionally to the square of the voltage. If the voltage decreases, the slip of the motor will increase, thereby reducing the speed of the motor. At the same time, the breakdown torque of the motor decreases as the square of the voltage. This places the motor in jeopardy of stalling while under load. The slip increases in proportion to any load change on an induction-type motor. Changing the impressed voltage across an induction-type motor to vary its speed is not recommended for this reason. This method may, however, be used in an emergency and in some special circumstances.

Another method of using slip for speed control is to change the amount of resistance in the secondary circuit, the rotor. This can only be done on the wound rotor-type motor. A squirrel cage rotor's resistance is fixed and cannot be varied for speed control.

The resistance method provides for variable speeds. Resistance is introduced into the rotor by connecting a three-phase rheostat through slip rings in series with the windings. Figure 8–11 depicts this condition.

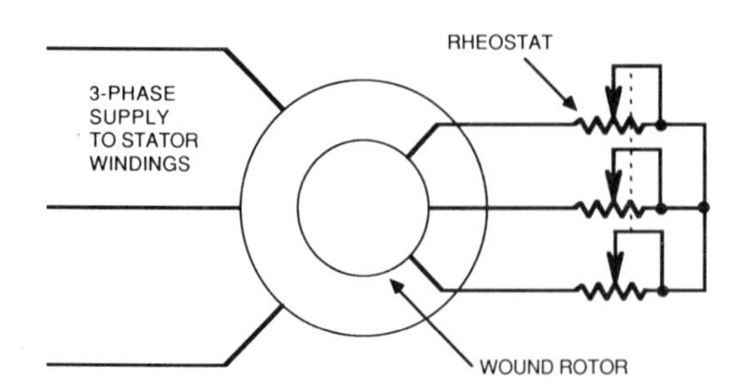

FIGURE 8–11. Speed control of a polyphase motor through use of resistance in series with the rotor windings.

Once set, the motor will run at a constant speed unless the load changes. Increased load will decrease the speed of the motor due to the slip. The load should be relatively constant for this method to effectively provide a fixed speed.

Additional resistance in series with the rotor windings will decrease the motors RPM if it is under load. The RPM will not change very much if the motor is not loaded, or under very light load.

This method lacks efficiency due to the power consumed by the resistance. The slower the motor runs, the lower the efficiency.

The introduction of a foreign voltage through slip rings into the rotor windings of a three-phase motor can also be used to control the speed. If this voltage opposes the induced voltage in the rotor, the RPM of the motor is reduced. When the voltage aids the induced voltage, the speed of the motor will increase. This method allows operating speeds at, above, and below the synchronous speed of the motor. This method has the advantages of higher efficiency, and the RPM of the motor is practically constant when the motor operates within its horsepower rating. Figure 8–12 illustrates one of these methods.

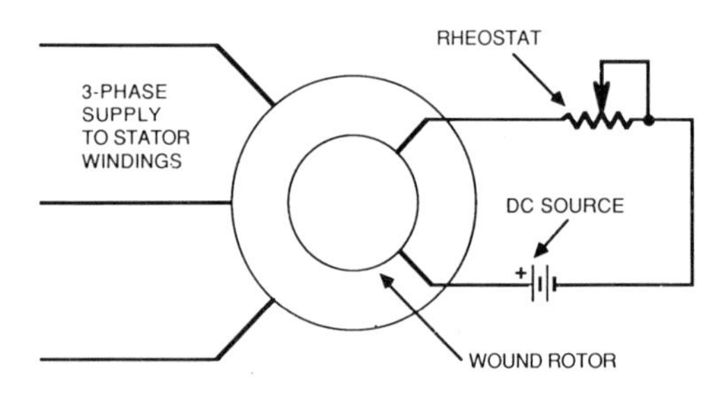

FIGURE 8–12. Speed control of polyphase motors using a foreign voltage in the windings of the rotor.

Electricians who work on motors must be aware that the number of leads on a three-phase motor may differ from the common nine-lead arrangement. The nameplate information and manufacturer's specifications must be studied to assure proper wiring for the speed, torque, and horsepower desired.

Direction of Rotation

Three-phase motors turn in the direction of the rotating magnetic field. The direction of rotation of these motors can be easily changed. This is accomplished by reversing the connections of two of the power lines to the motor leads. The result is the rotating magnetic field will reverse, and the rotor will turn in that direction.

When the application requires the motor to run in either direction, magnetic starters or contactors are generally used to reverse the connections between the three-phase power and the motor. Figure 8–13 shows a controller circuit that will cause the direction of rotation of a three-phase motor to reverse.

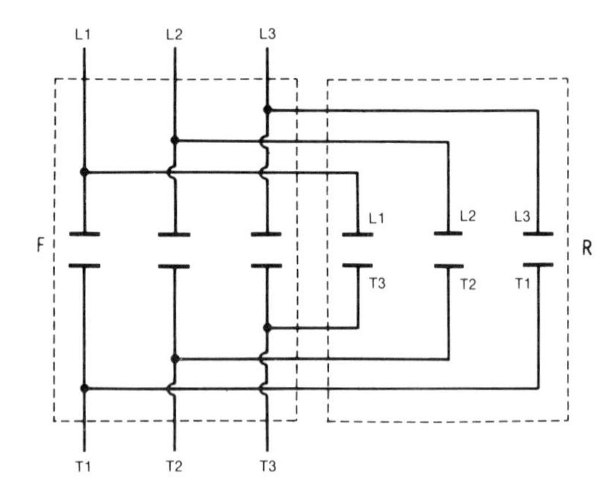

FIGURE 8–13. Magnetic controllers used to reverse the direction of rotation of a three-phase motor.

Types of Three-Phase Motors

There are three basic types of three-phase motors in use today. These are the squirrel cage, wound rotor, and synchronous motors. Each of these requires three-phase power for their operation. Stators for all three of these motors are essentially the same. The motors differ from each other due to the rotor design.

The squirrel cage and wound rotor types use the principles of induction and the rotating magnetic field to start and run. The synchronous motor in its standard design is not self-starting due to the dead weight of its rotor. Synchronous motors need special designs or provisions to come up to speed and to lock in sync with the rotating magnetic field. These motors make use of a foreign voltage on the windings of the rotor.

Three-Phase Squirrel Cage Motor

Three-phase motors meet most of the requirements of industry. Because of their simplicity of construction, low maintenance cost, and versatility, these machines are chosen for most applications. These motors have good speed characteristics, are adaptable to most mechanical loads, and they can be easily reversed. They are suitable for use in areas where the air contains explosive mixtures.

Of the three polyphase motors that are discussed in this chapter, the squirrel cage induction motor is the one most often chosen. To operate, this motor requires only the interaction of the stator and rotor fields through the principles of induction. It has no need for brushes or slip rings that require maintenance. This in turn increases the ruggedness of this type of motor. It is self-starting due to the rotating magnetic field and requires no special provisions required by any single-phase motor. Figure 8–14 depicts this type of motor.

The squirrel cage motor may be either wye or delta wired. It may be either a single- or dual-voltage type.

Rotor

The core of the rotor of the squirrel cage motor is constructed of steel laminates. The material is usually the same as that used for the stator. There can be less laminates, and they can be thicker, without producing excessive eddy currents and hysteresis losses. This is due to the lower frequency of the rotor's magnetic flux.

Slots on the rotor are normally arranged in a diagonal fashion, or *skewed*. This minimizes the noises due to the magnetizing forces and smoothes torque variations.

The number of slots is never the same as the number of stator slots. There may be more or less slots on the rotor, but the number will never be an equal multiple or division of the number of stator slots. This arrangement prevents *dead spots*, or points of zero torque.

Rotor conductors may be round, square or rectangular. They are made of copper, aluminum, or brass. The rotor conductors are inserted into the slots of the core and are not insulated from thc corc. Insulation is not necessary since the induced voltage into the conductors is seldom greater than 10 volts. In most cases it is considerably less.

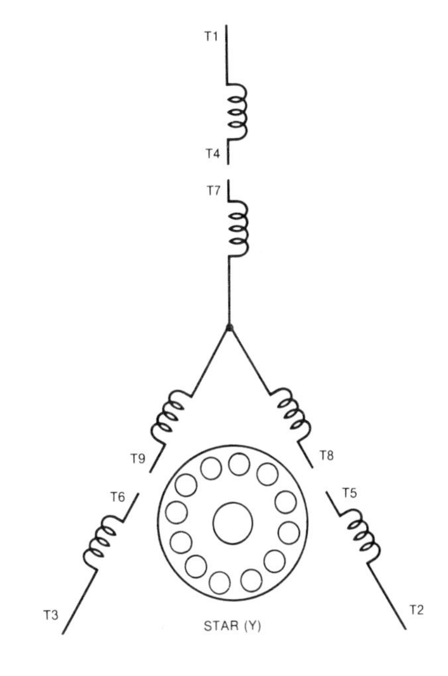

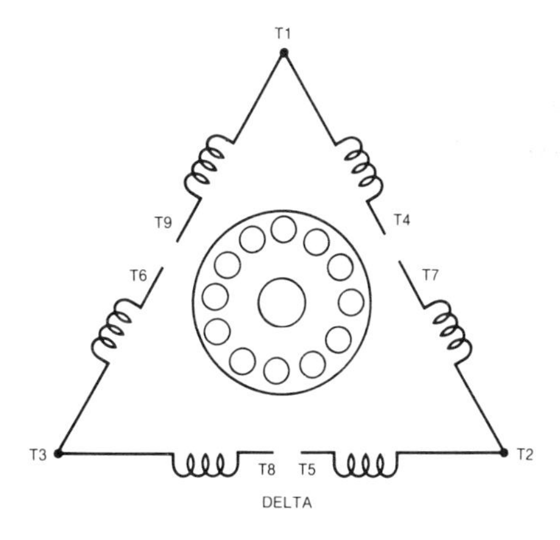

FIGURE 8–14. Three-phase, squirrel-cage induction motor.

End rings complete the circuits for the rotor's conductors. They are generally made of the same material as the conductors. The shape of the end rings differs between design and manufacturer.

Due to the large temperature change between no-load and full-load conditions, a good electrical connection is necessary between the conductors and end rings. Rotor bars are usually welded, brazed, or bolted to the end rings in order to achieve this condition. On small motors some manufacturers place the assembled rotor core in a mold and pour molten conductive material into the mold to make the conductors. Aluminum is normally the conductor of choice when this process is used. The mold also may include provisions for cooling fins.

Although the squirrel cage rotor is more rugged than the wound rotor, points of high resistance or *opens* sometimes occur where the conductors are joined to the end rings. When this happens, the motor will have reduced starting torque and will slow down when under load. Discoloration due to heat at the joint between the end rings and the rotor bar is evidence of this problem. Because brazing of these joints or replacing the bar requires special skills, the rotor must go to a motor shop for repairs.

Resistance and inductance of the rotor is not variable. It is determined by the design of the rotor and it is set at the time the motor is produced. The amount of rotor resistance and inductance is determined by the engineer based on the motor's application.

Rotor resistance is usually less than 1 ohm. The exact value depends on the physical characteristics of the conductors. Resistance varies with the type of material used, the length of the conductors, and the circular mil area.

Inductance of the rotor can be changed by using slots of different dimensions. A deep slot produces greater inductance than a shallow slot.

Airgap

The *airgap* is the radial distance between the rotor and the stator. Because the magnetizing field of the stator tends to confine itself to the iron of its magnetic circuit, it requires a magnetizing force (NI) several times greater to cross the air gap to the rotor. The increased current needed to produce this field lags the voltage by 90° and is mostly wattless power. This lowers the overall efficiency of the machine. For this reason, the *airgap* is kept as small as possible.

Measurements are made on the drive end of the motor. This end will have the greatest bearing wear. Most motors have openings for this purpose. Four measurements must be taken to insure the rotor is centered between the teeth of the stator. These are performed 90° apart. This check is essential on large motors and for those with sleeve bearings. Ball bearings normally fail before showing signs of wear. Records of the measurements are kept for comparisons so that changes in distances can be detected.

Operating Characteristics

Three-phase, squirrel cage motors being manufactured today that use NEMA standards are of four basic designs. These are the following.

1. *Design A*. This motor has normal torque and normal starting current. Design A is a constant speed motor with a slip from 2% to 5%. Locked-rotor torque ranges from 275% of normal running torque for a 1-horsepower motor with a synchronous speed of 1800, to as low as 70% for a 500-horsepower motor operating at 3600 RPM. The percent of torque decreases with increased horsepower and reduced RPM. Locked-rotor currents range from 500% to 1000% of full-load current. These motors often require reduced-voltage starting.
2. *Design B*. Design of this motor provides for normal starting torque and low starting current. The rotor has high resistance and high reactance when the motor is starting. These values become close to Design A values when the motor is running. Starting currents under full-voltage starts are within the limitations of most power companies for motors of 30 horsepower or less.
3. *Design C*. This motor develops high torque and has low starting current similar to Design B. It has two squirrel cage windings on the rotor, one placed above the other. During the starting, most of the current is carried by the outside windings which have a high resistance. The high resistance provides for the high torque. When the rotor comes up to speed, both rotor windings act equally in carrying the current.
4. *Design D*. The rotor for this motor has a high resistance for starting and running. Because the resistance of the rotor is high, starting torque is high, and the slip of the motor is greater. These motors are designed to have medium slip, 7% to 11%, or high slip, 12% to 17%. Most of these motors can be started under full voltage without drawing prohibitive currents.

The efficiency of the polyphase squirrel cage motor ranges from 60% for a one-half horsepower motor under one-half load, to 94% efficiency for a 300-horsepower motor under full load. The higher horse-power motors provide higher efficiency.

Reducing load decreases efficiency. The motor becomes more inductive-reactive under these conditions, and the power factor decreases. This in turn increases the losses of the motor. For this reason, the motor should be matched closely to the load. Avoid using motors too large for the intended load.

Standard squirrel cage motors, Design A, have a comparatively low starting torque which is in the range of 150% to 200% of the full-load torque. This is due to the low resistance and high reactance of the rotor windings. They must often be started using reduced voltages to lower the excessive starting currents which are in the range of 400% to 900% of the running current.

As the motor comes up to speed under a no-load condition, the slip decreases. The current decreases to its maximum no-load current, and the power factor improves. Only enough current flows to overcome bearing friction, windage, and power losses.

When the motor is placed under load, it must develop enough torque to overcome the mechanical load while turning at a constant speed. Current increases in the rotor to supply more torque. Adding more load will again cause the current to increase further. This action will continue until the motor has developed its maximum running torque. The increase in rotor current in turn causes a larger magnetic field which must be neutralized by the stator field. Stator current will increase just as the primary current of a transformer increases with the increased load on the secondary.

Maximum torque occurs when the rotor's reactance is equal to the rotor's resistance. If additional load is added at this point, the motor will tend to stall or stop. This is the point of breakdown torque or pull-out torque. The maximum torque the motor can develop is in the range of 150% to 225% of its full-load horse-power. A motor rated at 10 horsepower is capable of developing 15 to 22.5 horsepower maximum torque.

Pull-out torque occurs in the standard squirrel cage rotor motor at about 75% of synchronous speed, or at 25% of slip. Above this speed, the motor is stable. An increase in load will result in an increase in torque. Below this speed, an increase in load will result in decreased torque, and the motor will be forced to stop.

Full-load torque is established by the design of the motor. Usually this value is somewhat above the 75% of synchronous speed to allow for any minor overloads that may occur.

When the motor is operated at full load and at rated voltage, it develops its full horsepower at its rated full-load current. The power factor of the line will be within the stated power factor of the motor.

Three-Phase, Wound Rotor Motor

Wound rotor motors have the ability to provide high starting torques and are used for applications where high inertia loads must be started smoothly and effectively. They have relatively lower starting currents than a comparable squirrel cage motor. Because of this, full-voltage starting across the line is often possible.

Stator

The stator windings for the wound-rotor motor is the same as that of the squirrel cage type. This motor operates on the same principle of the rotating magnetic field. Calculations for its synchronous speeds are made in the same manner for both motors. The wound rotor motor has an additional advantage in that its speed can be adjusted. This is accomplished by varying the slip of the motor.

Rotor

The wound rotor motor differs from the squirrel cage motor in that it has insulated wire windings similar to the stator windings. These windings are normally wye connected with the ends connected to slip rings on the shaft. This arrangement allows resistance to be inserted in series with the windings through brushes that ride on the slip rings. Figure 8–15 illustrates this type of motor.

This design requires three slip rings to add resistance to the circuit. The number of rotor poles is always equal to the number of poles on the stator.

Wound rotors have a relatively low resistance. At normal running speeds, the external resistance is removed, and the ends of the wye windings are shorted together. Under these conditions, this type of motor behaves very much like the squirrel cage motor, and it has good speed regulation.

Operating Characteristics

Wound rotor motors are usually started with maximum resistance in the rotor. This provides maximum starting torque in the range of 225% of the full-load torque. This is in the range of the breakdown torque of the motor. At the same time, the resistance limits the starting current to about 150% of the full-load current.

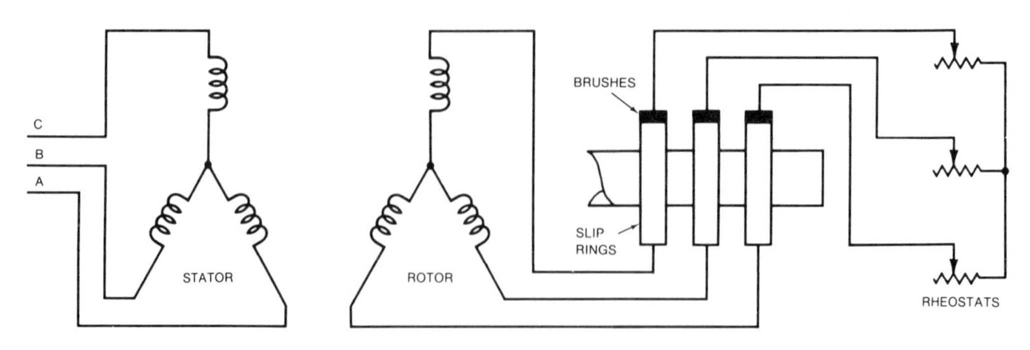

Figure 8–15. Three-phase, wound-rotor motor.

However, with optimum resistance selected, maximum torque will be developed per ampere of current.

The large resistance in the rotor's circuit causes the rotor's current to be almost in phase with the induced voltage. The resulting field is, therefore, almost in phase with the stator's field. Because the two fields are maximum at the same time, the magnetic reaction between them causes a strong turning force.

Figure 8–16 provides the torque curves for the wound rotor motor. This graph shows the torque curves for three values of resistance. With the high resistance inserted in series with the rotor's windings, maximum torque is developed when the slip is maximum, or 100%. This is the start condition. If the high resistance remains in the circuit, the torque will be reduced below the motor's normal full-load torque when the motor comes up to speed.

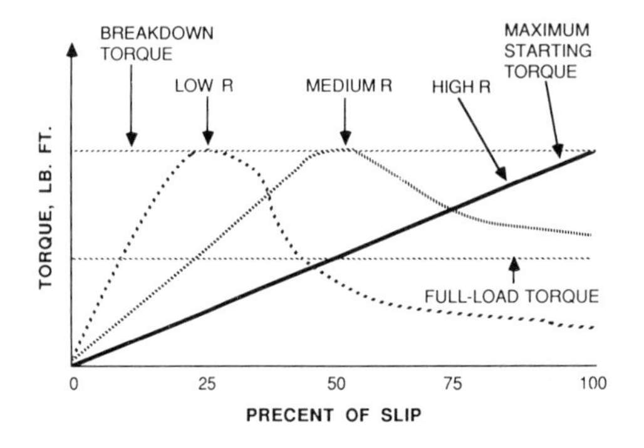

Figure 8–16. Torque curves for a wound-rotor motor.

If an attempt is made to start the motor without maximum resistance, the torque performance will be poor. The rotor circuit will be mostly inductive-reactance. This will cause the current in the circuit to lag the induced voltage by nearly 90°. The resulting flux field will lag the stator's field accordingly. The magnetic reaction between the two fields will be reduced. This in turn will reduce the torque. Under these conditions, the motor may not start if under load. Starting current in the range of that for the squirrel cage motor will also result.

In comparison, maximum torque for a medium amount of resistance will not be developed until the motor reaches approximately 50% of the operating speed. For low resistance this does not occur until about 80% of the rated RPM.

The use of resistance provides for most of the heat generated during the start cycle to be dissipated by the external resistors and not in the rotor. These resistors have high wattage ratings sufficient to handle the power to be dissipated. The rotor windings will be more inductive than resistive and the electrical energy will be transformed into a stronger magnetic field rather than heat.

Speed Control

The wound rotor motor like the squirrel cage motor has its base speed set by the number of stator poles and the frequency. It will operate at the synchronous speed minus the slip. Speed control is obtained with the wound rotor motor by increasing the slip from base speed.

Adding resistance to the rotor circuit when the motor is running will cause the rotor's current to be reduced. This will cause the motor to slow down. When the rotor's speed is reduced, more voltage is induced into the rotor's windings due to the increased slip. This in turn will increase current flow in the rotor's windings, thereby developing the necessary torque for the motor to run at the lower speed.

How far the RPM can be reduced depends on the characteristics of the load. Fifty percent reduction is usually permissible without the motor becoming unstable or stopping. Because of the higher currents involved with reduced speed, and the reduction in the normal ventilation, the motor will operate at a higher temperature.

When the motor is operated below full speed, it will run with reduced efficiency and horsepower. The resistance in the rotor consumes power which is wasted as heat. The horsepower is dependent on the RPM. Also at the lower speed, the motor is more susceptible to variations in speed due to load changes. Speed regulation is poor under these conditions.

Three-Phase Synchronous Motor

Synchronous motors are highly efficient machines for converting electrical energy into mechanical power. They can be used with most loads that require a constant speed. One factor that makes the synchronous motor popular is that this machine can provide a leading current and help correct the power factor on a system that has induction motors. Induction motors cause the current to lag the voltage. The currents of induction motors and synchronous motors can be 180° out-of-phase with each other. Under this condition, they will cancel each other's reactive current values on the power line. The remaining current will be purely resistive.

When full voltage is applied to a synchronous motor on the start cycle, the KVA required is often in the range of 550% to 700% on the normal running cycle. For this reason, reduced-voltage starting is required in many cases.

Efficiency

The efficiency of synchronous motors is higher than that of induction motor with the same horsepower and speed. Most medium and large motors have an efficiency between 88% and 96%.

Speed

Synchronous motors are classified according to their speed. They are either high-speed or low-speed machines. Those operating over 500 RPM are designated *high-speed motors*.

The speed of a synchronous motor is dependent on the frequency of the power source and the number of poles the stator has. RPM increases directly with frequency and inversely as the number of poles. The same formula used to calculate the speed of the induction motor is used to determine the speed of this type of motor. The motor will run at synchronous speed and will not have the slip required by the other three-phase induction motors. Varying loads within the rating of the motor will not cause the RPM to change.

Excitation Voltages

In order to obtain this constant speed characteristic, the synchronous motor requires two excitation voltages. The stator has three-phase AC power applied to its windings. This is the same as for the induction motor. The rotor of the synchronous motor also requires excitation. A DC voltage is applied through slip rings and brushes to the rotor. This, in effect, makes the rotor a polarized electromagnet. Figure 8–17 illustrates the required power sources for this motor.

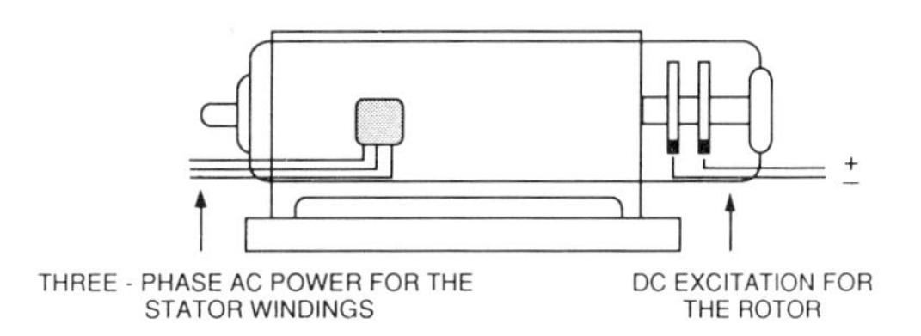

FIGURE 8–17. Power sources for the synchronous motor.

The rotor of the synchronous motor does not require the magnetic induction from the stator field for its excitation. The DC current connected directly to it creates its own field. Comparatively large airgaps between rotor and stator can be used. This allows for the production of low-speed machines with small horsepower ratings. They are generally more economical, and smaller, and easier to manufacture than the squirrel cage induction motor.

Other Classifications

These motors also have a classification according to the type of service. "General" types have a service factor of 1.15 and a rating of 40°C rise in temperature. The other classification for service is the "special" type. When a motor is built for a particular application that does not conform to the general rating, it is classified as a *special* motor. As an example, a special motor might have a service factor of 1.25 and be rated at 60°C.

The power factor is another way to classify this type of motor. Synchronous motors will either have a power factor rating of one (unity) or they will have a rating of

0.8 leading. Motors develop their horsepower ratings when operated with these power factors. The motor may be set at a different power factor during operation. It may have a lagging power factor depending on how it is operated.

Rotor

The rotor field of the synchronous motor usually has salient rather than cylindrical poles. These poles have definite projections on the face of the rotor. When energized by the direct current they have a definite magnetic polarity. Their windings are connected in series with adjacent poles having opposite polarity. The rotor must have the same number of poles as that of the stator.

Principles of Operation

Synchronous motors are not self-starting. The weight of the armature does not allow the rotor to respond quickly enough to the rotating magnetic field of the stator to begin moving in one direction before the field pulls it in the opposite. This results in zero starting torque for the motor. Some method must be designed into the motor to bring it up to speed before the excitation current is applied to the rotor.

Several methods are used to bring the motor to synchronous speed. One of these is to use a separate induction motor, or DC motor, to drive the rotor of the synchronous motor until an RPM near the synchronous speed is achieved. At that point, DC excitation is applied to the rotor's windings and it will lock with the rotating magnetic field of the stator.

Another method is to start the synchronous motor as an induction motor. A separate winding on the rotor called the *damper winding* is used to start the motor as a wound rotor motor. These windings are placed in the slots of the rotor. It is possible to add external resistance to the windings to provide the characteristics of a high starting torque of the wound rotor motor.

A squirrel cage winding is often used to start the synchronous motor. This winding is referred to as the "amortisseur winding." These windings are embedded into the salient poles. The motor is once again started as an induction motor using the principles of the squirrel cage rotor.

When either the damper or amortisseur windings are used to start the motor, they apply torque only during the start cycle. Once the rotor is locked to the synchronous speed, no voltage is induced into them. This is due to the fact that they are turning at the same speed as the rotating magnetic field. Therefore, the magnetic flux field of the stator does not cut through the windings.

As the rotor nears the synchronous speed, the DC source is applied to the field windings of the rotor. The magnetic field caused by this current locks to the rotating magnetic field of the stator. This brings the rotor up to the synchronous speed.

On some synchronous motors, the DC exciter is mounted on the shaft of the motor. In the start cycle, the DC exciter can be used as a DC motor instead of an alternator. There is usually enough residual magnetism in the pole pieces to start the exciter as a motor. As the synchronous motor comes up to speed, its rotor windings are excited and thus the motor is locked in sync with the rotating magnetic field.

While the motor is in the start cycle, the magnetic field of the stator windings also induces a voltage into the field windings of the rotor. Because of the large number of turns used for the field, and because these windings are connected in series with the voltages adding, an excessive voltage is developed across the windings due to transformer action. This voltage can damage the insulation.

To prevent this rise in voltage, the rotor's windings are connected to a resistor during the start cycle. This is the same procedure that was described for the current transformer in Chapter 1 of this text. The resistance remains across the windings until the motor reaches sufficient speed for the excitation current to synchronize the motor. Figure 8–18 illustrates this action.

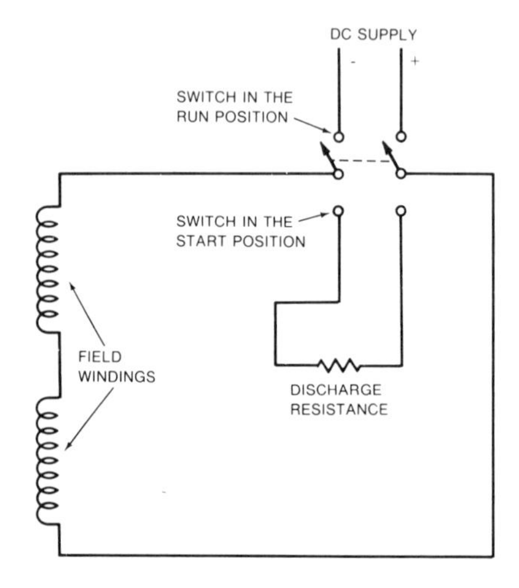

Figure 8–18. Equivalent circuit for the field discharge resistor.

When the DC excitation voltage is applied to the windings on the salient poles, these poles become electromagnets with fixed polarities. The fields of these magnets lock to the fields created by the stator and the rotor pulls into synchronous speed of the rotating field. Once the rotor is synchronized with the stator field, no further voltage is induced into the rotor's windings by the magnetic field of the stator.

The point at which the motor changes from an induction-start to a synchronous run motor is usually the most critical time for bringing the motor up to speed. The torque developed by the interaction of the two fields at this time is called the "pull-in torque." Salient poles synchronize more easily than the poles on cylindrical rotors. If the load is not too great during the start cycle, salient-pole motors will sometimes synchronize before the DC excitation is applied. This is due to the residual magnetism.

Synchronous motors have a pull-in torque of about 40% of the full-load torque. The value of the pull-in torque depends on the strength of the stator flux. The higher the applied voltage, the greater the torque. Full-voltage starting would provide greater torque than reduced-voltage starting.

Sometimes when the rotor synchronizes without the benefit of the DC excitation, the rotor will slip a pole. When this occurs, the magnetic field of the rotor opposes the magnetic field of the stator. The magnetic fields of alternate poles are opposite each other. A current surge occurs on the power system. This is due to the stator field drawing additional current to stabilize and maintain a constant magnetic flux. This surge is temporary, and the current falls back to normal as the rotor synchronizes. It is good practice, however, to apply the DC excitation within 1% to 5% of the synchronous speed to prevent this line surge. Most synchronous motors have automatic control starting to insure that the DC excitation is connected to the rotor at the most opportune time.

The synchronous motor is sensitive to any phase shift or disturbances on the power line. When these occur, the motor has a tendency to oscillate or "hunt." Under these conditions, the rotor tends to first run faster, and then slower than the rotating magnetic field. This condition in turn causes pulsations on the power system which can accentuate the hunting.

Motors having damper or amortisseur windings tend to correct the hunting problem. If the rotor's speed suddenly changes, and it is out of synchronization with the rotating magnetic stator field, a voltage will be induced into the windings used to start the motor. This produces currents in the windings which tend to pull the motor back into synchronization. If the motor is started by the use of another motor, it may be necessary to stop the motor and then start it again.

The torque angle between the stator and rotor fields will be nearly zero degrees when the motor is run without a load. As load is applied, the angle will increase. The rotor pole will lag the stator pole. The angle becomes greater with increase in load. As long as the motor is operating within its horsepower rating, the rotor will continue to turn at the synchronous speed. This operating characteristic is similar to having the rotor attached to the rotating field with a rubber band. The rubber band will stretch, but the rotor will continue to turn at the same RPM.

When the torque angle is zero, the CEMF of the motor is equal and opposite the voltage on the stator windings. As the angle increases under load, the CEMF changes in respect to the applied voltage. This in turn results in a higher stator current which develops the additional torque needed to maintain the constant RPM with the added load.

Exciters for Synchronous Motors

There are two basic methods of providing excitation current to the rotor of a synchronous motor. One method is to use an external source. The current is supplied to the field windings of the rotor through slip rings as illustrated in Figure 8–17. The other method is to have the exciter mounted on the common shaft of the motor. This procedure does not require brushes.

In the two examples that follow, the control contacts for the discharge resistor and for connecting excitation current to the windings of the rotor of the synchronous motor are not shown. Direct current excitation is directly connected to illustrate the condition when the motor is running at its synchronous speed. A control method is necessary during the start and run cycles of this motor for proper operation.

Figure 8–19 shows the diagram of a synchronous motor which has its rotor windings excited by an external source. The DC exciter may be motor driven by another motor, or its shaft may be coupled to one end of the shaft of the synchronous motor. Electronic power supplies are also often used for this purpose.

The DC generator is wired in the compound configuration. The variable resistor controls the amount of voltage applied to the rotor windings by adjusting the current in the shunt field of the generator. The amount of current flowing through the rotor windings is metered by an ammeter in series with the windings. The

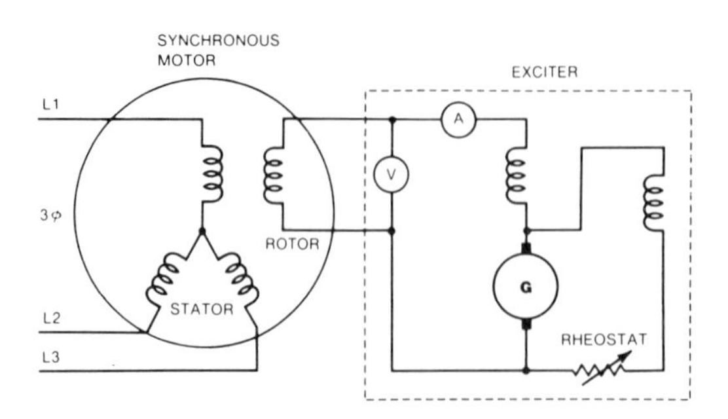

Figure 8–19. External DC exciter for a synchronous motor.

voltage across the windings is also metered by a voltmeter. This arrangement allows the operator to change the strength of the excitation current.

Figure 8–20 illustrates how DC excitation is produced using a generator mounted inside the motor on the common drive shaft. The motor is started using either a damper or amortisseur winding. The rotor of the exciter turns with that of the motor. Rectified single-phase power supplies the exciter's field windings to create a magnetic field. The strength of the field is adjustable by the rheostat.

The rotor of the exciter turns in this field generating a three-phase voltage. This voltage is then rectified by a three-phase bridge rectifier to produce the DC current for the rotor's field of the motor. The rotor's field current is adjustable by the rheostat in series with the field windings of the exciter.

With the appropriate controls, this system has the following advantages over the external method of excitation.

1. Slip rings and brushes are not required.
2. Field excitation is applied at precisely the right time.
3. Good torque efficiency is met.
4. Field control and excitation are maintenance free due to static controls.
5. Sparks are not generated between brushes and slip rings. This allows the motor to be used in hazardous locations.
6. Field excitation is automatically removed if the motor loses synchronization. This provides for resynchronization without having to stop the motor.

Levels of Excitation

Figure 8–21 depicts three different levels of DC excitation on the rotor of a synchronous motor. The load is held constant in each case. Because the load is the same for each level of excitation, the torque angle and the power delivered to the machine will also be the same for each level of excitation.

The top set of vectors shows the motor with the right amount of DC excitation for the motor to operate at 100% power factor. Note that the applied voltage and the line current are in phase with each other. The voltage developed across the rotor (E_R) is the vector sum of the applied voltage (E_A) and the CEMF. The line current lags the voltage across the rotor by about 90°. This is due to the inductive reactance of the stator windings.

Increasing the DC excitation on the rotor results in the rotor being overexcited. Its magnetic flux field will be stronger than it was at unity power factor. The larger

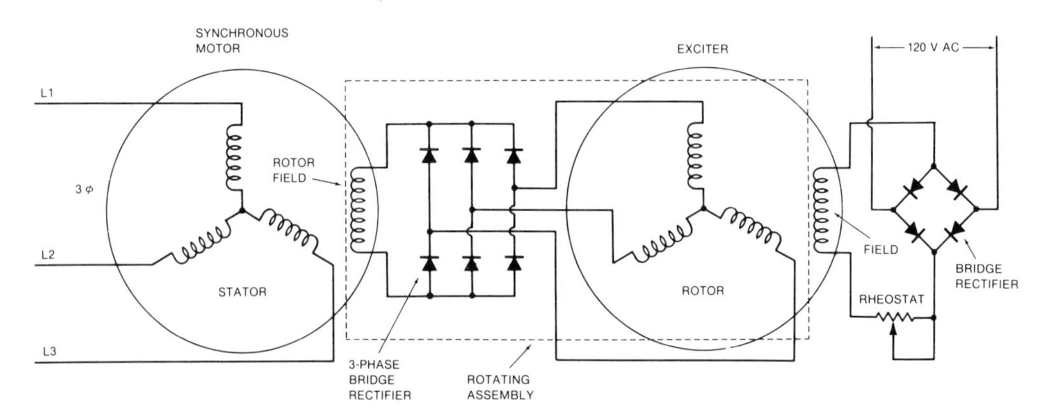

Figure 8–20. Internal DC exciter for a synchronous motor.

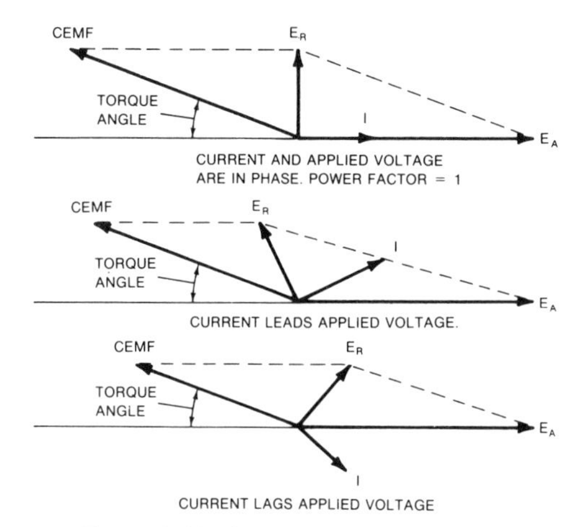

FIGURE 8–21. Current-voltage relationships in a synchronous motor.

flux field will cause the CEMF to become larger. E_R is shifted counterclockwise by this action. The current on the line will lag the rotor voltage by approximately 90°, and it will also shift counterclockwise. This shift carries the current to now lead the applied voltage. The line current will now be greater than it was at 100% power factor. These changes from unity power factor are shown by the middle set of vectors in Figure 8–21.

If the rotor's field is underexcited, the opposite effect will happen. The results are shown by the bottom set of vectors.

Decreasing the current through the field windings of the rotor will cause its magnetic field to decrease. This decrease, in turn, results in a decrease in the CEMF. The resultant vector (E_R) of the CEMF and applied voltage now moves in a clockwise direction. The line current is shifted in the same direction while remaining 90° behind the rotor voltage. This action places the line current at a lagging angle to the applied voltage.

The circuit is now inductive. Current lags the voltage. Once again the current is greater than it was at unity power factor due to the reactive current being added to the working current needed to turn the mechanical load.

Increasing the load on the motor will cause the torque angle to increase. E_R will increase accordingly as the vector sum of the CEMF and the applied voltage. This increase in E_R causes an increase in line current needed to turn the heavier load. If the imperfections of the motor are ignored, the amount of torque applied to the load will vary as the sine of its angle. Maximum torque occurs when the rotor lags the stator field by about 70 electrical degrees. The general use synchronous motor will develop a maximum torque in the range of 200% to 250% of its full-load rating.

This example shows that the synchronous motor has the ability to operate either with a leading or lagging power factor, or it can be set to unity power factor. These actions are possible by changing the DC excitation current through the field windings of the rotor. A more detailed explanation of power factor is given in Chapter 9.

Identifying Untagged Leads

The identifying tags on motors are sometimes removed or become obscured with dirt and oil so that they cannot be read. The motor may be perfectly good, and you may be called on to put it back into service. The nameplate on the motor indicates it is a three-phase, dual-voltage motor with each winding rated at 115 volts. You have nine black leads. How do you determine which leads are to be connected to the power source and which leads are to be connected together for proper operation of the motor?

Wye or Delta

You know from your previous study of motors that this motor has dual windings on each phase for dual-voltage connections. You also know that the motor is wired either wye or delta. You visualize these two types of wiring configurations and draw each of them on paper. Number the leads in the prescribed manner. Figure 8–22 illustrates your efforts.

Using a multimeter on the ohmmeter function, you can determine if the motor is wired wye or if it is wired delta. For the wye-wired motor, there will be continuity between three pairs of wires. You will also have continuity between one set of three wires. For the delta-wired motor, you will have continuity between three sets of three wires.

The wye-wired motor will have four circuits with continuity. The delta-wired motor will have only three circuits.

You now know if the motor is wye or delta connected. The next step is to determine the proper pole for each winding.

Identifying the Leads of a Wye-Wired Motor

Let us assume that you found the motor to be wye (star) connected. Your first step is to tag each wire of the set of three wires. Permanently attach the numbers T7, T8, and T9, respectively. Temporarily mark the pairs T1–T4,

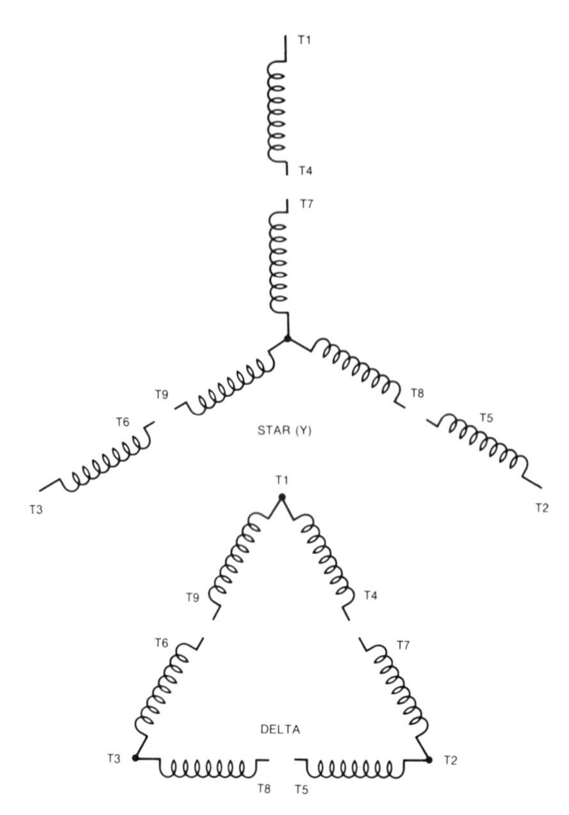

FIGURE 8–22. Wye- and delta-wired motors.

T2–T5, and T3–T6. At this point in the test, the relationships between the pairs of wires to the three wires marked T7, T8, and T9 is not known (Figure 8–23).

To determine which pair of wires belong to the phases marked T7, T8, and T9, the following method is used. Connect the circuit shown in Figure 8–24. Any low-voltage battery that will deliver the necessary current can be used for this test.

Current will flow through the windings marked T7 and T8. No current will flow through the coil marked T9.

The current should be set to a low value in the range of 1 to 5 milliamperes to begin the test. Given the current range and knowing the voltage of the battery, the value of resistance of the rheostat to limit the current to the desired level can be calculated using Ohm's law.

The meter should be set initially to its highest DC scale. While watching the meter, close the switch. Look for a deflection of the meter. If no deflection is observed, open the switch and once again watch for a deflection. A greater voltage will be induced when the switch is opened because of rapid deterioration of the current change (T = L/R). If once again no deflection is observed, continue the process with progressively reduced voltage ranges on the meter.

If you are not successful in this effort, then change the meter connection to a different pair of wires and repeat the process. If no deflection is observed, the current will have to be increased. Reduce the resistance in the test circuit to accomplish this. Repeat the process. You want to get a discernible deflection on the low range of the voltmeter. If the deflection is downscale when the switch is closed, the leads of the meter should be reversed.

Once you obtain the deflection when the switch is closed, and before opening the switch, the meter should either be set to its highest range or removed from the circuit. This is to prevent the meter from being pegged when the switch is opened. The deflection will be in the opposite direction when the switch button is released. If the pointer hits the stop very hard, damage will result.

Test each of the three pairs of wires. The pair that has minimal or no deflection belongs to the pole associated with the lead marked T9. The temporary tag on this pair of leads should read T3–T6. If not, change the temporary tags to reflect this association of leads.

Change the circuit connections at this point so that the current source is now connected between T8 and T9. Repeat the process as before. The pair of leads of the two remaining pairs that has the smallest deflection will be T1–T4. Under these conditions, there will not be any current flow through the coil that is attached to T7, and no voltage will be induced in the coil windings attached to T1–T4. The last pair of leads will be T2–T5.

At this point the three pairs are associated with their proper poles on the motor. It is now necessary to know the polarity of each of them. T1, T2, and T3 connect to the three phases of the power line for proper wiring. If the power lines are connected to the opposite ends of one or more of these windings (T4, T5, and T6), the magnetic fields of the windings will cancel the magnetic flux of the second set of windings associated with them. Under these conditions the motor will either not run at all, or it will run with reduced horsepower.

Using the same test setup as in the last procedure, the polarities of T2–T5 and T3–T6 can be determined. First, note that when the switch is closed, the current flows from the negative side of the battery to T9. The current passes through the two sets of windings, exits at T8, and flows back to the positive side of the battery.

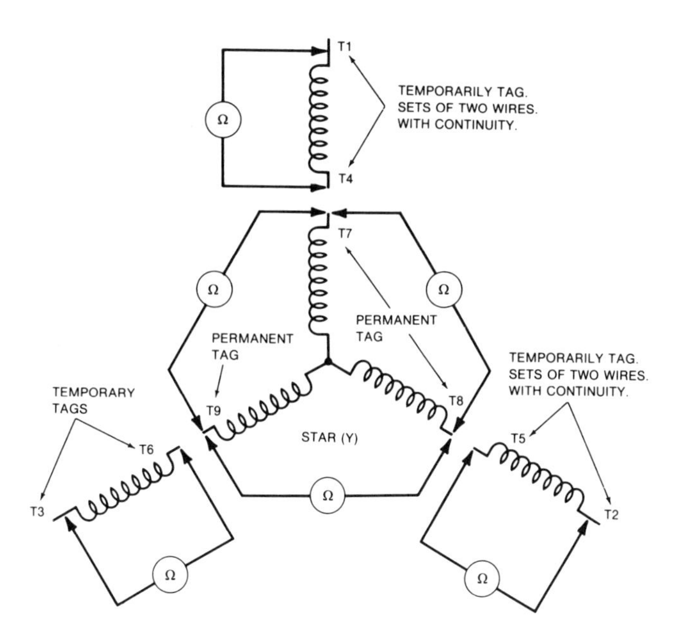

FIGURE 8–23. Tagging the leads of the wye-wired motor.

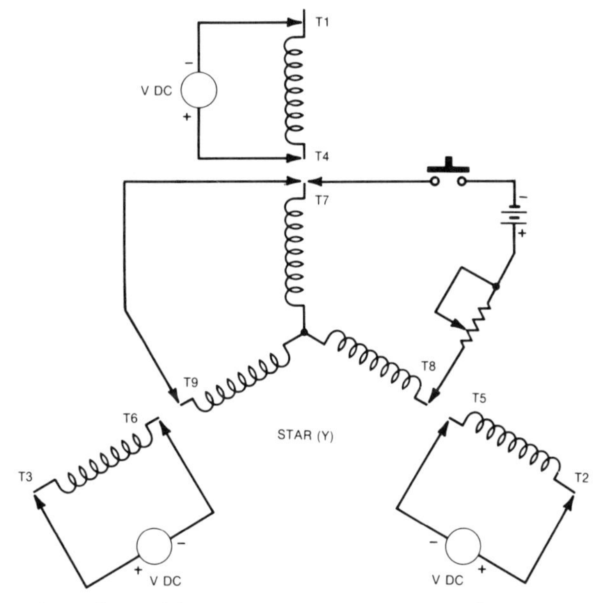

FIGURE 8–24. Determining pole and phase relationships of motor's windings.

This will cause the polarity of the induced voltages in the pairs to be negative on leads T3 and T5, and positive on leads T6 and T2. You will obtain an upscale deflection when the meter is connected for this polarity. If a downscale deflection occurs, the numbers on the leads are reversed.

Change the circuit back to the arrangement where the negative side of the battery is connected to T7, and the positive side to T8 as shown in Figure 8–24. The polarity of T1–T4 can now be determined in the same manner. When the switch is closed, the meter will deflect upscale if T1 is negative and T4 is positive. All six leads can now be permanently identified with their proper letter number designator.

An alternate method for determining the proper relationships between the windings is also possible. This can be accomplished by connecting 208 volts/three phase to T7–T8–T9 which were identified using the ohmmeter as previously described. Each of the wye-connected windings of the motor will have 120 V across them. This is equal to 208 V divided by 1.73, or the square root of 3. The open coils of the motor will have 120 V induced in them through the 1:1 ratio and transformer action.

Connect the lead temporarily marked T6 to T9. Take AC voltage measurements between T3 and T7, and between T3 and T8. If the windings of T3—T6 are of the proper polarity and associated with the same pole as T9, the meter will indicate approximately 318 volts for both readings. This value is determined by adding the in-phase voltages of 120 V to 120 V to obtain 240 V between the center point of the wye and T3 to 120 V at leads T7 or T8 which are 120° out of phase. The other two sets of windings can be determined in a like manner. Figure 8–25 illustrates this test arrangement. A vector diagram of the voltages is also depicted.

If your initial readings are 120 V, the coil you selected is associated with the correct pole, but its polarity is reversed. The reason for this is that the voltage induced into the windings is 180° out of phase with the applied voltage. These voltages on the same pole will cancel each other leaving 120 V across the third coils associated with T7 and T8. Under these conditions, merely reverse the connection and 318 V will be seen on the meter. Continue this process until the leads of the other two coils are properly identified.

When your initial readings indicate two unequal voltages when measuring between T3 and T7 or T8, the coil you have selected is not associated with the pole of the windings attached to T9. Select another pair of terminals and repeat the measurements until one of the two conditions showing proper pole association is achieved. At this point, the terminals can be identified as previously described.

Identifying the Leads of a Delta-Wired Motor

The same principles used to identify the leads of a wye-wired motor can be used to identify the leads of the delta-wired motor. The first step in this case is to find the center lead of each of the sets of three wires. This is done using an ohmmeter. Figure 8–26 shows the test arrangement.

Between the temporary marked leads T4–T9 you will measure twice the resistance as between T1–T4

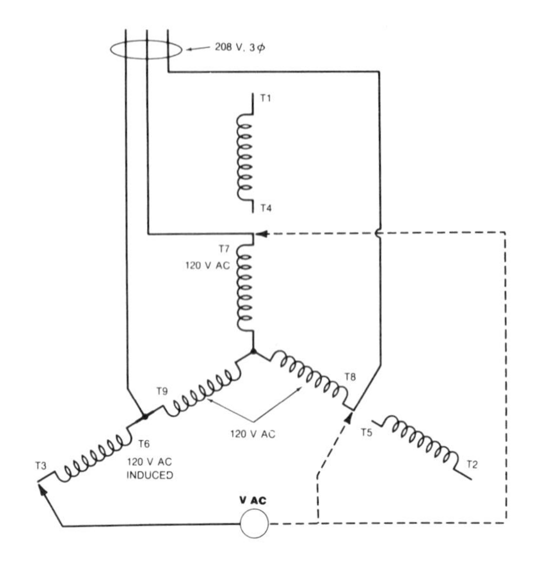

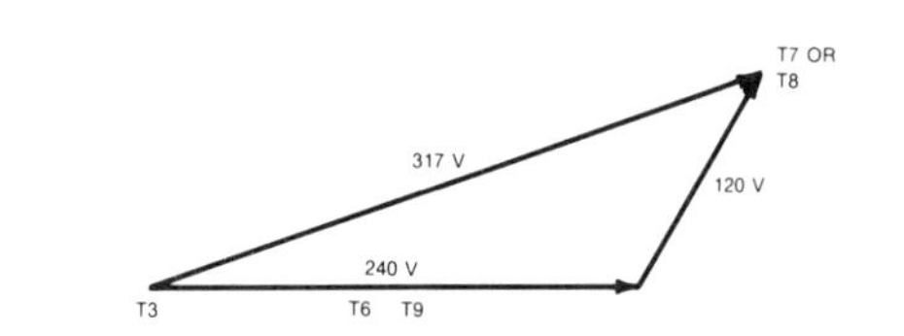

FIGURE 8–25. Using alternating current to determine the proper pole and phase relationships of a motor's windings.

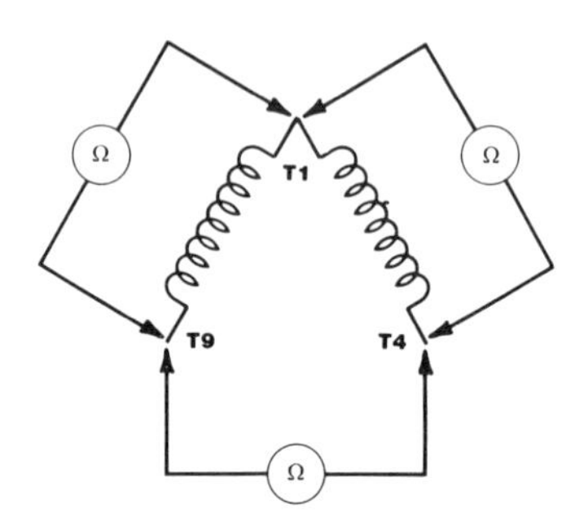

FIGURE 8–26. Identifying the center wire of each set.

and T1–T9. Identify T2 and T3 in a like manner. T1, T2, and T3 may be marked permanently at this time.

Either DC or AC can be used to identify the other leads associated with the center point. Figure 8–27 shows the test setup using DC voltage.

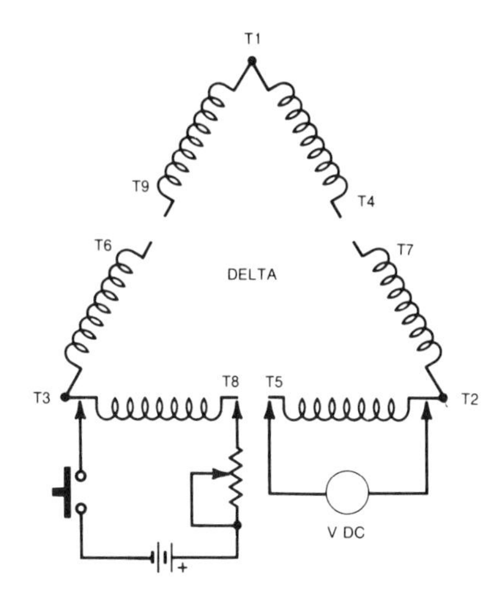

FIGURE 8–27. Identifying associated windings on a pole of a delta-wired motor using DC.

Current is applied to terminal T3 and the temporarily marked terminal T8 by closing the switch. Four readings across terminals T1–T9, T1–T4, T2–T5, and T2–T7 are made. These measurements can only be made as the switch is closed and the current through the windings of T3–T8 is changing. One of these readings will be larger than the other three. If the maximum reading is obtained between T2 and one of its related terminals, then you have guessed right about T8. If the maximum reading is between T1 and one of its related terminals, then the temporary markings of T6 and T8 should be switched and made permanent.

Either T5 or T9 can also be permanently marked based on the maximum deflection and their association with T1 or T2. The other leads can now be identified using the same method.

The leads can also be identified using AC. Figure 8–28 illustrates the test arrangement.

With AC of 120 V applied between leads T3 and T8 which are temporarily marked, meter readings are taken across the four possible combinations of T1–T4, T1–T9, T2–T5, and T2–T7. Three of these readings will indicate zero or very low voltage. The fourth reading will be 120 V AC. If the 120 V AC reading is found between T2 and one of its related leads, then T8 is marked correctly. If not, and the 120 V AC reading

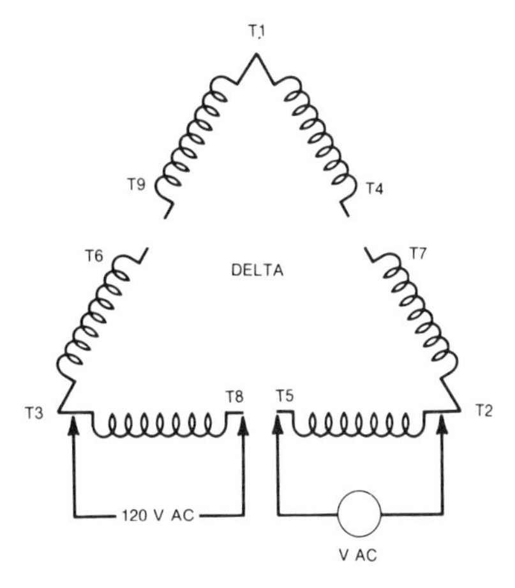

FIGURE 8–28. Identifying associated windings of a delta-wired motor using AC.

is found between T1 and one of its leads, then the markings of T6 and T8 will need to be switched. Either T5 or T9 can now be identified and permanently marked. The same method can now be used to identify the remaining terminals of the motor.

Testing the Motors

After the type of wiring and the leads have been identified and marked, the motor is tested for proper operation. Connect the motor for low-voltage operation according to the standard wiring diagram for the wye- or delta-wired motor. If the identification of leads has been conducted properly, the motor will be able to turn its normal load at the rated speed.

Current readings need to be taken on each phase. The current values should be within the normal range and approximately equal on each of the lines. If a large imbalance is noted, then the terminals have not been numbered correctly. The identification of leads will need to be repeated.

Questions

1. Two-phase motors have their voltages displaced by ______ degrees, and three-phase motors have ______ degrees displacement.
2. Why are three-phase motors superior in operation characteristics to those of single-phase motors?
3. What are the two major components of a three-phase motor?
4. How is the rotor's impedance determined?

5. The frequency of rotor current is equal to ______.
6. What factors determine the amount of turning force a motor is producing?
7. Three-phase motors are either wired ______ or ______.
8. Which of the two systems used to wire three-phase motors requires the higher current to run the motor?
9. To what terminals of a three-phase, nine lead motor are the power lines connected?
10. How many circuits are found for a nine-lead, delta-wired motor?
11. On a wye-wired motor, what is the tag number of the conductor attached to a coil whose opposite end is tagged T2?
12. How are the coils of a wye-wired, nine-lead motor connected for low-voltage operation?
13. To connect a nine-lead, delta-wired motor for its high voltage operation, the coils will be connected in ______ with each other.
14. Does the three-phase induction motor run at synchronous speed?
15. List the factors that determine the speed of a three-phase induction motor.
16. Do three-phase motors always have nine leads?
17. A three-phase motor with a wound rotor may have its speed varied by increasing or decreasing the ______ in the rotor winding circuit.
18. How do you reverse the direction of rotation of a three-phase motor?
19. Which of the three-phase motors studied in this chapter is the most reliable and requires the least maintenance?
20. Why are the airgaps of induction-type motors kept small?
21. Is a 1-horsepower or a 100-horsepower motor more efficient?
22. When is it desirable to use the wound rotor motor in place of the squirrel cage type?
23. What is the effect of leaving the starting resistance in the rotor's winding circuit while the motor is running?
24. Which type of motor can provide a leading current on the power line to help correct the power factor?
25. What is the operating speed of a four-pole, synchronous motor?
26. What are the power requirements for a synchronous motor?
27. How are synchronous motors classified?
28. There are two basic ways of starting a synchronous motor. Name them.
29. It is good practice to apply the excitation current to the rotor of the synchronous motor when the RPM is within ______ to ______% of the synchronous speed.
30. What action do the damper or amortisseur windings have on a synchronous motor in the event it begins to "hunt"?
31. Do all synchronous motors have slip rings and brushes? Explain.
32. If the rotor of a synchronous motor is underexcited, the motor's current in respect to the voltage applied will be (leading, lagging, or in phase).
33. You have a six-lead, three-phase motor with the windings rated at 230 volts. How many ways can you connect the motor? What would the supply voltage equal in each case?
34. How many circuits would you detect using an ohmmeter on a dual-voltage, wye-wired motor? Name them.
35. What are the tag numbers of the other two leads associated with T3 on a dual-voltage, delta-wired motor?
36. Should AC or DC be used to identify the related coils and their polarities?
37. What precaution should be taken with the voltmeter when using DC to test for polarity of a coil?
38. After putting a motor in service, what is the purpose of measuring the line current on each phase?

CHAPTER 9

Installation

The installation of a motor starts when the engineer specifies the motor's characteristics and the order is placed to the manufacturer.

Installation Procedure

When the motor arrives on site, the electrician's job begins with receiving the machine, inspecting it, and in time, putting it in service. The electrician must possess a broad range of knowledge in order to perform the tasks associated with the efficient and proper installation of a motor. Depending on the circumstances, this knowledge is not limited to just the electrical wiring tasks required by the blueprints, specifications, and addenda.

Receiving the Motor

Check the invoice against the information on the shipping container. This is to assure the correct machine was shipped and received at the worksite. If an error is discovered at this time, it will save the effort and cost to crate the motor to ship it back to the manufacturer.

Store the motor indoors if possible. Package materials are often not suitable for exposure to the weather. If indoor storage is not possible, cover the package with a tarpaulin or polyethylene sheet.

Allow a sufficient amount of time to let the motor reach the temperature of the atmosphere in which it is stored before the machine is unpacked. If the machine is cooler than the surrounding air, sweating will occur. This accumulation of moisture can cause rust and corrosion on the machine's surface. The electrical windings may also be damaged.

Carefully examine the motor when it is unpacked. Report any physical damage immediately to the carrier and to the manufacturer.

Handling the Motor

Use the hooks or slings in the lifting lugs that are on many motors to move the motor. The lugs have the capacity to carry the weight of the motor. Do not attempt to lift additional weight with them. Remove heavy attachments such as gear boxes and pumps. Never attempt to lift the motor by its shaft extensions. This can bend the shaft or cause damage to the bearings.

Checking Electrical Characteristics

Check the nameplate on the motor. The model number must correspond to the one shown on the invoice. The electrical characteristics of the motor must correspond to the distribution system from which it will receive its power. System voltage must be within 10% of that on the motor's nameplate, and the frequency must be within 5%. Does the motor meet all the requirements of the blueprints, specifications, and any changes in an addendum? The time to discover any conflicts is now and not after the complete installation of the motor.

Checking for Environmental Problems

Check the motor for any signs of rust or moisture. Most manufacturers will coat bare metal on the machine with a rust preventive before shipment.

If the motor is to be installed immediately, remove the rust preventive with a suitable solvent such as mineral spirits. Most of these solvents are flammable and may be toxic. Use them in well ventilated areas and take the necessary precautions to avoid fire or explosion. Avoid excessive contact with the skin when handling these solvents. Wear rubber gloves and safety glasses.

In the event the motor is to be stored, the rust preventive is to remain on the motor until it is ready for installation. If the rust preventive no longer covers the exposed machine parts of the motor, apply more rust protection.

Remove any rust with a fine abrasive paper. Once rust forms, it will continue to make progress even if the rust preventive remains on the metal. Touch up the painted areas where the bare metal shows.

Check the motor's cover to be sure it is acceptable for the environment in which it will operate. If the motor is going to be mounted other than on the deck, drip holes are to be at the bottom of the motor when the motor is orientated in its specified position. This may require the rotation of the motor's cover. In some cases this may be impossible because of other attachments to the motor. In this case, return the motor to the manufacturer and provide the corrected information.

Even with the most careful planning, these problems still arise. It is best to catch them before the actual installation. An inspector may reject the job after it is complete. This would result in increased costs and frustration on everyone's part.

Does the motor's design meet the requirements for the area in which it will operate? Some of the environmental questions to be checked are

1. Will the motor be indoors or outdoors?
2. Will the area be wet or dry?
3. Is the area classified as hazardous as defined by the NEC? It is the user's responsibility to specify the class and division in which the motor is going to be installed.
4. Will the motor be exposed to liquids such as oil and other chemicals? Are these corrosive? The hardware for standard motors is designed to operate under common environmental conditions. For unusual conditions, stainless steel hardware may be required and should have been specified on the order. Special paints, such as catalyzed epoxy enamel, will protect motors exposed to a highly corrosive atmosphere.
5. Is the motor going to operate in extreme temperatures? The temperature rise is based on a 40°C ambient temperature. Specify higher ambient temperatures on the order.

Motors that operate in extreme conditions require special lubricants. Standard motor grease is suitable for temperatures of –30°C to +120°C. Special steel for the shaft may also be required.

6. Will the motor operate at a high altitude? A special motor may be needed under these circumstances. This information is given on the nameplate.

Standard motors are designed to operate up to 3300 feet. The air density is less at high altitude, and the heat transfer from the motor's surface is reduced. Temperature rise increases 1% for each 330 feet above 3300 feet. Motors with service factors above 1.0 can be operated up to 9900 feet if the horsepower rating is not exceeded.

Another approach is to operate the motor at a lower ambient temperature than 40°C. If operated at a 30°C, the maximum altitude is increased to 6600 feet. At 20°C, the motor can operate safely up to 9900 feet.

7. Will airborne particles with magnetic and/or abrasive qualities be present? These will collect on the motor's windings? Over a given time, the movement of these particles on the windings can wear away the insulation.

If these various conditions are specified when the motor was ordered, the manufacturer would have taken them into consideration. Some of the special features which are available to protect the windings are the following.

1. Application of additional insulating varnish or protective coatings. Several extra dips and bakes will increase the moisture resistance quality of the windings.
2. Motors may be encapsulated with vacuum pressure–impregnated windings or be totally enclosed.

3. Abrasive resistant windings that are totally enclosed are available.

Checking Bearings, Seals, and Shaft

Check the bearing seals and the shaft extensions. Use special care in cleaning these areas to avoid damage to the surfaces. These items are precision ground, and it is very difficult to remove rust and other corrosive formations without causing further damage. Keep these areas protected. Remove any burrs or bumps due to careless handling with a fine file or scraper.

Many manufacturers will ship the motors without oil in the reservoirs to prevent spillage during transit. An oil film remains on the bearings to protect them. Check the levels and fill if necessary. Be careful to use clean equipment when filling. Dirt and grime introduced into the oil reservoir can cause premature bearing failure. Use oil or grease with the proper weight and viscosity. Refer to the lubrication nameplate or the instruction manual for this information.

If the motor is going to be put in service immediately, fill the reservoirs to the recommended level. Some motors will have sight glasses or gauges for this purpose. Do not overfill. When the motor comes up to operating temperature, the oil will expand. This will force the oil over the oil sleeve and into the motor. Overfilling the reservoir prevents the bearing of clearing itself of excessive oil. This will cause churning and foaming of the oil which will result in extra losses due to higher temperatures. Under these conditions the oil will oxidize.

Checking for Mechanical Freedom

Turn the shaft by hand. It should rotate easily without rubbing or binding. If the shaft does not move easily, identify the cause and correct the problem. If the motor has an open enclosure, check for packing materials or other debris which may have got between the rotor and the stator poles. Look carefully once again for any signs of physical damage.

For motors that have brushes, check to see that the brush rides smoothly on the slip rings or commutator. There should not be any rough spots.

Insulation Tests

Test the insulation of the windings using a megger that will deliver a minimum of 1000 volts or more. The minimum acceptable resistance for this test is 1.5 megohms.

If a high-potential (hipot) test is required, only qualified personnel are to conduct the test. The maximum voltage to be applied is

$$85\% \times (2 \times \text{Rated Voltage of Windings}) + 1000\ \text{V}$$

The factor of 85% of twice the rated voltage is for a new motor that has not been put in service. For motors that are in service, a 50% multiplier is used for this test.

Before applying the voltage for the hipot test, the windings are connected to the frame of the motor for a short time. The purpose of this procedure is to discharge all residual electrostatic charge that may have accumulated on them. Remove the ground before applying the test voltage.

The minimum acceptable resistance for this test is calculated using the following formula.

$$R_w = V_w/1000 + 1$$

where,

R_w is the resistance of the winding in megohms at 40°C.

V_w is the rated voltage of the machine.

A more detailed description of the hipot test is in Chapter 3 of this text. Recommended practices for testing insulation resistance of rotating machines is in Publication No. 43 of the IEEE. These procedures are followed by most U.S. manufacturers of motors.

If the insulation resistance is lower than the recommended minimum value, moisture may have accumulated in the motor. In this event, dry out the motor.

Drying the Motor

Use an oven that has circulating air to dry the motor. Regulate the temperature between 95 and 115°C. When the motor is heated above 90°C for at least 4 hours, increase the temperature to 115–135°C. Make periodic checks with the megger to determine if the resistance of the windings is increasing. Continue to heat the motor until the insulation resistance is about the same value at the beginning and end of half-hour intervals.

If an air-circulating oven is not available, cover the motor with canvas or similar types of material to enclose it. Leave a hole in the top of the covering so that moisture can escape. Place heating units or lamps under the cover to increase the temperature of the air around the motor. Be sure there is sufficient area around the heating units or lamps so that a fire is not started. Do not allow the heaters to come into contact with the motor's windings. They can cause damage to the insulation. Heat the unit until megger checks show the resistance is constant for half-hour intervals.

Another method of drying the motor is to apply about 10% of the rated voltage to the motor. Lock the rotor so it cannot turn. Increase the current flow until the winding's temperature reaches 90°C. Do not exceed this value. If the motor has heaters, energize them to help raise the temperature of the windings.

Continue this heating process until the resistance of the windings is constant for half-hour periods. Regardless of the method used to heat the motor, take care to prevent an excessive temperature rise that will damage the insulation quality of the motor.

Auxiliary Equipment

Auxiliary equipment for the installation of the motor may arrive at the same time as the motor or it may be shipped separately by a different manufacturer. When these items arrive, inspect them for any damage. Compare the invoice to the original order for any discrepancies. Report any difficulties immediately.

Auxiliary equipment must have enclosures suitable for the environment in which they are designated. Verify the suitability of the types of enclosures. Figure 9–1 illustrates some NEMA-type enclosures.

NEMA Type 1 enclosures are designed primarily to prevent accidental contact with live parts and to prevent falling dirt and dust from entering the box. The enclosure, however, is not dustproof. It is a general purpose cover for indoor use under normal conditions.

Type 3 enclosures are dusttight and raintight. They are intended for outdoor applications. Besides the protection provided by Type 1 enclosures, Type 3 enclosures also protect against falling liquids and light splashing as well as wind blown dust. They are sleet resistant, but the external mechanism may not be operable when the enclosure is completely covered by ice. Common applications are at construction sites, on loading docks, and in subways and tunnels. The Type 3R enclosure is also weatherproof but is not protected against blown dust. It protects against interference in operation of the contained equipment and has provisions for drainage.

NEMA Type 4 enclosures are watertight. These are designed to meet the hose test. This test requires a 1-inch nozzle to deliver a minimum of 65 gallons of water per minute, at a distance of not less than 10 feet, for at least 5 minutes. The water can be directed at the enclosure in any direction during the test, and there must be no evidence of leakage under these conditions.

Type 4X boxes are also corrosion resistant. They are suited for applications in paper mills, meat packing plants, chemical and fertilized plants, and where the contaminants normally present in these environments would destroy an ordinary steel enclosure over time.

Enclosures NEMA Types 7 and 9 are suitable for use in hazardous locations. Type 7 is designed for Class I locations where flammable gases or vapors are, or may be, present in the air in quantities sufficient to produce explosive or ignitable mixtures. Type 9 enclosures provide protection in Class II locations where combustible dust is or may be present. The number on these boxes may be followed by a capital letter. This letter references the enclosure for use in an area subjected to a particular group, or groups, of gases, or where various types of dust are present.

Type 12 enclosures are for industrial use. They prohibit the entry of such materials as dust, lint, fibers, and filings. They also exclude oil or coolant seepage.

NEMA Type 13 enclosures are also dust and oiltight. They are of cast construction with a gasket for use in the same areas as Type 12. Type 13 enclosures have a cast housing as an integral part of the device for the conduit. The mounting arrangement on this device is by means of blind holes rather than mounting brackets on Type 12 enclosures.

Mounting Hardware

Inspect the mounting hardware that was shipped with the motor to determine if it is appropriate for the installation. Does the invoice match the purchase order?

Motors may be mounted with various orientations from being on the deck, on a wall, or on a ceiling. If the proper mounting hardware is not available when the motor is ready for installation, considerable delays can be encountered. Keep the hardware with the motor so it is readily available for the installation.

Storage

Take the following precautions when storing a motor before it is to be put into operation:

1. Store indoors rather than outside. If the machine must be outside, cover it completely to protect it from weather and other environmental hazards. In either case, select an area that is clean and free of contaminants. Do not place the motor directly in contact with the earth or a concrete floor.

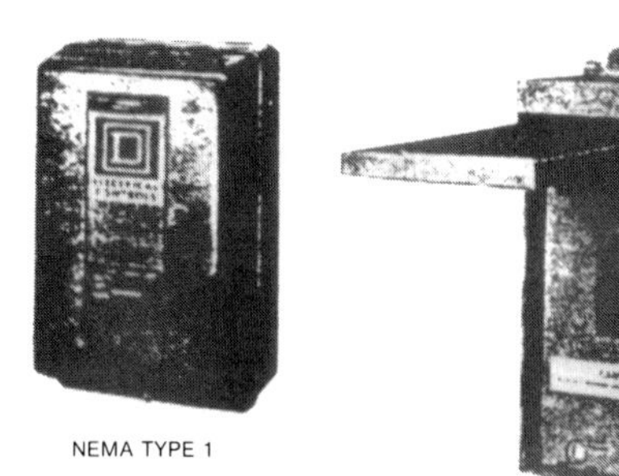

NEMA TYPE 1

NEMA TYPE 3R

NEMA TYPE 4X

NEMA TYPE 7 & 9
SPIN TOP

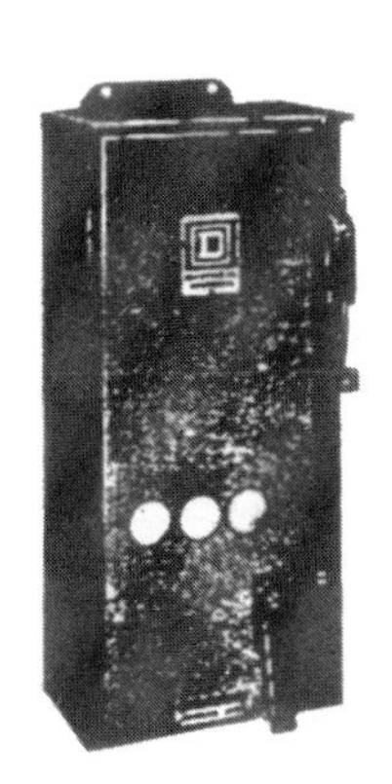

NEMA TYPE 12

NEMA TYPE 3

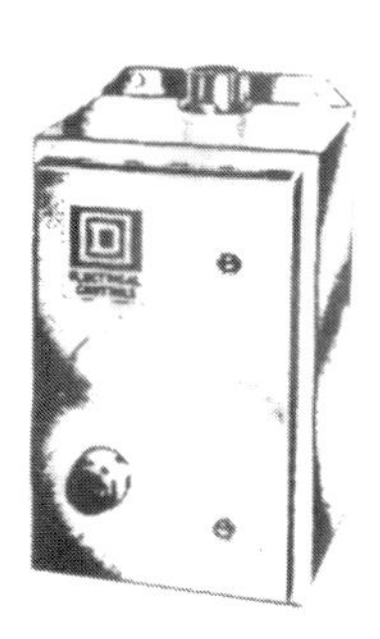

NEMA TYPE 4

NEMA TYPE 4 & 5

NEMA TYPE 9

NEMA TYPE 12K
WITH KNOCKOUTS

NEMA TYPE 13

FIGURE 9–1. NEMA-type enclosures. (*Courtesy Square D Co.*)

Use some scrap wood as a spacer so the air can circulate completely around the machine.

2. Protect the motor from becoming wet.
3. Do not expose the motor to extreme temperatures.
4. Fill oil reservoirs to completely cover bearings. This level will usually be above the normal recommended level. Tag the motor as to this condition. This is to warn the electrician who puts it into service to drain the excessive oil before operating the motor.
5. Remove the tension on brushes if the motor has them. Corrosion is apt to occur between the brush and commutator or slip ring. This can result in a flat spot which will result in uneven wear and possible sparking.
6. Keep the temperature of the motor slightly higher than the surrounding air to prevent condensation. If the motor has space heaters, activate them while the motor is in storage. Be sure combustible materials do not come into contact with the heaters' covers. The temperature may be high enough to cause a fire. If space heaters are not part of the motor, use thermostat-controlled heaters or lamps to maintain the desired temperature.
7. Leave the rust preventive protection on the exposed machined parts. Slush these areas if they are not completely covered.
8. Megger the windings at regular intervals. Keep a log of these readings to detect any significant decrease in the insulation resistance. If moisture is starting to accumulate, the temperature range on the thermostat may need increasing. It may also be necessary to dry the motor.
9. Rotate the armature every few months. This is particularly applicable to motors that use grease for the bearing lubricant. The grease will drip down over time and not completely protect the bearing and races. This can result in corrosion or rust.
10. If the motor is in storage for over a year, consult the manufacturer's representative for a competent technical field inspection to ensure that the storage process has not harmed the motor.

Mechanical Considerations

When the motor is brought into the area in which it is to be installed, there are several considerations which must be given to its proper installation. Article 110 of the NEC sets forth the general requirements. Among these requirements is the statement that all electrical installations shall be installed in a neat and workmanlike manner. An electrician who is competent in the profession, and who takes pride in the work performed, will adhere to this statement.

Location

Normally, the blueprints will specify the exact location of a motor. Given a choice, pick a location that will have the least amount of dirt, dust, liquids, and other contaminants. Allow enough space around the motor for the free flow of air to maintain an ambient temperature of 40°C or lower.

There must also be sufficient space for the installation and maintenance of the motor. Do not use these spaces for storage. The motor under certain circumstances must be guarded by one of the recommended methods prescribed in the NEC to protect personnel against accidental shock hazards.

Guards are also required on the exposed moving parts of a motor. This is especially important on motors that are remotely or automatically started or stopped. This recommendation extends to those motors that have automatic resets on overload protection devices.

Where the general requirements for a motor installation are not met, the necessary modifications to the motor and the area in which it is installed shall be followed. These changes are necessary to meet the requirements of the NEC, local codes, and the manufacturer's recommendations.

Noises due to the motor's operation may be a factor in some cases. Motors operating in homes and public buildings such as schools, hospitals, and theaters require some form of sound isolation. Some motors are quieter than others and can be used in public areas. Special sound absorbing mounts are used on the motors. Other motors require soundproofed equipment rooms. You may want to review some of the recommendations made in Part I of this text for the sound reduction methods used for transformers. The same principles apply to motors.

Mounting

Most motors come with a mounting base. The mounting dimensions are usually based on NEMA standards.

These are set according to the frame size of the motor and how the motor will be mounted. Motors with the same number normally have interchangeable, base mounting dimensions depending on how the motor is to be mounted. Specifications for detailed dimensions on any given motor should be consulted. These will be given in the manufacturer's printed matter on the motor.

The vast majority of motors are the horizontal type that are floor mounted. There are, however, many different possible configurations for mounting horizontal motors which the specifications of a particular job may require. The motor may have to be mounted on the ceiling or on a wall. It may be mounted in a vertical position as well as horizontal. Figure 9–2 shows the many different possibilities. Unless the purchase order specifies how the motor is going to be mounted, the motor will arrive at the site for the F-1 mounting.

Standard motors may be supplied for F-2 mounting. When the motors are mounted on the ceiling or a wall, the ventilation openings are located downward to make the motors dripproof. Likewise, vertical mounted motors, which have ventilation openings in the bell housing, will have these openings on the end of the motor that points down. The other end of the motor will be closed to prevent the entrance of liquids.

Usually the smaller motors designed for horizontal mounting can handle external thrust loads in either direction within certain design limitations. The manufacturer's specifications for the particular motor must be consulted to assure that these limits will not be exceeded. Larger motors require special lubrication systems, and the purchase order must specify if the shaft will point up or down.

When motors are to be mounted on walls or ceilings, additional lifting lugs may be required. Standard lifting lugs may not be located for this purpose. Where these conditions exist, employ experienced riggers to put the motor in place. This additional expense is worthwhile to avoid possible injury to personnel and equipment.

Mount the motor on a firm, level foundation. The foundation must have sufficient strength to carry the load. Fasten the motor securely to the foundation.

Medium to heavy loads cannot safely be fastened to soft masonry materials such as plaster or plasterboard. Use special concrete bases for horizontally mounted motors. The size and strength of the base for a particular motor is not simple to calculate. Where this is critical, the specifications are the responsibility of the engineering department.

Generally, a base with a thickness of 2 to 4 inches, and of sufficient size to comfortably hold the motor, will have the strength to keep the motor in place during its operation. A 1:3:6 mixture is normally recommended for the concrete base. This consists of one part cement, three parts sand, and six parts gravel.

Masonry plugs and bolts are normally used to fasten the motor to the foundation. Their size and number are specified by the manufacturer of the motor. Due to varying conditions of the mechanical loads, a safety factor of 4:1 is recommended for the plugs and bolts for static loads. That is, they will withstand four times the requirements of the design load. Dynamic loads, such as motors, require a larger safety factor. When motors are wall or ceiling mounted, or when there is considerable vibration with resulting shear and twisting forces, the ratio requirement is as high as 10:1.

Loads on the masonry anchors are transmitted to the material in which they are installed. The industry's standards recommend a spacing between anchors of ten times the diameter of the anchor to obtain 100% efficiency. They also recommend that the anchor is not placed closer than five times its diameter from the edge of the material. Anchors may be located closer to each other by as much as 50%, but there is a proportionate reduction in the efficiency of the installation.

Before the introduction of the many fasteners now in use, wood plugs were often used. These do not have the holding power of that of the modern anchors, and they are no longer acceptable according to the NEC.

Coupling

The type of drive to select depends on many factors. Some of these include the type of motor, requirements of the mechanical load, the power supply, space limitations, safety requirements, working conditions, and the initial and operating costs of the system. The types of couplings are

1. Direct
2. Belts and pulleys
3. Chain and sprockets
4. Gears
5. Magnetic clutch

1. Direct coupling is used when the speed requirement of the machine is the same as that of the motor. Some common examples are fan blades, pumps, and motor-generator units. One of the other methods of

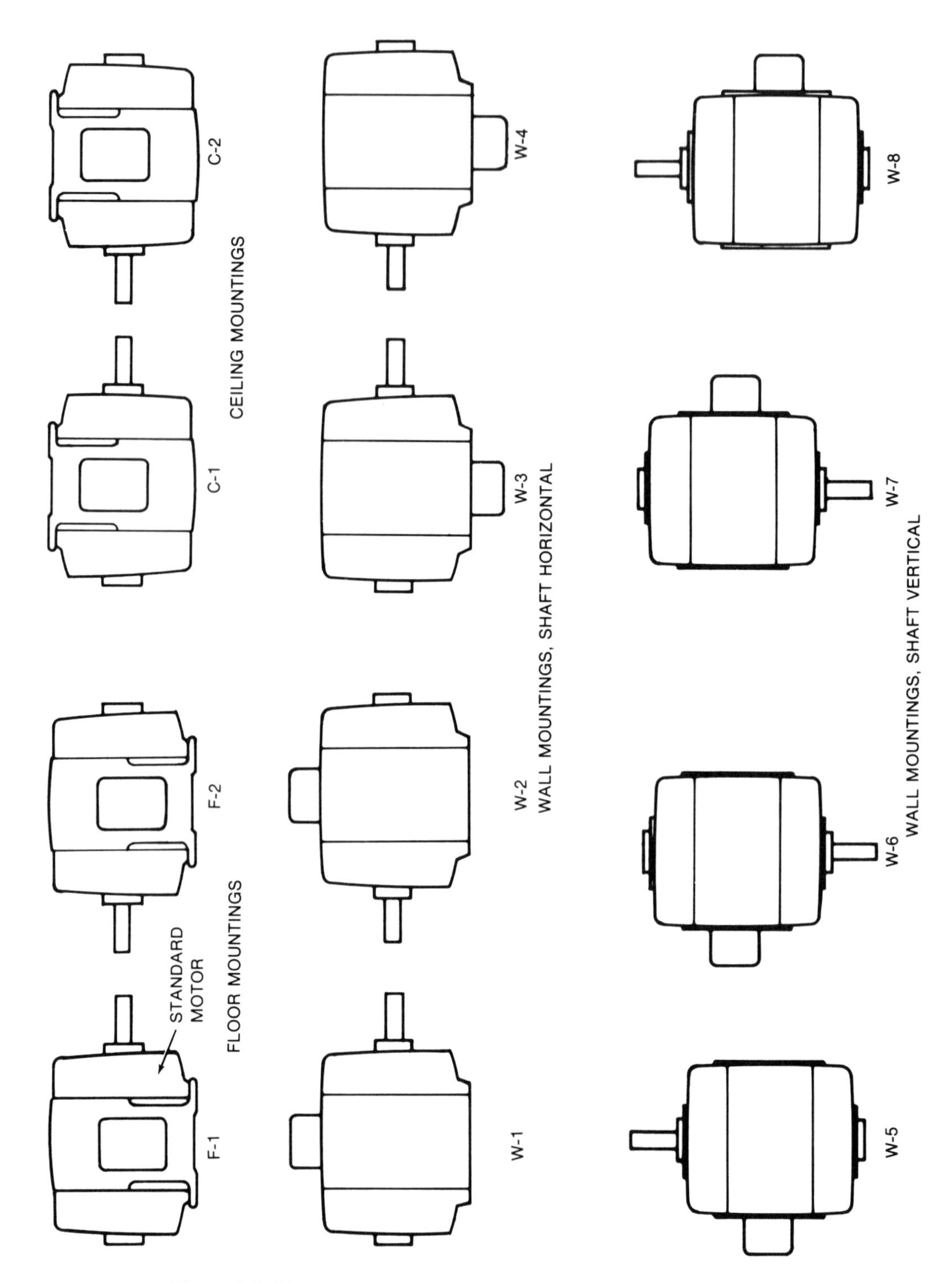

FIGURE 9–2. Motor mounting configurations. *(Courtesy General Electric Corp.)*

coupling is used when the motor's speed differs from the requirements of the driven machine.

In the cases of the fan blades and pumps, these can be connected directly to the shaft of the motor using a couple with set screws. Another method is for the shaft to be threaded so the devices can be attached in this manner.

In both of these cases, self-alignment is assured. The installer must make sure that the driven body is firmly secured by the set screws to the motor shaft when the coupler is the selected method.

When the threaded shaft is used, the possibility exists in some cases that an impeller will unscrew if the motor is run in the reverse direction. Precautions to prevent this must be designed into the system to prohibit the accidental reversal of the motor. Threaded shafts require either left- or right-hand threads depending on the direction of rotation of the motor. When the motor is in operation, it must turn so that the force on the driven body will maintain a force to keep the threads tight.

The possibility also exists that when a motor loses power, the driven body will cause the motor to spin in the reverse direction. Water and oil pumps are good examples. As the liquid flows back into the well, the impeller is driven in the reverse direction. If power is restored during this time, the motor may start in the reverse direction.

One approach to prevent the impeller from separating from the shaft is to use a pressure-tight pin driven through the shaft and impeller. The mechanics of this system will prevent the two from coming apart.

In other cases, the accidental reversal of the motor can cause severe damage to the driven equipment. A time-delay relay in the control circuit will prevent the motor from being started until enough time has lapsed for the impeller to stop.

There are other methods that will prevent the motor from turning except in one direction. Centrifugal switches are one possibility. Three-phase motors may be equipped with anti-phase reversing relays.

When a coupler is used between the motor and the driven machine, such as a motor-generator unit, care must be taken to align the two. Failure to do so will result in excessive vibration. This in turn causes premature bearing failure due to the stresses of misalignment.

The manufacturer of the motor-machine unit will have prescribed methods for making this alignment. The use of a straight edge with feeler gauges is usually not good enough to meet the alignment requirements. Dial indicators are recommended by many of the manufacturers. The General Electric Corporation limits the deviation to 0.002 inch for measurements made on all four quadrants between the faces of two couplings of a motor-generator unit. The bases of the two units have provisions for the adjustments to achieve the necessary alignment.

Once alignment is complete, and the machines are firmly bolted into position, grout the bases around and under the units. This prevents any further movement of the units in respect to each other. Use a cement sand ratio of 1:2 or 1:3 for this purpose. Add just enough water to keep the mixture firm and stiff. A minimum of 1 inch thickness is necessary to provide enough strength for the grout to be effective.

Direct couples may be rigid or flexible. Flexible couples are generally preferred. Use flexible couples when alignment is extremely critical. They allow for slight misalignment and minimize the wear on the bearings. They must be used when the motor and the driven unit have four bearings or more. Three-bearing designs normally require rigid coupling.

2. Belts and pulleys are the simplest and least costly method of coupling the motor to the load when the RPM of the motor does not meet the requirements of the load. Speed changes is accomplished through the ratio of the diameters of the pulley on the motor and that of the load. This is expressed mathematically as

$$\text{Speed of Motor}/\text{Speed of Machine} = d/D$$

where,

D is the diameter of the pulley on the motor,
d is the diameter of the pulley on the machine.

If the diameter of the machine's pulley is greater than that of the motor, its speed will be lower than that of the motor. For example, a motor operating at 1800 RPM with a pulley of 2 inches is driving a 6-inch pulley on the machine. The machine's RPM will be 600.

$$1800/X = 6/2$$
$$6X = 3600$$
$$X = 600 \text{ RPM}$$

The maximum practical speed ratio obtainable is dependent on several factors. These include the standard size of pulleys available, distance between pulleys, and the arc of contact on the smaller pulley. The practical ratio decreases with the increase in horsepower of the motor. Small FHP motors may have ratios of as much as 10:1. Motors over 50 horsepower have a limit of 4:1.

Because belt drives cause a bending force on the shaft of the motor, locate the motor's pulley as close to the motor's bearings as possible. This reduces the leverage on the shaft and increases the life of the bearings and the shaft. Overtightening the belt to avoid slippage will result in the early failure of belts, shafts, and bearings.

The motor will normally have an adjustable base to allow the belt(s) to be tightened. Adjust the tension on the belt just enough to prevent slippage when the maximum torque is being delivered by the motor to the load. A rule of thumb for proper tension is that the top of the belt will sag about 1/8 inch for each foot of length. This will vary somewhat with different types of belts. Follow the manufacturer's recommendations for the belt in each case.

The length of the belt (L) can be determined by the following formula.

$$\text{Length} = (D + d)1.57 + 2L$$

Measure the diameter of the two pulleys and add them together. Multiply the sum by 1.57. To this add two times the distance between the centers of the pulleys. If the diameter of the pulleys and the distance between centers are in inches, divide the final figure by 12 to convert the total length into feet.

When the belt is not tight enough, an excessive amount of slip will result. The slip causes friction between the pulley and the belt. Both the belt and the pulley will take on a shiny appearance. This is in contrast to a dull, smooth appearance when both are in good condition. The belt loses its traction with the pulley. The belt may begin to burn under this condition. Smoke is often present. The heat will destroy the belt within a short time. In severe cases the belt will come apart.

Other factors than proper tension on the belt can cause excessive slip. If the load becomes too great, the slippage will increase. A common cause of a relatively fixed load increasing its slip is due to bearing failure on the driven machine. Other causes include improper maintenance on the belts. Keep them clean of oil and other contaminants. Some systems have improper designs.

Allow a minimum distance of 2.5 times the diameter of the largest pulley between the centers of the driving and driven pulleys. Allowances of three to five times is better. With shorter belts, the flexure occurs more frequently at each cross-section. This means the belt is bent and stressed more often for the smaller distances. Also, the arc of contact between pulley and belt is reduced on the smaller pulley. This results in less power being transmitted between motor and load.

Belt speed is a limiting factor in the transmission of the power. Flat belts are normally limited to 5000 ft/min, and "V" belts have upper limits of 6500 ft/min. Above these speeds, contact is lost between the belt and pulley, and the centrifugal and twisting forces tend to throw the belt.

The maximum allowable speed varies with the manufacturers of pulleys and belts and the system design. It is best to run the belts at a higher speed rather than a lower one when using smaller pulleys. The belt adheres to the pulleys better and is less likely to slip. Refer to the manufacturer's specifications to assure the maximum speed is not exceeded. Calculate the belt speed using the following formula:

$$\text{Ft/Min} = (3.14 \times \text{Diameter of Pulley} \times \text{RPM})/12$$

Locate pulleys in the horizontal plane with their centers aligned whenever possible. This provides the most efficient type of drive. The lower side of the belt should be taut with the sag at the top. This arrangement increases the arc of contact between the pulleys and the belt. Avoid having the belt sag at the bottom. Proper tension on the belt is achieved when the sag is about 1½ inches for each 10 feet between the centers of the pulleys. Figure 9–3 illustrates this arrangement.

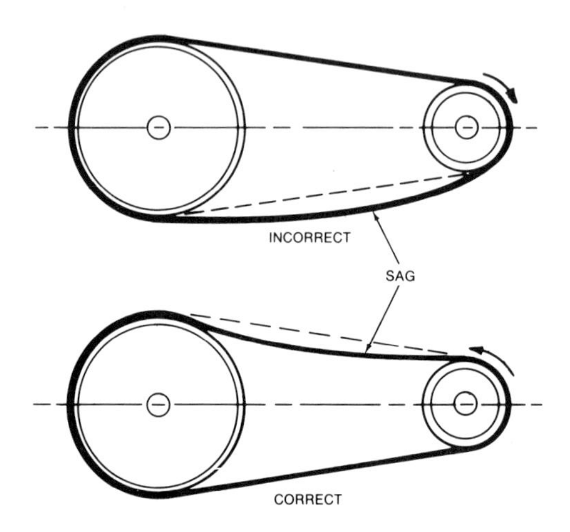

FIGURE 9–3. Horizontal pulley drive.

If possible, avoid vertical drives. The arc of contact is reduced when one pulley is directly above the other. This is particularly true if the smaller pulley is on the bottom. Under this condition only about 120° of contact is achieved between the small pulley and the belt. By putting the small pulley on top, the angle of contact is increased to 155°. This arrangement is recommended when vertical drive must be used. An angle of 45° or less between the pulleys provides for better operating efficiency (Figure 9–4).

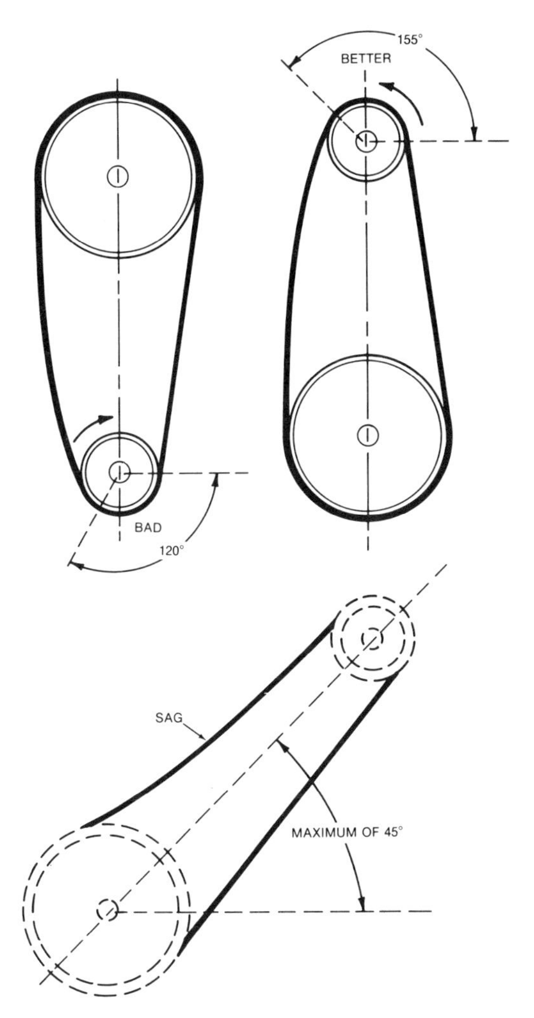

FIGURE 9–4. Vertical pulley drive.

Once the motor is properly mounted, the next step is to align the driven pulley with the motor's pulley. A misalignment between the two pulleys will cause the belt to wobble and flap. This causes strain on shafts and bearings and stretches the belt. In some cases the belt will come off the pulley. Figure 9–5 illustrates one procedure to assure the pulleys are in line with each other.

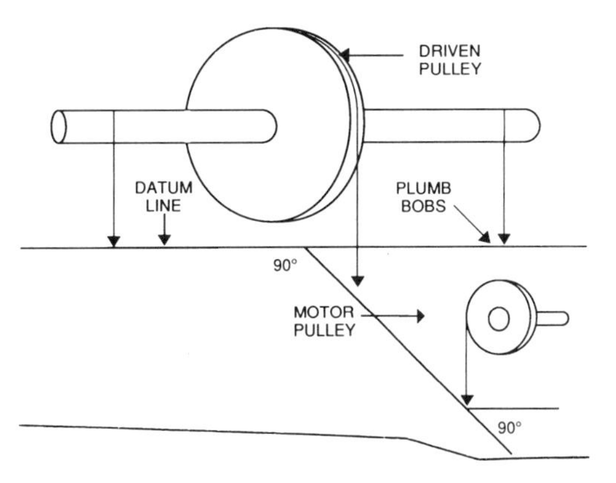

FIGURE 9–5. A method for aligning two pulleys.

The tools needed for this method are a plumb bob, square, and a level. A combination square and level may be used.

The first step in alignment is to determine if the motor and driven machine are level. This alignment is to be set first. Shimming one or the other machines may be necessary. If one of the devices cannot be brought into a level position, align the other to the same position.

To align the two pulleys so they are parallel and centered with each other, establish the datum line first. Use the plumb bob for this purpose. Drop the plumb bob from the shaft on each side of the driven pulley. Mark the points where the plumb bob barely touches the floor. Draw the datum line between these two points. This places the datum line parallel with the shaft of the pulley.

The next step is to drop the plumb bob from the center of the driven pulley. Once again mark the point on the deck. Using the square on the datum line, align it with the center point of the driven pulley. Draw a straight line. The datum line and the pulley's center line are at 90° to each other.

Next, drop the plumb bob from the motor's pulley. Align the motor so that the plumb bob barely touches the center line of the two pulleys. Perpendiculars can be drawn with the square from the pulley's center line though the centers of the motor's mounting holes. Check the level once again between the two devices. Use shims if necessary to bring the two into level.

Install the belt between the motor and the load. Draw up the belt so there is about 1/8-inch sag for each foot of belt. Tighten the leg bolts and nuts. Check the alignment to be sure it has not changed.

Belts tend to stretch somewhat during the first 24 to 48 hours of operation. Check the tension frequently during this period. Note any unusual movement of the belts or wear on them.

3. Chain drives are often used when belt slippage cannot be tolerated under varying loads. Although belt and pulley systems can be designed so that the slippage between the two pulleys is in the range of 2%, this may be too much for some systems.

Speed ratio of chain drive is calculated in the same manner as for pulley drives. This is based on the RPM and the motor and the diameters of the two sprockets. If the motor sprocket is larger than the driven sprocket, the load will turn at a higher speed based on the ratio of the diameters of the sprockets. Conversely, if the motor's sprocket is smaller than that of the driven load, the load will turn at a lower speed.

For chain drives, the maximum speed ratio is limited to about 8:1. In addition, the maximum speed of a chain is in the order of 1500 ft/min.

Chain drives are noisier than belt drives. In some cases this may be a consideration that needs to be taken into account. For quietness and smoothness of operation, chain speeds are usually limited to 600 ft/min. The length and speed of the chain can be calculated using the same method as that for belts and pulleys.

Alignment of the sprockets is a little more critical than that of the belts and pulleys. The same procedure is used for both systems. A little more care is taken with the chain and sprocket drives.

4. Gear drives are the most expensive of any system used to couple a load to the motor. They are used when an absolute positive drive is required to reduce slip to zero. They are also the most desirable method of coupling in the cases where a belt or chain break may cause a dangerous situation.

Gear drives also provide much greater speed ratios than either belts or chains. Although for a single set of gears this ratio is limited to about 10:1, any ratio is possible by the application of several sets of gears. Gears are often used when the load must turn at a very low RPM. This reduction in speed can be accomplished in a much smaller space than the previously discussed methods.

The speed ratio of gear drives is based upon the number of teeth on the pinion gear and the driven gear. The pinion gear is the one with the least number of teeth. For speed reduction, which is normally always the case when gear drives are used, the pinion gear is attached to the motor's shaft.

If an 1800 RPM motor has a pinion gear with 16 teeth driving a gear with 48 teeth, the load will turn at 600 RPM. In this case there are three teeth on the driven gear for each tooth on the pinion gear. The pinion gear must make three complete rotations for each rotation of the driven gear. The mathematical formula for determining these values is

$$\text{RPM Motor}/\text{RPM Load} = \text{Teeth Load}/\text{Teeth Pinion}$$
$$1800/X = 48/16$$
$$48X = 16 \times 1800$$
$$X = 600$$

Gear drives are noisier than either belt or chain drives. To reduce this noise level, pinion gears in some cases are made of a hard fiber or plastic material. This not only reduces the noise level, but in the case of a jam, only the pinion gear will be damaged. Normally, this gear will be the easiest one to remove and replace.

Gears are usually enclosed and submerged in an oil bath. This keeps the gears lubricated and reduces friction. The oil removes the heat generated by the gears, increasing the life of the gears. The oil bath also absorbs the noise energy, thereby reducing the noise level of the system.

5. Although not frequently used, magnetic coupling has some distinct advantages over all of the other types. The primary advantage is this method needs no mechanical means for the motor to turn the load. The motor can be running at full RPM without the load. By applying current to an electromagnet, a magnetic force is created which provides torque to the load.

The current to the magnet can be increased in increments so that a sudden turning force is not applied to the load. The load can be slowly brought up to speed. On the other hand, full torque can be applied to the load by the motor on start by allowing the maximum current to flow through the electromagnet.

By using various levels of magnetic force, the load can be made to turn at exact speed. Reducing the magnetic force will increase the slip, and the load will turn at a lower speed. Increasing the magnetic force will cause the load to turn at a higher RPM. When magnetic coupling is used, the RPM of the load cannot exceed the RPM of the motor.

Because no friction is present with this system, the slip between motor and load will not generate heat. There are no mechanical parts to wear out, and the maintenance on the coupler is minimal. Overall, magnetic coupling is the most efficient of all of the systems.

Electrical Considerations

All electrical wiring, accessories, and circuit protection for motors and their controllers shall comply with the NEC and any additional requirements of the local jurisdiction in which the motor is to be installed. The codes prescribe the electrical requirements for the installation of motors. Manufacturers' recommendations are to be considered in each case.

Article 430 of the 1987 NEC is the basis for most of the statements and calculations made in this section. Because the NEC is revised every 3 years, specific sections of Article 430 that are identified need to be reviewed for any changes in future editions of the code. Local jurisdictions do not always adopt the NEC when a new edition is published, or they may differ on some of the provisions. The local code will take precedence under these conditions.

Figure 9–6 shows the major sections of a motor system. Each section has certain requirements. The basic rules and calculations are discussed for most common problems associated with the installation of motors. Not all exceptions and varying circumstances are attempted, and you will need to refer to the exact code requirements for actual conditions on the job.

Power Supply

The requirements of the motor must conform to the voltage and frequency of the power supply of the plant. The power supply must have the required ampacity to provide the current drawn by the motor.

A transformer may be the power source when the voltage ratios must be changed to meet the requirements of the motor. At the same time, the motor's voltage may conform to the distribution voltage of the plant. In this case, a transformer is not needed.

Branch Circuit Conductors

The gauge of branch circuit conductors is based upon the type of loads they supply and the amount of current necessary to supply the particular load. In the case of a single motor load the ampacity of the conductor must be 125% of the full-load current (FLC) of the motor. The primary reason for this is the overload relay can be set to operate at this higher value, and the conductors can carry the higher current for a long period of time. This limits the FLC of the motor to 80% ampacity of the conductor. Under this condition, the conductor will not overheat and the stress is reduced under normal operation.

If the motor is a noncontinuous load, the conductors would be sized according to the time of the duty cycle and the classification of service of the FLC on the nameplate of the motor. Because the motor does not operate continuously, the conductors have time to cool. Unless it is impossible for a motor to operate continuously under any condition, it is classified as a continuous load. This is the case for the majority of motors.

The FLC stamped on the nameplate of the motor is *not* used for sizing the branch circuit conductors for a continuous duty motor. Instead, the FLC is taken from the tables in Part M of Article 430 of the NEC. Four tables are provided for this purpose. A different table is given for DC motors, single-phase motors, two-phase motors, and three-phase motors. In each of these tables, the higher the horsepower and the lower the voltage, the greater the FLC.

Apply the previous rules for conductors to the following problem. Determine the size of THW copper conductors rated for 75°C for a 50-horsepower, three-phase motor rated at 460 V to be operated continuously. The FLC on the nameplate is 60 amperes. Figure 9–7 illustrates this problem.

If your answer is AWG No. 4 conductors, you are correct. Here are the steps you should have taken.

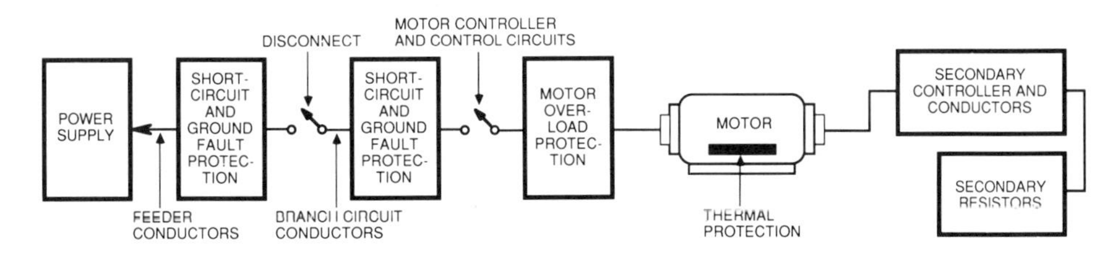

FIGURE 9–6. Sections of a motor system.

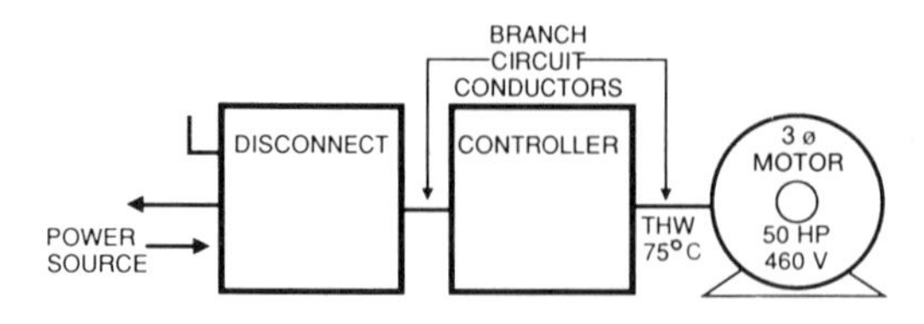

FIGURE 9–7. Branch circuit conductors for a single motor.

1. From Table 430-150 of the NEC you find 65-A FLC listed for a 50 HP, three-phase, 460-V motor.
2. Because the load is continuous, multiply 65 amperes by 1.25. This equals 81.25 amperes.
3. From Table 310-16 of the NEC, you determine that No. 4 AWG is needed for the conductor specified.

If the horsepower rating of the motor is not listed in the tables, the FLC can be determined by the following method.

1. Use the FLC for the motor listed with the next lower horsepower rating.
2. Divide the FLC by the horsepower rating.
3. Multiply the results obtained in step 2 by the horsepower rating of the unlisted motor to obtain its FLC.

Using this information, determine the FLC for a three-phase, 12.5-horsepower motor rated at 230 volts. If your answer is 35 amperes you have correctly solved the problem.

The steps in solving this problem are

1. Locate the 28-A FLC of the three-phase, 10-horsepower motor rated at 230 volts in Table 430-150.
2. Divide 28 amperes by 10 horsepower to obtain a multiplier of 2.8.
3. Multiply 2.8 times the unlisted 12.5-horse-power motor to obtain the 35 A.

Multispeed motors draw different FLC for each speed. The lower the speed of operation, the higher the current. Branch circuit conductors on the line side of these motors must be 125% of the nameplate current of the highest FLC shown. Between the controller and the motor, the conductors are sized according to the *name-plate* FLC times 125%.

A motor that is not used for continuous duty does not have to have branch circuit conductors rated at 125% of the FLC. Because the motor is started and stopped, varying heat loads are placed on the conductors. Conductor sizes are based on the length of time of the duty cycle and on how the motor is used. The nameplate FLC is multiplied by the factor given in the NEC, Table 430-22(a) Exceptions. If the motor is running for 15 minutes, it must have a rest period of 45 minutes, unless the nameplate of the motor indicates otherwise. The run time is subtracted from 60 minutes.

Feeder Circuits

The conductors of the feeder circuit supplying several motors are sized in accordance with Section 430-24. To determine the gauge of wire for the conductors, the FLC for each motor is added together. An additional 25% of the highest rated FLC is added to this value. Using this rule, determine the size of the conductors for the feeder for the motors illustrated in Figure 9–8.

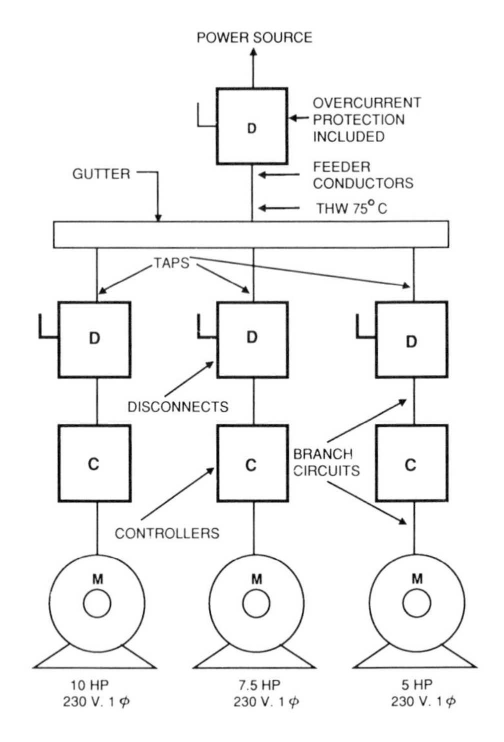

FIGURE 9–8. Feeder conductors for multiple motors.

If your answer is AWG No. 1/0, you are correct. The steps to obtain this answer are

1. FLC for each of the single-phase motors is found in Table 430-148. These values are 50, 40, and 28 amperes for the three motors based on their horsepower ratings.
2. Multiply the largest value (50 A) by 1.25 to obtain the value of 62.5 amperes. Add the 40 and 28 amperes FLC of the other motors to this figure. The result is 130.5 amperes which the feeder conductors must be able to deliver safely to the loads.
3. Use Table 310-16 to determine the wire size as AWG No. 1/0.

When the demand factor is less than the continuous load, the feeder conductors may be smaller with permission of the local inspector. They must have ampacity sufficient for the duty cycles, type of loads, and conditions of operation.

Taps

In the illustration shown in Figure 9–8, the feeder conductors are brought to an auxiliary gutter and taps are made to supply the three motors. The taps must terminate in the branch circuit protective device. In this case, the disconnects must have fuses. If not, a panel with either circuit breakers or fuses is necessary to feed and protect the individual branch circuits.

Protective devices are not needed at the source if the taps meet one of the following conditions.

1. The taps have the same current carrying ability as the conductors of the feeder.
2. They have an ampacity of at least one-third that of the feeder conductors, are protected from physical damage, and have a length not in excess of 25 feet.
3. The length of taps do not exceed 10 feet, and they are protected by a raceway or an enclosed controller.

There are exceptions to these three conditions, and the appropriate section of Article 430 should be consulted when conditions differ.

Conductors for Wound Rotor Secondary

The rotor of the wound rotor motor is connected through slip rings and conductors to a bank of resistors. The conductors for continuous duty operation shall have 125% of the ampacity of the FLC of the secondary circuit. FLC of the secondary will be found on the nameplate of the motor or it will be provided by the manufacturer.

For varying duty cycles and applications, the conductor's ampacity may be other than 125% of the FLC. Table 430-22(a), Exceptions, is used to determine the percent of the multiplier for the secondary circuit conductors as well as that of the primary circuit. The percentages range from 150% for 30- and 60-minute-rated motors, to a low value of 85% for 5- and 15-minute-rated motors.

Conductors for Capacitors

When capacitors are separate from the motor, the conductors that connect the motor to the capacitors shall not be less than 135% of the rated current of the capacitors. In addition, the current carrying capacity of the conductors must be at least one-third as great as the branch circuit conductors supplying the motor.

In summary, main, feeder, and branch circuits must have conductor ampacity that is at least as great as the sum of

1. 125% of the continuous nonmotor load.
2. 100% of noncontinuous, nonmotor load.
3. 125% of the largest motor's FLC plus the FLC of any other motors. If there are two motors with the largest HP rating, one of them is considered to be the larger.
4. 100% of all other motors' full-load current.

Disconnect

There are exceptions in the NEC to many of the provisions listed for the disconnecting means. Each installation requires a review of the code to assure compliance. Appropriate sections of the NEC include 430-81, 430-83, 430-109, 430-110, and 430-111. A summary of the requirements for the switches used as disconnects for motors follows.

1. A disconnect shall be used to remove power from both the motor and the controller.
2. The disconnect shall have sufficient number of poles to open the circuit to all ungrounded power supply conductors.
3. The grounded conductor is permitted to be opened by the disconnect provided the ungrounded conductors are simultaneously opened.

4. The disconnect shall be clearly marked as to its on-off status.
5. The disconnect shall be within sight of both the motor and the controller, or be capable of being locked open. "In sight" means that it is visible from both the controller and the motor and within fifty feet of each of them.

Although exceptions are made in the NEC, in general, each motor shall have an individual disconnect means. The disconnect shall be a motor-circuit switch rated in horsepower, a circuit breaker, or a molded-case, nonautomatic, circuit interrupter switch.

Stationary motors with voltage ratings of 0 to 300 volts have the following requirements for the disconnect.

1. For a motor rated at 2 horsepower or less, the options are
   a. A switch rated in horsepower
   b. A general use switch with an ampere rating of at least 200% the ampere rating of the motor
   c. Use an AC (only) snap switch with a rating of at least 125% of the motor's current rating
2. Motors having ratings in a range greater than 2 HP up to and including 100 horsepower shall have the disconnect switch rated in horsepower.
3. Motors larger than 100 horsepower shall have a minimum ampere rating of 115% of the motor's nameplate current rating. If the switch is used for control purposes, it shall have a horsepower rating.

Motors having voltage ratings between 301 and 600 volts shall have the disconnect switch rated in horsepower for motors less than 100 horsepower. Motors above 100 horsepower shall have a minimum rating of 115% of the motor's nameplate current rating. When used as the controller, the switch shall be rated in horsepower.

For portable motors the attachment plug and receptacle is permitted to serve as the disconnect. For motors under 1/3 horsepower, the attachment plug and receptacle may also serve as the controller. Motors above 1/3 horsepower shall meet the same requirements as a controller for stationary motors.

For motors rated 600 volts or less, the ampere rating of disconnect shall be 115% of the FLC as determined from Tables 430-147 through 430-150. In the case of torque motors, the ampere rating shall be 115% of the nameplate current. One hundred percent of FLC is permitted to be used when the disconnect is rated for continuous operation at 100% of its rating.

Where two or more motors, or a combination of a motor and heating and lighting circuits, are served by a single disconnect, the ampere rating of the disconnect shall not be less than 115% of the summation of the full-load current of the branch circuit. Article 220 and other appropriate sections of the NEC are used to determine the current of the heating and lighting loads. The FLC of the motor(s) obtained from the tables in Article 430 is added to these loads.

Locked-rotor currents (the current drawn by the motor when the rotor cannot turn with full voltage applied) for each motor obtained from Table 430-151 shall be added to the currents of the other loads to obtain the equivalent full-load, locked-rotor current for the combined load. Locked rotor currents are about six times the FLC for a given motor.

Controller

A *controller* is a device that starts and stops a motor. It may be as simple as a single-pole, single-throw switch. Under other requirements, the controller may include provisions for reduced voltage starts, acceleration, speed control, reversing direction or rotation, jogging, inching, plugging, or braking. The controller delivers the required power and frequency to the motor to accomplish these purposes. Under these conditions, the starter or controller, may be very complex (Figure 9–9). In general, the controller must have a horsepower rating equal to or greater than that of the motor it controls.

Motor controllers are manual, magnetic, or solid state electronic devices. They may be a combination of all three.

Manual devices require the use of the hand to open or close a switch, press a button, or operate a drum controller. The basic functions of performing these tasks with the magnetic controller are accomplished using coils and solenoids.

In the case of solid state electronics, the device itself acts as a switch and resistance in applying the required current and conditions required by the motor. Silicon controlled rectifiers are often used for this purpose. By electronically timing the gate pulse to the

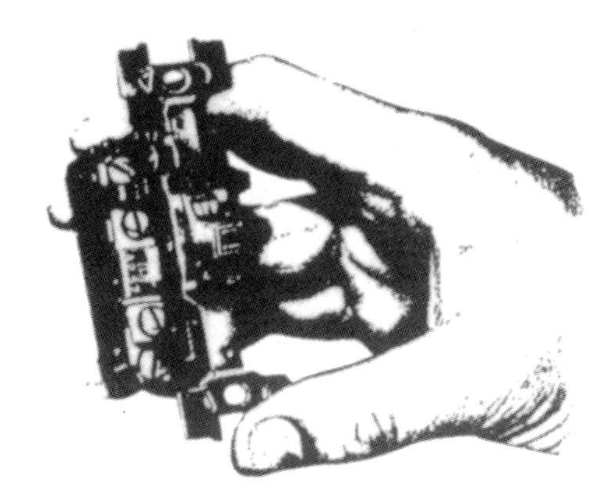

(A) Simple controller.

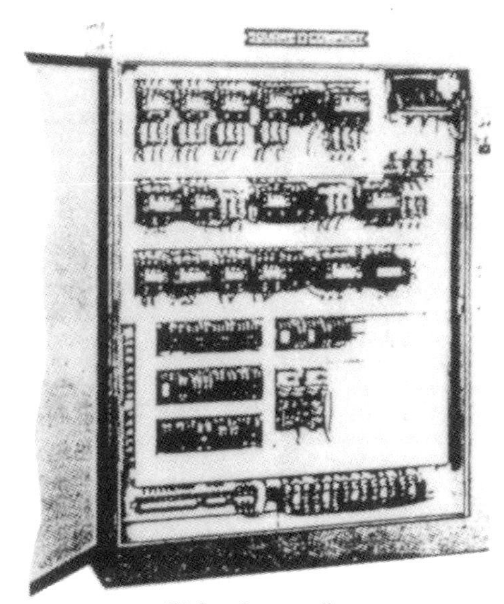

(B) Complex controller.

FIGURE 9–9. Simple and complex controllers. *(Courtesy Square D Co.)*

SCR, the time that current is applied to the motor can vary from the full 360° to zero current at 0°.

Magnetic and solid state devices allow for remote operation. The devices can be located close to the motor to connect the high currents required. Manual control may require the currents to flow over great lengths of conductors. For example, in the case of an oil processing plant and storage tank farm, miles of large conductors would be required if manual control was used.

Starters using magnetic and solid state devices also provide isolation for the operator from the high currents and voltages associated with the motor. An operator may control a 575-volt, 100-ampere motor with a control circuit rated below 100 volts and with less than an ampere of current.

Motor controllers may be fully manual, semiautomatic, or fully automatic. The diagrams in Figure 9–10 illustrate each of these conditions.

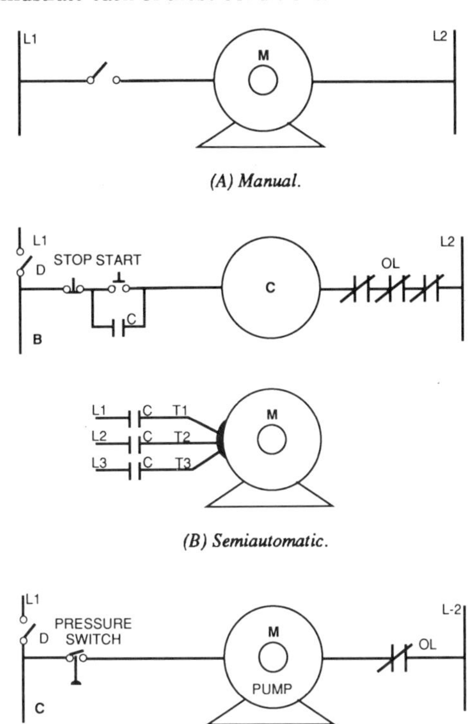

FIGURE 9–10. Motor controllers.

Figure 9–10A depicts a motor that is manually controlled by a simple hand switch. When the switch is closed the motor will run. Although not shown, an overload which will open if the motor draws too much current, should be included.

Figure 9–10B shows a semiautomatic system. To start the motor, the start pushbutton must be pushed manually. When this is done, coil *C* will activate and close the four *c* contacts automatically.

There is one contact in series with each of the three-phase conductors going to the motor. These will close automatically and apply power to the motor.

The fourth *c* contact across the start switch will close shorting the normal open contact of the start button. This will provide memory and keep the motor running after the start button is released. This arrangement is known as a *three-wire control system.*

When the motor is running under normal conditions, it can be stopped by pushing the stop button. As in the case of the overload opening, this will break the current path to the coil causing the *c* contacts to open. When the stop pushbutton is released, the motor will not automatically restart due to the *c* memory contact across the start button being open at this time.

Pilot devices that start a motor running are normally open contacts. Those that stop the motor are normally closed contacts. Switches and relay contacts are usually shown on the line diagram in the condition, open or closed, as they sit on the shelf of the distributor. When this is not true, notations will be made on the diagram that will describe if the contacts are energized or not. Timing diagrams are sometimes provided to show the periods that contacts will be open or closed.

The line diagram in Figure 9–10B is a common wiring scheme for control circuits. Figures 9–11 and 9–12 depict the wiring that is performed on the motor starter to obtain this circuit.

Figure 9–10C illustrates a fully automatic system. Once the disconnect is closed, the pump will turn on and off due to the force exerted on the pressure switch. This is similar to a typical water pump in a home which obtains its water from a well. Manual operation of the switch is not necessary in order to maintain water pressure. The only time human intervention must occur is for maintenance or repair.

Overloads

In series with the start-stop circuit and the coil are three overload contacts. These contacts are normally closed. The current to the motor passes through heater coils that will cause these contacts to open if the current goes above a predetermined level for a given interval. This will break current flow though the coil, and its armature will release, opening the power to the motor. Usually, these contacts must be reset manually. A sufficient time lapse must occur for the motor to cool before the overloads can be reset, and before the motor can be restarted. Figure 9–13 shows the mechanics and operation of thermal overload relays.

Overload devices are designed to protect the motor. Motors of more than 1 horsepower used for continuous duty are required by Section 430-32 to be protected against overloads. The overload shall be a separate

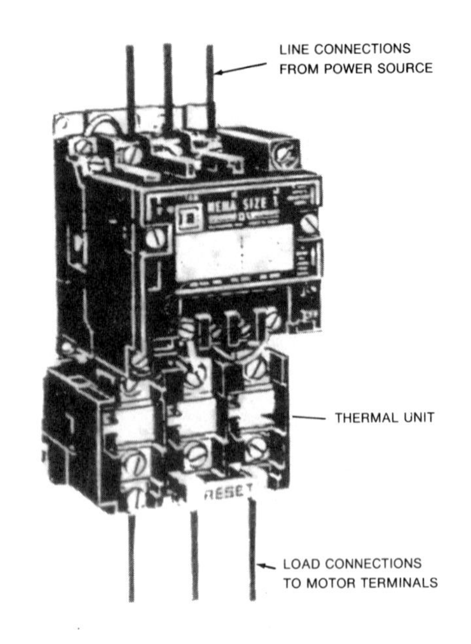

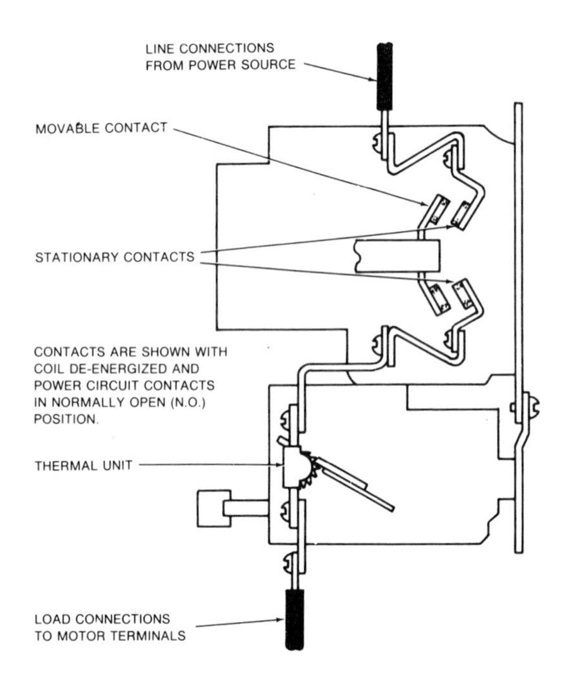

FIGURE 9–11. Magnetic starter power circuit. (*Courtesy Square D Co.*)

device that is activated based on the nameplate FLC. Motors marked with a service factor not less than 1.15 or a temperature rise not to exceed 40°C will have the

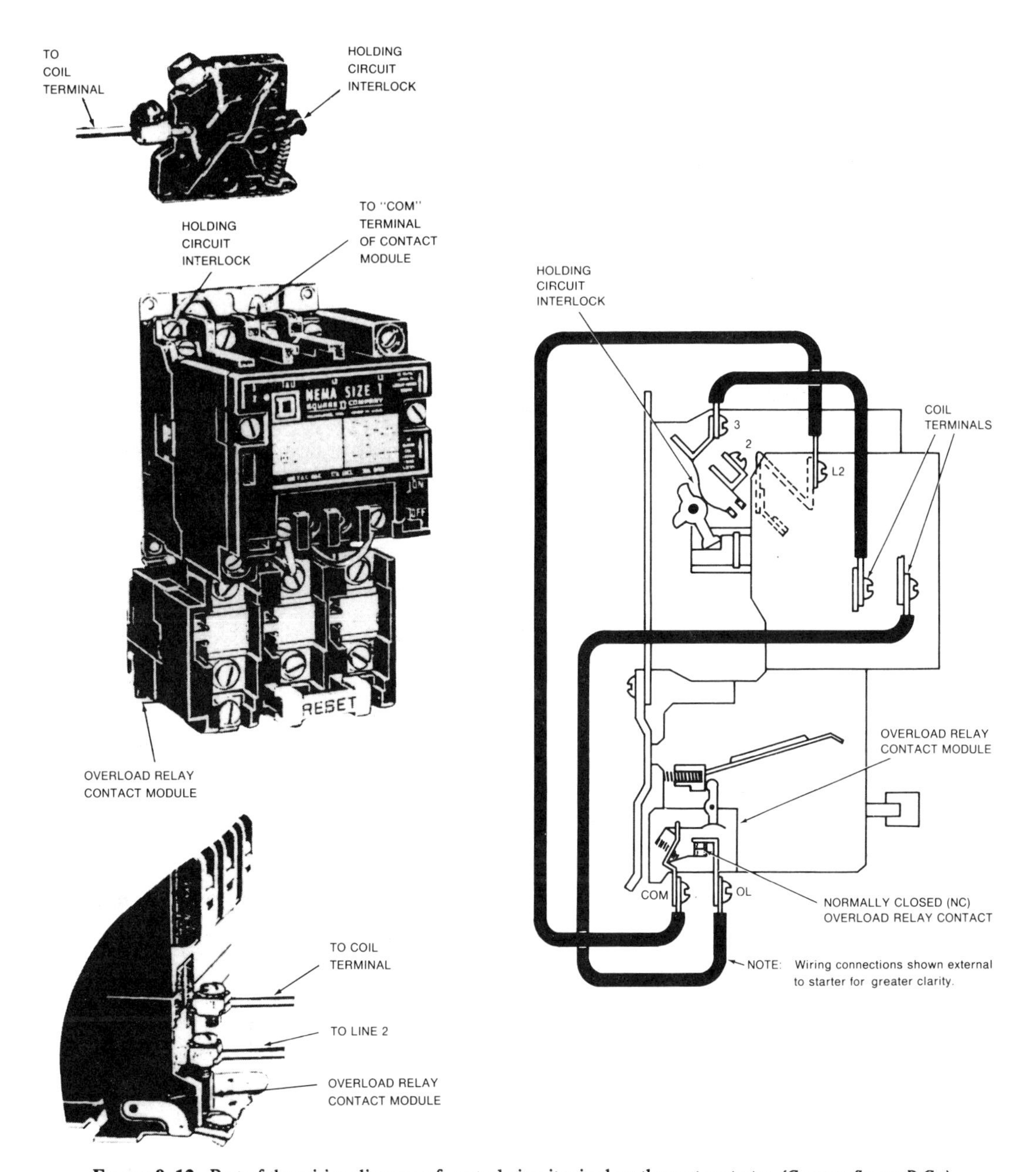

FIGURE 9–12. Part of the wiring diagram of control circuit wired on the motor starter. (*Courtesy Square D Co.*)

overload set to trip at no more than 125% of the nameplate FLC. All other motors are to be set at not more than 115% of FLC.

The size of the overloads for a motor is normally given on the cover of the magnetic starter of the motor control center. If this information is not available at this source, the manufacturer's specification sheet or catalog will provide the information.

When the specifications of Section 430-32 are not sufficient to start the motor, Section 430-34 allows for the next higher size of overload relay. This section limits the setting to 140% of the FLC for motors with

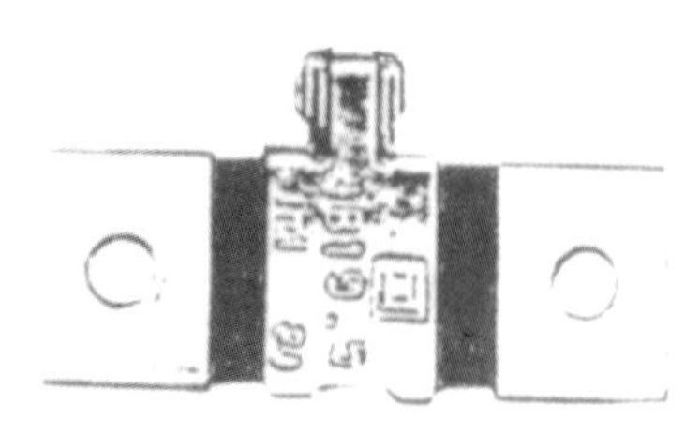

ONE PIECE THERMAL UNIT

HEATER WINDING (HEAT PRODUCING ELEMENT) IS PERMANENTLY JOINED TO THE SOLDER POT, SO PROPER HEAT TRANSFER IS ALWAYS INSURED. NO CHANCE OF MISALIGNMENT IN THE FIELD.

SOLDER POT (HEAT SENSITIVE ELEMENT) IS AN INTEGRAL PART OF THE THERMAL UNIT. IT PROVIDES ACCURATE RESPONSE TO OVERLOAD CURRENT, YET PREVENTS NUISANCE TRIPPING

SINGLE POLE
MELTING ALLOY
THERMAL OVERLOAD RELAY

THREE POLE
MELTING ALLOY
THERMAL OVERLOAD RELAY

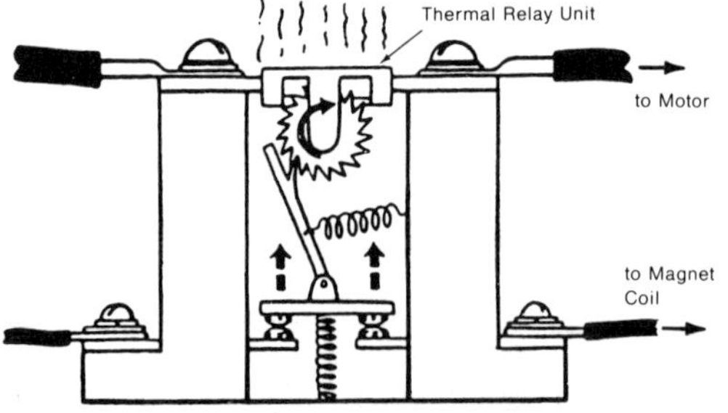

Drawing shows operation of melting alloy overload relay. As heat melts alloy, ratchet wheel is free to turn—spring then pushes contacts open.

FIGURE 9–13. Mechanics and operation of a thermal overload relay. (*Courtesy Square D Co.*)

service factors not less than 1.15 and with temperature rises not over 40°C. All other motors are limited to 130% of the FLC.

Thermal Protection Devices

In lieu of, or in conjunction with, the overload protection on the starter other thermal-sensitive devices can be used to protect the motor from overheating or failure to start. These include such devices as thermostats, thermistors with positive temperature coefficients, resistance temperature detectors (RTD), and thermocouples. Some of these require special circuits that will interrupt the control circuit of the motor and cause the power to be disconnected. Others may automatically disconnect power from the motor without disrupting the motor control circuit.

Some of these devices must be manually reset, while others automatically reset themselves after they have cooled. If the latter is used, and the control circuit is not interrupted, care should be taken by the electrician when working on the motor due to the possibility that it may suddenly start. A motor with automatic restart thermal protection must never be used where an unexpected start may injure someone or cause damage to property.

The trip current for thermal protection devices that are an integral part of the motor is based on the FLCs given in Tables 430-148, 430-149, and 430-150. These devices are set to operate under the following conditions. For motors not exceeding 9 amperes, a multiple of 170% of FLC is allowed. From 9.1 up to and including 20 amperes, 156%, and for motors with FLC above 20 amperes, 140%.

Thermal devices other than those on the motor starter are normally embedded in the stator windings or the frame of the motor. They are mounted in such a way that heat transfer from the motor to the thermal sensor is effected.

Other Protective Devices

Other types of protection can be added to a motor in order to protect it from damage. These devices will normally perform their function in conjunction with the control circuit. They will either prevent the motor from being started under adverse conditions, or they will stop the motor if an abnormality exists which may cause damage to the motor or related equipment.

When a three-phase system loses one of its phases, it is known in the industry as *single-phasing*. This is not an unusual occurrence, and can be caused by the malfunction of a single-phase transformer in the bank, or the opening of an overcurrent device in one phase.

A three-phase motor with an FLC of 40 amperes would suddenly draw 69.2 amperes on the other two phases if single-phasing were to occur (40 A X 1.73 = 69.2 A). This will cause the motor to overheat.

One could rely on the overload protection to take the motor off the line when the temperature reached a certain level. Another method, which would react instantly to the loss of a phase, would be a circuit which would sense the absence of one phase. Figure 9–14 illustrates such an arrangement.

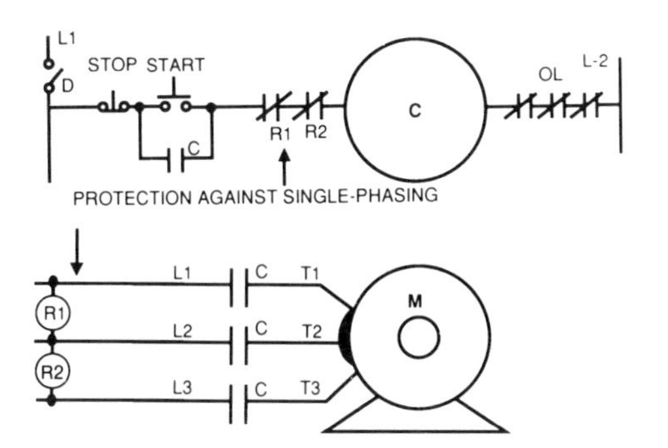

FIGURE 9–14. Protection against single phasing.

In this case, two relays, R1 and R2, are connected across the three phases of the power line to the controller contacts of the motor. If all three phases are present, the relays will be energized, and their contacts will be held close. The contacts are in series with the coil of the controller. Under this condition, the motor will start and continue to run when the start button is pushed.

If one of the phases, L1 or L3, is not present, the single-phasing relay connected to that line will not be energized. When L2 drops out, there is not sufficient voltage to energize either of the relays. Under these conditions, the relay's contacts will remain open, and the motor cannot start.

If the motor is running when the phase is lost, the associated single-phasing protection relay will lose power with its associated contact opening. This will remove power from the controller's coil, and the motor will stop. The motor cannot be started until power is restored on all phases.

This circuit could also serve as protection against low voltage. The relay coils for the single-phasing protection would be selected so that they drop out if the voltage across them fell to a predetermined level.

Another set of relays could be used to protect against high line voltage. These would have normally closed contacts placed in series with the control circuit. If the voltage on the line reached a predetermined level, the armature would be activated, and the contacts would open, interrupting the control circuit.

Motors may also be protected from excessive currents by the addition of current sensing devices. Current transformers are commonly used to sense the current on each of the phases. The output of these transformers is connected to relays which have normally closed contacts in series with the control circuit or the memory contact. The relays have pull-in current ratings which will provide a limit on the maximum current the motor can draw.

Protection against excessive vibration of the motor can be provided in a similar manner. The vibration sensor would be mounted on the motor. It would have a set of normally closed contacts. These contacts could be wired either in series with the controller's coil or in series with the memory contact *c* of the coil. When a certain level of vibration is reached by the motor, the contacts will open. This will cause the motor to stop. The motor may or may not be able to be restarted without some manual intervention. This depends on the design of the circuit and/or the vibration detector.

Thermal sensors are sometimes installed to sense the temperature of the bearings of a motor. When a predetermined temperature is reached that would damage the bearings, the sensors activate a coil to open a set of normally closed contacts that stop the motor. This arrangement is normally designed to prevent the motor from being restarted until the temperature on the bearings is lowered to a safe value.

Centrifugal switches, tachometers, photoelectric methods, and Hall sensors can all be used to control the speed of a motor. These devices in conjunction with associated control circuits can protect the motor from over or under speed operation. They can also be used to maintain a constant or predetermined RPM.

Motors and their related circuits may also be protected by surge and lightning arresters. High voltages and spikes on the power lines can cause damage to insulation on conductors and motor windings. Solid state control devices are particularly receptive to damage due to these occurrences.

Overcurrent Protection for the Controller

A motor control circuit must be protected from overcurrents in accordance with Section 430-72. If tapped from the load side of the motor branch circuit overcurrent protective device, the circuit is not considered to be a separate branch circuit. It may be protected by a supplementary or the branch circuit overcurrent device. When the control circuit does not obtain its power in this manner, Section 725-12 or 725-35 would apply.

Maximum ratings for conductor protection is given in Table 430-72(b). Section 430-72(c) sets the conditions for protection of the circuit when a control transformer is used.

Grounding

Motors, related equipment, and conductors will conform to the general requirements for grounding as provided in the NEC, Article 250 and Article 430, Part L. The primary purpose is to prevent accidental voltages above ground on frames and enclosures which may be hazardous to the health of personnel. A second purpose is to provide a complete circuit so that the overcurrent device will interrupt power to the equipment in the case of an accidental short to ground. Under certain conditions, isolation, guarding, or insulation are exceptions used in place of grounding requirements. Under the following conditions, frames of stationary motors shall be grounded

1. When supplied by metal enclosed wiring.
2. In wet locations when not isolated or guarded.
3. In hazardous locations as defined in Articles 500 through 517.
4. When any terminal on the motor is above 150 volts.

Controller enclosures shall be grounded regardless of the level of the voltage. The only exceptions to this are in the cases of controller enclosures which are part of ungrounded portable equipment, and lined covers used with snap switches.

Portable motors that operate over 150 volts to ground shall be either guarded or grounded. The grounding conductor shall be either green or green with yellow stripes.

Many of the portable tools under 150 volts to ground are encased in double insulation to prevent accidental shock and do not require grounding. There are many others, however, produced with metal frames which do require the frame to be grounded. These are often equipped with the three-wire plug with the round connector being attached to the grounding conductor. When these are plugged into three-wire receptacles, grounding is automatically effected.

There are many two-wire receptacles still in use, and these can present a problem for the three-wire plug on the tool. Adapters with provisions for completing the ground circuit may be used with receptacles which have only two connections. The adapter's green lead must be grounded. The grounding prong on the plug must never be broken off merely to accommodate a two-wire receptacle. This is a very dangerous practice which results in several deaths by electrocution each year.

Overcurrent Protection

Overcurrent and overload protection are required for circuits that serve motors. Overcurrent protection is provided by fuses and circuit breakers that are installed in series with the ungrounded conductors to protect against short circuits and ground faults. These devices must be sized to protect the circuit conductors, while at the same time they must not open with the inrush of the starting currents.

Single-element, nontime-delay fuses or instantaneous circuit breakers do not protect the motor or its circuits from a continuous overload. A separate device, such as a thermal overload relay, is used to provide for overload protection. Exceptions to this are the dual-element fuse or the inverse time circuit breaker. They will provide protection for both overcurrent and overload when the recommended multiples of FLC are applied. Even when these types of overcurrent protection are used, they may serve as a backup to other means of overload protection.

Figure 9–15 provides a diagram for determining the size of fuses to provide short-circuit and ground fault protection for motors and other loads. The diagram provides general guidelines as specified by Bussman for selecting ampere ratings for fuses for the most common circuits. Specific applications may require different fusing levels, and the NEC is to be consulted in these cases.

Table 430-152 of the NEC gives the actual multiples of FLC for the maximum rating of motor branch circuit short-circuit and ground-fault protection devices. This table includes the multiples for nontime-delay and time-delay fuses as well as instantaneous trip and inverse time circuit breakers.

The percentages given in Table 430-152 are to be used when the motor will start without interrupting the overcurrent device for the each type of motor. Where greater currents are required to start the motor, the NEC in Sections 430-52 through 430-54 allows greater percentages as multipliers. These include

1. When nontime delay fuses are used that do not exceed 600 A, the percentage may be increased to 400%.

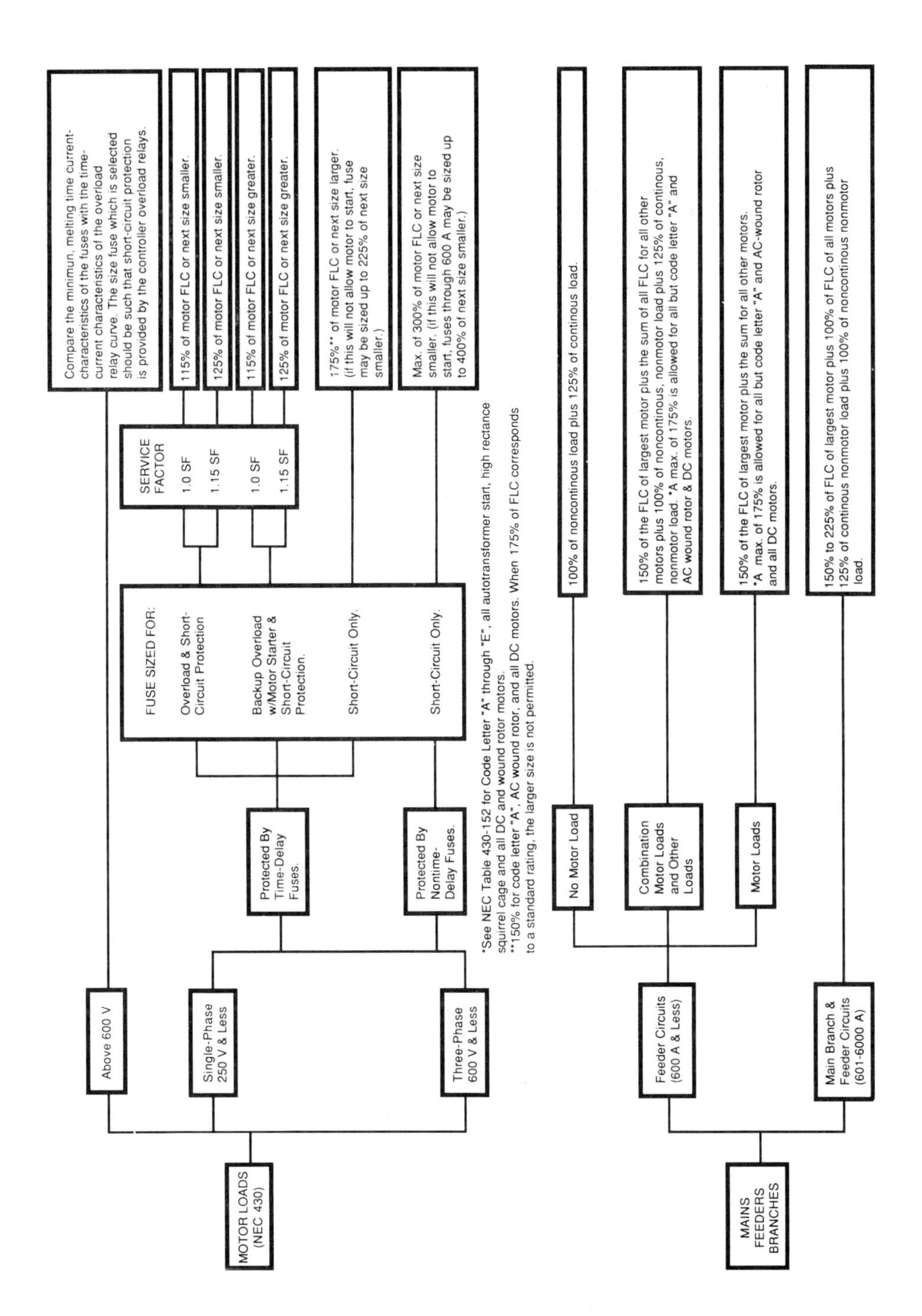

FIGURE 9–15. Overload and overcurrent protection for motors and other loads. *(Courtesy Bussman Division, McGraw Edison Co.)*

2. Time-delay fuses may have their values increased to 225% of the FLC.
3. Current ratings for instantaneous trip circuit values may be increased to 1300% of FLC.
4. Ratings for inverse-time circuit breakers can be increased to 400% of FLC up to 100 amperes. If greater than 100 amperes, the ratings can be increased to 300% maximum.

Overcurrent protection at the higher values in each case may not exceed the value calculated using the higher multiple. The next smallest standard fuse shall be selected.

Nontime-delay fuses detect shorts in the circuit. They operate on thermal principles and will usually hold 500% of their rated current for about one-quarter of a second. Some of these fuses are designed to hold this level of current up to 2 seconds. The motor must start and reach its running speed within this time interval when nontime-delay fuses are used for overcurrent protection.

Time-delay fuses are the dual-element type with thermal and instantaneous trip features. These fuses hold 500% of their rating for 10 seconds. This amount of time will allow almost any motor to start and reach its running speed within this interval.

Instantaneous trip circuit breakers respond to short-circuit currents, ground faults, and locked-rotor currents. Some of these devices have adjustable current ratings. These breakers will trip at about three times their ratings on the low setting and at ten times the value on the high setting. They are designed to hold the starting current of a motor. In compliance with Section 430-52, they shall not be set at more than 1300% of the FLC of the motor.

If the motor will start and run up to speed, the maximum setting is 700% of FLC as required by Table 430-152. Always set the ampere rating as low as possible for the motor to perform properly. Instantaneous trip circuit breakers do not provide for overload, and separate overload protection shall be provided.

The inverse time circuit breaker responds both to overcurrent and overload problems. It has both thermal and magnetic elements to remove power from the motor if either of these problems is present. Heat buildup will be detected by the thermal element, and short circuits or ground faults will be detected by the magnetic action of the breaker. The inverse time circuit breaker will maintain 300% of the load for variable intervals from 4 to 40 seconds depending on the level of the current setting and the voltage applied to the circuit. This type is the breaker of choice in industry today.

Figure 9–16 depicts a distribution system with several different motor loads as well as other types of loads found in a typical plant. This diagram illustrates some examples for the calculations of fuse and circuit breaker sizing for the different combination of loads under different circumstances.

Problem 1—What size time-delay fuse is required for the 200-horsepower, 230-volt, three-phase, squirrel cage induction motor that is supplied by fuse No. 2?

1. Determine the FLC from Table 430-150 (480 A).
2. Determine the multiplier for the FLC from Table 430-152 (150%).
3. Multiply 1.5 by 480 amperes (720 A).
4. Look up the standard size fuses in Section 240-6. The standard sizes found in this range are 700 and 800 amperes. Exception No. 1 of 430-52 allows the larger value of 300 amperes.

If the motor will not start without blowing the 800-ampere fuse, what is the maximum ampere rating allowed for the fuse?

Exception 2(d) of 430-52 allows up to but not to exceed 300% of the FLC for fuses in the range of 601 to 6000 amperes.

1. Multiply 3.00 by 480 amperes (1440 A).
2. Determine the standard size fuse from the list in 240-6. Standard sizes close to 1440 amperes are 1200 and 1600 amperes. Because the code allows up to 300% of FLC, the smaller value must be selected (1200 A).

Problem 2—In Figure 9–16, if the continuous load is 225 amperes, and the noncontinuous load is 75 amperes, what sizes of nontime-delay fuses are needed for fuses 3, 9, and 10?

Fuse size for the continuous load must be at least 125% of the continuous load. For noncontinuous loads, the fuse size must be a minimum of 100% of the load. In each case, the fuse may be larger than that which is calculated. At the same time, it must be small enough to protect the conductors of the circuit.

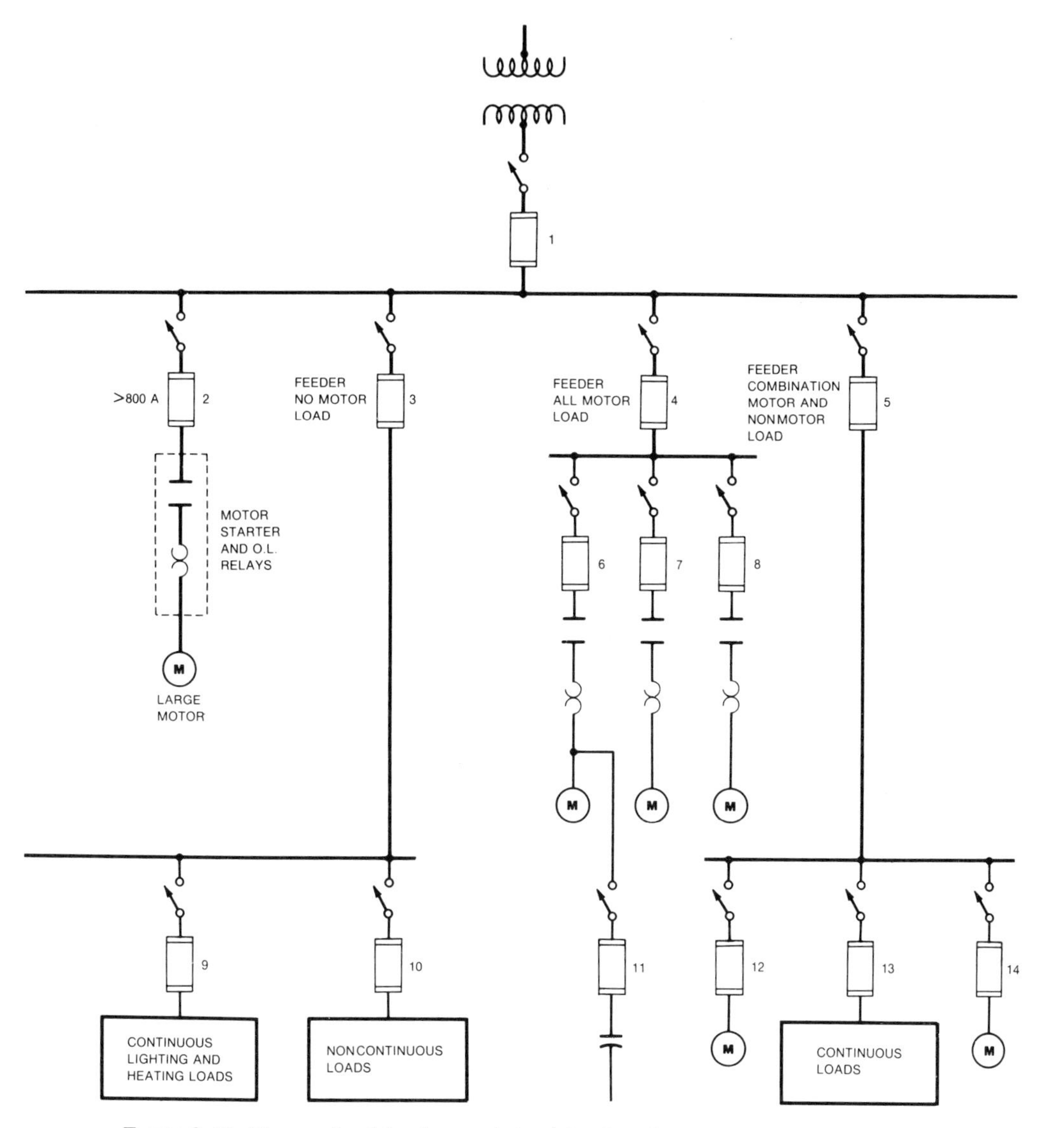

FIGURE 9–16. Diagram for sizing fuses and circuit breakers for a power distribution system.

1. Multiply 1.25 by 225 amperes to determine the size for fuse No. 9 for the continuous load (281.25 A). The standard size fuse is 300 amperes.
2. Multiply 1.00 by 75 amperes to determine the size for fuse No. 10 (75 A). The standard size fuse is 80 amperes.
3. To determine the size of the fuse in the feeder, add the results of the calculations in steps 1 and 2 (281.25 A + 75 A = 356.25 A). The standard size nontime-delay fuse is 400 amperes for the feeder supplied by fuse No. 3.

Problem 3—The motor supplied by fuse No. 6 is a 10-horsepower, 230-volt, three-phase motor with a service factor of 1.15. The fuse is a dual-element, time-delay fuse. What is the ampere rating of the fuse?

The motor has both overcurrent and overload protection with backup overload protection by the dual-element fuse. The steps to determine the fuse size are

1. Determine the FLC of the motor from Table 430-150 (28 A).
2. Multiply 28 A by 1.25 (35 A).
3. Look up the standard size fuse in Section 240-6 (35 A).

Problem 4—Fuse 7 feeds a 50-horsepower, 230-volt, three-phase, synchronous-type motor. The fuse is to be replaced with an inverse time circuit breaker that will provide overcurrent protection only. What is the maximum size circuit breaker allowed?

1. Determine the FLC for Table 430-150 (104 A).
2. Table 430-152 provides a maximum value of 250%. Exception c of Table 430-52 allows up to 400% of FLC for circuits under 100 amperes, and 300% for circuits over 100 amperes. A notation under Table 430-152 limits the maximum value to 200% under certain conditions. Because the maximum size circuit breaker allowed was requested in this problem, multiply 104 A X 3.00 = 312 A.
3. Consult Section 240-6 for the standard size circuit breaker that is closest to but does not exceed 312 amperes (300 A).

Problem 5—Fuse 8 is a nontime-delay fuse that feeds a 5-horsepower, 230-volt, three-phase induction-type motor with Code Letter A. What is the ampere rating of the fuse?

1. Determine FLC for Table 430-150 (15.2 A).
2. Table 430-152 limits the multiplier to 150% for Code Letter A. Multiply 1.5 X 15.2 (22.8 A).
3. Look up the standard size fuse in Section 240-6. Exception 1 of 430-52 allows the next higher standard rating for the fuse (25 A).

Problem 6—Determine the size of dual-element, time-delay fuse for the all motor load (Fuse 4). Section 430-62(a) is applicable for this purpose.

1. Determine the highest rated overcurrent device used with the three motors in this problem (300 A).
2. Add the FLC of the other two motors to this value. 300 A + 15.2 A + 28 A (343.2 A).
3. The standard size fuse must be equal to or the next lowest value. Fuse 4 would be rated at 300 amperes.

The ampere rating of the fuse must not exceed a value that will not protect the feeder conductors. The recommendation in Figure 9–15 is to set the fuse rating at 150% of the FLC of the largest motor plus the sum of the FLC of the other motor loads. When the conductors are capable of carrying larger loads in anticipation of changes or for future additions, the fuse may be set at the higher value.

Problem 7—What would the size of fuse No. 12 be using a dual-element, time-delay fuse to protect a motor rated at 5 horsepower, 115 volts, single-phase? The nameplate on the motor indicates it has a service factor of 1.15 with a Code C, and an FLC of 50 amperes.

Given these values and conditions, the fuse must provide both short-circuit, ground-fault, and overload protection. Section 430-32 requires continuous-duty motors of more than 1 horsepower to have overload protection.

1. The FLC given on the nameplate of the motor is used for this calculation (50 A).
2. Determine the multiplier from Section 430-32 (125%).
3. Multiply 1.25 X 50 A (62.5 A).
4. Find the standard fuse rating in Section 240-6. Because the next size smaller must be used, a 60-ampere fuse is selected.

Problem 8—Determine the size of an inverse time circuit breaker to replace fuse No. 14. The motor is a 2-horsepower, 115-volt, single-phase machine. It has a service factor of 1.0 with a Code F. The nameplate indicates it is thermal protected with an FLC of 17 amperes. The motor will not start using the 115% multiple recommended for overcurrent and overload protection with backup. The breaker will provide only overcurrent protection.

1. Find the FLC from Table 430-148 (24 A).
2. Find the multiplier from Table 430-152 (250%).
3. Multiply 24 A X 2.50 (60 A).

4. Determine the standard size of the next smallest rated circuit breaker from Section 240-6 (60 A).

Problem 9—What would the ampere rating be for fuse No. 5 which feeds the combination motor and nonmotor load?

Use the following steps to solve this problem.

1. Multiply 150% times the FLC of the largest motor. The 5-horsepower motor has an FLC of 56 amperes as per Table 430-148 (1.5 X 56 A = 84 A).
2. To this value, add the FLC of the other motor (84 + 24 A = 108 A).
3. To this number add 125% of the nonmotor, continuous load (125% X 100 A + 108 A = 243 A).
4. Find the next lowest value of the standard size fuse in Section 240-6 (225 A).

Problem 10—Determine the ampere rating for the overcurrent protection for the main, fuse No. 1.

The primary purpose of the main fuse is to protect the conductors supplying power to the system from overcurrents. It may have a higher rating than the calculated load if the conductors have the capacity to supply higher currents for additional loads. Articles 210 and 220 of the NEC are applicable for the nonmotor loads and Section 430-62 for the motor loads. Using this information, the steps in calculating the current rating for the fuse are

1. Multiply the continuous nonmotor load by 125% (1.25 X 325 A = 406.25 A).
2. Add to this value the noncontinuous nonmotor load (75 A + 406.25 A = 481.25 A).
3. The ampere rating of the largest motor fuse is then added to the total of the nonmotor loads (800 A + 481.25 A = 1281.25 A).
4. To this value add the FLCs of all the other motors (1281.25 A + 28 A + 130 A + 15.2 A + 56 A + 24 A = 1510.45 A).
5. Section 240-6 shows standard values of 1200 and 1600 amperes. The 1600-ampere value is permitted if the conductors can carry the load.

Bussman recommends that the overcurrent protection for the main be a factor of 150% times the FLC of the largest motor when using dual-element, time-delay fuses. If nontime-delay fuses are used, the multiple factor is 300%. The factors for the other loads remain the same.

In the first case (1.5 X 480 A = 720 A), the total calculated above would be reduced by 80 amperes. The total current would then be 1430.45 amperes. Using these calculations would not have changed the value of the main fuse.

For the nontime-delay fuse (3 X 480 A = 1440 A), our calculated value would be increased by 640 amperes. This gives a value of 2150.25 amperes. The standard fuse sizes are 2000 and 2500 amperes in this range.

From the calculations and previous discussions on overcurrent devices, it becomes an obvious necessity to consult the NEC when installing and wiring any motor. Circumstances vary for each case, and it is unlikely that most electricians will memorize all the rules and tables necessary to correctly wire a motor. In any case, the various values may undergo changes in future editions of the NEC.

Although the interrupting capacity of the fuses and circuit breakers were not discussed in this section of the text, this value must also be included in your decision for selecting the proper fuse or circuit breaker for any given circuit. A review of this feature about fuses may be necessary. The information is in Chapter 3 of this book and available from manufacturers of overcurrent protective devices.

Putting the Motor into Operation

Before energizing a motor for the first time, or in the event the motor has been idle for a long interval, it is advisable to check the motor's insulation resistance. These procedures were outlined earlier in this chapter and in Part I of this text.

Make sure all power is removed from the machine for these tests. Windings must first be grounded to remove any residual electrostatic charge. Failure to do this may result in personal injury. Dry the motor if the resistance measures on the low side. Follow the procedures previously outlined.

Drain oil reservoirs and fill with fresh oil. Reservoirs may have been filled above the recommended level when the motor was put in storage. Use the specifications for oil viscosity, grade, and weight provided on the lubrication nameplate. Check the tightness of the filler caps and drain plugs. Look for oil leaks.

When possible, check the interior of the motor for any foreign matter that may have collected in it during storage or installation. Turn the rotor. Check for freedom of movement. Note any unusual force necessary to turn it. Listen for unusual noises.

If possible, turn the mechanical load by hand also. The load should be free to turn without undue force or unusual noise. Motor and load must be aligned for the type of coupling used.

Check the ventilation system to assure its proper operation. If the motor has provisions for water or oil cooling, measure the flow and temperature of the coolant. Measure the ambient temperature if the area appears to be too warm. Be sure ventilation holes are not blocked and there is a free flow of air to and around the motor.

Measure the voltage of the power supply. It should be within 10% of the motor's rated voltage. Three-phase voltages should be balanced. Measure each phase to ground for the wye-connected system. Measure voltages between all three phases. If the voltages of the power supply are unbalanced, the deficiency must be corrected before applying power to the motor.

Use the wiring diagram to confirm proper connection of the motor. This is a good time to check all accessible connections made by the manufacturer and those made during the installation of the motor.

Without applying power to the motor, check the control circuit for proper operation. Observe the proper sequence of functions and the timing of the operations. Operate pilot devices to determine if they are functional. Check the various features of the controller such as reduced voltage starting, reversing, plugging, or breaking.

After performing the above tests and checks, you are ready to apply power to the motor. If possible, on the initial start, run the motor without its load. Note any unusual noises or vibrations. When the load is connected to the motor, it may mask the motor noise.

Note if the shaft of the motor is turning in the prescribed direction. Some loads may be damaged if they require a particular direction of rotation. Take corrective action if any of these deficiencies are found.

Measure line voltage and the current drawn by the motor. Take these readings for each setting of multi-speed motors. Voltage should be within 10% of the motor's rating, and current on a three-phase system should be balanced. Compare these values to the motor's ratings. If abnormal readings are obtained, stop the motor and correct the problem.

Test the various functions that the motor must perform through the controller commands. Operate the pilot devices that send information to the controller. Note the motor's operation in response to each of these commands.

In the event of excessive vibration or unusual noise, stop the motor immediately. Check the motor mounting for tightness. Listen carefully as you turn the motor by hand. Noises coming from the motor can be amplified by putting your ear to the handle of a plastic screwdriver and placing the metal tip against the bearing housing. Check the shaft of the motor for proper end-play. Correct the problem before placing the load on the motor.

After these tests have been completed, couple the load to the motor. Check for proper alignment again. This is particularly true if you have taken corrective action to eliminate vibration on the unloaded motor.

Follow the regular procedure for starting the motor. If it is possible to initially operate under reduced load, do so. Observe any unusual noises, vibrations, or hot spots on the motor. If the motor has space heaters, be sure they are not powered during normal operation of the motor. Increase the load in increments and make your observations at each level of increasing load until the full load has been applied. Check the motor under full load for at least 1 hour.

Test the electrical characteristics of the power system under the full-load condition. Line voltage should be balanced and within 10% of the motor's rated voltage while the motor is being run under these conditions. Line current must not exceed the FLC on the nameplate times the service factor when the motor under continuous load.

Check for temperature rise and hot spots on the motor. Do not depend on the sense of touch to determine if excessive temperatures are present. Use the motor's measuring devices such as the outputs of RTDs and thermocouples if the motor is so equipped. Otherwise use a thermometer to take the readings.

If there are any questions about the levels of temperature at any point on the motor, take the motor out of operation. Determine if the operating temperatures are within the tolerance of the motor. Heat is still the most destructive factor to any electrical component or apparatus. It is estimated that a 10°C rise above maximum operating temperature of a motor will reduce the motor's life by one-half.

In the event the cause of the temperature rise cannot be determined, call the manufacturer's representative. Be prepared to give the full information on the nameplate of the motor. The level of heat may be normal. If not, the representative of the company will be more familiar with the various causes.

Overheating may be caused by the way the motor is operated. Repeated starts and stops for jogging or inching cause motor temperature to rise. Improper ventilation or a high ambient temperature are common factors. Excessive dirt, oil, or grease on the motor housing and cooling fins may prevent the heat from escaping from the motor. Unbalanced line voltage and current, low or high voltage, high current, or in the case of a variable-speed motor, harmonics in the power supply can cause excessive temperatures. High temperatures will also be caused by mechanical binding due to bad bearings, improper alignment between motor and load, or excessive load.

On large, sophisticated motor systems, instrumentation is usually available on the controller panels to provide the various items of information for determining if a motor is running within the correct parameters. Voltages, currents, and temperatures at various points of the motor can be read directly from the panel. The panel may also indicate which pilot devices are operative. RPM and direction of rotation can also be given along with several other measurements such as kVA, kilowatts (kW), and power factor.

For the majority of motors, however, little or no instrumentation will be included. The electrician will need to supply the necessary meters and temperature measuring devices to make the readings.

Start the motor's record card during this initial start-up procedure. Record all pertinent data for future maintenance and repairs of the motor. This information is also important in preventing motor failure by detecting abnormalities and correcting them before damage occurs. The equipment information card should be kept up to date over the life of the motor.

Power Factor Correction

When the source supplies power to a load that is not purely resistive, the current and the voltage will be out of phase with each other. In the case of motor loads, the current will lag the voltage due to the inductive nature of the machine. This results in the kVA supplied by the power source being greater than the kW consumed by the load. The difference between the two is sometimes referred to as "wattless" power.

Definition

Power factor is defined as the ratio of watts to volt-amperes. This is expressed mathematically as

$$\text{Power Factor} = \text{Watts/Volt-Amperes}$$

This ratio corresponds to the trigonometric function of the cosine of Theta (θ). (Values of the trigonometric functions are given in Appendix B.) Theta is the angle between the adjacent side of the triangle and the hypotenuse (Figure 9–17). Power factor is the cosine of the angle.

$$\text{Power Factor} = \text{Cosine } \theta$$

Power factor is a numerical reference to the electrical efficiency of the circuit. The range can be from 0 to 1, or from 0% to 100%.

It can be seen as the kilo-volt-amperes-reactive (kVAR) gets larger in respect to the kW of the circuit, more kVA must be supplied for the magnetizing force of the inductor. This condition is known as *poor* or *low power factor*. As the angle becomes larger and larger and approaches 90°, the power factor becomes smaller and smaller and approaches zero. At this point, the circuit would act totally inductive, and no real power would be consumed. The cosine of 90° is equal to zero.

As the kVAR is reduced, the angle becomes smaller. The power factor increases until it equals unity, 100% at zero degrees. At this point, the circuit is purely resistive. The true power consumed will be maximum and equal to the kVA.

True power dissipated in a circuit can be calculated by the following formula.

$$\text{kW} = \text{kVA} \times \text{PF}$$

Cause of Low Power Factor

Motors operating at less than their full load are the primary cause for a plant to have a poor power factor. Typically, motors that drive such machines as saws, grinders, and presses do not have continuous loads on them. This results in a high magnetizing current when these machines are unloaded with power factors in the range of 30–50%.

One solution is to keep the machines fully loaded at all times. This is not practical in many industries and for certain processes such as plastics, rolling mills, chemical processing, foundries, and granaries. Intermittent or cycle process is required. To keep the motor loaded would merely consume power without receiving useful work in return.

Power Triangle

Figure 9–17 shows the impedance, voltage, and power triangles for an inductive circuit. The current flow through the resistance of the circuit is in phase with the voltage dropped across it. This condition is set as the reference point of the circuit at zero degrees. The current flow through the inductance lags the voltage across the inductance by 90°. The vector sums of the voltage across the resistor and the voltage across the inductor equal the source voltage.

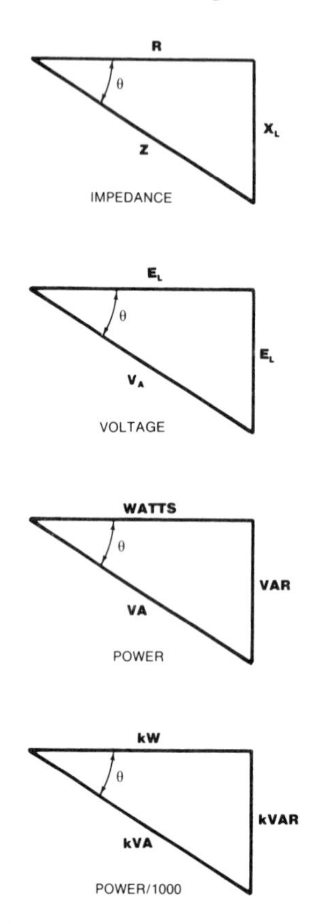

FIGURE 9–17. Impedance, voltage, and power triangles for an inductive circuit.

Multiplying the individual parts of the impedance triangle by the current (I) will give the values for the voltage triangle. $I \times R = E_R$, $I \times X_L = V_L$, and $I \times Z = V_A$, or the source voltage. Multiplying the values in the voltage triangle by I, or multiplying the values in the impedance triangle by I^2 will give the values in the power triangle.

The true power dissipated by the resistance of the circuit is given in watts. The "wattless" power of the inductor is in volt-amperes-reactive (VARs). Power supplied by the source, apparent power, is measured in volt-amperes (VA). Dividing the values in the power triangle by 1000 will change the units to kW, kVAR, and kVA.

Cost of Poor Power Factor

Poor power factor equates to poor efficiency, and poor efficiency translates to higher costs. Many utility companies penalize the customer that has a poor power factor. Even though the total kVA is not used by the customer, the utility must have the capacity to deliver the full amount. This is the reason that transformers are rated in kVA rather than kW.

Some utilities bill the customer for the full kVA delivered to the plant. This can be very costly for the customer. Suppose a utility charges $5.15 per kVA demand. The customer has a 1000 kVA demand, at 480 V, three-phase, with a power factor of 80%. The electric bill would be equal to

$$\$5.15/\text{kVA} \times 1000\ \text{kVA} = \$5150$$

But the customer is using only 800 kW of true power.

$$800\ \text{kW} = 1000\ \text{kVA} \times 0.80$$

If the power factor was corrected to 100%, then only 800 kVA would be necessary to power the plant. The cost for 800 kVA would equal

$$\$5.15/\text{kVA} \times 800\ \text{kVA} = \$4120$$

The power factor correction results in a savings of $1030. Without power factor correction, 20% of the cost of power is going for "wattless" power with no useful work being done for the customer.

Other utility companies collect a penalty for poor factor based on the kVAR demand. This is done in conjunction with the kW demand. The utility will provide a waiver up to some percentage of the kW demand before charging for the kVAR. For example, there may be a 40-cent charge per kVAR in excess of 35% of the kW demand. A plant using 100 kW would be charged 49 cents per kVAR over 35 kVAR. With poor power factor this can also become very expensive.

Most utilities that charge for poor factor will bill for the kW used with an adjustment for poor power factor. For example, the utility may not penalize the plant if the power factor is 90%. If the power factor falls below this value, a multiplier is applied to the kW billing based on how much below the allowable power factor.

Using our previous example where the plant is using 800 kW at $5.15 per kVA with a power factor of 80%, the total bill to the plant would be multiplied by the factor

Poor PF Multiplier = Allowed PF/Actual PF
1.125 = 90%/80%

Therefore the bill would be calculated as follows:

Total Power Cost = Cost/kW X kW X Multiplier
$4635 = $5.15/kW X 800 kW X 1.125

Savings of $515 could be made by bringing the power factor to 90%.

Additional Benefits of High Power Factor

Besides reducing utility costs, increasing the power factor of a plant may provide additional benefits. Some of these are

1. Additional loads can be placed on the present system.
2. The voltage conditions in the plant will be improved.
3. Power losses are reduced.

Suppose a plant has a 1000 kVA, 3% impedance, transformer operating near capacity (Figure 9–18). The plant draws 960 kVA at 480 V, three-phase. The power factor of the plant is 70%. The true power being used by the plant is

$$\begin{aligned} kW &= kVA \times 0.70 \\ &= 960\ kVA \times 0.70 \\ &= 672\ kW \end{aligned}$$

The plant wishes to increase production by 30%. This would put an additional 201.6 kW load on the electrical system. Assuming unity power factor for the additional load, the electrical system would have to deliver 1161.6 kVA. This is beyond the capacity of the system.

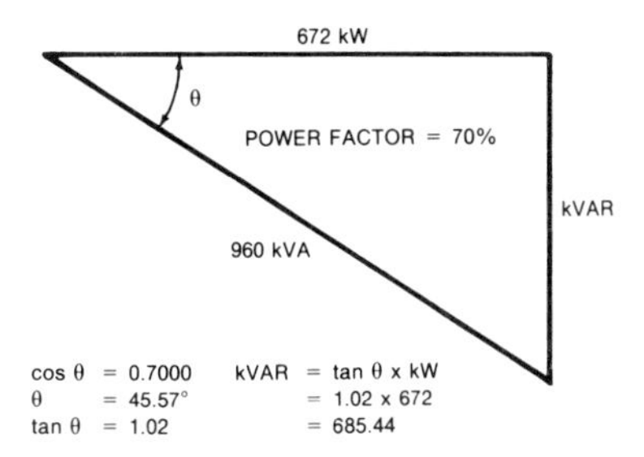

FIGURE 9–18. Determining the kVAR.

One solution is to add a new transformer to the system. If the power factor remains at 70%, the minimum rating of the transformer would be

Minimum Transformer Capacity
= (672 kW + 201.6 kW)/0.70
= 1248

The next standard size transformer would most likely be 1500 kVA.

In addition, the capacity of conductors, switches, and overcurrent devices would need to be increased. It would not take a lot of calculations to determine that this approach would be very expensive.

By correcting the power factor to unity, there would be 328 kW of capacity available instead of the 40 kW with a 70% power factor. With 100% power factor, the present transformer would provide the needed capacity. Depending on the location of the corrective devices used to raise the power factor, other components in the circuit may or may not need an increase or decrease in their current carrying capacity.

Low voltage due to excessive current drawn from the transformer can cause the motors to overheat and operate inefficiently. By increasing the power factor on the present load, the voltage would be increased. The percent of voltage rise can be determined by

$$\%VR = \frac{(kVAR\ (Reduced) \times \%Z)}{kVA\ of\ Transformer}$$

The kVAR must first be determined to solve this problem (see Figure 9–18). We know that the power factor is equal to 70%. This is the cosine of the angle. The angle whose cosine is 0.7000 is equal to 45.57°. The tangent of 45.57° is equal to 1.02. The tangent of an angle is equal to the opposite side (kVAR) divided by the adjacent side (kW). Therefore,

$$\text{Tangent } 45.57° = \frac{\text{kVAR}}{\text{kW}}$$

or

$$\begin{aligned} \text{kVAR} &= \text{Tangent } 45.57° \times \text{kW} \\ &= 1.02 \times 672 \\ &= 685.44 \end{aligned}$$

Substituting the kVAR reduction into %VR formula, it is determined that

$$\%VR = 2.05\% = (685.44 \text{ kVAR} \times 3\%)/1000 \text{ kVA}$$

For a 480-V system with 465-V at the terminals of the transformer, this means that the voltage rise at the terminals would increase by

$$(465 \text{ V} \times 0.0205)\ 9.53 \text{ V}$$

The terminal voltage of the transformer would now equal 474.5 volts:

$$474.5 \text{ V} = 465 \text{ V} + 9.5 \text{ V}$$

Power factor correction will also reduce the losses due to the reactive current flowing through the resistance of the circuit. The kW loss is equal to I^2R. R is equal to the resistances of the feeder and branch circuit conductors.

The percent reduction of the losses is equal to

$$\% \text{ Reduction Losses} = 100 - 100\ (\text{Original PF/New PF})^2$$

In the case of the problem above, the PF was corrected from 70% to 100%. The losses would be reduced 51%.

$$\% \text{ Reduction Losses} = 100 - 100 \times (70\%/100\%)^2$$

Although the dollars saved by correcting the power factor for this purpose is relatively small, the payback for the installation to reduce these losses alone can usually be accomplished in 3 to 5 years.

How Power Factor Is Improved

The previous examples illustrate how inductive loads such as motors increase the kVA loading of a circuit in respect to the kW actually consumed by the load. By adding capacitance in parallel with the load, or by using synchronous motors that are over-excited, a leading power factor can be created that will cancel the lagging currents caused by the inductance. Figure 9–19 depicts these conditions.

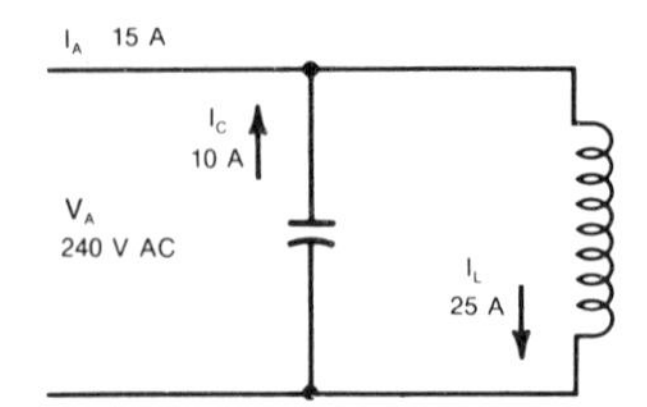

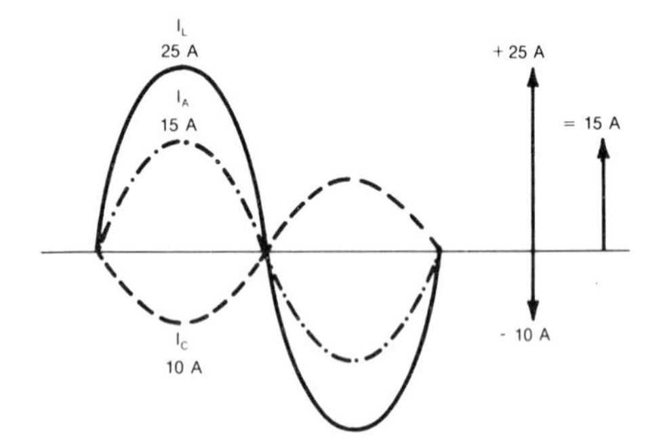

FIGURE 9–19. Capacitive current cancels inductive current.

Using the source voltage as the reference, the current through the inductor (25 A) will lag the voltage by 90°. The current through the capacitive branch (10 A) will lead the applied voltage by 90°. These two currents are 180° out of phase with each other as shown by the vectors and the sine waves. The resultant line current (15 A) will be equal to the difference between the two currents. Ten amperes will circulate between the capacitor and the inductor.

Measuring Power Factor

The first step in improving the power factor on a motor or in a plant is to determine what the power factor is. The power factor may or may not need to be corrected. After the motor is put into operation, make the following measurements (Figure 9–20).

1. Power consumed by the circuit
2. Line voltage
3. Line current

Figure 9–20A shows the instrumentation necessary to obtain these three measurements on a single-phase motor. Some installations may already have the instru-

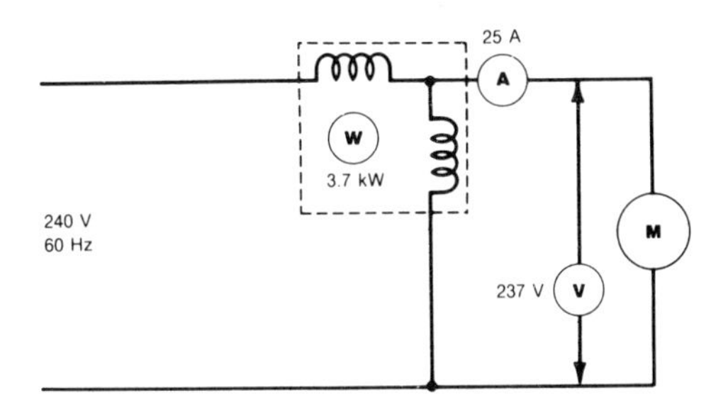

(A) Single phase.

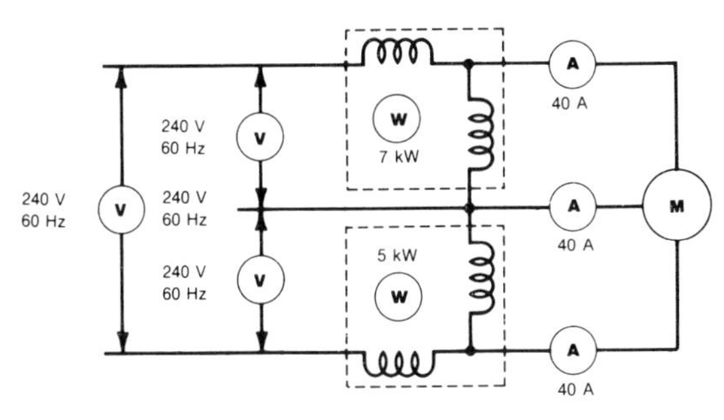

(B) Three phase.

FIGURE 9–20. Determining the power factor of a circuit.

mentation necessary to obtain these readings. This may even include a power factor meter.

If not, a combination watt/volt/amp meter is ideal for this procedure if one is available. Otherwise, three separate meters can be used.

If clamp-on probes are available for the current readings, the circuit will not have to be broken to insert the ammeter and the wattmeter in series with the line. If not, the simplest place to break the line connection is at the overcurrent device. The series winding of the wattmeter and the ammeter can be placed in series with the fuse for the necessary readings. The wattmeter must sample both the current and the voltage and will have three leads for this purpose.

Given the values shown in Figure 9–20A, the kVA can be calculated.

$$\text{kVA} = (\text{V X I})/1000$$
$$5.925\ \text{kVA} = (237\ \text{V X}\ 25\ \text{A})/1000$$

The wattmeter indicates 3.7 kW of power being consumed by the motor. With this information, the power factor can now be calculated.

$$\text{Power Factor} = \text{kW/kVA}$$
$$= 3.7\ \text{kW}/5.925\ \text{kVA}$$
$$= 62.45\%$$

Figure 9–20B shows the arrangement necessary to determine the power factor for a three-phase motor. A single-phase wattmeter can be used for this measurement, but two readings will have to be taken. Usually, a three-phase wattmeter will have two meters and will be connected as shown. The total power is equal to the algebraic sum of the two readings.

At unity power factor with balanced loads, the two readings will be identical and will merely add together. If the power factor is 50%, one meter will indicate zero watts, and the other will show the full wattage used by the circuit. For power factors between 50% and 100%, one of the meters will indicate a higher power reading than the other. Below 50%, one meter will give a negative power reading and the other a positive one. A toggle switch on the meter allows the reversal of current through the meter with the negative reading so that the meter will deflect up-scale. In this case the negative power would be subtracted from the positive power to obtain the kilowatt reading for the circuit.

The voltages and currents shown in Figure 9–20 are balanced. In the case of unbalanced parameters, the values of each can be added together and averaged for the calculations. Voltages should not be off more than 10%. If so, this condition needs to be corrected before proceeding with the measurements.

Using the values given in Figure 9–20B, the kVA of the circuit can be determined. For three-phase circuits, the multiple of the average voltage and the average current must be multiplied by a factor of 1.73, or the square root of 3.

$$\text{kVA} = (\text{V X I X}\ 1.73)/1000$$
$$= 240\ \text{V X}\ 40\ \text{A X}\ 1.73)/1000$$
$$= 16.61$$

The total wattage in the circuit is equal to the algebraic sum of the two readings on the wattmeters.

$$\text{kW} = \text{W1} + \text{W2}$$
$$= 7\ \text{kW} + 5\ \text{kW}$$
$$= 12$$

This information can now be used to determine the power factor for the circuit.

Power Factor = kW/kVA
= 12 kW/16.61 kVA
= 72%

Practical Examples of Power Factor Correction

It is seldom feasible or practical to correct the power factor to unity. To do so, in most cases, would result in a leading power factor at times. This would increase the kVA in respect to the kW just as much as a lagging power factor. Neither is desirable. A power factor in the range of 95% will provide most of the economic advantages of a 100% correction.

Problem 1—Using the values given in Figure 9–20A, draw the power triangle for the circuit. What is the CkVAR (Capacitance kilovolt ampere reactive) rating for the capacitor needed to correct the power factor to 95%? (See Figure 9–21.)

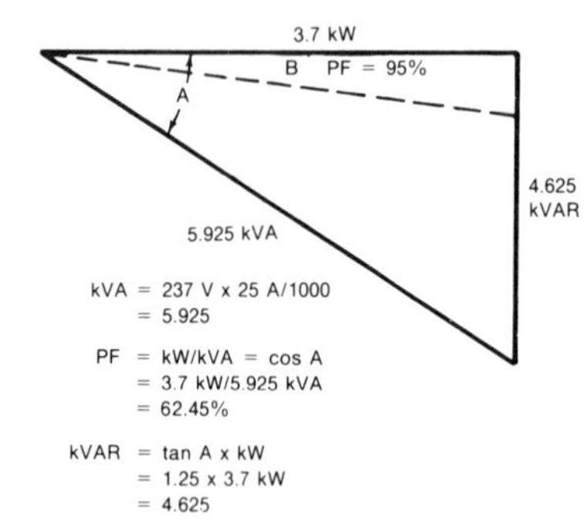

FIGURE 9–21. Power triangle for Figure 9–20A.

The steps necessary to solve this problem are

1. Determine the kVAR for the power factor at 62.45% at Angle A.

Cos A = 0.6245

Using the trigonometric functions, find the angle. The angle whose cosine is 0.6245 is 51.35 degrees. The tangent of 51.35° is 1.25. The kVAR for Angle A is equal to

kVAR = Tan A X kW
= 1.25 X 3.7 kW
= 4.625

2. Determine the kVAR for a power factor of 95% at Angle B.

Cos B = 0.95

The angle whose cosine is 0.95 is equal to 18.19°. Tan B is therefore equal to 0.3287. The new kVAR when the power factor is corrected to 95% would equal

kVAR = Tan B X kW
= 0.3287 X 3.7 kW
= 1.216

3. To determine the size of the capacitor in kVAR, subtract the kVAR of Angle B from that of Angle A.

CkVAR = kVAR(A) – kVAR(B)
= 4.625 – 1.216
= 3.409

4. From a manufacturer's chart, select a standard size capacitor to correct the power factor. The chances are either a 3 or 4 CkVAR will be available.

The manufacturer of the capacitor will usually have a table that limits the size of capacitor that should be used with a given type of motor design, the horsepower, and the RPM of the motor. In this case, we will assume the maximum value to be 3 CkVAR.

Problem 2—The 3 CkVAR capacitor has been installed across the motor. What is the new power factor of the circuit, and how much line current does the motor now draw?

Use the following steps to solve this problem.

1. Determine the new kVAR rating of the circuit. To calculate this value, subtract the CkVAR value of the capacitor from the kVAR for the original power factor of 62.45%.

kVAR = 4.625 – 3
= 1.625

2. Determine the tangent of the new angle (C).

Tan C = kVAR/kW
C = 1.625/3.7
= 0.4392

3. Determine Angle C. Using the trigonometric functions, the angle whose tangent is 0.4392 is 23.71°.

4. The cosine of Angle C is equal to the power factor. Again use the trigonometric functions and find the cosine of the angle. .9156
5. Calculate the new kVA for the circuit for a power factor of 91.56%.

$$\text{Cos C} = \text{kW/kVA}$$
$$\text{kVA} = \text{kW/Cos C} = 3.7\ \text{kW}/.9156 = 4.041$$

6. The new line current can be found by dividing the new kVA by the line voltage.

$$\text{I} = \text{kVA/V} = 4.041\ \text{kVA}/237\ \text{V} = 17.05\ \text{A}$$

Therefore, the current to the motor has been reduced by nearly 8 amperes.

Problem 3—What is the value of the 3 CkVAR capacitor in microfarads?

Use the following steps to solve this problem.

1. The current through the capacitor branch (I_C is equal to

$$I_C = \text{CkVAR/V} = 3000/237\ \text{V} = 12.66\ \text{A}$$

2. Capacitive reactance (X_C would equal

$$X_C = V_C / I_C = 237\ \text{V}/12.66\ \text{A} = 18.72\ \text{A}$$

3. Then C would equal

$$C = 1/(2\pi f X_C) = 1/(6.28 \times 60 \times 18.72) = 141.7\ \mu\text{F}$$

Problem 4—Draw the power triangle for Figure 9–20B. Using these values, determine the CkVAR necessary to correct the power factor to 95% (Figure 9–22).

Solve this problem using the following steps.

1. Determine the kVAR for the circuit. The power factor equals 72%, therefore, Cos A equals 0.7200. Angle A can be determined from the trigonometric tables as being 43.94°. Tan A equals 0.9638, which is equal to the side opposite the angle divided by the adjacent side.

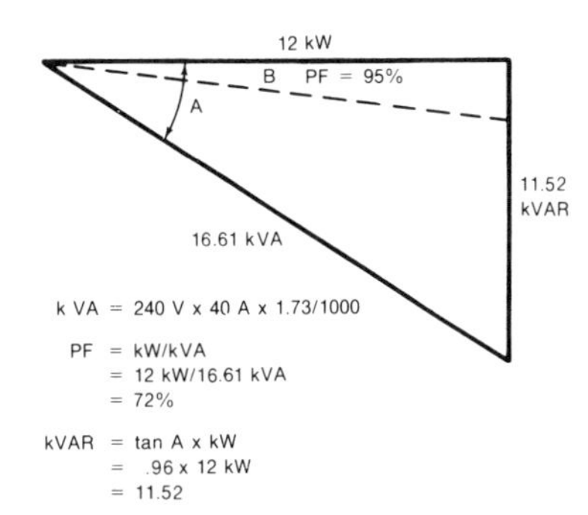

FIGURE 9–22. Power triangle for Figure 9–20B.

$$\text{Tan A} = \text{kVAR/kW}$$
$$\text{kVAR} = 0.9638 \times 12 = 11.57$$

2. Calculate the kVAR for a power factor of 95%. The angle whose cosine is 0.95 equals 18.19°. Therefore, the kVAR at 95% power factor will equal

$$\text{kVAR} = \text{Tan } 18.19^\circ \times \text{kW} = 0.3287 \times 12 = 3.944$$

3. Subtract the kVAR for 95% power factor from the kVAR for 72% power factor to determine the amount of CkVAR.

$$\text{CkVAR} = \text{kVAR(A)} - \text{kVAR(B)} = 11.57 - 3.944 = 7.625$$

4. Standard size capacitors are available in 7.5 and 8 CkVAR. In this case, let us use the 8 CkVAR capacitor.

Problem 5—What is the new power factor for the circuit?

1. Determine how much kVAR remains after the circuit is corrected with the 8 CkVAR capacitor.

$$\begin{aligned} kVAR(C) &= kVAR(A) - CkVAR \\ &= 11.57 - 8 \\ &= 3.57 \end{aligned}$$

2. Find the new angle. The tangent of Angle C is equal to

$$\begin{aligned} \text{Tan } C &= kVAR(C)/kW \\ &= 3.57/12 \\ &= 0.2975 \end{aligned}$$

The angle whose tangent is 0.2975 is 16.57°.

3. The cosine of Angle C is the new power factor. 95.8%.

Problem 6—With the power factor corrected to 95.8%, how much current does the circuit now draw?

1. Calculate the new kVA for the circuit.

$$\begin{aligned} \text{Cos } C &= kW/kVA \\ kVA &= kW/\text{Cos } C \\ &= 12/0.958 \\ &= 12.526 \end{aligned}$$

2. We know that the kVA for a three-phase circuit is equal to

$$\begin{aligned} kVA &= (V \times I \times 1.73)/1000 \\ I &= (kVA \times 1000)/(V \times 1.73) \\ &= 12{,}526/(240 \times 1.73) \\ &= 30.17 \text{ A} \end{aligned}$$

In this case the current in the circuit has been reduced by nearly 10 amperes on each of the three-phase conductors.

The kVA of a motor is usually in the range of the kW used under full-load conditions. The kW at full load can be approximated by the following formula

$$kW \text{ (Motor Input)} = (HP \times 0.746)/\% \text{ Efficiency}$$

Problem 7—What would be the approximate kVA of a motor rated at 10 HP with 80% efficiency?

$$\begin{aligned} kVA &= (10 \text{ HP} \times 0.746)/.80 \\ &= 9.325 \end{aligned}$$

Location of Capacitors

Capacitors used to correct the power factor can be applied at the individual loads, on the feeder, or the main bus. A combination of placements may also be used with capacitors located at all the possible points in a given plant. There are advantages and disadvantages to each of these approaches. Figure 9–23 illustrates the different possible locations for the capacitors to be placed for power factor correction. Capacitors located at Points A, B, and C are installed on the load side of the circuit.

From an engineering point of view, power factor correction at the individual loads is technically the best and has the greatest flexibility. This approach has the following benefits.

1. Capacity is on the line only when the load is energized.
2. Separate switches are not required by the code when the capacitors are switched with the load.
3. Better power utilization is effected, and the motor runs with improved performance due to the lack of the voltage drop on the line.
4. If the motor is relocated, the power factor correction goes with it. This provides for greater flexibility for moving motors to different areas in a plant where they are needed.
5. Because a single load is being corrected, the selection of the capacitor can be more precise in meeting the requirement for a particular motor.
6. Maximum utilization of plant kVA capacity is possible for additional loads which may be needed in the future.
7. The voltage on the motor's branch circuit will increase due to the reduced current through the conductors. The branch circuit may be able to carry an additional load.

Capacitors located at Point A reduce the current flow through the overloads. It is necessary to change the size of the overloads to reflect the reduced current flow to the motor. The current demand on the transformer and the current flow through the feeder and branch circuit is reduced by installing capacitors at this point. The original current will flow only on the conductors between the capacitors and the motor.

Article 430-32(a) of the NEC requires maximum sizing of overloads to be from 115% to 125% above the FLC marked on the nameplate of the motor. When this allowance is not sufficient to start the motor, Article

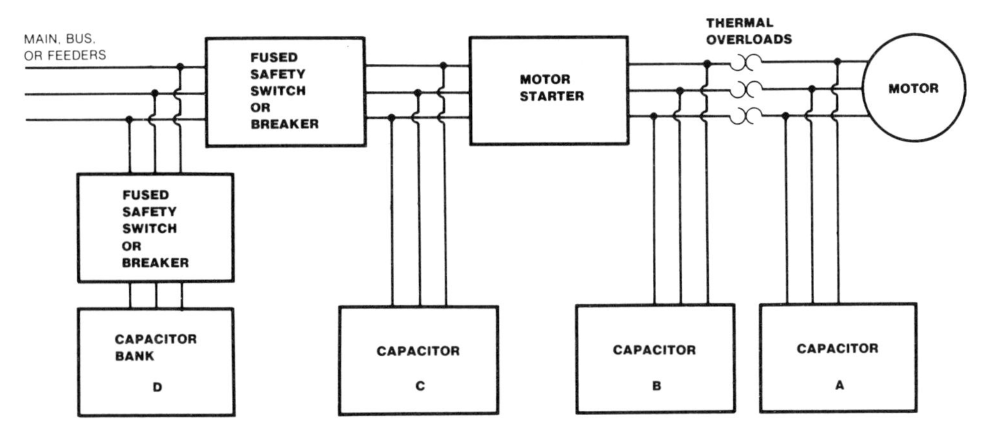

FIGURE 9–23. Location of capacitors for power factor correction.

430-34 provides for an increase in the value of the overload ratings to 130-140% of the FLC. On an existing motor installation, the capacitors would be located at Point B to take into account the increased ratings.

Capacitors may be installed at Point C, when a separate disconnect exists. This point for power factor correction shall be used under the following conditions.

1. When the motor is to be jogged, plugged, or reversed
2. For all multispeed motors
3. When open transition starts are used
4. Any time the capacitors are to be momentarily disconnected and then connected during the cycle.

It is recommended that Point C also be used with motors which have a high inertia and require a long period of time before stopping. With capacitors connected across the motor under these conditions, the motor may become a self-excited generator and cause motor damage due to excessive currents.

When the capacitors are connected at Point C, they will remain in the circuit even if the motor is off unless the disconnect is opened. As long as the capacitor remains on the line, it will provide power factor correction for the total system. Overcorrection is possible under this condition depending on the size of the capacitors, capacity of the plant, and the overall power factor condition.

Capacitors installed at Point D have some advantages over installation at the load. These are

1. Larger capacitors can be used reducing the cost per CkVAR over installations at the individual motors.
2. Total labor and material costs are generally lower.
3. Power factor is corrected for the total plant and charges for low power factor are reduced or eliminated.
4. Power factor correction is easier to maintain under different operating conditions.

If the capacitors are connected on the main bus or a feeder, a fused safety switch or breaker shall be used to disconnect the capacitor bank and to protect against overcurrents. The NEC requires the switch to have a capacity of 135% of the current rating of the capacitors. Manufacturers generally recommend 165% for the current rating of the disconnect.

The code also requires that the capacitors be fused at as low a value as is possible. This is usually in the range of 150% to 175% of the current rating of the capacitors. Molded case circuit breakers are in the same range.

Even though the NEC permits capacitors to be protected by the motor's overcurrent devices when they are installed at Points A, B, and C, it is recommended that they still be fused. This is particularly desirable when the capacitors are installed indoors and personnel are working around them. In the event of high leakage current, they can overheat and explode. Because many of these capacitors contain liquids such as oil, an accident may cause severe burns or death.

Capacitance is defined in electrical work as the ability to hold a charge. A capacitor can hold a charge for a long period of time. In order to protect personnel who may work on them, the NEC specifies a specific amount of time for the capacitor to discharge. For capacitors 600 volts or under, the charge must be reduced to less than 50 volts within 1 minute after de-energizing. For values over 600 volts, 5 minutes is allowed for the capacitor to discharge to 50 volts or less. The capacitor's nameplate must state if it contains discharge resistors. The smart electrician will use a discharge rod to ground each hot terminal before beginning work on a capacitor that has the power disconnected.

If the capacitors are to remain on the line, check the system to assure that a leading power factor is not introduced. Some utility companies charge for leading power factor as well as for lagging power factor. Other utilities that do not charge for low power factor do not want capacitors on the line. When a leading power factor is introduced due to the capacitors, they may cause some line disturbance for other customers and interfere with their operations.

When capacitors are selected for permanent connection to the line, the CkVAR rating of the capacitors should not exceed 25% of the kVA rating for the plant. This will help prevent line interference, and the likelihood of harmonic resonance will be lowered. Odd harmonics cause the most problems and should be avoided. The manufacturer's representative will provide guidance if this problem arises. The 25% ratio will also keep voltage spikes to an acceptable level.

Automatically switched capacitor banks are becoming common in the industry. These devices switch 50 to 60 CkVAR on and off the line as the power factor of the plant varies. This limits the undesirable effects of excessive correction when the load is light and makes the utility company happy. The cost, however, is higher than that of a fixed bank of capacitors that are not switched.

Comments on Control

A complete study of motor controllers is beyond the scope of this text. I would be remiss in not mentioning the importance for the electrician of understanding this part of the motor circuit. Without a good working knowledge of control work, the electrician will be ill prepared for working on motors. It is an important topic that needs further study by the electrician who wishes to stay informed and up to date with the industry.

The continued growth in the complexity of motor control circuits is advancing on almost a daily basis. Not long ago, numerical control was hailed as the ultimate in motor control. We are now witnessing an increased usage of solid state devices which use digital logic in these control mechanisms. At first, discrete components such as transistors were introduced for these purposes. These individual components are swiftly giving way to integrated circuits. The integrated circuit may have hundreds of transistors with related circuit components to be used as switches. Yet, all these parts are contained in a chip no larger than the size of a thumbnail.

In the mid-1950s, while at the University of Illinois, I was privileged to hear Norbert Wiener, the father of cybernetics. Dr. Wiener, who died in 1964, was a mathematician and logician. He had a major role in developing the high-speed computer. He received his bachelor's degree from Tufts University at the age of 14. He received his first doctorate from Harvard University at age 18.

Dr. Wiener had completed university requirements for doctorate studies in several fields. I do not recall exactly the areas or number, but I believe they included work in the areas of mathematics, electrical engineering, chemistry, physics, and medicine.

Dr. Wiener coined the term "cybernetics" and used it as a title of one of his books which was published in 1948. *Cybernetics* is a branch of science that combines the fields of study of communications and the mechanics of devices that perform work.

A major part of Dr. Wiener's research was with feedback of information from the load to the controller to enhance the adjustments of the mechanisms. He proposed the combined study of communications and control under the discipline cybernetics.

In general, workers in the two fields had little or no knowledge of the other. An ultimate objective was the marriage of the fully automatic plant with the computer. Robotics used in manufacturing is an outcome of Dr. Wiener's research.

Computers in the 1940s and in the early and mid 50s mostly used vacuum tubes. (The transistor was invented in 1947.) They were large, cumbersome, and used considerable electrical power. Dr. Wiener's topic on the day I heard him speak was the biological computer. Dr. Wiener noted that the internal control and communications system within the human being were similar to what we attempt in a very primitive way with machines. For example, most humans can pick up an egg without breaking it. For a machine to perform the same function, the same process becomes very complex.

Dr. Wiener suggested that if the human brain could be kept biologically alive after the death of a person, it might be possible to use it as a biological computer that could store information, analyze it, and send control signals to mechanical devices to effect desired changes.

This is what the human operator of any machine does in a routine fashion. The human brain used as the control unit of a computer would be much smaller and more powerful than any computer in existence at that time or possible in the foreseeable future.

After Dr. Wiener's introductory remarks, I must admit that I understood very little of what he had to say. He was attempting to explain to us his mathematical system of probability which could predict the signal output from the human brain when it was given a single, DC input pulse. At the end of his lecture Dr. Wiener asked, "Gentlemen, do you know what the probability is of predicting the output of the human brain for one simple input?" He went on to answer his own question, "Gentlemen, it is 16 factorial (16 X 15 X 14 . . . X 1) raised to the 23 power. Gentlemen, I predict there ain't no such damn number."

The biological computer may or may not become a reality. But the electronic computer has become a reality, and its applications and prevalence continue to grow and to affect our everyday lives in many ways. The speed and efficiency of computers, and their ability to solve very complex problems, are advancing daily. In some operations computers already work better than the human brain. They can store much more information and retrieve that information better. More importantly, their outputs are predictable.

A big push in computer technology today is artificial human intelligence programs that emulate the human brain. Have you ever played a game of checkers or chess against a computer?

Solid state programmed controllers are good examples of how industry is making use of cybernetics today. These devices are computers that readily work with motors that are protected or controlled by multiple pilot devices. These devices replace the switching functions of the manual and magnetic controls with solid state devices and incorporate the power of computer logic to insure the proper operation of the motor at the proper time. As the complexity of these modern circuits increases, the level of knowledge and understanding of the electrician must also keep pace in order to install, maintain, repair, and replace motors used with computer-controlled programmers.

Questions

1. What is the first thing to be done when a motor arrives at the worksite?
2. Why should a motor not be lifted by its shaft?
3. What relationship must exist between the nameplate voltage on a motor and the voltage of the system that will supply it?
4. Are there any special provisions for a motor to be operated in a corrosive atmosphere?
5. Under what conditions may a motor be allowed to operate at its rated temperature rise if the installation is above 3300 feet?
6. Should oil reservoirs be drained if a motor is to be stored for an extended period of time?
7. What is the maximum voltage allowed on a new motor rated at 460 volts when conducting a high-potential test?
8. What is the minimum acceptable ohms for a high potential test on a motor rated at 600 volts?
9. If the insulation resistance of a motor is below the minimum acceptable value, what must be done?
10. List the basic characteristics of a NEMA Type 3 enclosure.
11. What is the primary limiting factor in mounting a motor in any direction?
12. Masonry plugs and bolts should have a safety factor of ______ when anchoring large, dynamic motor loads.
13. A motor operating at 1200 RPM has a 3-inch pulley which drives a 9-inch pulley on a machine. What is the rotation speed of the machine?
14. Is it preferable to locate the driven pulley close to the motor's body or near the end of the shaft? Why?
15. What is the general rule for tightening the belts on a pulley-driven system?

16. Two pulleys of 12 and 6 inches have their centers 3 feet apart. What is the length of the belt in inches?
17. A 6-inch pulley is rotated at 1725 RPM. What is the belt speed?
18. What is the advantage of using a chain drive in place of a belt drive?
19. Why are pinion gears sometimes made of hard fiber or plastic?
20. Magnetic clutches do not require a ______ connection between the motor and the load.
21. A motor operating continuously has an FLC of 68 amperes as specified in Table 430-151 of the NEC. What must the ampacity of the conductors be?
22. The FLC stamped on the nameplate of a motor is to be used to size the ______.
23. Three motors have FLCs taken from the tables in the NEC as 21, 27, and 40 amperes. A single feeder is used to supply these motors. What is the minimum ampacity of the feeder?
24. A power factor correction capacitor bank draws has a rated current of 35 amperes. What is the minimum ampacity of the conductors to this bank?
25. Is the following statement true or false? According to the NEC, in general, all motors must have a disconnect means.
26. Locked rotor currents are about ______ times the FLC for any given motor.
27. A controller is a device which ______ and ______ a motor.
28. On a three-wire push-button station, what is the purpose of the set of contacts connected in parallel with the start button?
29. Pilot devices which stop a motor are normally ______.
30. Does the FLC of a motor pass through the contacts of the overload relay? Explain.
31. List several devices used to provide thermal protection for a motor.
32. What does the term "single-phasing" mean?
33. Besides overcurrent and overload protection, list several other types of protection a motor may have.
34. All ______ conductors must be provided with overcurrent protection.
35. Can a fuse provide both overcurrent and overload protection? Explain.
36. Why are overcurrent devices sometimes rated at several hundred times the FLC of a motor?
37. How is the FLC of a motor determined for the purposes of providing overcurrent and overload protection?
38. Fuse size for a continuous heating and lighting load must be ______% of the load. For noncontinuous loads it is ______%.
39. What are the primary purposes of overcurrent and overload protection?
40. What is meant by the interrupting capacity of an overcurrent device?
41. Before starting a motor for the first time, list the items that need to be checked.
42. After the motor is started, what are the primary indicators that it is running properly?
43. Define the term "power factor."
44. If a motor is drawing 50 kVA and its power factor is 80%, what is the true power used by the circuit? How much "wattless" power is there?
45. How can the power factor for the motor in Question 44 be corrected to 95% power factor? Explain. Give value.
46. Why is it undesirable for a plant to have a low power factor?
47. What is the primary cause for low power factor in a plant?
48. List the advantages of correcting the power factor of a plant.
49. List two other ways without using capacitance by which the power factor of a plant may be improved.
50. What is the advantage of correcting the power factor at the motor? At the main?

CHAPTER 10

Maintenance and Troubleshooting

Motor maintenance begins when the motor is installed. Records on the motor are started, and notations on the conditions of operation are made. Each motor has its own individual record for maintenance and repair. These records are not only important to the electrician who must maintain the machine, but also to management. Managers can make cost-effective decisions using these records.

Maintenance Requirements

Motors require more maintenance than transformers. This is because motors have more parts. Some of these additional parts move. When parts move against each other, friction is created which results in wear and an increase in temperature. If the temperature rises above rated levels, damage results to the motor.

Types of Maintenance

There are two types of maintenance programs—corrective maintenance and preventive maintenance. *Corrective maintenance* is performed when a problem arises. Replacement of defective parts is a common task during these activities. *Preventive maintenance* reduces the need to replace parts by providing timely care for the machines.

Do not limit the maintenance program to the repair and replacement of parts. The primary purpose of any maintenance program is to prevent the premature failure of the device. Schedule a preventive maintenance program to inspect, clean, lubricate, adjust, and test all accessible motors. Too often planned maintenance is considered to be unnecessary and too expensive. This false economy is not recognized until a serious preventable problem occurs.

A comprehensive planned inspection and preventive maintenance program with a fixed schedule may seem to be unnecessary. The problem is that when this is not part of the work assignment, some motors will receive greater attention than others. Some motors will rarely be serviced, and others that are difficult to access or operate in undesirable conditions for the electrician's comfort, will receive no consideration.

The tendency to gamble with preventive maintenance does not show good common sense and defies logical reasoning. The question is not *if* a motor is going to fail, but rather *when* it will fail.

A well-planned preventive maintenance program will occupy the work force in constructive tasks during periods when repairs or installation tasks are limited. Since the corrective maintenance crew is idle during these times, preventive maintenance makes even more sense, and the plant receives a return on the salaries paid during these times that can be nonproductive.

Causes of Motor Failure

Like transformers, the primary enemies of motors are high temperatures, dirt, and grime. Foreign materials that collect on a motor reduce the effectiveness of the motor's ventilation system. This in turn causes the operating temperature to rise. The higher the temperature at which the motor must operate, the shorter its useful life. These undesirable conditions can result in either mechanical or electrical failure of the machine.

Dirt that collects in the moving parts of a motor will cause premature wear on those parts. This will cause the motor to be out of operation more often than necessary. Metallic and abrasive particles can cause electrical failures as well. Moisture and chemicals can destroy the electrical and mechanical parts of the motor.

Scheduling Maintenance

A well-planned preventive maintenance program will help limit the number of outages during production peaks. Defective equipment will be repaired before problems occur and at times that are least disruptive to plant operation. Taking motors off the line during production times results in reduced production and higher costs.

Observe the daily operation of the motors in your care. In addition, establish an inspection schedule for the motors based on their service conditions. The frequency of inspection can eventually be accurately timed based on the downtime experienced by any given motor. Periodically check all motors for

1. General cleanliness.
2. Electrical conditions.
3. High ambient temperatures and proper ventilation.
4. Tightness of mounts and alignment with the load.
5. Proper lubrication and bearing wear of motor and load.
6. Deterioration of wiring insulation.
7. Condition of the field windings.
8. Condition of the armature or rotor.
9. Condition of slip rings, commutator, and brushes.
10. Wear on switches.
11. Deterioration of capacitors.

Preventive Maintenance

A good routine preventive maintenance program takes place continuously with specific tasks scheduled weekly, quarterly, semiannually, or annually. Experience gained with each motor dictates the frequency of schedule. These programs include

1. Observations of changes in the motor's operating conditions.
2. Keeping the motor and the area around the motor clean.
3. Providing for adequate ventilation.
4. Lubricating machinery on schedule.
5. Making small repairs before major problems occur.
6. Measuring electrical characterisics of the motor and comparing them over a period of time to detect any changes that may be detrimental to the motor.
7. Keeping records of the maintenance checks and any repairs made to the motor.

The following recommendations for routine maintenance are based on normal operating conditions of most motors. Those with severe and heavy duty will obviously require more frequent attention, and those with very light duty may require only routine or casual checks.

Daily Observations

The electrician needs to be aware on a daily basis of the clues that indicate a motor may need attention. Motors that have problems usually provide advance warning. The electrician or the operator of the machine is in the best position to detect these clues and to act. These observations are made on a daily basis and in an informal way. When the warning signs are heeded and corrective action taken, the preventive maintenance program is greatly enhanced.

Our senses are constantly feeding information to our brain which analyzes these data and causes us to act. We may ignore this information which is the case almost 99% of the time. On the other hand, the brain must determine what is important in order for us to act. What is important are values that we have programmed the brain to call our attention to when certain conditions are present.

In order for us to take positive action to correct a motor problem, a certain level of knowledge must already exist in the brain to compare the information being transmitted to it by the senses. This knowledge is mostly gained by study, experience, and hands-on, job-related work.

Some potential problems can be observed by our sense of vision. Look for the presence of lubricants leaking from the bearings or excessive arcing of brushes on a commutator. Make a notation when dirt and grime are accumulating on the motor. Where instrumentation is available, glance at the readings as you pass the meters. Abnormal readings will register as experience is gained. Some meters even have red lines to help get your attention. The sight of smoke usually gets everyone's attention.

Our sense of hearing is a good source for detecting potential troubles. Listen for unusual noises. Bearings make a different sound when they are worn and need replacement. Metallic parts rubbing together can be heard. Brakes squeal with a metallic squeak when they are badly worn.

Use the sense of touch by being aware of the temperature in any given area. Touch the motor casing and around the bearing housings. Be cautious in this procedure in the event the area is hot enough to burn and cause blisters. Experience will allow you to evaluate the appropriate level of temperature. Check the temperature with a thermometer if your sense of touch tells you the space or a part of the motor is too warm.

Vibrations can be felt. Excessive vibrations can also often be heard, and in many cases, can be seen. When large motors are involved, the whole building may vibrate when a problem is present.

The sense of smell can often detect smoke before the smoke can be seen. The smell of burnt varnish indicates problems with the breakdown of insulation on the windings. This would also indicate that the temperature of the motor has exceeded its rating. Toxic fumes given off by burning insulation of overloaded conductors are evident. Ionized air due to the ozone present has a distinct smell indicating the possibility of high-level electrical arcing. The sense of taste can often detect high ozone concentrations.

To summarize, use all of your five senses on a daily basis to detect the beginning of any problem. Investigate the cause. Take corrective action before an emergency arises and a major problem presents itself.

Weekly Maintenance

Preventive maintenance on a weekly basis includes

1. Start each motor to determine if it comes up to speed within the normal time frame.
2. Check the line voltage to see if it is within tolerance and is balanced on a three-phase system.
3. Measure line current and compare to previous records and if it is within the FLC of the motor's nameplate rating. Line current should be balanced for a three-phase system.
4. Listen to each motor for any unusual noises.
5. Inspect switches, fuses, starter, and any electrical controls.
6. Observe any excessive sparking of brushes.
7. Check lubricant levels of bearings and look for any leaks.
8. Look for corrosion on slip rings and commutators.
9. Check brushes for excessive sparking while the motor is running.
10. Check oil rings for rotation with the shaft for the motors that have this arrangement.
11. Use a thermometer to measure the temperature of frames and bearings.

Quarterly Maintenance

Besides performing the weekly maintenance and checks, conduct the following procedures each quarter of the year.

1. Clean motor thoroughly to remove dirt from the ventilation ducts and insulation.
2. Measure and record the insulation resistance.
3. Operate motor at its normal load and measure the temperature. Compare reading to previous values.
4. Inspect for mechanical tightness of mounting frame bolts of motor and load. Do the same for belt guards, end-shield bolts, and other fasteners.
5. Check condition of pulleys, belts, sprockets, and chains. Check tension of belts and chains.
6. Check the end-play of the rotor shaft.
7. Check the clearance between bearings and shaft for motors with sleeve bearings.
8. Measure the airgaps when possible on motors using sleeve bearings. Compare measurements to previous readings.
9. Inspect brushes for breaks and wear.
10. Check capacitors for leakage or swelling of the case. Look for corrosion around the

electrical connections. Test any capacitor that appears suspicious.

Semiannual Maintenance

1. Clean motor thoroughly. Disassemble motor if a thorough cleaning cannot be accomplished in any other way.
2. Drain and flush oil reservoirs on motors having sleeve bearings. Replace oil with recommended type. Fill to proper level.
3. Replace grease in reservoirs of ball and roller bearing type motors. Flush old grease.
4. Inspect lubricant seals and gaskets. Replace them if they are worn or leaking.
5. Check bearings.
6. Check electrical connections for good contact and any corrosion. Clean and tighten.
7. Measure insulation resistance.
8. Inspect the ends of squirrel cage rotors for overheating and loose or broken bars.
9. Inspect wound rotors for hot spots and condition of insulating varnish.
10. Remove brushes and inspect them for cracks or breaks. Check tension on the brushes.
11. Check coupling system for alignment and mechanical tightness. Tighten journal setscrews. Check shaft keys for looseness. Look for worn gears, chains, and belts.
12. Tighten other hardware on the motors such as the mounts, end-shields, and guards.
13. Check all safety provisions to determine if they are in place and in good working condition.
14. Inspect fans for loose or broken blades. If the motor is belt driven, check and lubricate bearings.
15. Operate the motor. Check for smooth running and the level of vibration. Check the operating speed(s). Activate control devices to determine if the motor responds correctly.
16. Take temperature measurements after motor has run under load for an adequate period.
17. Measure line current and voltage. Compare to previous readings and the nameplate ratings.

Annual Maintenance

Annual preventive maintenance is a repeat of the weekly, quarterly, and semiannual procedures. In each case, the intent of the scheduled preventive maintenance is to extend the life of the motor and to reduce the number of unexpected outages. A well-planned program provides for

1. Scheduled inspections.
2. Keeping the equipment clean and well ventilated to limit temperature rise and electrical failures.
3. Application of the required lubrication at the proper time.
4. Making small repairs before they become major repairs.
5. A system of record keeping that is adequate to record the history of the motor.

As a general rule, motors need a complete overhaul every 5 years. This procedure will be more frequent if the motor operates under severe conditions or in unsuitable environments. This practice helps avoid untimely breakdowns of the motor.

Disassembly of a Motor

How often a motor must be disassembled for routine preventive maintenance depends on the environmental conditions in which it operates. Effective maintenance programs provide for the complete disassembly of some motors on a scheduled basis. The size of the motor, usage, and the environment in which it operates suggests the frequency for major maintenance to the prudent supervisor. Corrective maintenance almost always requires at least partial disassembly in order to make the repairs.

General Rules

Before attempting to disassemble a motor, remove power from the machine. Disconnect the power and control wiring from the motor. Record the connections and keep in a safe place. Remove any auxiliary equipment that may prevent access to the motor. The motor may, or may not, have to be removed from its mounting.

Follow the procedures in the manufacturer's instruction manual if it is available. If not, the following discussion will usually suffice for the disassembly of most common types of electric motors.

Select a clean area for the job. The addition of dirt and grime can cause future problems. If the motor is disassembled while still mounted in place, thoroughly clean the area around the motor before taking it apart.

Handle all parts with care. Mark and tag the parts. Store the parts in an orderly fashion in a safe place. Lost parts increase one's frustration level when attempting to put the motor together. The electrician who fails to observe these simple procedures will appreciate their value too late.

Marking Bell Ends and Frame

The first step in taking the motor apart is to mark the frame and bell ends of the motor (Figure 10–1). Use a different set of marks on each end of the motor in the event the bell ends are identical. Switching bell ends, or aligning them in a different position than they were originally, will make assembly difficult, if not impossible. Also a change in their positions may cause stresses between the rotor's shaft and the bearings.

FIGURE 10–1. Mark the bell housings and frame of the motor so that their positions will be the same when assembled.

Precautions

After marking the bell ends and frame of the motor, you are ready to begin taking the motor apart. Use the following precautions.

1. Do not use a steel hammer directly on any part of the motor. The impact can break or crack castings. Parts of the motor that have been forged or punched will dent. This will distort their shapes and cause a mismatch of the mating parts.
2. Do not use a screwdriver to force mating edges apart. This will cause nicking and burring of mating edges.
3. Be prepared to record your disassembly procedure and make arrangements for storing the parts in an orderly fashion.
4. Have a pencil and notebook ready to record any interior wiring connections or any unusual mechanical arrangements.

Procedures

To disassemble the motor

1. Remove the fastening screws and bolts.
2. If the motor has brushes, remove them from their holders. Neglect in doing this can result in damage to the brushes and/or the commutator if the rotor is removed.
3. You are now ready to part the end bells from the frame. As soon as these parts separate, the rotor or armature will rest on the stator. Take precautions that will prevent damage by using supports to prevent the rotor from crashing into the stator poles. If the bell housing is open, it may be possible to place thin strips of waxed hardwood between the rotor and the stator. The larger the motor, the more chance of damage.
4. Use a machinist's hammer and block of wood to part the end bell from the frame (Figure 10–2). The wood block protects against damage to the mating parts.
5. Remove the bell housing from either end of the shaft very slowly. Parts in the bell housings may be connected by wires to parts in or on the frame. These can easily be damaged. Figure 10–3 illustrates the centrifugal switch mounted in the bell housing of a split-phase motor. One of the wires on this switch will go to the start winding on the stator. The other wire will either be connected to one side of the run winding or one side of the power source.
6. As you continue the process of disassembly, record the mechanical and electrical connections of all internal mechanisms. Make

FIGURE 10–2. Parting the end bells from the frame using a machinist hammer and a block of wood.

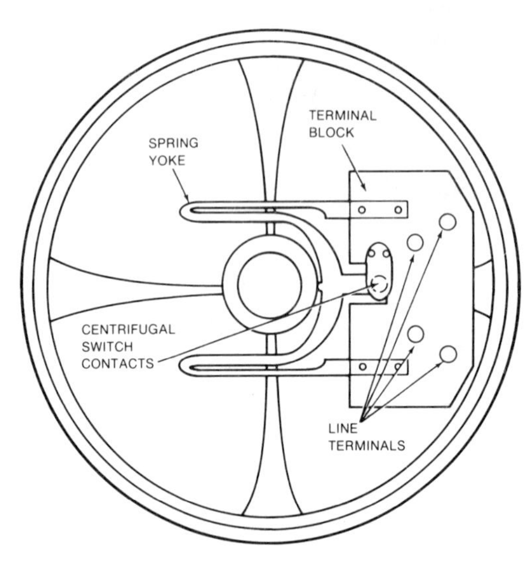

FIGURE 10–3. Centrifugal switch in the bell housing of a split-phase motor.

a wiring diagram. List the color codes and/or numbers attached to the wiring.

7. Once the end bell is removed from the shaft end of the motor, the rotor can usually be pulled out. Take care not to hit the rotor against the stator poles or windings in this process. Support the rotor (Figure 10–4). Very large motors may require special holders or slings to support the rotor as it is removed. Figure 10–5 shows special brackets for holding a rotor as it is being moved lowered by a sling into place. These brackets can be mounted on casters for moving the rotor to other areas.

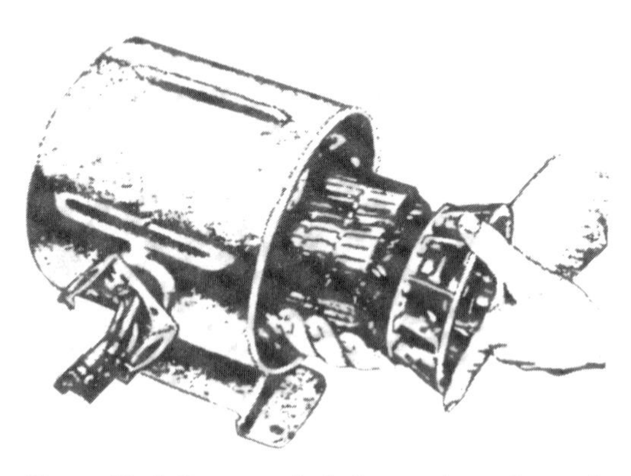

FIGURE 10–4. Proper method of supporting and removing a small rotor or armature to prevent damage.

8. Record the order of disassembly. When putting the motor back together, the reverse order will be used.

Cleaning the Motor

Some motors operate in more hostile environments than others. These motors require greater and more frequent attention than those that perform their tasks under more suitable conditions.

All motors should be disassembled and cleaned as needed or on some fixed schedule. Oil, grease, dirt, dust, metallic, and chemical contaminants build up over time and reduce or block the required ventilation. If this is allowed to happen, eventually the operating temperature of the motor will rise beyond its maximum rating and damage will occur.

Metallic and chemical contaminants along with abrasive dust and dirt will attack the electrical insulation qualities of the motor resulting in early failure. Most insulation failures can be blamed on these types of debris collecting in the motor.

Motors operating in wet and humid environments will collect moisture. This will also cause insulation

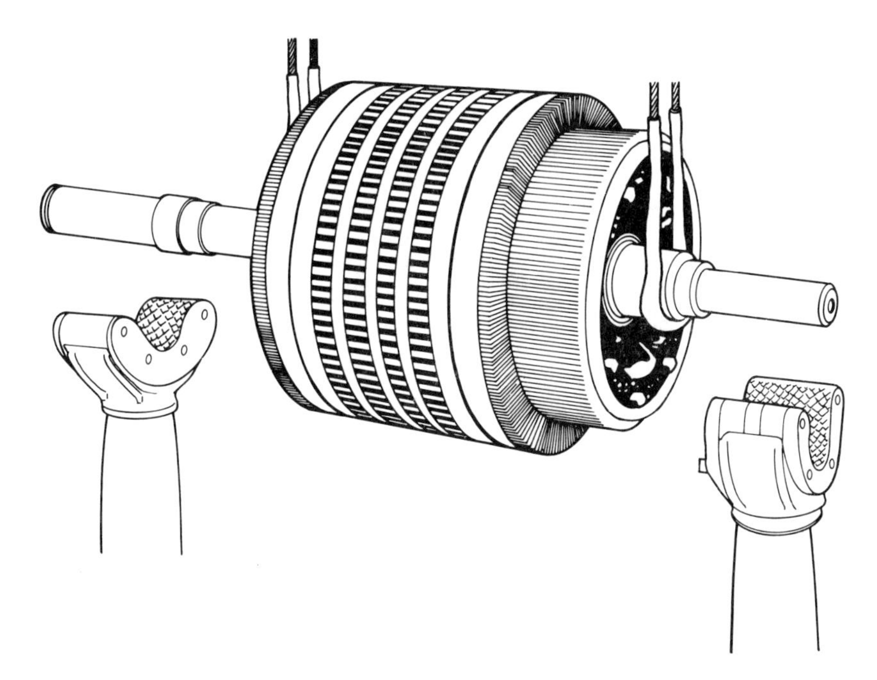

FIGURE 10–5. Special supports for removing large rotors.

breakdown and result in parts of the motor rusting. Completely enclosed motors or humidity control may be necessary under these conditions.

Air Pressure and Vacuum

Compressed air can be used to remove dust and light dirt from the exterior and interior of a motor. The air should be dried to remove the moisture content.

Because of flying particles caused by the compressed air, it is recommended that the electrician wear safety glasses for eye protection. Be sure adequate ventilation is present and respiratory protection is worn to prevent the breathing in of the contaminants. In addition to the many harmful contaminants present in some plants, motors may have mechanical brakes that contain asbestos particles. Asbestos has been proven to be harmful to the lungs. The operator must never use the compressed air to remove dirt and dust from his or her person or clothing.

Air pressure should be in accordance with the OSHA standards. Check these regulations for the particular job you are performing.

Air pressure usually is not to exceed 30 psi. High pressure will force stubborn particles deeper in crevices and the windings causing future problems. Compressed air is never to be used when metallic dusts such as copper, iron, or carbon are present.

The preferred method is to vacuum the motor rather than using compressed air to blow the debris out of it. A soft brush on the end of a hose will help loosen stubborn particles for removal. Vacuuming eliminates the introduction of moisture, forcing of particles deeper into the windings and crevices, and reduces the health risk to the operator.

Compressed air can be used to clean ventilation systems and air passage ducts. Be sure to change all dirty air filters associated with these systems. A dirty filter will reduce air flow and cause the motor to overheat.

In the case of air handlers, dirty filters can either increase or decrease the load on the motor. When the load is increased, the current will increase causing the temperature to rise. When the load is decreased, the motor will run faster but draw less air across its surface due to the dirty filter. Reduced air flow in this case will result in less heat being removed from the motor. Both cases are detrimental to the life of the motor.

Cleaning with air will not remove grease, oil, and many other contaminants. Further cleaning may be done using water with detergent, or using a cleaning solution. These methods are not to be used with DC motors or those having brushes, commutators, or slip rings. Do not use them on other motors that are exposed to conducting dusts. Liquid solutions carry conducting contaminants deep into the motor to produce shorts and grounds at a later time. Components on these types of motors may be further cleaned using a moist but not wet rag dampened with the cleaning solution.

Cleaning with Water and Detergent

After vacuuming and using compressed air to clean out loose particles, use a low pressure steam jenny for further cleaning the motor. Do not exceed 30 psi maximum for the steam flow. To prevent possible damage to the varnish and insulation, use a fairly neutral, nonconducting detergent. General Electric Company recommends Dubois Flow for this task. Mix a pint of detergent with 20 gallons of water.

Apply the solution using warm water and a spray gun if a steam jenny is not available. Do not exceed the maximum pressure when using a spray gun. A soft bristle bush may also be useful to loosen the dirt and debris in the windings and crevices.

Rinse the motor with clean clear water or low-pressure steam to rid it of any detergent solution. Use a megger to measure the windings' resistance. It is advisable to dry the motor. Use one of the methods previously described.

Cleaning with detergent and water is the preferred method of cleansing a motor. If drying facilities are not available, a cleaning solvent can be used.

Cleaning with Solvent

Many cleaning solvents are toxic or flammable. Use safety glasses and appropriate respiratory protection when applying these solvents. Avoid contact of the solution with the skin. Many of them cause rashes or can be absorbed by the body through the skin. Perform the cleansing operation in a well-ventilated area. Keep the solution and its vapors away from open flames or sparks. Failure to observe these precautions can result in personal injury and possible fire or explosion.

Some manufacturers recommend trichloroethane as the solvent of choice for cleaning motors. Although this chemical is considered to be nonflammable with a low level of toxicity, use it only in a well-ventilated area free of open flames. It is best not to expose yourself to any prolonged periods of breathing the vapors of any chemical solvent.

Carefully dry the motor after cleaning it with a solvent. Do not allow the temperature of any part of the machine to exceed 125°C. Do not energize the machine until it is thoroughly dried and the resistance checks are acceptable.

Varnishing Windings

After several cleanings with either solvents or water and detergent, varnish the windings. Use the manufacturer's recommendations as to the type of varnish. Apply the varnish when the motor is warm. The varnish can either be brushed or sprayed onto the windings. For small motors, it is best to dip the rotor and stator windings into a varnish vat and clean the adjacent metal parts with a varnish solvent. Figure 10–6 illustrates a method for dipping the armature.

FIGURE 10–6. Method of dipping armature into varnish vat.

After varnishing, bake the windings at 100–150°C for about 4 to 7 hours. Temperature and time will vary with the type of varnish. Follow the instructions. Under some circumstances, it is wise to apply a second coat of varnish and repeat the baking process.

If an oven for baking is not available, or the motor must be returned to the line without delay, there are commercial varnishes that can be used without baking. In any event, allow an adequate time for the varnish to dry. Keep the machine in a low humidity atmosphere while this process is taking place. Never energize the motor until the varnish is completely dry.

Inspection

Any time a motor has been disassembled for major maintenance is a good time to check all the parts. They are easier to see and any problem will normally be more evident. Inspection can take place while the cleaning operation is in process.

1. Check the bell housings for cracks or breaks. Remove any burrs or nicks from mating parts.
2. Replace worn bearings, brushes, and switches.
3. Look for deterioration of insulation and megger the resistance of the windings. Check all windings for burned spots. Correct any problems.
4. Look for evidence of the armature rubbing against the stator poles. Replace bearings if this condition exists. The armature may have to be turned if it is out of round.
5. Measure the concentricity of commutators and slip rings.
6. Look for rust and damage to the surface of the motor. Remove any rust and touch up the paint on the motor.

Bearings

Bearings are an important part of a motor. They maintain the rotor in a centered position in respect to the stator. These devices allow the rotor to turn with little friction. Because the rotating body of the motor applies the most force on the bearings, the type of bearing to use is dependent on the type of force exerted on it.

The rotor of an electric motor may subject the bearings to different types of loads. These loads are called radial, thrust, and angular.

Radial loads are those that apply force perpendicular to the shaft. Use radial bearings when the motor is mounted horizontally.

Thrust loads are forces applied parallel to the shaft. These occur when the motor is mounted in a vertical position. Use thrust type bearings under these conditions.

Angular loads are combinations of radial and thrust loads. Forces are applied both perpendicularly and parallel to the shaft. Use angular-type bearings when this condition exists.

Bearings are one of the few parts that wear. They are lubricated with grease or oil to reduce friction which causes heat. Too high a temperature will cause the bearings to fail.

Ball bearings are used when the motor drives loads where heavy axial and radial thrusts are present. Normally, ball bearings are packed in grease at the factory, and greasing is not required before putting a motor into service.

Modern motor designs provide for an adequate supply of grease or oil to be kept in a reservoir which is dusttight and oiltight. Greasing or oiling usually is not required for several months or even years.

Lubrication Schedule

Oil and grease will ultimately become depleted, however, and lubrication is a necessity if the motor is to achieve its normal life expectancy. Greasing periods vary depending on the type of service and the horsepower of the motor. Table 10–1 sets forth General Electric Company's recommendations for lubrication.

TABLE 10–1. Recommended Greasing Periods*

Usage	Horsepower		
	0.5–7.5	10–40	Over 40
Easy. Motor operates infrequently (1 hr/day).	10 yr	7 yr	5 yr
Standard. Machine tools, fans, and pumps are good examples.	7 yr	5 yr	3 yr
Severe. Motor operates continuously in locations subject to severe vibration.	4 yr	2 yr	1 yr
Very severe. Motor is subjected to a dirty environment and vibration. The shaft of the motor becomes hot.	9 mo	8 mo	6 mo

**Courtesy of General Electric Company.*

Type of Grease

Follow the manufacturer's recommendations as to the type of grease to use. Normally this will be a lithium-based grease specifically designed for ball bearings. Never use a grease that is abrasive, acidic, or alkaline. The temperature rating of the grease must always be greater than that of the operating temperature of the windings of the motor, preferably over 150°C. When under operating conditions or in storage, the grease should not separate into its oil and soap components.

A special grease is needed if the motor is to operate at an unusually high or low temperature. Avoid mixing different types of grease. Completely purge the reservoir of old grease when changing to a different type of lubricant. Mixtures of different types of grease sometimes cause chemical actions that result in the formation of acids or alkalies. Either of these will attack the bearings and cause premature failure. Ball bearing grease normally has an oxidation inhibitor.

Procedure

Due to the inherent dangers of rotating machinery and electrical shock, whenever possible, always stop the motor before lubricating it. If the motor cannot be taken off the line for maintenance, exercise the necessary precautions to prevent injury to yourself and others.

Clean the grease fitting to prevent dirt entering the grease reservoir. Remove the lower grease relief plug. Clean the relief hole of any caked or hardened grease (Figure 10–7). Some manufacturers provide a special brush for this purpose. Do not use the brush if you are attempting to lubricate the motor while it is running.

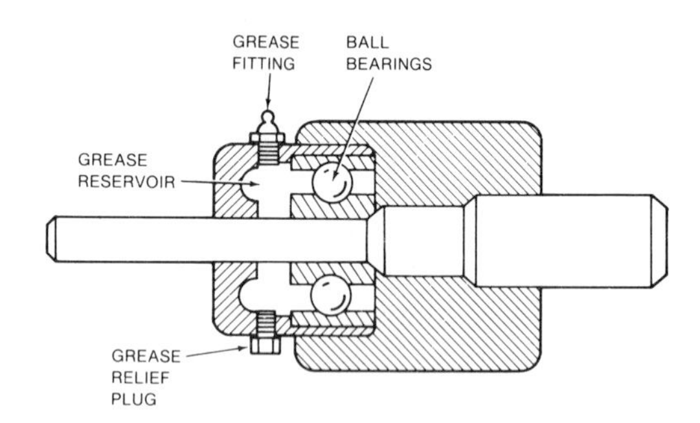

FIGURE 10–7. Lubricating ball bearings.

Pump grease into the fitting until clean grease appears through the relief hole. This will clear all the old grease from the reservoir.

Use a hand-operated grease gun only for this task. Do not use power guns. Excessive pressure might rupture the bearing seal.

Allow the motor to run for about ten minutes with the grease relief hole open. This will clear the excessive grease. Too much grease can cause the bearings to overheat.

Replace the plug in the grease relief hole. Clean all excessive grease around the grease fitting and relief hole. The excess grease may be carried into the machine and attached itself to the windings. The grease will collect dirt and hasten the deterioration of the windings.

Motors that do not have the grease fitting and relief plug cannot be greased in this manner. To grease these bearings, remove the bearing housing and clean out the old grease with a cleaning solvent. Apply new grease either by hand or from a tube. Fill the cavity about one-third to one-half full. More than this amount will be excessive.

Cleaning

Occasionally it may be necessary to disassemble a motor for cleaning. Clean the bearings and bell housing at this time. Take the following precautions.

1. Unless absolutely necessary, do not remove bearings which are in good condition. When bearings are removed, store them in waxed or oiled paper.
2. Keep the bearings clean. Handle them only with a clean, lint-free cloth or canvas glove.
3. Use a cleaning solvent such as trichloroethane which will remove dirt and grease. Do not use a cleaning solvent on sealed or semisealed ball bearings. The solvent may remove the slushing oil inside of the bearing.
4. If the bearings are accessible after cleaning, soak them in mineral oil heated at about 100°C. Thoroughly dry the parts before assembly.

The following method can be used to clean the bearings and the reservoir without disassembling the motor.

1. Wipe the housing, relief fitting, and relief plug clean to prevent entry of dirt or particles. A small piece of abrasive material can quickly damage a bearing.
2. Remove the fittings by unscrewing them from the casing.
3. Remove any hardened grease around the fittings using tools supplied by the manufacturer or with the blade of a clean screwdriver. Take precautions not to damage any part which the screwdriver comes into contact.
4. While the motor is running, inject a suitable cleaning solvent into the top fitting hole using a syringe. The cleaning solvent will thin the grease and it will run out of the bottom hole of the reservoir. Continue to add solvent until a clear mixture is seen exiting the drain.

5. Replace the relief plug and add solvent until it can be seen splashing in the top hole. Allow the motor to run for a few minutes and open the drain. If the mixture shows signs of dirty grease, replace the drain plug and repeat the process until a clear mixture exits the reservoir.
6. Replace the drain plug and add a light lubricating oil to the reservoir. Run the motor for a few minutes and drain the mixture of oil and solvent. The oil will flush any solvent that may have remained in the reservoir.
7. Turn the motor off and replace the grease fitting and relief plug. Grease the bearing according to the previously described method.

Sleeve Bearings

Most single-phase, low-horsepower motors use sleeve bearings. These bearings are either babbitt-lined, porous bronze, or brass. The babbitt-lined bearings have the least friction and the longest life.

Oil is normally the lubricant for sleeve bearings. Follow the manufacturer's recommendations as to type and viscosity. The same rules apply to oil as to grease.

1. Temperature rating of the oil must be higher than that of the windings.
2. Never mix different types of oil.
3. Use special lubricants for motors operating in extreme temperatures. Preheating the oil is a method sometimes used in very cold conditions.
4. Never use oils that are acidic or alkaline, or those that contain abrasives.

Sleeve bearings normally have specifically designed grooves which permit the transfer of oil from the reservoir to the shaft through a wool felt wick or packing of wool felt or cellulose fiber. Packing as a means of transferring the oil to the shaft is seldom used on modern fractional horsepower motors. When packing is used, change it at least once each year. Figure 10–8 illustrates the lubrication system using a felt wick to transfer the oil from the reservoir to the shaft of the motor.

The design includes a tube with a cap to add oil and a drain plug to remove contaminated oil. The reservoir is large enough to provide adequate room for dust, dirt, and oil sludge to collect at the bottom.

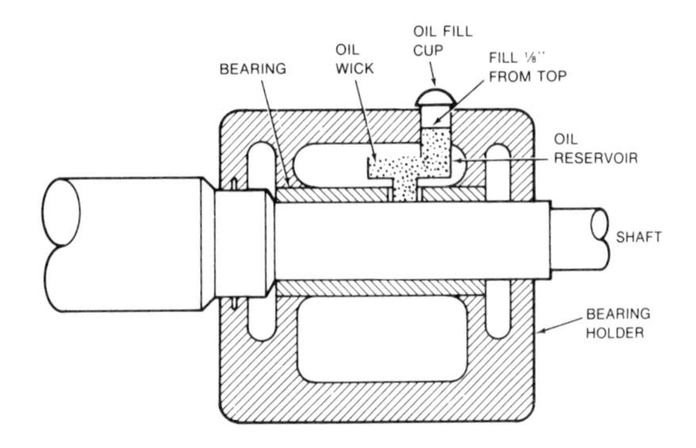

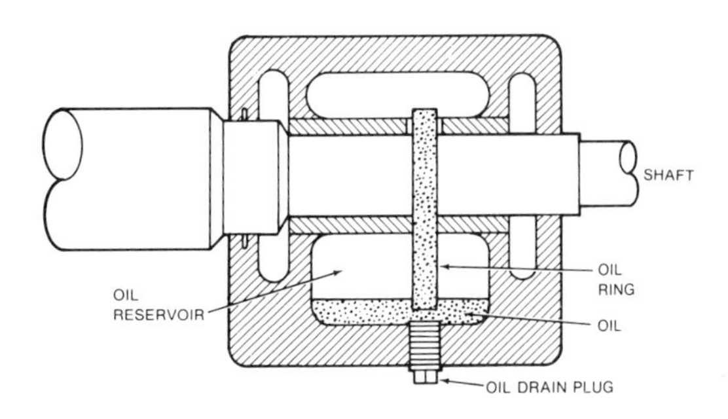

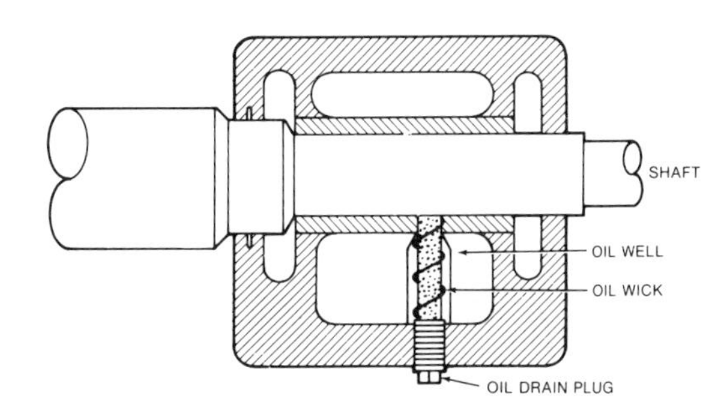

FIGURE 10–8. Sleeve bearing lubrication system.

Follow the recommendations of the motor's manufacturer as to how often the oil needs to be changed. Heavy use in high ambient temperatures or dirty conditions require more frequent oil changes.

Procedure for Oiling

When replacing the oil, allow the oil to run out of the drain plug until the oil is clean. Replace the drain plug and fill the reservoir until the oil is within 1/8 inch of the top of the fill tube.

It is common for sleeve bearing types of motors to run from 1 to 3 years without the need for the oil to be replenished. If the oil discolors or oxidizes, change it immediately.

Overheating

Motor bearings overheat for the following reasons:

1. No lubricant. To remedy this condition, add oil or grease. Do not overfill.
2. Too much lubricant. Drain excessive oil until the oil falls to 1/8 inch below the top of the fill tube. Open the drain plug on the grease fitting and allow the motor to run for about ten minutes. Excessive grease will drain. If not, the grease is caked and should be removed and the ball bearings greased.
3. Oil not reaching the shaft. Check the wick and means of holding the wick against the shaft. Replace the wick if necessary. If ring construction is used, be sure the ring is free to rotate.
4. Worn bearing. Replace the bearing. When replacing one bearing, replace all of them even though there is no evidence of wear. Use exact replacements. It is best to have the bearings replaced by a motor shop or machinist. Special tools such as bearing pullers, extractors, and bearing arbors with presses are needed in many cases. These are not standard tools that electricians carry on the job. Shaft wear can be corrected by metallizing, spraying with molten metal, and turning and smoothing the shaft to make it true. Another method is to reduce the diameter of the shaft on a machine lathe. This operation will require a change in the size of the bearing. The machinist uses inside and outside micrometer calipers to make these measurements.
5. Excessive belt tension. Adjust belt tension so that the sag is about 1/8 inch for each foot of belt.
6. Excessive end thrust or side pull. Check belt tension. Check the motor operating characteristics against the type of load it is driving. Misapplication can be the problem. Check pulley alignment.
7. Shaft current. These stray currents are a form of eddy currents which are caused by the magnetic field cutting through the metal shaft. They are normally caused by irregularity of the magnetic flux field. For example, an uneven airgap between rotor and stator poles may be the culprit. Some current in the shaft is an inherent characteristic of an AC machine and is not always harmful. Figure 10–9 shows the path of the shaft current.

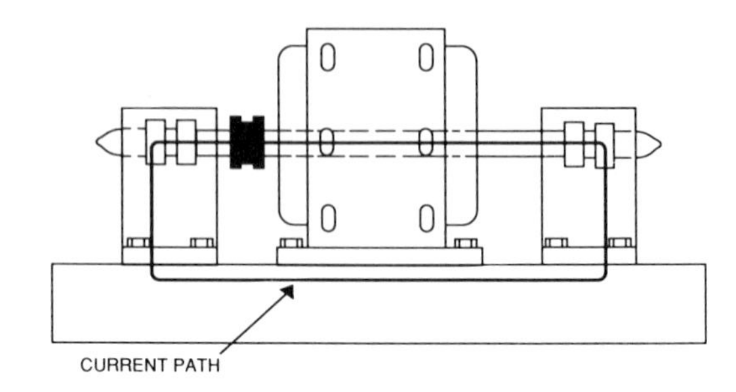

FIGURE 10–9. Path of the shaft current.

When conditions are exaggerated, however, and large shaft currents are present, they can cause damage to the bearings. These conditions are more likely to occur in large induction-type motors.

When contact between the shaft and bearing is good, a ball bearing is more likely to conduct the current than a sleeve bearing. The bearing may melt due to the heat created by the current.

If bearings on a motor are receiving proper attention, but are failing on a regular basis, check for shaft current. Shaft current can be determined by shorting the ends of the shaft with a low-resistance copper wire, and measuring the current flow through the wire with a clamp-on ammeter.

Shaft currents can be eliminated by breaking the current path. This can be done easily for motors that have bearing pedestals. Insulate the bearing pedestal opposite the coupling end of the motor. On motors with the bearings supported in brackets, the insulation is placed between the bearing and the bearing housing.

Although the problem of shaft current does not occur frequently, when it does occur it can present some serious problems. It is not always easy to reduce it to acceptable levels.

A company that I worked for years ago had a contract to maintain the motor-generator units in several military missile silos. These units had a large flywheel. The bearings on the flywheel were burning out on a regular basis. It never occurred to us at first that this problem was electrical and not mechanical in nature.

Magnetic flux from the motor generator unit was leaking. The flywheel as it turned was cutting the flux field. This caused very large currents to be induced into the flywheel. This in turn was causing arcs in the bearing housing.

We attempted many remedies to correct the problem. This included shielding the flywheel from the flux field which was not possible without redesigning the unit. Insulating the bearings of the flywheel also failed due to the weight involved. We were unable to find a suitable insulation material that would withstand the stresses involved.

Our final attempt was to provide an alternate current path. Large carbon brushes were used to make contact with the flywheel with grounding leads attached to the brushes. This attempt, although not completely satisfactory, solved the problem.

Checking Bearing Wear

Worn bearings normally tell you their condition by the noise they make when the motor is in operation. Listen to them when they are in operation.

Because worn bearings increase the friction between shaft and themselves, the temperature will rise above the normal level. Although feeling for "hot bearings" has been acceptable in the past, too many bad bearings are missed in this manner. Use a thermometer or thermocouple to make the measurement. These methods are much more reliable. Establish base temperatures on the motor's record card when the motor is put into operation. Don't forget that low or dirty lubricant can be the cause of the higher temperature. Check this condition first.

Bearing wear can also be determined by attempting to move the shaft of a motor when it is at rest. Increased end-play and movement up and down or sideways are good indicators of bad bearings. Figure 10–10 illustrates this procedure.

Badly worn bearings allow the rotor to come into contact with the stator core. Evidence of this will be present on the rotor and stator cores as scrape marks. The increased noise of operation will also be evident. If this condition continues for even a short period of time, the motor will be destroyed due to the heat generated by the friction between these two parts.

The best procedure for determining wear on sleeve bearings is to use a feeler gauge to measure the distance between the rotor and the stator in each of the four quadrants. One measurement is to be made in the direction of the load.

The distance between them should be equal. Measurements taken at the time of installation or the manufacturer's specifications are to be used as the base figures. Taking these measurements during regular

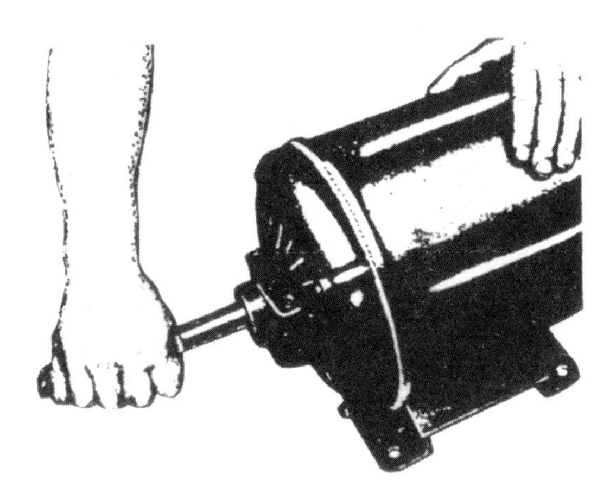

FIGURE 10–10. Checking shaft bearings and end play by manipulating the shaft.

preventive maintenance intervals will give adequate warning as to bearing deterioration before serious damage is done to the motor. Figure 10–11 shows a set of feeler gauges.

This method is not effective for roller or ball bearing motors. These bearings normally fail before there is a noticeable change in distance between the rotor and stator poles.

Replacing Bearings

Special tools are necessary for removing the bearings from a motor. Regardless if the bearing is a sleeve or roller type, an even pressure must be exerted on the bearing in order for the bearing to be removed without doing damage to the casing or shaft of the motor. Never attempt bearing removal without the special tools. Removal procedures varies greatly between different types of motors. Use the manufacturer's handbook on the specific motor for the correct procedure. Only some of the tools, general precautions, and methods are included in the following descriptions.

Figure 10–12 shows a set of arbors for removing sleeve-type bearings. These tools come in various sizes with a single arbor having as many as five different diameters. The tool is designed so that the outside diameter will pass through the bearing holder of the casing without doing damage to it. At the same time, the next diameter is equal to that of the bearing and will rest evenly on it.

Make sure the working surface on which the end bell rests is smooth, firm, and even. Figure 10–13 illustrates the placement of the arbor before removing

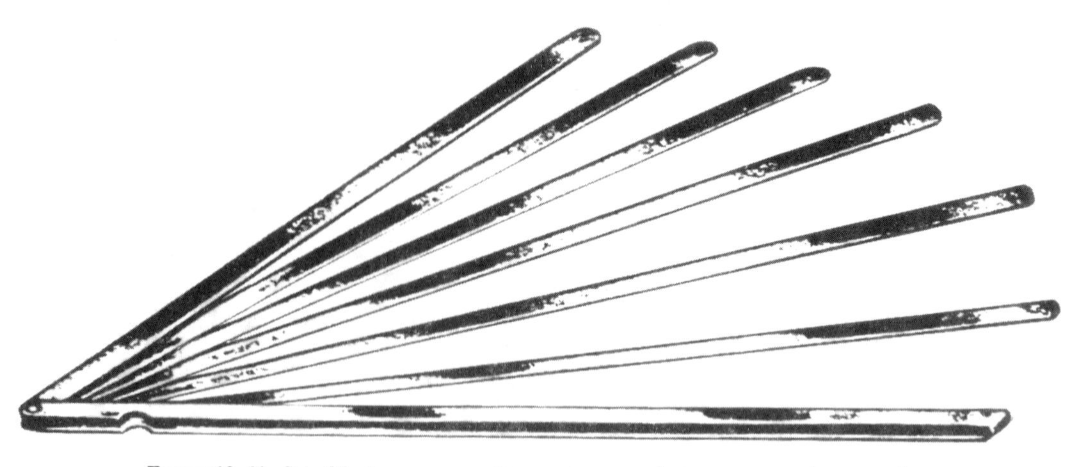

FIGURE 10–11. Set of feeler gauges used to measure space between rotor and stator poles.

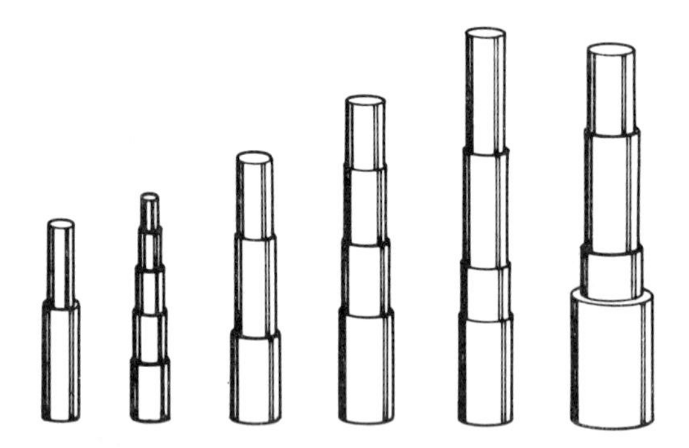

FIGURE 10–12. Set of sleeve-bearing arbors.

the bearing. Figure 10–14 shows the removal of the bearing using the arbor and a machinist hammer.

A preferred method to removing the sleeve bearing is to use an arbor press. This machine eliminates the chance of accidentally striking the housing which is normally made of cast aluminum or iron. Sharp blows can crack or break the housing. An additional benefit is that bearings removed in this manner can be reused if they are still good. Figure 10–15 shows an arbor press.

If an arbor press is not available, a set of extractor and inserter tools can be used to remove the sleeve bearing. These tools are designed to put even pressure on the bearing either when removing it or replacing it. This method is better than the arbor and hammer technique.

New sleeve bearings can be reinserted in the bell housing using any of these methods. The open end of the housing will face up for this operation. The bearing is placed into the opening and then forced either by tapping the proper size arbor with a hammer or providing even pressure using a press or inserter.

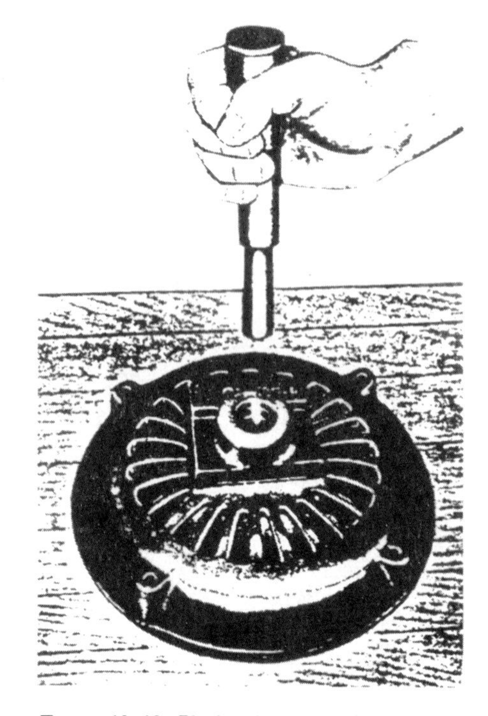

FIGURE 10–13. Placing the arbor before removal of a sleeve bearing.

When removing ball bearings from a shaft or the bell housing, apply a steady, even pressure to the shaft. Special tools are made for this purpose. Figure 10–16

FIGURE 10–14. Removing a sleeve bearing using an arbor and machinist hammer.

FIGURE 10–15. Removing a sleeve bearing with an arbor press.

shows the removal of a bearing using a hook-type puller. The tool shown has only two hooks, and it needs to be centered perfectly for the pull. Pressure may be applied to the outer race of the bearing, but it will destroy the bearing. If the bearing is good and is to be used again, apply the pressure to the inner race if possible. Similar tools have three hooks, and a more even pull is easier to achieve with them.

Damaged ball bearings that are particularly tight can be removed by first separating the bearing. Force the outer race over the balls leaving only the inner race on the shaft. Use a torch to apply quick heat to the inner race while applying a steady pulling force with the hook-type puller. The heat will normally loosen the bearing on the shaft and make it easier to pull. Do not apply too much heat or damage may occur to the shaft.

Use a plate- or bar-type puller when the construction of the bearing assembly is such that the inner cap and ball bearing remain on the shaft after the end bell and outer cap have been removed. Figure 10–17 provides three illustrations of how the bar-type bearing puller operates.

Usually replacement bearings are of the same type and number as the originals. Install them in the same relative position as the old bearings. Angular-contact bearings that are to be stacked together must have their high points of eccentricity lined up. This point is indicated by a burnished spot on the inner race.

It cannot be emphasized enough the need for proper maintenance on the bearings of a motor. Bearings are the single most likely component to fail when the motor is operated within its electrical ratings.

Commutators

Commutators are used on DC machines and on the starting windings of some AC machines. When used on AC machines they are called *rotary commutators*. They are also used on auxiliary machines to provide excitation for synchronous motors.

Motors are useless without good commutation. Good commutation is noted by the absence of excessive sparking or burning of the commutator bars and brushes. To meet these requirements, a continuous and close contact is maintained between the commutator and the brushes. The commutator must be kept clean.

The surface of the commutator must be smooth and concentric. The mica insulation must be properly undercut. Brushes must be properly shaped and have the proper pressure applied. The armature must be well balanced and the bearings in good condition.

Another requirement of good commutation is the formation of a uniform copper oxide–carbon film on its surface. Normal oxidation of the copper along with heating and ozone action on particles of the graphite from the brushes cause this film to form. The film acts as a lubricant over the commutator surface and extends the life of the brushes and the commutator.

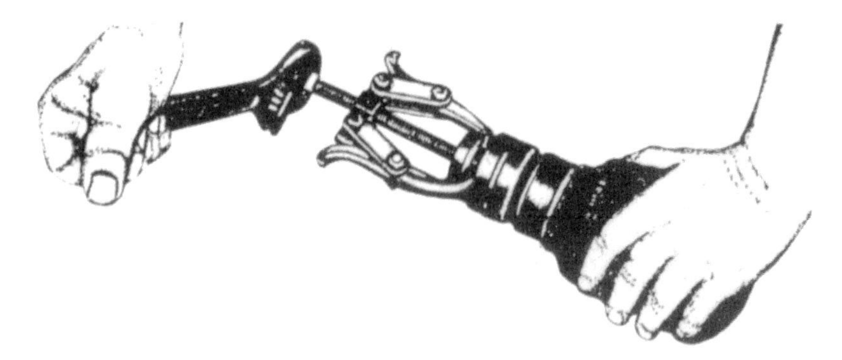

Figure 10–16. Removing a ball bearing from a shaft using a hook-type puller.

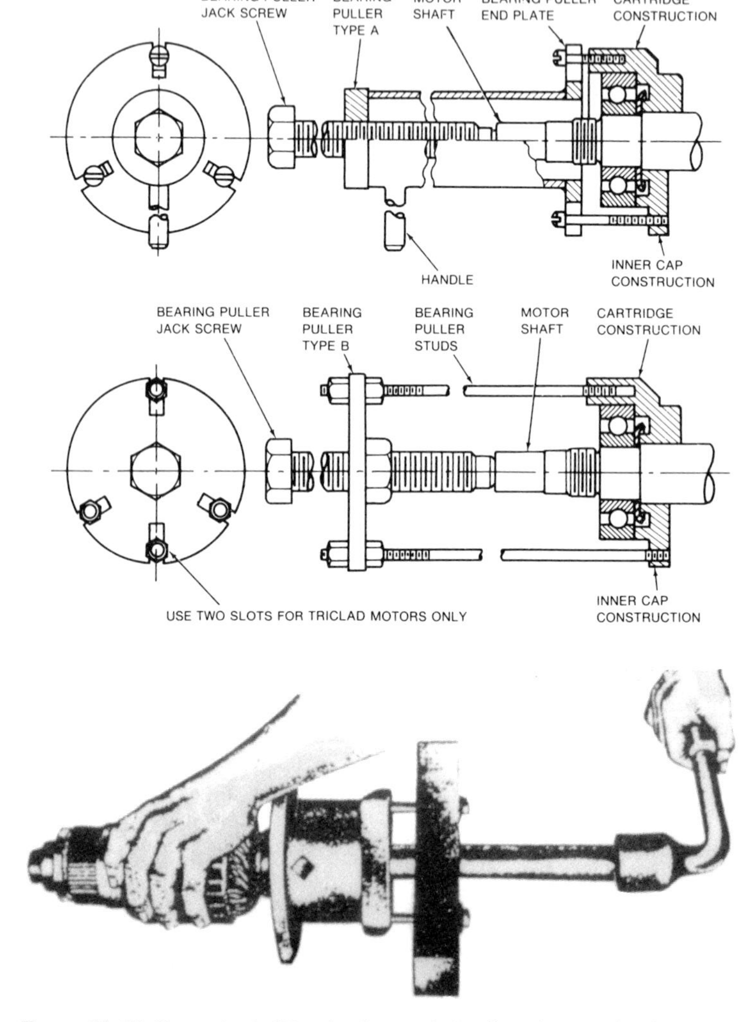

Figure 10–17. Removing ball bearing from a shaft using a bar-type bearing puller.

The color of the film varies from copper to straw, or light chocolate to almost black. The surface appears highly polished and smooth. There should not be any color shading around the brushes. Chlorine fumes will cause a green discoloration. Sulfuric fumes will turn the area blue. A streaked or gummy-looking surface is the result of oil, oil vapors, grease, or the presence of some other types of contaminants. Clean the commutator when any of these conditions exist. This is the key to extending the life of the machine.

Commutators normally have a polished brown color where the brushes ride on them. A bluish burned color indicates overheating of the connection between the surface of the commutator and the brush.

Cleaning

Usually these parts require only occasional wiping with a piece of canvas or lint-free cloth. Be sure no threads become lodged between the commutator and the brush. This can increase wear and cause arcing. Figure 10–18 illustrates the cleaning of a large commutator.

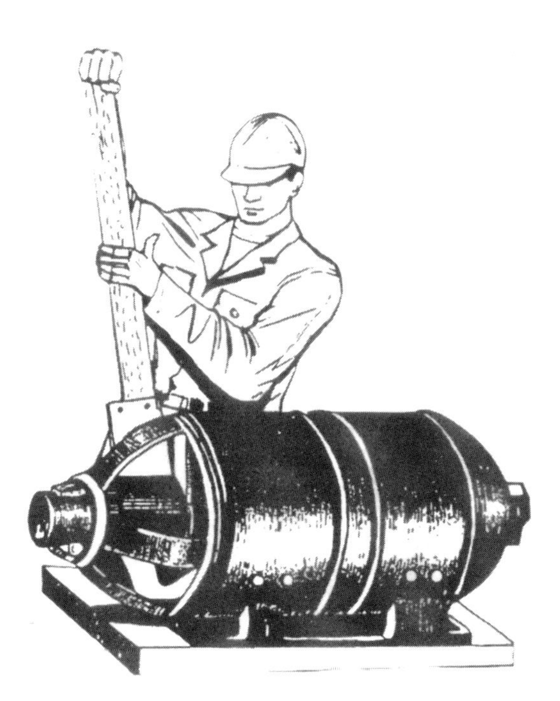

FIGURE 10–18. Using a wooden stick and canvas to clean a large commutator.

Never use any lubricant on a commutator while cleaning it. Avoid using silicone-based materials such as used for gaskets in the vicinity of a commutator. Silicones cause abnormal wear of the brushes.

Approved solvents can be used to remove oil and grease. Be sure the commutator is completely dried of the solvent before the machine is put into operation.

When foreign matter sticks to the commutator and cannot be easily removed by wiping, use a fine grade of sandpaper. Do not use coarse-grained sandpaper or emery paper. The coarse sandpaper will scratch the surface of the commutator, and emery paper has metallic particles that will collect in the slots and cause shorting of the bars.

Use a wood sanding block to hold the sandpaper while performing this task. The concave surface of the block must fit the convex surface of the commutator. Sand perpendicular to the axis of the slots and very slowly to prevent scratching the surface (Figure 10–19).

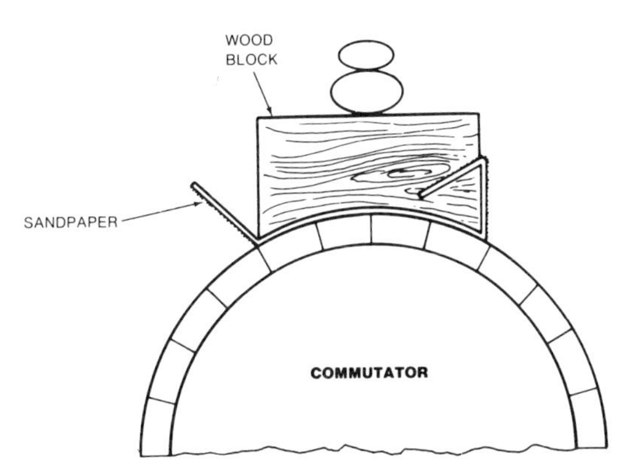

FIGURE 10–19. Cleaning the surface of a commutator using a sanding block.

No matter how careful you are with the sandpaper, it will put small invisible scratches into the surface of the segments. It is wise to burnish the commutator after cleaning it with sandpaper, or as explained later, reestablishing the concentricity by turning, grinding, or sanding.

Completely dried hardwood makes an excellent burnishing tool. Shape the hardwood like the sanding block in Figure 10–19. While the motor is running, move the block back and forth across the surface of the commutator. The hardwood will remove the small scratches from the surface.

Another method of using hardwood for the burnishing is to shape the hardwood in the shape and size of the brushes. Replace the brushes with the hardwood. The motor will not run under this condition, and the rotor will need to be turned by another motor in order for the hardwood to do its job.

Checking Concentricity

The commutator must be nearly round to prevent brush bounce and excessive arcing. The higher the speed of commutation, the greater the need for more perfect concentricity. A commutator with speeds of 5000 feet per minute needs to be concentric within 0.001 inch on its diameter. At speeds above 9000 feet per minute, the requirement is for the commutator to within 0.0005 inch.

Movement of the brush up and down on the commutator can be felt in some cases. Place an insulated probe or the rubber eraser of a pencil on the top of the brush while the motor is running. Any rough spots will be felt.

If no unusual movement is noticed using the probe, and you still suspect the commutator is not in round, use a concentricity gauge to measure the variation. Figure 10–20 shows this arrangement.

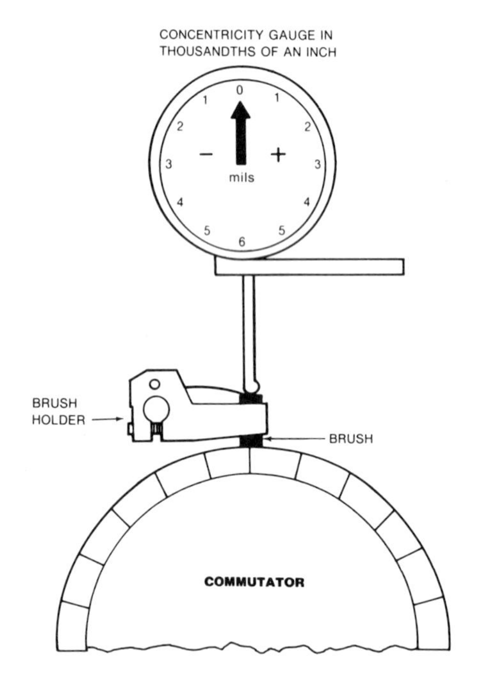

Figure 10–20. Measuring concentricity of a commutator.

The gauge is set on top of the spring holding the brush in the brush holder. Adjust the pressure on the gauge so that it barely touches. Adjust the indicator so that the gauge will read zero. Rotate the commutator by hand, segment by segment. Note the high and low points. If the concentricity of the commutator is not within tolerance, corrective action is indicated.

The preferred method of making the commutator concentric is to send it to the motor shop and have it turned on a machine lathe. In the absence of this tool, the second preferred method is to use a grinding stone and to turn the commutator. This operation would also be performed in the motor shop.

If the commutator is not too far out of round, and the motor needs to be back on the line quickly, a handstone specified for smoothing commutators can be used. The material in the stone consists of a synthetic abrasive that has the property for the stone's particles to tear loose as soon as the particles become dull. The stone does not become clogged and smooth, thereby always presenting a cutting action.

The stone is formed as the sanding block in Figure 10–19 so that its surface fits evenly on the circumference of the commutator. While the motor is running the block is moved slowly back and forth across the commutator's surface. Use a coarse stone for the initial smoothing, and replace it with a fine stone for finishing. Never use oil or other lubricant with the handstone. This will cause the stone to clog and become smooth, thereby reducing the stone's ability to grind.

In the absence of a handstone, sandpaper and a sanding block may be used. This is the least preferred method. The same procedures outlined for the handstone would apply to the sanding block.

Before using any of the methods described, check the bearings. If the bearings are worn, the commutator is going to turn off center and cause problems. Also check the commutator for loose segments. If a segment has loosened, it may be either higher or lower than other segments. Place the bar in its proper position and tighten the fasteners which hold the segments. Mica above the level of the segments may also be the cause of the commutator not to be concentric.

Mica Undercutting

High mica in the slots is a frequent source of brush noise, streaking, and rapid deterioration of the brushes. Check the slots for high mica any time the commutator appears rough. Figure 10–21 illustrates a slot that has been properly undercut 1/32 inch and a condition where the mica is high.

Any time the mica is high in the slots, it must be undercut below the level of the segments. The depth of the undercut will vary depending on the different sizes

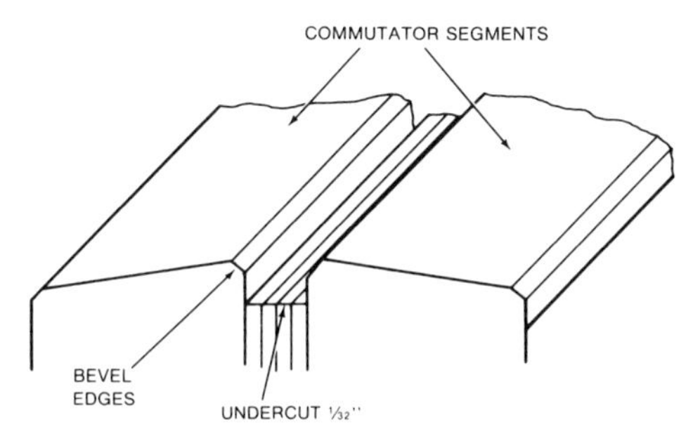

(A) Correctly undercut.

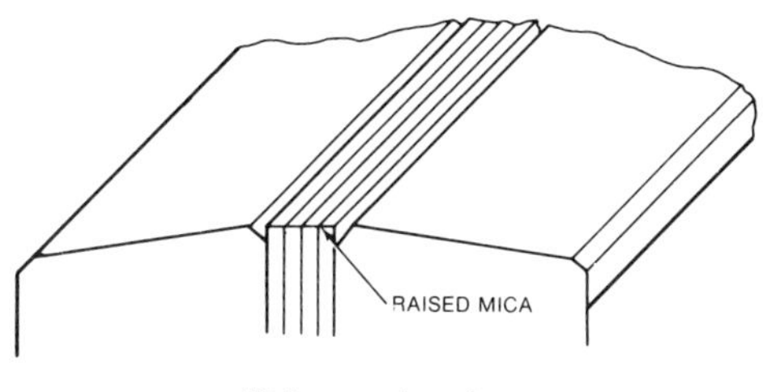

(B) Incorrectly undercut.

FIGURE **10–21.** Commutators correctly and incorrectly undercut.

of commutators. For medium-width commutator bars, the undercut is in the range of 1/16 inch plus or minus 1/64 inch.

When the machine operates at a low speed or in a dirty environment, the slots are likely to collect dirt. In either of these cases, a V-shaped slot may be more satisfactory than the square or U slot shown in Figure 10–21. The V slot collects less debris than the U slot. The U slot has the advantage in that the commutator bar can wear down to the bottom of the slot before it is necessary to undercut it. On the other hand, wear on the segment can only be to the top piece of mica on the V-type slot.

Undercutting consists of removing the mica from the slot so that there is no contact between the mica and the brushes. There are several tools available for this purpose.

The preferred method is to use a flexible, shaft power driven, saw-type undercutter. When this tool is available, it is not necessary to remove the armature from the machine. Only the brush rigging needs removal in order to saw out the slots.

If the armature is removed, a mica undercutting tool such as that shown in Figure 10–22 is used to remove the mica from the slots. This tool has adjustable V blocks to hold the armature. The saw is mounted on a motor whose height is adjustable. The saw can move horizontally the length of the slot to remove the mica. Figure 10–22 depicts this tool. There are also portable power saws designed for undercutting the mica.

In most cases a power tool will not be available to the plant electrician for undercutting mica on the commutator. To make immediate repairs, a hacksaw blade can be shaped as shown in Figure 10–23. The teeth of the hacksaw blade are ground off in the area where the tool comes into contact with the commutator. This reduces the chance of burring the commutator bars. Wrap tape around the end opposite the cutting edge to prevent injury when using this tool. If a taped handle is not satisfactory, make a wood handle to hold the blade. There are commercially manufactured slotting files on the market that can be purchased.

When power tools are used to undercut the mica, rough edges are often developed on the segments of the commutator. If left there, they will cause poor commutation. The standard procedure for removing the burrs is to bevel the edges of the commutator segments. The process provides smooth transition for the brushes from one segment to the next.

Any time the mica has been undercut on an armature, check each slot for proper depth of cut and uniformity. V cuts may be off center. U cuts may have raised mica on one side where the saw has not removed all the mica in the slot.

Never use any lubricant when cutting the mica. Be sure all loose pieces are removed from the slots before putting the machine in operation. Insulate the undercut with an air-dry varnish. Remove any excess varnish from the segments. Always keep the slots clean to assure good performance of the commutator.

Slip Rings

The principles of maintaining the surface of the commutator applies equally to slip rings. The same uniform copper oxide–carbon film is necessary for good electrical contact with the brushes. Check slip rings for contaminants, flat spots, streaks, and grooves. Clean them if necessary. Remove minor grooves or flat spots with a stone or by sanding. If the rings are too far out of round, send the armature to a motor shop to have the surface of the rings machine ground or turned on a lathe.

Brushes

The contact between the commutator of the motor and the external circuit is through the brushes. The brushes must be able to

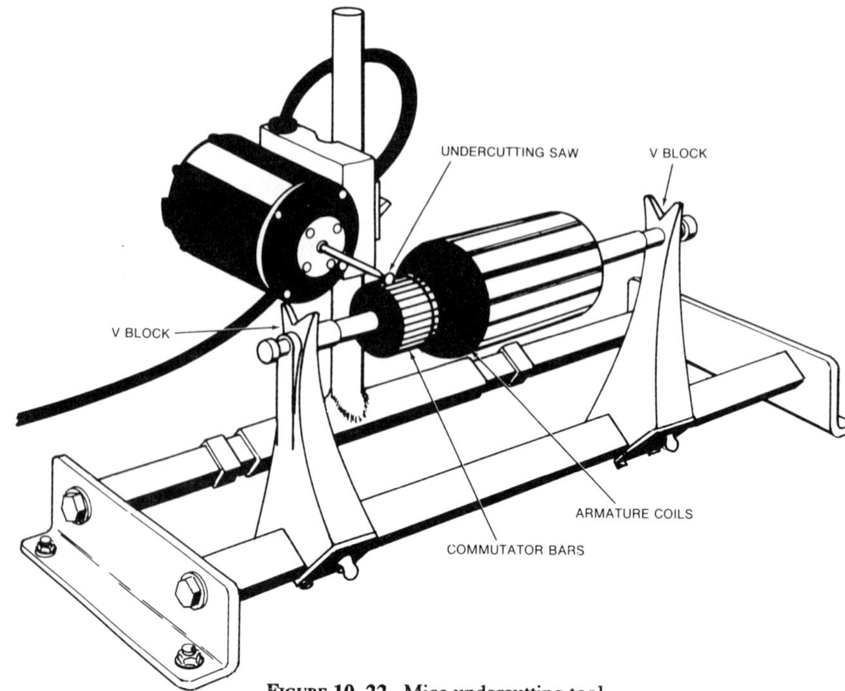

FIGURE 10–22. Mica undercutting tool.

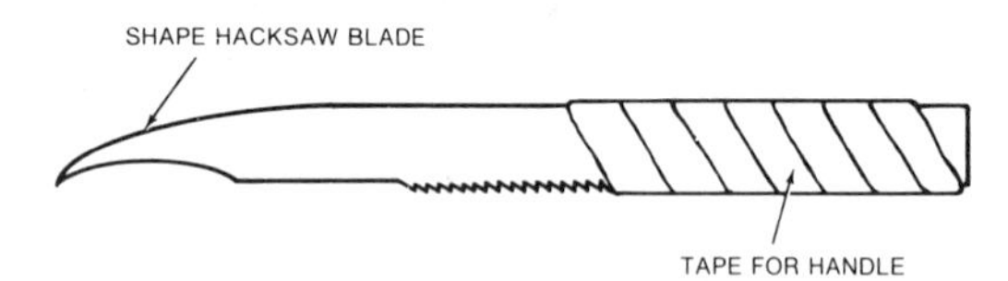

FIGURE 10–23. Slotting file made from hacksaw blade.

1. Carry the load current.
2. Act as lubricated bearings on the commutator or slip rings.
3. Withstand very high levels of short-circuit currents and peaks.

There are several types of brushes in use. They differ depending on the application. It is extremely important that the same type of brush is used as a replacement for a worn brush. Follow the manufacturer's recommendation.

Types

The types of brushes are

1. High-grade carbon. These brushes are used for their low cost, and they are restricted to low currents and speeds.
2. Electrographite. The carbon is processed at high temperatures in an electric graphitizing furnace. This results in a carbon with reduced hardness and increased electrical and thermal conductivity. The brush has a low friction, high resilience (hugs the commutator without bouncing), high current capability, and is not abrasive.
3. Natural graphite. These brushes are not as strong as the previously listed types. They can be identified by their silvery appearance and soft flaky structure. They are used because of their good lubricating qualities. These brushes are prone to selective action and shunt burning at high currents. Applications are limited. With proper allowances, however, they have a long life with a minimum of maintenance.
4. Copper graphite. Powered copper and graphite are pressed together and baked at a relatively low temperature to form these brushes. They are primarily used on slip rings and commutators where high current densities are present.

The average current carrying capacity of the first three brushes is in the range of 40 to 60 amperes per square inch. Metallic graphite brushes have capacities up to 150 amperes per square inch.

Causes of Brush Failure

A brush may fail due to normal wear. A good preventive maintenance program will overcome motor failure due to this cause. Replace brushes when they wear down to one-half of their newly installed length. It is good practice to change all the brushes at the same time. It is normal to have a fairly rapid rate wear on new brushes until they are firmly seated on the commutator.

The causes of premature failures are

1. Brush chatter. A rough or eccentric commutator, high mica, excessive vibration due to imbalance, and loose or worn bearings will create the conditions for brush chatter. Figure 10–24 shows a set of brushes that have been fractured due to brush chatter. The 4 brushes on the left show normal wear, while the remaining 12 show noticeable damage.
2. Spring tension. Either too much spring tension or too little can cause abnormal wear on the brush. Set tension according to the manufacturer's recommendations. The range is normally from 1¾ to 2½ psi for the contact pressure for light-metallized, carbon, or graphite brushes. Heavy metallized brushes require between 3 and 5 pounds per square inch. Figure 10–25 illustrates brush damage due to improper spring tension.
3. Arcing. Both improper brush pressure and chatter will cause excessive arcing. In addition, improper setting or position of the brushes on the commutator will also result in this condition.

The cure for brush chatter was described above in the section on maintenance of the commutator. Setting the spring tension and the proper positioning of the brushes follows.

Setting Spring Tension

Measure the spring tension on each brush when performing preventive maintenance or when replacing a brush. Adjust the pressure according to the manufacturer's specification. Failure to do this will result in unnecessary corrective maintenance and motor failure at inopportune times. Figure 10–26 illustrates the measurement of brush spring pressure with a spring balance.

FIGURE 10–24. Set of brushes which have been fractured due to brush chatter.

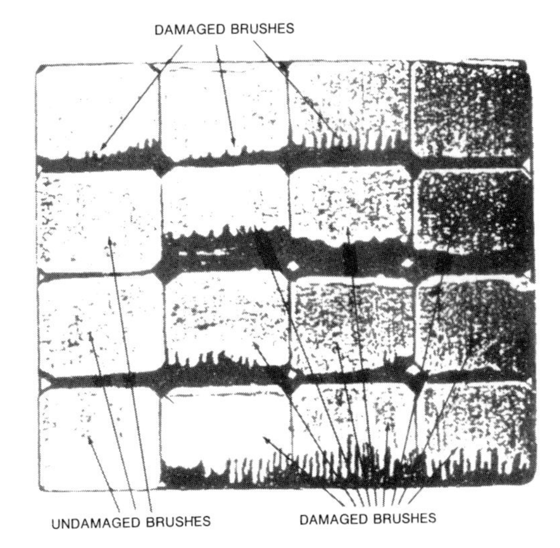

FIGURE 10–25. Brushes damaged by electrical arcing due to improper spring tension.

Uneven tension will cause selective action. One may assume that low tension will reduce brush and commutator wear. In reality, low tension often results in streaking, arcing, and grooving of the commutator. One may also assume that greater pressure will reduce chatter. If chatter is caused by a rough commutator, flat spots, or high mica, this condition will only be aggravated by the increased tension. It is highly important to maintain proper brush tension.

FIGURE 10–26. Measuring brush spring pressure with a spring balance.

Use the following procedure to adjust the brush pressure on the commutator (Figure 10–27).

1. Raise the brush and insert a piece of paper between the brush and the commutator.
2. Attach the spring balance to the brush-spring arm.
3. While pulling lightly on the paper, raise the spring balance perpendicular to the brush until the paper begins to slip free.
4. Read the pressure at this point.
5. Take two or three readings and average the values. Divide the reading by the contact area of the brush. The result will give the brush pressure in pounds per square inch.
6. Adjust the spring tension if the average reading is not within tolerance. Some brush holders have provisions for making the adjustment. On others, the brush spring is bent for the proper force.
7. Repeat the steps until the proper tension is set.

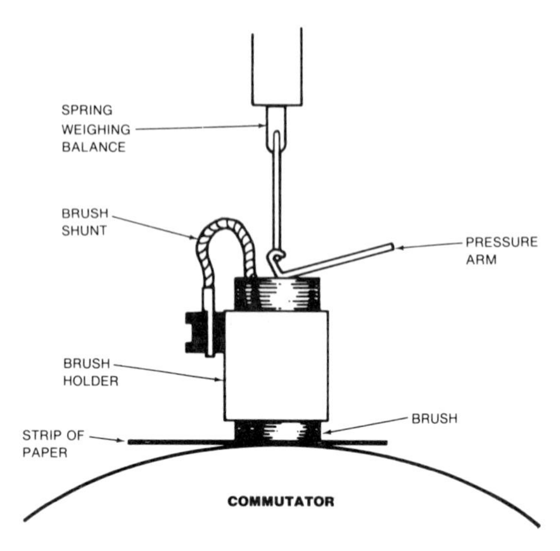

FIGURE 10–27. Adjusting the brush pressure with a strip of paper and a spring balance.

Angle of Brush Setting

The angle of the brush setting on a DC motor may be radial, leading, or lagging. Figure 10–28 illustrates these three conditions.

Radial settings are perpendicular or at a right angle to the commutator. When the motor can be reversed, this position is usually considered to be the best selection. Some reversible motors have one brush set at a leading angle with its counterpart set to lag.

The leading setting forms an angle of less than 90° ahead of the direction of rotation of the armature. This setting is best of the three because of less brush-in-holder friction and the ability to better follow the irregularities of the commutator. All the pressure is used to hold the brush against the commutator. Whereas, in the other two methods a greater side thrust is exerted by the brush against the holder.

For low commutator speeds, 5500 feet per minute or less, set the brushes to a lead angle of about 20°. For higher speeds, set the angle between 30° and 40°. Adjust the angle in each case for the best commutation, or the lowest amount of arcing. The formula for determining the surface speed of the commutator is

$$\text{Feet per Minute} = \frac{\text{Diameter (inches)} \times 3.14 \times \text{RPM}}{12}$$

As a general rule, the brushes have either been permanently fixed in position by the manufacturer, or the neutral point is marked. Mechanical neutral is

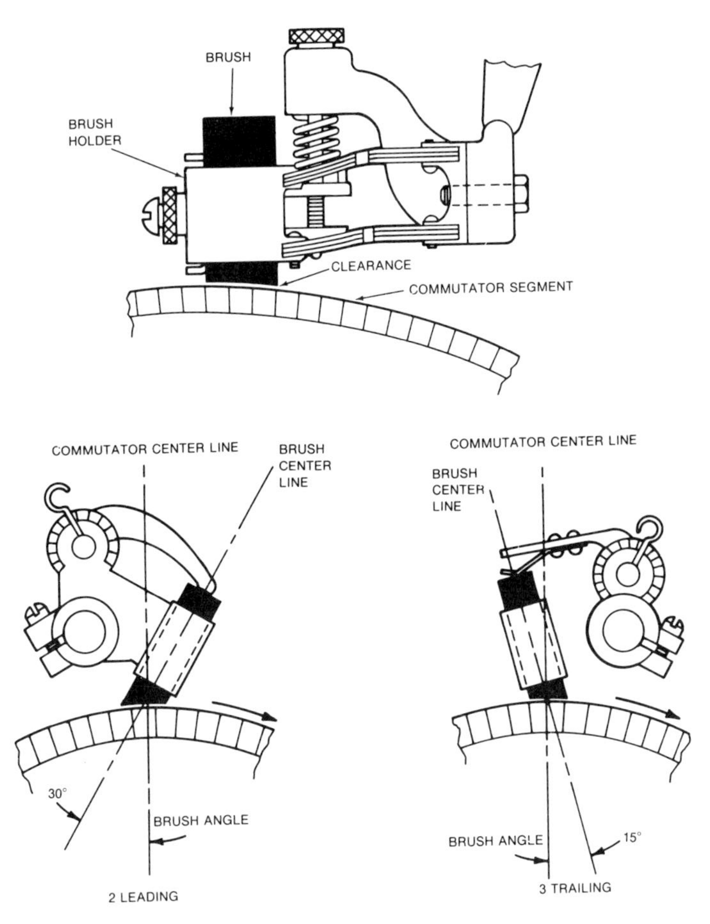

FIGURE 10–28. Angle of Brush Setting.

found at the midpoint between the center lines of the main poles. Often the brushes are slightly off this point in order to be at electrical neutral for best commutation. This is particularly true on machines that do not have commuting poles.

When provisions are available for adjusting the brushes, as in the case of a brush yoke on the motor, use the following procedures for setting the brushes (Figure 10–29).

1. Disconnect the power from the machine. Never attempt to shift the brushes when power is connected to the motor.
2. Loosen the brush yoke adjustment screw.
3. Move the yoke slightly in the direction of rotation for a leading angle and counter direction for lag angle. Make incremental adjustments using a screwdriver or a bar for leverage.

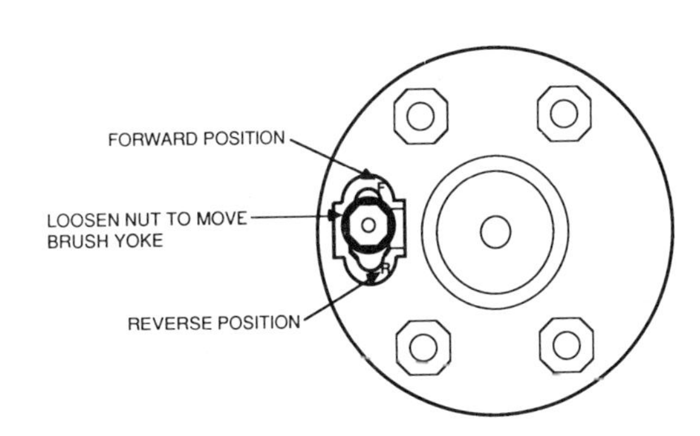

FIGURE 10–29. Brush yoke adjustment.

4. Tighten the brush yoke.
5. Apply power and full load to the machine.
6. Observe if commutation has been improved due to the adjustment.
7. If arcing is worse, remove power, and repeat Steps 2 to 6.

Brush Holders

Check the brush holders for cleanliness and freedom from obstructions. Clean them with a solvent if they appear dirty or have grease and oil in them. Remove any stubborn spots or burrs with No. 00 sandpaper. Insert the brush and check it for freedom of movement. Check the side play. If the brush does not move freely without wobble, replace it. Sand the sides of any brush that fits too tightly with a medium-coarse sandpaper. The fit must be close while giving the brush freedom of movement in the holder.

Set brush holders at a maximum of 1/8 inch above the commutator. Distances greater than this allow too much movement of the brush. If the brush holder is set too close, there is danger of the brush holder rubbing against the commutator and causing damage. If the manufacturer specifies a distance, set the holder accordingly.

Use the following procedure for setting the distance of the brush holder from the commutator (Figure 10–30).

1. Loosen the fastener that secures the brush holder.
2. Insert a fiber strip or gauge with the proper thickness between the brush holder and the commutator.
3. Move the brush holder down until it rests firmly on the gauge.
4. Tighten the fasteners.
5. Remove the gauge.

Fitting Brushes to the Commutator

Sand new brushes to the shape of the commutator to assist the brushes in seating themselves. Use a medium-coarse grade of sandpaper for this purpose. Follow the procedure below (Figure 10–31).

1. Use a strip of sandpaper that is wide enough to cover the brush completely.

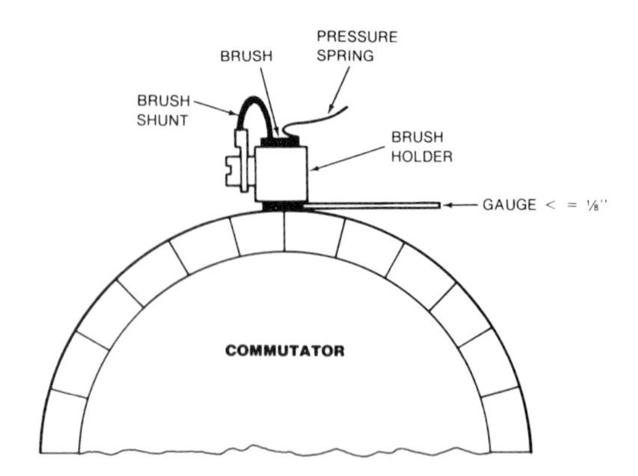

FIGURE 10–30. Spacing brush holder from commutator.

2. Place the sandpaper strip between the brush and the commutator with the sanding surface against the brush.
3. While holding down on the brush with one hand, pull the sandpaper through with the other hand.
4. Sand in the direction that tends to push the brush up into the holder for leading and lagging brushes. The direction of sanding would be reverse that shown in Figure 10–31 for a lagging brush. Radial-mounted brushes can be sanded in both directions.
5. Lift the brush and put the sandpaper back in place. Eliminate this step for radial-mounted brushes since the sandpaper can be worked in both directions.
6. Continue the process until a good fit is achieved between the brush and the commutator.
7. When the machine has been put in operation, apply a brush seating stone to the commutator to improve the fit.
8. After the brushes have been firmly seated, clean the commutator of any residue resulting from the sanding. If possible, when using compressed air, direct the air from the back of the commutator and away from the motor.

Staggering Brushes

Current flow is electron flow. Polarity of current is from negative to positive for the external circuit. Wear on the commutator differs for current flow from brush to commutator as opposed to current flow from com-

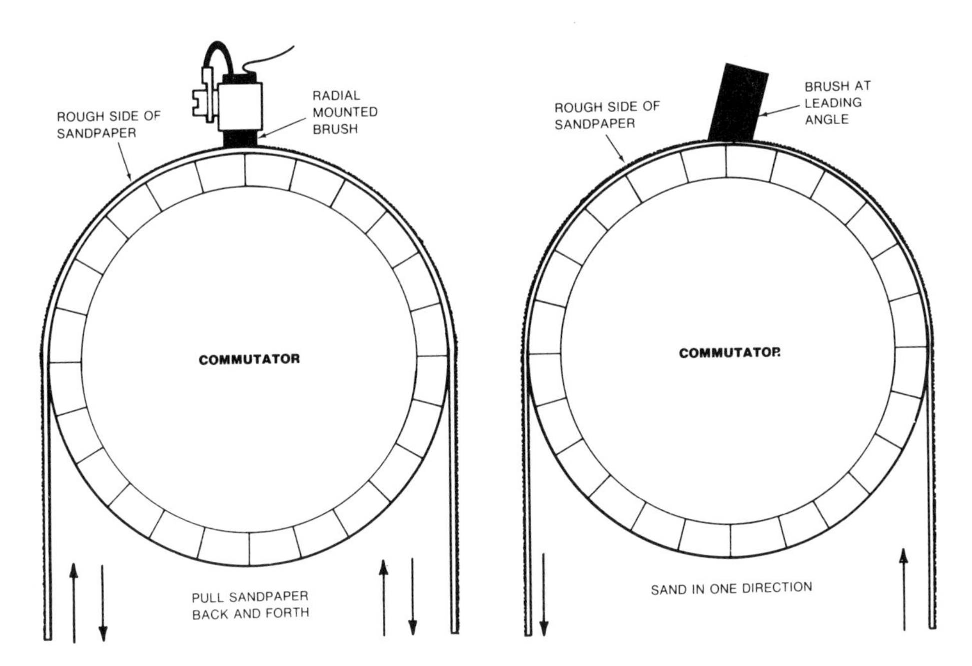

FIGURE 10–31. Shaping brush to commutator surface.

mutator to brush for a DC machine. Brushes are staggered on the commutator to provide for even wear under the brushes.

Figure 10–32A illustrates the correct method for staggering the brushes. Brushes of opposite polarity track each other. This procedure helps prevent grooving of the commutator.

The method of staggering brushes in Figure 10–32B is incorrect. In this case, brushes with the same polarity follow each other. This will aggravate the pitting and wear on the commutator.

It is not possible to stagger all the brushes in accordance with Figure 10–32A when the machine has a number of poles equal to two times an odd number. Under these conditions, stagger all but one set of positive and negative brushes according to the correct method. This set can then be aligned with any two adjacent rows.

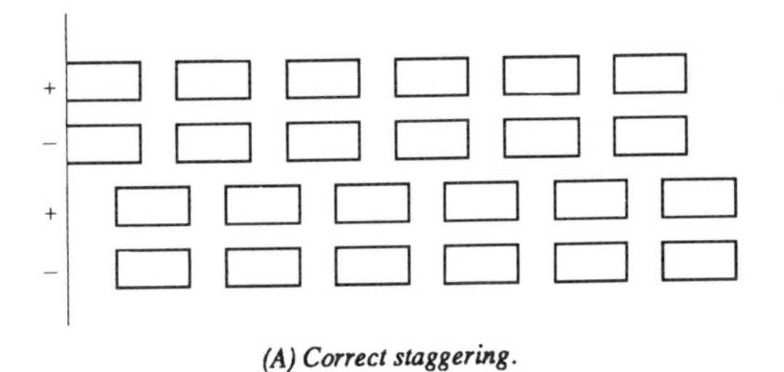

(A) Correct staggering.

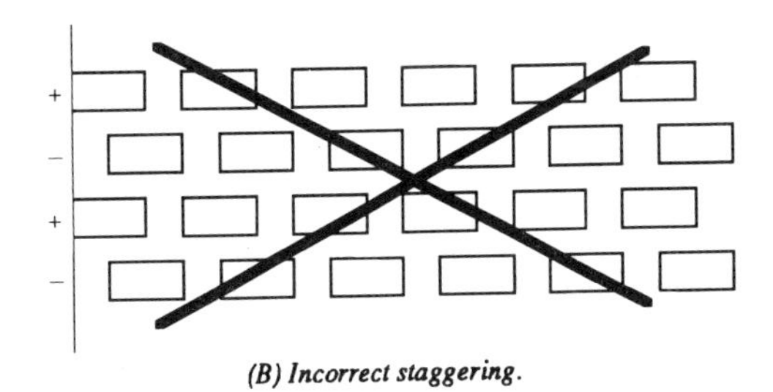

(B) Incorrect staggering.

FIGURE 10–32. Commutator brush positions.

Spacing Brushes around the Commutator

Brushes must be evenly spaced around the commutator in order for the machine to operate effectively. When spacing is uneven, excessive currents are often present which will cause overheating of the armature and destruction of the brushes.

Any time brushes are replaced, or the studs and holders are adjusted, check the brushes for proper spacing. Small changes in the brush rigging can be accumulative which result in serious error in the brush spacing.

To check for proper spacing, wrap a piece of paper

around the circumference of the armature. Carefully mark the point where the paper overlaps and where the toe of each brush group contacts the paper (Figure 10–33).

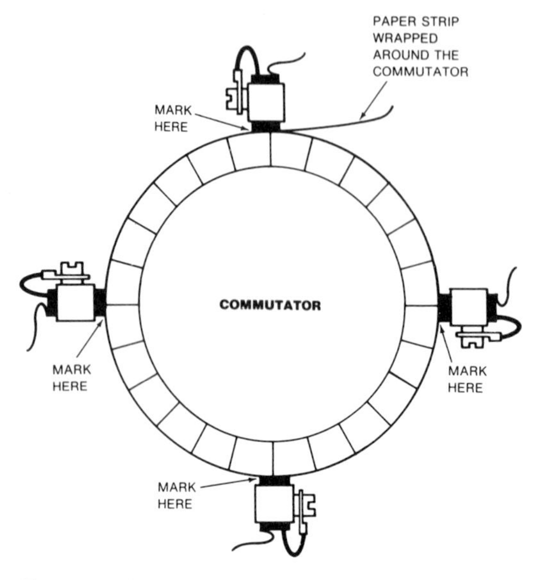

FIGURE 10–33. Spacing brushes around the commutator.

Remove the paper. Cut the paper at the lap. Measure the length of the paper. This will be the circumference of the commutator. Divide the circumference by the number of brush groups. The result is the equally spaced distance which should be between the toe marks.

Measure the distances between the toe marks. If the distances correspond to the results of the calculations, no further action is necessary.

If the distances between toe marks are not equal, perform the following procedure.

1. Divide the distance of the circumference into as many equally spaced segments as there are brush groups. This can be done either by using a ruler and pencil or by folding.
2. Draw a line squarely across the width of the paper at each mark or crease.
3. Replace the piece of paper under the brushes.
4. Align the paper so that the toe of the first brush group is in line with one of the lines.
5. Check for the proper alignment of the other brush groups. If the toes of the other brush groups are not on their corresponding mark, adjust the brush holders.

Electrical Tests of Windings

Before assuming a motor is defective and consuming considerable time performing electrical tests, take the time to observe the conditions of the circuit. If the motor is still running, measure the line current using a clamp-on ammeter. The use of this instrument eliminates the need to disconnect the leads in order to connect an ammeter in series with the line. A typical AC split-core multimeter with clamp-on ammeter capabilities is shown in Figure 10–34.

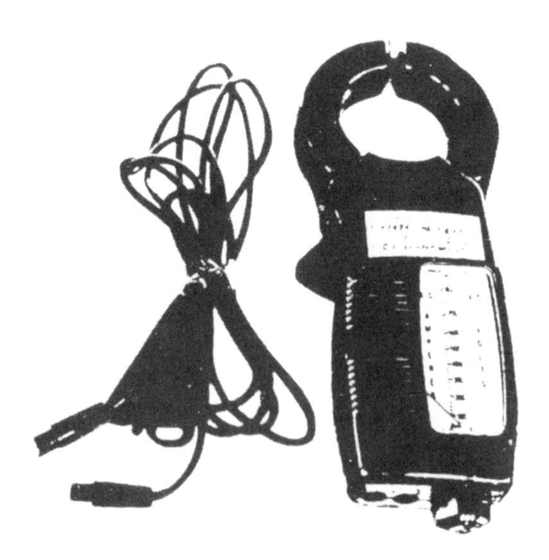

FIGURE 10–34. Clamp-on ammeter.

Compare the ammeter readings with the full-load current rating on the nameplate of the motor and the values recorded when the motor was put into service. Current should be within 10% of the rating with the motor under full load. In three-phase systems the current should also be balanced. Unbalanced current may indicate a problem with one of the motor windings.

If these values vary significantly, measure the voltage(s). They should be within 10% of their ratings. Three-phase voltages should be about equal in value. The clamp-on ammeter in Figure 10–34 is also an AC voltmeter and can be used to make these measurements.

When the current is high and the voltage low, the motor may be the cause. Disconnect the motor from the line and measure line voltage. If the voltage supplied to the motor is either too high or too low, this condition will need to be remedied before proceeding with electrical tests on the motor.

If the voltage rises to its rated value with the motor disconnected, the chances are the motor will be at fault. Check first for an increase in the mechanical load before assuming an electrical problem. Increased load may result from a mechanical fault with the motor such as a bad bearing or the depletion of the lubricant, as well as an increase in the driven load. In any of these cases, the motor will draw too much current with a resulting decrease in line voltage.

Electrical faults in a motor are mostly due to failure of the insulated windings. Windings will fail due to the motor operating at a temperature above its rating. This condition can be caused by an overload or poor ventilation. Exposure to moisture, corrosive atmospheres, dust, or metallic filings will provide the circumstances for premature failure. In addition, an electrical spike on the power line may cause damage to the windings.

Before beginning time-consuming electrical tests, use your senses of smell, vision, and touch to locate the problem. When windings burn out, they often omit the odor of burned varnish. Smoke may be present or the windings may be charred and blackened. The area around the fault can have a much higher temperature than the surrounding metal. Some failures are not immediately detectable in this manner, and electrical testing may be necessary.

Run the motor unloaded first. Carefully observe its performance. Look for unusual noise, vibration, changes in speed, and overheating.

Failure of the insulated windings will cause one or a combination of three types of faults. These are ground fault, short between windings, or an open winding. A ground fault or shorted windings are usually the cause for a winding to open.

Several test methods may be used to determine electrical faults with the rotor and windings of a motor. Which of these you use is dependent on the availability of the test instruments and your ability to correctly make application. The following methods are presented for your consideration.

1. Test lamp method
2. Ohmmeter method
3. Megger method
4. Growler method
5. Millivoltmeter method
6. Voltmeter method
7. Impedance method

Examples of how to make these various tests on a machine are explained in the following sections of this text. No attempt has been made to cover all possible tests for every conceivable defect. At least one example is given for each method. Two or more examples are provided using some of the methods.

Test Lamp Method for Finding Ground Faults

It is simple to make a test lamp for testing a motor. Figure 10–35 depicts such a circuit. Although you can use the bare wires for making measurements, line voltage may be present between the terminals. For safety and convenience sake, use an insulated alligator clip on one wire and a probe on the other end. Use a small wattage lamp of 10 to 25 watts. This will limit the current flow. Save the test lamp for use on future problems.

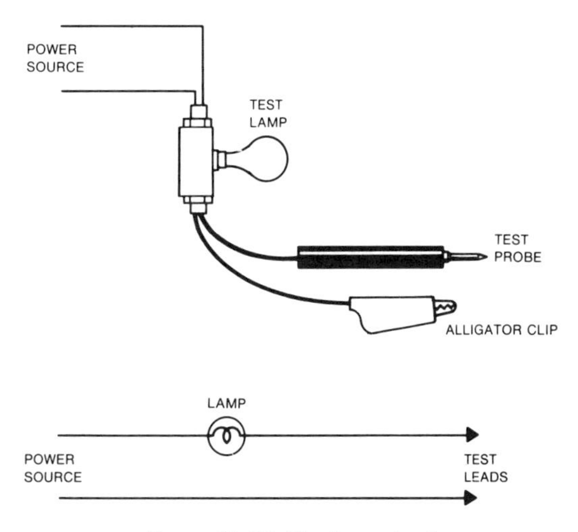

FIGURE 10–35. Test lamp circuit.

Connect the test lamp as shown in Figure 10–36 for finding shorts to ground on an armature of a motor. Use the following procedure in making this test.

1. Connect the alligator clip to the shaft of the armature.
2. Apply power to the system.

3. Touch the probe to the shaft to determine if the lamp will light.
4. Touch the probe to each successive bar of the commutator.

The lamp should not light if the bar is insulated from ground. It will light if the bar has a short path to ground.

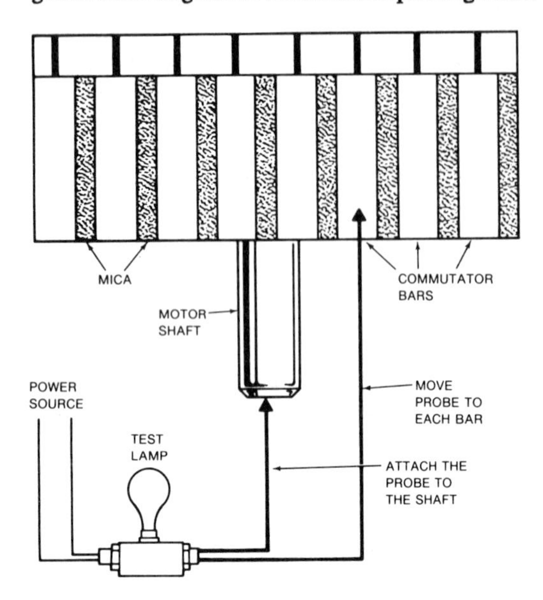

FIGURE 10–36. Testing armature for grounds using a test lamp.

This test can be used to check wound rotors as well as armatures. The only difference is that the probe will be placed on the slip rings rather than the bars of the commutator.

Stator windings can be checked for grounds in a similar manner (Figure 10–37). The alligator clip is attached to the stator's frame. The probe is connected to each of the windings. If the lamp burns with full intensity, there is a direct short between the winding and ground. If the lamp burns with less than full intensity, there is a high resistance ground. The winding may be shorted to the frame in the middle of the winding to provide enough opposition to current to reduce the current flow through the lamp.

It is not necessary to disconnect the star point to determine if the stator windings are shorted. This would be done only if you were interested in locating which of the windings was shorted. Under short to ground conditions, in most cases, the machine would be sent to a motor shop for repair, or you would replace

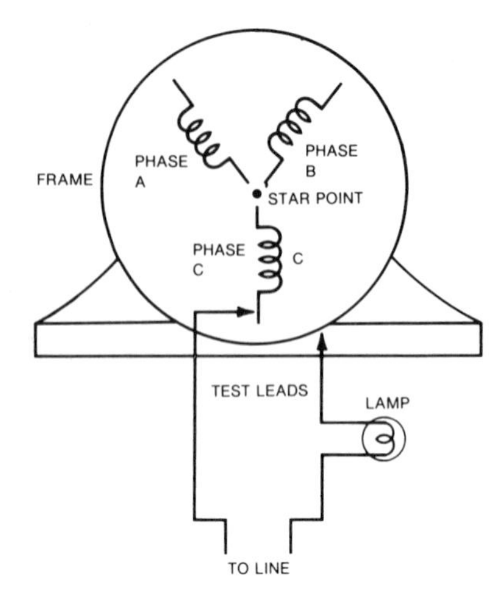

FIGURE 10–37. Testing stator windings for grounds using a test lamp.

the motor. Therefore, there is no need to locate the defective winding.

Ohmmeter and Megohmmeter Tests for Ground Faults

To make ground fault checks using the ohmmeter or a megohmmeter, connect the meters in the same manner as shown for the test lamp setup shown in Figures 10–36 and 10–37. The lamp circuit is replaced by either the ohmmeter or the megohmmeter. One lead of the meter is attached to the shaft of the armature, and the other lead is moved from one commutator bar to the next in an orderly progress.

Neither of these meters requires an external source of power. The ohmmeter has an internal battery for this purpose, and the megohmmeter has an internal generator.

A deflection by the ohmmeter on any bar indicates a short path. Unless the short is visually observable and easy to remedy, the armature will be sent to the motor shop for repair or be replaced.

Because the ohmmeter uses a low-voltage source, the lack of deflection does not absolutely rule out a grounding path. If a higher voltage is applied, a high-resistance ground may be detected. For this reason, it is better to use the megohmmeter for this test.

The megohmmeter applies a voltage of about 500 to 1000 volts between the commutator bars and the shaft. These higher voltages may flash a short to ground which the low voltage of the ohmmeter was not capable of detecting. A reading of zero ohms indicates a short, and you would have probably found the same condition with the ohmmeter. On the other hand, the megger may indicate a high-resistance short between 0 and 500,000 ohms. The armature will need corrective maintenance or replacement if this condition is found to exist.

Growler Method for Finding Ground Faults

A *growler* is nothing more than an inductor whose field is allowed to pass through an armature or the stator fields. It gets its name from the fact that when used to test for short circuits a vibrating noise is heard.

There are two types of growlers. These are the external growler which is used to test armatures, and the internal growler which is used to test the stator windings of the motor. Figure 10–38 shows the external growler, and Figure 10–39 depicts the internal growler.

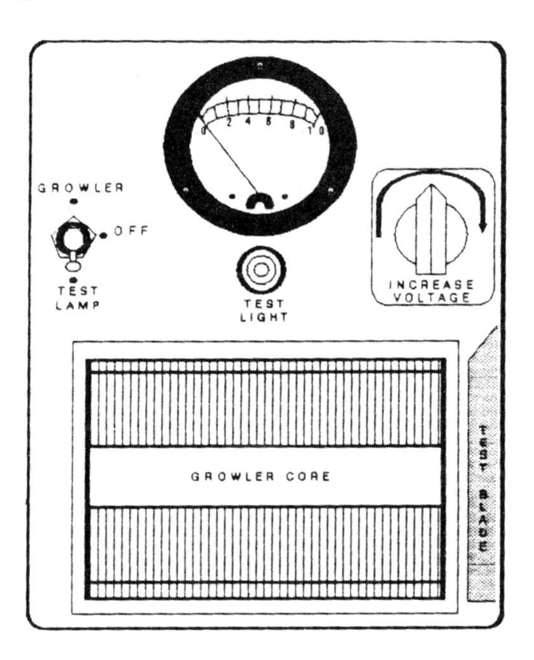

FIGURE 10–38. Typical external growler.

FIGURE 10–39. Typical internal growler.

The external growler can be operated with the test light as well as a growler. Its operation under these conditions is the same as for the test lamp circuit which was previously discussed.

The amount of test voltage is variable to provide more positive readings. A test blade is furnished for some of the tests. The instrument also has test leads attached to it which are not shown.

Figure 10–40 shows a more detailed drawing of the growler core. The core has the coils wrapped around it. It is laminated to reduce eddy currents. It can be removed from the instrument to make the testing of some armatures easier. The distance between the two jaws is adjustable to accommodate different sizes of armatures.

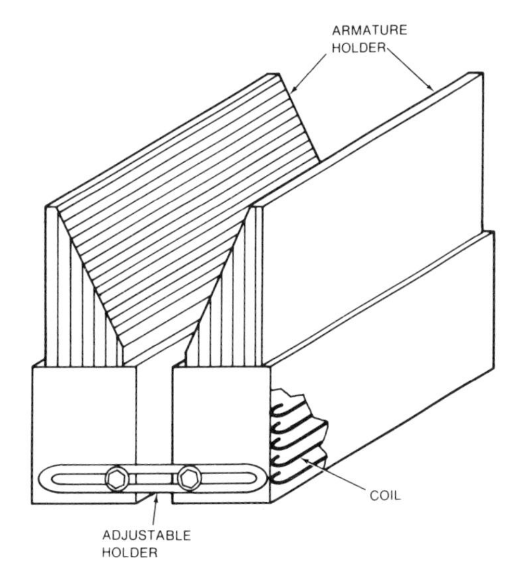

FIGURE 10–40. Growler core.

To test an armature for ground faults, perform the following sequence.

1. Place the armature in the jaws of the growler core.
2. Adjust the jaws for a proper fit.
3. Connect the instrument to an AC source of power.
4. Connect one test lead to the shaft of the motor.
5. Touch each commutator bar individually with the other test lead.

If there is a spark when making and breaking contact with the commutator bar, then a ground fault is present. Lack of a spark indicates a no-fault condition. Figure 10–41 depicts this test procedure.

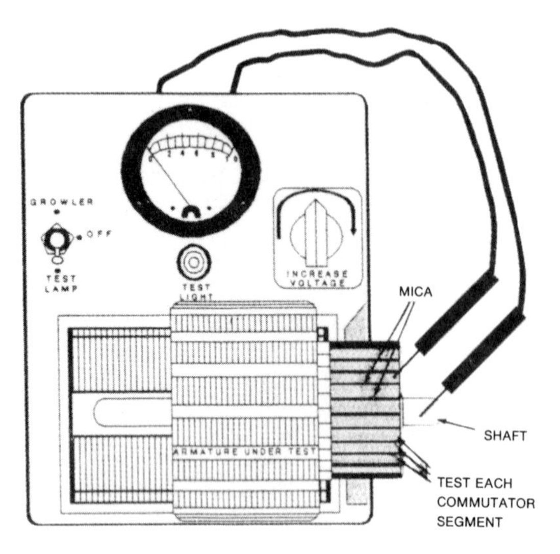

FIGURE 10–41. Checking armature for ground fault using a growler.

Millivoltmeter Method of Finding Ground Faults

Figure 10–42 depicts the test setup for using a millivoltmeter to check for ground faults in an armature. To make this test, connect one side of a 1.5-volt dry cell to one of the commutator bars, and the other side of the battery to a rheostat. Connect the other side of the rheostat to a second commutator bar several segments away from the first connection. A strong rubber band can be used to hold the wires against the selected commutator bars. Set the rheostat to its highest resistance.

Use the following procedure in making the test.

1. Connect the millivoltmeter across the output of the battery circuit. Observe polarity. You may want to use a voltmeter to measure the output of the circuit before connecting the millivoltmeter to assure yourself the voltage is lower than the full-scale rating of the millivoltmeter.
2. Adjust the rheostat so that the meter deflects about three-quarter scale.
3. Disconnect the meter from the output of the battery.
4. Connect one lead of the meter to the shaft of the motor to obtain a good ground connection.
5. Touch the other lead of the meter to each segment of the commutator between the battery connections until all the bars have been tested. A zero deflection of the meter indicates that no ground fault exists.
6. Move the battery connection to the next group of commutator bars. Repeat steps 5 and 6 until all segments have been tested.

Voltmeter Method for Finding Ground Faults

A voltmeter can be used to find ground faults. Figure 10–43 depicts the circuit. Make all connections for this test with power removed from the circuit.

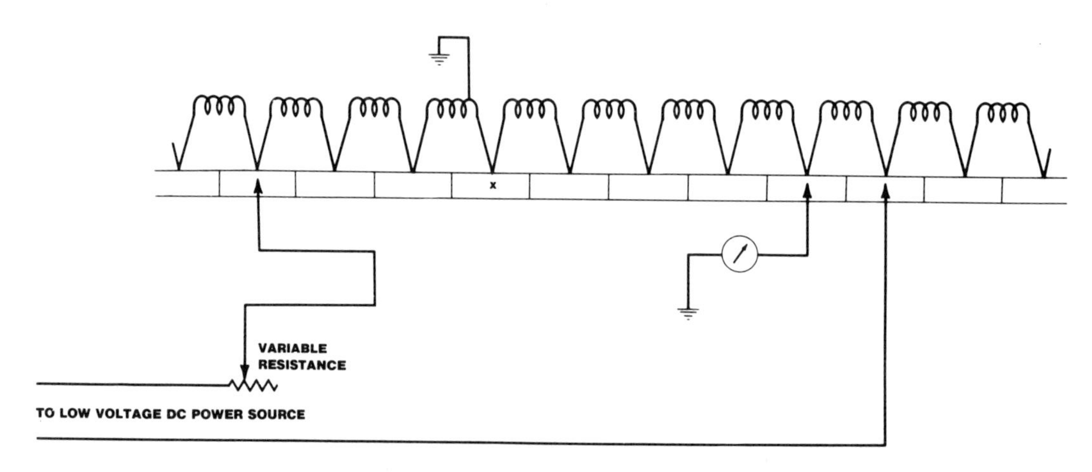

FIGURE 10–42. Checking armature for ground faults using the millivoltmeter method.

Connect one lead of the voltmeter through a fuse to the ungrounded side of the power line. Be sure the meter's voltage range is set to an AC range greater than the line voltage. Connect the other lead of the meter to one lead of the stator's windings. The grounded lead of the power line is attached to the metal frame of the motor. This arrangement will place the high resistance of the meter in series with the stator windings in the event a short circuit to ground is present. Apply power to the circuit.

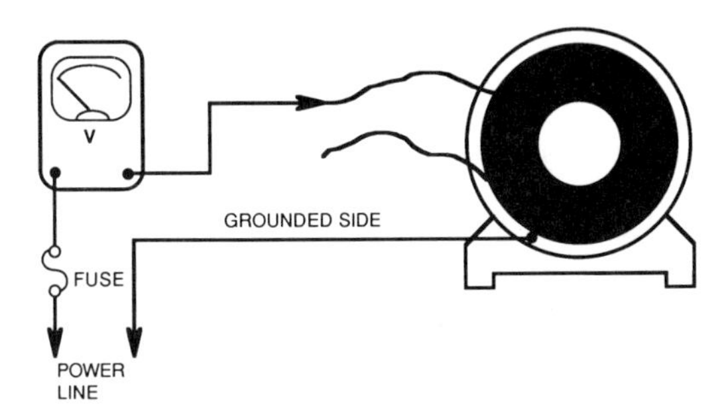

FIGURE 10–43. Voltmeter method for finding ground faults.

If the meter reads zero volts, the windings are not grounded. A reading of full, or nearly full, line voltage indicates that the windings are grounded. A partial reading of the source voltage, no matter how small, indicates a high resistance ground.

To locate the exact shorted winding on an AC machine, it is necessary to disconnect all the windings and test them individually. On a DC machine, the field is checked first. Some of the circuit coils on the motor may be intentionally grounded. These coils must be disconnected from ground to make this test valid.

All the previous methods described for finding ground faults in armatures, except the millivoltmeter test, are applicable to the stator windings. The voltmeter method is only recommended if you do not have an ohmmeter or megger.

Testing for Open, High-Resistance, or Shorted Windings

The simplest method for testing for open or high resistance windings is to use the ohmmeter. Readings for each winding can be compared to the others to detect differences in the readings. When there is no deflection by the meter, the winding is open. Figure 10–44 illustrates this method.

Because the shunt, series, and armature windings will all have different values of resistance, they cannot

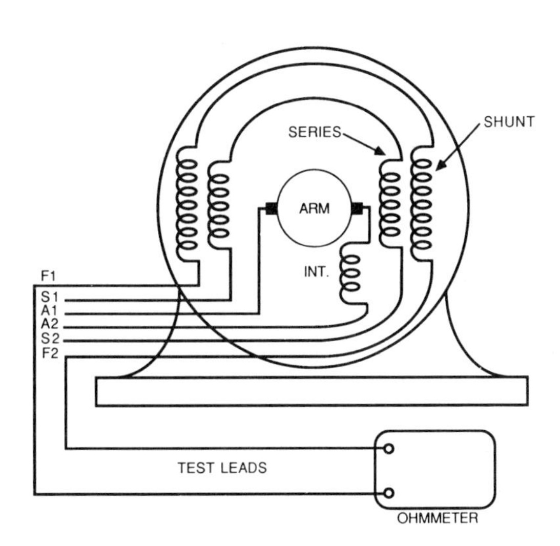

FIGURE 10–44. Locating open faults in the windings of a DC compound motor using an ohmmeter.

be compared for high-resistance faults. The shunt winding will have the highest resistance of the three circuits, and the armature the lowest. If the connection between the windings in the series and shunt fields can be open, then the individual windings of each field can be compared to the other respectively.

If the armature winding is showing a fairly high reading, check the conditions of the commutator and the brushes. Correct any problems before continuing the measurements.

Check the resistance of individual armature windings by rotating the commutator segment by segment and taking an ohmmeter reading for each segment. Each reading should be equal. If you experience difficulty using this method, remove the brushes and make readings between each adjacent commutator bars.

A reading between zero ohms and the value of the other windings would indicate a partial shorted winding. A high reading indicates a high resistance problem. This problem normally occurs at the solder joint on the commutator and can be repaired by soldering the joint.

Other test instruments are also appropriate for determining an open fault. The test lamp will not light if a winding is open. It will burn brighter in the event of a shorted winding.

The megger will indicate a very high resistance if the winding is open. It will deflect to zero ohms with the least turn of the hand crank generator if the circuit has continuity. Meggers are of little use in determining short-circuit conditions because of the low resistance of the windings. They are excellent, however, in determining if there is a short between the windings.

The millivoltmeter method for testing the armature can also be used. The ground lead of the meter as shown in Figure 10–42 would now be attached to the adjoining commutator bar to take the reading across a single winding. If the readings are too far downscale, decrease the amount of resistance in the test circuit, or reduce the number of segments that are under test at any given time.

If the armature winding is open, the meter would read the full three-quarter scale deflection that was set at the beginning of the test. The full voltage will be across the open circuit. After taking the reading on one winding, the leads of the meter would be placed across the next winding. Readings across each winding will be equal if the windings are good. A zero reading would indicate a shorted winding.

Figure 10–45 illustrates the use of the millivoltmeter to locate open faults in the stator windings of an AC machine. Use the following procedure for making the measurements.

1. Using a 1.5-volt battery, a rheostat, and a millivoltmeter, wire the circuit.
2. With maximum resistance, short the leads of the test circuit and adjust the millivoltmeter so that it reads about three-quarter scale deflection.
3. Check similar windings in the circuit and compare the values.

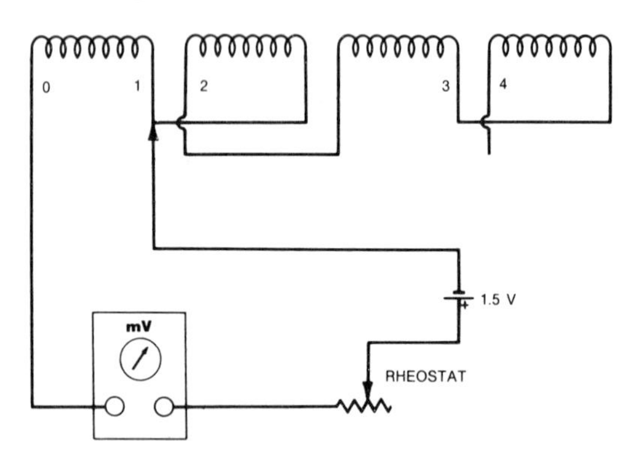

FIGURE 10–45. Locating faults in stator windings using a millivoltmeter.

No reading of the millivoltmeter indicates an open winding. A low reading indicates a high-resistance condition. A high reading indicates shorted turns in the windings.

An external growler can also be used to check for open and shorted windings on the armature. This test is the exact same as for using the combination of the test lamp and the ohmmeter when the test leads are used. The test leads are connected to each set of adjacent bars. The lamp will light or the meter will deflect if there is continuity of the winding circuit. The lamp will not light or the meter will not deflect if the winding is open. The lamp will not tell you if there are shorted turns, but the meter will. Compare the readings for each winding. The values should be the same. A high reading in respect to the other readings indicates a partial open winding. A low reading indicates a partial short.

The external growler can also be used to determine if there is a short between the windings of an armature. In this case the armature is placed in the jaws of the growler. The growler is energized. The test blade or a hacksaw blade is placed on the top slot of the armature. Move the blade to the slot on the left, and then return it to the top slot. Then move it to the slot on the right. Turn the armature to the right so that the next three slots to the left are on the top. Repeat the blade test for these slots. Continue the procedure until all the slots have been tested. If there is a short circuit between the coils, the blade will vibrate noticeably (Figure 10–46).

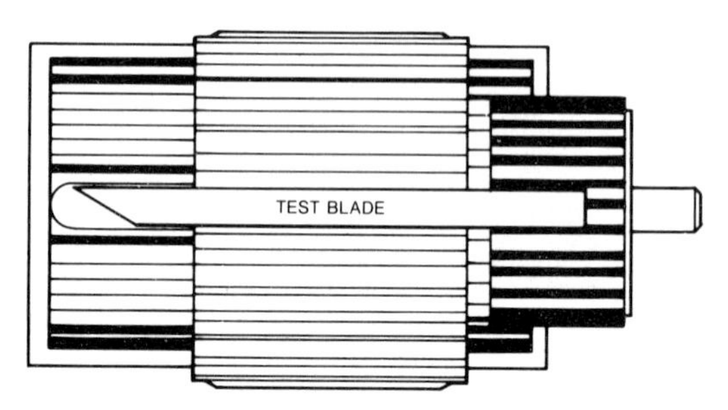

FIGURE 10–46. Using external growler to determine shorts between windings on an armature.

An armature that has cross-connections or equalizers cannot be tested using the test or hacksaw blade. The blade will vibrate at every slot giving a false reading that the coils are shorted. Other types of wired coils may give indications of shorts over more than one slot. When you have doubts about the results of your test when using a growler, use a meter test that will give positive results.

The internal growler method is practical only for AC stators. Figure 10–47 illustrates this method.

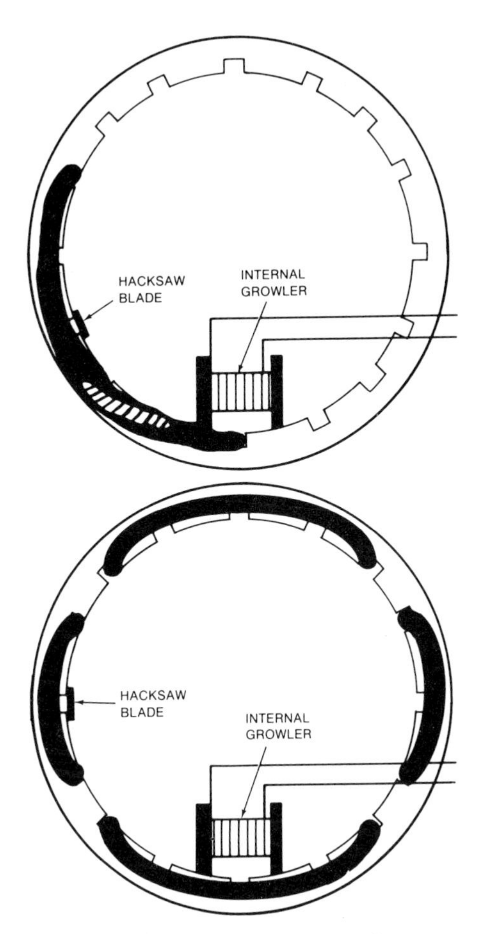

FIGURE 10–47. Testing for shorts and opens using an internal growler.

With the windings open, place the growler over a slot. Move it to each slot until each has been checked. A growler with a built-in feeler-blade will give off a noticeable buzz if any of the windings is shorted. If the growler has a meter, it will deflect on the shorts.

To check for open circuits, short-circuit each individual coil onto itself. Under this condition, the growler must buzz, or the meter deflect, at each of the slots. The absence of a buzz or meter deflection indicates an open coil.

If a motor has parallel windings, the windings must be separated from each other. Growlers are not effective when the windings are connected in parallel with each other.

A good way to test for shorts in an armature of AC motors that uses brushes to complete the coil circuits is to remove the brushes. Apply power to the stator. Because the armature is open, it will not turn. Turn the armature slowly by hand. It should turn freely. If there is humming or the armature resists being turned, a short exists in the armature coils.

Impedance Test for Wound Rotors

The following procedure is good for testing wound rotor motors up to one horsepower. Perform the steps listed below for this test.

1. Clamp the rotor so that it cannot turn.
2. Lift the brushes from the slip rings.
3. Short-circuit the slip rings.
4. Wire the test circuit as shown in Figure 10–48.
5. Connect the test leads to each phase in turn. Limit the current to the FLC on the nameplate of the motor.
6. Slowly turn the rotor a complete 360 degrees opposite the normal direction of rotation for each phase.
7. Record maximum and minimum values of current for each phase.

Phase voltages and currents should be close in value to each other. Lack of balance indicates a short circuit in the rotor. If the rotor tends to lock in position at any point, an open winding is indicated.

This test setup can also be used to check the condition of the stator windings.

1. Remove the shorting connector from the slip rings.
2. Connect the test leads to each phase in order.
3. Record the voltage and current for each phase. Limit the current to the FLC value.
4. Determine the impedance of each phase by dividing the voltage by the current (E/I).
5. Compare the values.

A difference in impedance indicates a problem with one of the windings. Open circuits and high-resistance joints will result in a high impedance calculation. Shorted coils or turns will provide a lower than normal impedance.

Once the rheostat is set, the voltmeter will indicate about the same value for each test. If this value varies

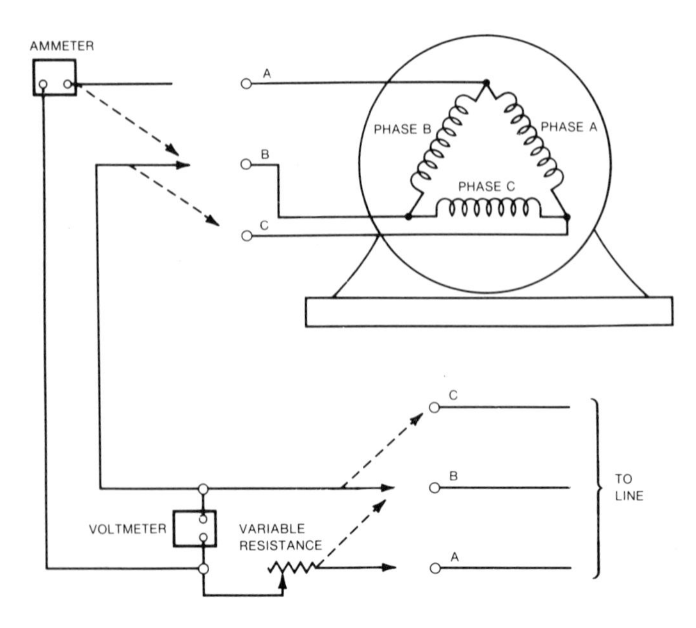

Figure 10–48. Stationary impedance test for locating shorted or open coil on a wound rotor.

more than 10%, disconnect the test setup and measure the voltage on each phase. This should have been done before starting tests on any motor. The problem can be the source voltage rather than the motor.

Tests for Motor Capacitors

Outside of keeping capacitors clean, they require little or not preventive maintenance. Do not allow dust, dirt, grease, oil, or any metallic particles to collect between the terminals. This can result in insulation breakdown between the terminals and cause arcs. Keep the cases clean so that the heat generated by the capacitors can transfer to the surrounding air. Most motor capacitors have about 60,000 hours of life for continuous operation at the rated voltage and temperatures not in excess of 70°C.

Capacitors do need to be observed and checked from time to time as part of the routine maintenance program. Remember that a capacitor can retain its charge even though power has been removed from the circuit. Always discharge capacitors with a grounding rod before working on them.

Note the motor's operation. If the motor is coming up to speed, developing normal torque, and is operating at speed, the capacitor is probably all right. If not, a further check of the capacitor's condition is indicated.

Observe the capacitor for swollen case or leakage of its electrolyte. If either of these problems exists, replace the capacitor.

Check for a shorted capacitor using an ohmmeter. Be sure the capacitor is discharged before connecting the meter. A capacitor can store enough energy to destroy the meter.

Set the ohmmeter on its highest multiple. Connect the leads across the capacitor. On a normal capacitor, the meter will deflect upscale and quickly fall back to a very large ohmmic value. If the capacitor reads zero ohms or a very low value of resistance, it is bad. Replace it. Full-scale reading on the standard ohmmeter represents 0 ohms (Figure 10–49).

If the capacitor fails to deflect upscale when the ohmmeter is set to a high multiplier, the capacitor is probably open. Replace it. With very small capacitors [picofarads (pF)], you may not get a deflection. This is normal. All capacitors used with motors, however, are much larger. If you repeat the test because of not watching the meter closely, be sure to discharge the capacitor. It will charge to the potential of the battery voltage of the meter.

Neither of these tests is absolute because of the low voltage utilized by the ohmmeter. The short test may indicate the capacitor is good, but when the AC line voltage is applied, it has a large current leakage. In

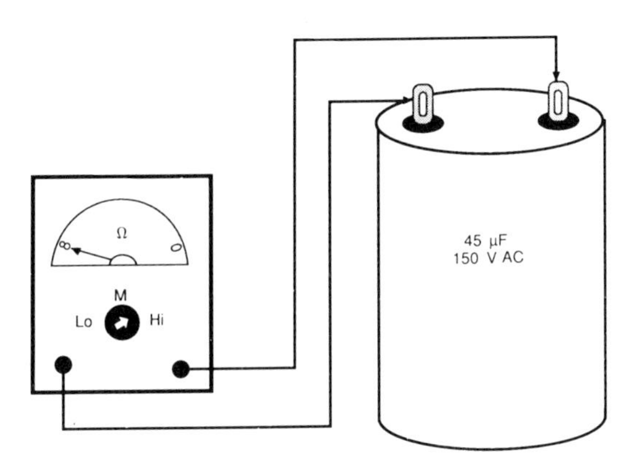

FIGURE 10–49. Testing a capacitor for shorts and opens using an ohmmeter.

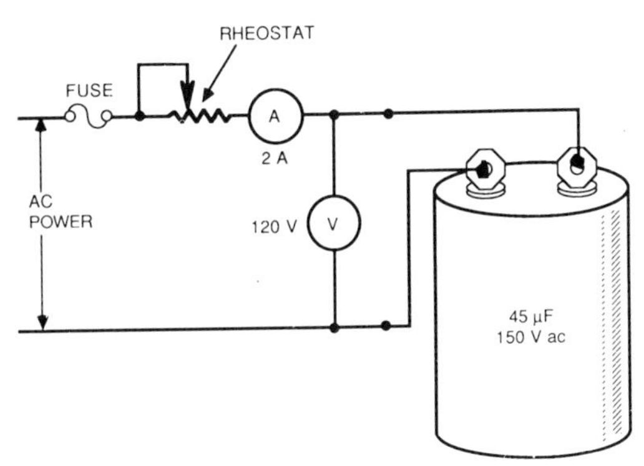

FIGURE 10–50. Test circuit for capacitors.

addition, the ohmmeter test will not tell you if the capacitor has changed in value.

There are commercial capacitor testers on the market. These testers allow for rated voltage testing of the capacitor while measuring its current leakage. In addition, these instruments have a capacitor bridge circuit which allows for the determination of the farad value of the capacitor. When this type of device is available, learn to use it. In the majority of the cases, you will not have a capacitor checker, so another method is needed.

Set up the circuit as shown in Figure 10–50. It is a good idea to fuse the circuit in the event maximum resistance is not in the circuit when it is energized, and the capacitor is in a shorted state.

Disconnect the capacitor from the motor circuit while making this test. Most manufacturers of motors use brown insulated conductors to connect the capacitor to the circuit. One of the brown leads may have a tracer color running its length. Set the rheostat so the maximum resistance is in the circuit before applying power.

If the current flow through the capacitor and the voltage across it are known, the value of capacitance in microfarads can be calculated using the formula

$$C = IK/V$$

K is a constant equal to

$$K = 1/(2\pi F \times 10^{-6}) = \frac{1,000,000}{6.28 \times 60}$$

For 60 hertz, K equals 2650. This constant is derived from the capacitive reactance formula. The value of K will change with a change in frequency.

Assuming 120 V AC across the capacitor and a current of 2 amperes as shown in Figure 10–50, the value of the capacitor would equal

$$C = (2\text{ A} \times 2650)/120\text{ V}$$
$$= 44.16\ \mu\text{F}$$

Most motor capacitors have a tolerance of 20%. If the experimental farad value of the capacitor is not within 20% of its rated value, replace the capacitor. The acceptable range of the capacitor in this example is plus or minus 9 microfarads, or from 36 to 54 microfarads.

Troubleshooting Techniques

During my working career, I have known electricians and technicians who seldom touch a piece of test equipment when troubleshooting a defective piece of equipment. Without exception, each of these people made exceptional application of their senses and was able to translate any given set of symptoms into the component part which was defective. Most of these electricians and technicians worked on a limited amount and type of equipment. Over a period of time, experience taught them well about the equipment for which they were responsible.

Placed in the same circumstances, you too will be able to correct problems in the same manner. This does not relieve you of the responsibility of knowing how to use test instruments and being able to interpret the information the instruments provide. The proper use of test instruments is necessary because electricity cannot be seen. Its effects can, however, and the electrical parameters can be measured.

Besides the common test equipment described in this text, other types are available with which you may need to become familiar. There are special motor test sets which are designed to perform electrical tests on a wide range of motors or on a particular motor. Motor control centers may have special instrumentation for the motors under their control. Brush recorders and oscilloscopes are necessary at times to record or observe transients on the power line which may cause some difficulties to motor operation. Solid state control devices are often susceptible to power surges and transient voltages. Harmonics of the power frequency may cause difficulties with some motor operations. These can only be detected using a device such as a scope or recorder.

All of the exceptional troubleshooters I knew also used test instruments at times. This was particularly true in determining the cause of a given failure of a single component that recurred at some regular interval, or when a new or different piece of equipment became their responsibility. Each of these people had developed a logical sequence that they used to locate electrical problems. They were students of their field of expertise and not mere changers of parts and equipment.

Determining if Motor Is at Fault

A motor system consists of four major components.

1. Power supply
2. Controller
3. Motor
4. Load

When a motor problem occurs, it is first necessary to determine which of these components is at fault. Power supplies and controllers can fail at an equal rate if not more frequently than the motor itself. Mechanical loads can increase due to an increased size of the load the motor is driving, or failure of bearings or coupling mechanisms.

Use the following procedure to determine if the motor is at fault.

1. Disconnect the motor from the controller. Be sure to tag all the leads so that they can easily be replaced when putting the motor back into operation. In this process you may find the problem to be a loose connection that was previously overlooked.
2. Operate the controller to start the motor.
3. Measure the input voltages to the controller. If these voltages are not present, or they vary more than 10% of the rated voltage of the motor, the power supply is the probable cause. Correct this problem before proceeding with further tests.
4. Assuming the power source is acceptable, activate the controller to start the motor. Measure the output voltages of the controller. Proper voltages need to be present at the appropriate time intervals. If not, the controller will be at fault. It is beyond the scope of this text to discuss control circuits. If you have doubts about whether the controller is the problem, proceed with the next steps to determine if a motor fault exists.
5. Check the mechanical coupling between motor and load. Disconnect the coupling. Turn the rotor of the motor to see if it moves freely.
6. Connect the motor directly to the power source and bypass the controller. Do not attempt to operate a DC series motor without a load.
   a. Measure the input voltages. If the load voltages have dropped more than 3% from the no-load voltages, the supply system does not meet the requirements of the NEC. The motor may be drawing more current than the circuit rating.
   b. Measure the input currents with a clamp-on ammeter. Disconnect the motor immediately if the currents are above the FLC ratings. An unloaded motor should draw one-quarter to one-third its rated FLC. If the currents are much greater than this, the motor is the probable cause. Currents should be balanced on three-phase motors. Unbalanced currents indicate a motor problem.
   c. Correct the deficiency in the power system if the motor is within current rating and the voltage drop is still more than 3%.
   d. Assuming proper voltages, currents, phases, and frequency, listen for unusual noises and vibrations. These can be caused by either mechanical or electrical problems. Some noises are normal. AC motors have a characteristic hum which is not objectionable.

Air for ventilation contributes to the motors' noise. Experience will help you to determine when either of these noises are abnormal.

e. Feel the motor for hot spots. Check the motor's temperature. Poor ventilation and electrical faults can cause temperature rise above the motor's rating.

f. Check the motor's RPM with a tachometer. Unloaded induction motors will run near the synchronous speed. If this value is too low, the motor has a problem. Synchronous motors will either run at the synchronous speed or not at all.

Isolating the Defective Component

Once you have determined for sure that the motor is at fault, it is necessary to locate the problem with the motor. Some faults can easily be corrected, while others are major, and the motor will be sent to the shop for repairs or replaced.

The first step in isolating a defective component is to analyze the information you already know about the motor. The knowledge you have gained in your basic studies about motors will serve you well in this process. Compare this information with the data you have collected so far in the troubleshooting process. Consult the equipment's maintenance card for the history of the motor and for further clues.

Different motors have different components that present common symptoms. Any motor may not start when power is applied. The cause may be a defective centrifugal switch on a split-phase motor, an open capacitor on a capacitor-start motor, or a broken brush on a DC motor. At the same time, there are common components that affect all motors the same. If the bearings are frozen on any motor, the motor is going to exhibit bad behavior and not start.

The next step is to separate the mechanical problems from the electrical problems. You have already gone a long way in performing this task by freeing the motor from its load and checking the freedom of the rotor. When a mechanical binding is found, correct this problem first.

Most mechanical and electrical problems can be observed readily. Use your sensory powers to isolate the one from the other. Both may exist. A mechanical problem may have caused an electrical one.

Once you have reasonably assured yourself that no mechanical problem is present, or you have corrected it, try running the motor again. If a problem persists, electrical tests are in order. Know your test instruments and follow the manufacturer's recommendations for the safe use of the devices.

The order of the electrical tests will depend on the symptoms and the type of motor. Most electrical faults can be found using a test lamp, ohmmeter, or megohmmeter. Check the motor components for opens, ground faults, and shorts between components.

It has been my experience that the most difficult problems are those that are intermittent, or when conditions change during the troubleshooting process. The motor will test well on the bench, but when it is put back into operation, the problem reoccurs.

Intermittent problems are usually caused by a dirty contact or a loose connection that has been overlooked. Conditions can change. For example, servicing a vertically mounted motor in a horizontal position is not the same as when the motor is in operation. A short that exists in one position may not exist in the other. A dirty contact or loose connection may be in the controller and not the motor. A good motor is almost impossible to repair. This can be frustrating at a minimum, and may invoke some strong language when the problem is finally corrected.

Troubleshooting Data for Motors

For the purpose of this section, we will assume the complete system: power supply, controller, motor, and load are all connected. The symptom is listed in Chart 10–1, along with the possible causes of the malfunction, and the probable remedy is given. Many of the causes for motor circuit malfunctions are common to all motors. The causes and remedies are given in these scenarios first. Following these common factors, the data about components for particular types of motors are listed.

CHART 10–1. Motor Troubleshooting Data

Possible Cause	Remedy
1. Motor Fails to Start	
	All Motors
Blown fuse or open breaker	Check fuse or breaker. They must have a minimum rating of at least 125% of FLC.
Low voltage	Check with voltmeter. Determine cause of the low voltage. If other electrical machinery is on the same line, turn them off one at a time to determine if one of them is the cause. Check transformer or the source. If feeder conductors are too small, reduce the electrical load or replace the conductors.
Frozen bearings	Replace bearings and recondition the shaft.
Mechanical overload	Reduce the load on the motor. Check the air gap. Look for a bent shaft, misalignment of coupling, worn or loose bearings, and loose hardware in general. Reduce the friction caused by one or more of these conditions.
Open or shorted field	Make electrical tests on field windings. If the open or short can be easily located and repaired, do so. If not, send the motor to the shop for repairs or replace it.
Open or shorted rotor	Make electrical tests on the rotor. If the open or short can be easily located and repaired, do so. If not, the rotor will need to be rewound or replaced.
Overload open	Reset overload. If the motor causes the overload to open after a short period, check for shorts and grounds. Check FLC of motor and compare to the setting of the overload.
Controller defective	Check voltage into and out of controller. If voltage is not present at the output, repair or replace controller. Clean and file the contacts on the controller relay. Dirty contacts can also cause a low voltage at the motor.
	Split-Phase Motor
Centrifugal switch	If the motor hums, and it will run normally if started by hand, the chances are the centrifugal switch is not operating properly. Disassemble the mechanism. Clean the contacts. Adjust spring tension. Repair or replace.
	Capacitcr-Start and Capacitor-Run Motors
Capacitor	Test the capacitor(s). Replace if necessary.
Centrifugal switch	Same as for the split-phase motor.
	Repulsion Motors
Centrifugal mechanism	Check the mechanism for proper operation. Disassemble, clean, inspect, and adjust. If defective, replace.
Brushes and commutator	Check brushes for wear and deterioration. Replace if necessary. Check brushes for proper pressure and placement from the neutral plane. Clean commutator and check for wear and high mica.
	AC-Wound Rotor Motors
External resistors	Test the resistor circuit. Clean slip rings and check the brushes for wear and proper pressure.
	AC Synchronous Motors
Exciter circuit	Exciter voltage is being applied when the motor is attempting to start. Make sure the field contacts are open and the field discharge resistors are connected.
	DC and Universal Motors
Control resistors	Check resistor circuits for continuity and rated values.
Brushes and commutator	Same as for repulsion motors.
2. The Motor Is Noisy When Running	
	All Motors
Bearings	Check bearings for wear and proper lubrication. Replace worn or loose bearings. Replace dirty or worn-out oil or grease.
Coupling mechanism	Check for bent shaft on motor or load. Straighten if necessary. Measure the alignment of the couplings. Realign if necessary. Tighten the coupling.
Excessive end-play	Adjust end-play take-up screw, or add thrust washers to the shaft.

CHART 10–1.—Cont. Motor Troubleshooting Data

Possible Cause	Remedy
Loose hardware	Tighten all loose components on the motor and load. Check fasteners on the motor and load mounts. Motors with centrifugal mechanisms, brushes, slip rings, and commutators can cause noise due to wear and looseness of the mechanisms.
Unbalanced motor or load	Balance motor or load.
Airgap	Check the airgap for proper centering of the rotor. Replace bearings if necessary. A dirty airgap can cause noise. Clean the motor.
Laminations loose	Tighten laminates. Dip in varnish and bake.
Ventilation system	Clogged filters and blocked ventilation systems can cause unusual noises. Check for free flow of ventilating air. The motor fan may break or be unbalanced.
3. Motor Overheats	
	All Motors
Load	Check ammeter readings against full load current rating of motor. Reduce load or replace with larger motor.
Windings	Test windings for shorts, opens, and ground faults. If not easily repaired, send to shop to be rewound or replace the motor.
Ventilation system	Remove any obstructions and clean.
Ambient temperature	Temperature of area above ambient. Lower the temperature by improving the ventilation system.
Bearings and alignment	Bad bearings or poor alignment of the motor with the load can increase friction and cause motor to overheat. Replace the bearings and correct the alignment of the motor and load. If the rotor is striking the stator poles, the motor will quickly overheat. The end-bells of the motor may not be aligned properly.
Source voltage	If the voltage is too high or too low, the motor will operate at a higher temperature. Correct voltage to within 10% of the motor's rating.
Rotor	Rotor frozen or rubbing against stator poles. Check bearings and binding load.
	Split-Phase and Capacitor-Type Motors
Centrifugal switch	Switch does not open when motor is up to speed. Start winding remains in the circuit. Repair or replace switch mechanism.
Capacitor	Check capacitor. Replace if bad.
	Repulsion Motors
Centrifugal mechanism	Check, clean, and repair the centrifugal mechanism. Replace if too badly worn.
	AC Synchronous Motors
Exciter	The motor is underexcited. Correct voltage by adjusting value or repairing the exciter.
	DC and Universal Motors
Brushes	Check condition of brushes, shunts, and pigtails. Replace brushes if necessary. Set proper pressure. Equalize brush tension.
4. The Motor Produces a Shock When Touched	
	All Motors
Stator coil	Test for grounded coil. Check motor frame connection(s) to ground. Clean and tighten.
Static charge	Check ground connections on the motor.
5. Motor Stops After Running a Short Period of Time	
	All Motors
Overload relay	Take current readings and compare to FLC of the motor. Check overloads for proper setting. Reduce mechanical load on the motor or replace with larger motor.
Thermal protectors	See causes and remedies in Section 3 above.
Supply voltage	Replace fuses or reset breaker.
Ambient temperature	Ambient temperature is too high. Increase the ventilation to the area.

CHART 10–1.—Cont. Motor Troubleshooting Data

Possible Cause	Remedy
6. Motor's Bearings Overheat	
	All Motors
Lubricant	Check lubricant. It should be clean and at the proper level or fill. May be too much lubricant as well as too little. Replace dirty or contaminated lubricant. Follow manufacturer's recommendation as to type of lubricant. Clean bearings.
Bearings	Check and replace defective bearings.
Lubricating system	Oil may not be reaching the shaft. Check wick or freedom of ring to rotate on ring type of bearings.
Cooling system	Fan not working or ventilation blocked. Some motors have special systems to circulate a coolant.
Shaft and alignment	Check for bent shaft and proper alignment. Correct any deficiencies. Bearings may have excessive end thrust or side pull. Check application of the motor for the mechanical load.
7. Motor Has Excessive Arcing at the Brushes	
	Most Motors Using Brushes
Brushes	Check brush wear, pressure, and fit to commutator or slip rings. Be sure brush moves freely in its holder. Check brush position in respect to the neutral plane.
Overload	Measure current to motor and compare to FLC rating. Reduce the load.
Bearings, coupling, and load	Check for any condition which will cause the armature to bind and increase the load. Correct any deficiencies.
Commutator	Clean commutator and burnish. Check to see if it is in round. Remove any rough spots. Look for high mica.
Armature	Test for open and shorted winding(s) in the armature. Repair or replace armature.
Field windings	Test for shorts, opens, and ground faults. Correct or replace motor.
8. Motor Runs Too Slow	
	All Motors
Overload	Check all factors such as the mechanical load, bearings, coupling, and alignment of the motor to the load which will increase the friction. Correct any of these problems.
Rotor	Check for open circuit or high resistance in the rotor circuit. This applies to squirrel cage rotors as well as wound rotors. Solder, welded, or brazed joints can open or develop high-resistance points. Rotor will have to be replaced if open occurs in a die-cast type that has an open bar or defective ring.
Supply voltage	Low supply voltage will cause the motor to run slow and to overheat. Correct voltage condition before continuing motor operation.
	Synchronous Motors
Exciter control	Field is excited too soon. Adjust time-delay relay so the exciter current is not applied until rotor nears synchronous speed.
	AC Induction Motors
Motor	All AC induction motors with the exception of the universal motor run at synchronous speed minus slip when unloaded. If the motor will not come up to speed with voltage and frequency of correct value when not loaded, the motor has a definite fault. Check for opens, shorts, and grounds.

Summary and Conclusions

When troubleshooting motors, a logical and sequential procedure is to be used. Without this approach, the least suspected part is often overlooked when troubleshooting any circuit.

Learn to use test instruments correctly. Understand what these instruments can tell you about any circuit. Study your system so you know what to expect in terms of voltage, current, and resistance readings provided by the meters.

Many of the basic principles for locating motor faults are explained in this chapter. Common symptoms and possible remedies to correct motor faults were outlined for your consideration. They will never be a substitute for the experience you will gain in troubleshooting your own systems. Become a student of the subject. It will pay you good dividends in the future.

Keep in mind that temperature, moisture, harmful chemicals, dust, and the lack of preventive maintenance are the factors which destroy motors.

A good preventive maintenance program will greatly reduce the hazards and thereby the need to troubleshoot a fault. You will find many potential problems before they cause the system to go down. Keep records of all maintenance schedules and procedures performed on the motors in your care. Take meter readings when the motor is installed and while it is operating correctly. If this is done, you will be better prepared to perform corrective maintenance in a more orderly, efficient, and economical manner.

Questions

1. Which electromagnetic device requires more maintenance, motors or transformers? Why?
2. Name two types of maintenance programs.
3. List several factors that can cause motor damage.
4. Preventive maintenance should take place ______.
5. Why are motor records important?
6. What factors determine how often a motor must be disassembled for preventive maintenance?
7. List the basic rules you should follow when taking a motor apart.
8. List two precautions to take when removing the bell housing from the frame of the motor.
9. List at least two reasons why motors must be kept clean.
10. The best way to clean dirt and debris out of a motor is to use high-pressure air and blow it out. Why?
11. May water and detergent be used to clean a motor? If so, what precautions must be taken before putting the motor back in service?
12. After a motor has been cleaned several times, what should be done to it?
13. Name the three types of loads exerted on the bearings of a motor.
14. Is it true that the frequency of maintenance of a high-horsepower motor is less than that of a low horsepower motor when all other factors are the same?
15. Why are hand grease guns recommended over power grease guns for lubricating the bearings of a motor?
16. List several reasons why bearings may overheat.
17. What is the effect of bad bearings in a motor?
18. What is the best type of tool for removing a sleeve bearing? Why?
19. Bearing pullers are used to remove what type of bearing? Sleeve, ball, or roller?
20. List the factors necessary for good commutation.
21. List the three primary requirements for brushes.
22. List four types of materials used to manufacture brushes. Which of these has the highest current carrying capability?
23. To shape the brush to the curvature of the armature, ______ is used.
24. What is the effect of brush chatter and electrical arcing on brushes?
25. List all instruments and materials needed to measure brush pressure.
26. Brush holders are positioned about ______ inch above the commutator.
27. Why should brushes of different polarity be staggered on the commutator?
28. What type of electrical problems can be determined using a test lamp?
29. To determine if a high resistance short exists between windings, the most appropriate instrument is the ______.
30. Can a motor's current be measured without disconnecting leads? Explain.
31. List three factors that may cause breakdown of windings' insulation.
32. What is the first order of troubleshooting?
33. How many types of growlers are there? Name them. What are their purpose?
34. Why would an impedance test be better than an ohmmeter for detecting a few shorted turns in a winding?

35. What type of test instrument allows you to measure voltage, current, and resistance with a single device?
36. With the brushes removed from the slip rings of a wound rotor of an AC motor and power applied to the stator windings, the rotor can be turned by hand without any opposition or buzzing. What does this indicate?
37. Why is a capacitor checker a better instrument to check capacitors than an ohmmeter?
38. How do you go about repairing a bad capacitor?
39. Are there any special precautions you should take when testing a capacitor?
40. Using the special circuit in Figure 10–50, you apply 115 V DC and the ammeter at first deflects and then falls back to zero. What is the problem?
41. List the four major components of a motor system.
42. Which of these major components would you test first when troubleshooting a motor problem?
43. A motor continually has its overload relay operate after a short period of time and take it out of operation. What is your first step in correcting the problem?
44. You receive an electrical shock each time you touch the enclosure of a motor. List the probable causes.
45. Under normal conditions, if a ground fault develops between a stator winding and its pole, what should happen?
46. A capacitor-start motor does not start turning when power is applied. It merely sets and hums. If the rotor is turned by hand it runs normally. What components in logical order are most likely the causes?
47. The exciter voltage is not present on an AC synchronous motor when you attempt to start it. What should you do?
48. A repulsion motor will not start unless you shift the brushes on the commutator. What is the probable cause?
49. Why will a motor overheat when the bear ings are causing increased friction?
50. A motor is connected directly to the power line though a toggle switch. It runs for about 20 minutes and turns off. You go to get a meter to begin troubleshooting the motor, and when you get back, the motor is running normally. This cycle repeats. What is happening?
51. Why is a preventive maintenance program important?

PART

III

Appendices

APPENDIX A

General Electric Model List Nomenclature

The following nomenclature is the basis for the majority of GE integral-horsepower motor and generator model numbers. The model number format shown below is held to 15 digits maximum and provides basic descriptive information as follows:

Model:	5	K	215	A	L	205	A	E
Description:	1	2	3	4	5	6	7	8

1. The numeral *5* is used to designate rotating electrical machinery.
2. *Electrical-Type Letters*
   a. Polyphase
      - K NEMA designs A & B
      - KAF Adjustable frequency
      - KG NEMA design C
      - KGS NEMA design C energy saver
      - KL Hydraulic elevator
      - KOF Oil well pumping
      - KR NEMA design D
      - KS NEMA design A & B energy saver
      - KW High efficiency
      - KY Special characteristics
      - SR Synchronous

   b. Single Phase
      - KC Capacitor-start
      - KCJ Capacitor-start, high torque
      - KCY Special capacitor-start

   c. Wound Rotor
      - M Continuous rated
      - MR Short-time rated

   d. Generator
      - LS Motor generator set
      - SJ Synchronous generator
3. NEMA frame size
4. First form letter
   a. Group 1. The following take precedence over all Group 2 letters (except explosion proof)
      - M Marine motors
      - N Navy motors
      - P Partial motors
      - T Textile motors
      - S Severe duty motors

   b. Group 2 letters are shown in Table A–1.

TABLE A–1

Type	DP	TEFC	TENV or TEAO	Explosion Proof	WPII
Footed frame and no customer mounting rabbets or auxiliary devices such as brakes	A	B	H	C	E
Round frame and no auxiliary devices such as brakes	D	F	W	G	—
All others	J	K	K	R	—

* "S" is used for Severe-duty

5. Second form letter
   - C Cast iron construction—sourced motors
   - L Aluminum alloy construction
   - N Cast iron construction—old design horizontal
   - P Aluminum alloy construction–vertical P base
   - T Cast iron construction–vertical P base
   - W Rolled steel construction—180 degrees diameter
   - S Cast iron construction—current design horizontal
6. Form numbers
   a. Number of Poles. For horizontal motors, the first digit in this series represents the number of poles divided by 2:
      1 = 2 pole
      2 = 4 pole,
      etc.
      The number 9 is reserved for multispeed.
   b. Input Voltage. The last digit in this series designates input voltage for standard horizontal motors only.
      4 = 575 volts
      5 = 230/460 volts
      6 = 200 volts
      8 = 460 volts
7. First suffix letter
   Denotes interchangeable design change of the complete motor.
8. Second suffix letter
   Denotes special item or thermal device
   a. Special items
      E = Special varnish treatment
      H = Precision balance
      L = Special balance
      K = Stearns brake
      N = Dings brake
      F2, W1, W2, W5, W6, W7, W8, C1, C2 = Special mounting
   b. Thermal devices
      S,T,X = Klixon automatic reset
      U,W,Y = Klixon manual reset
      J,V = Thermotector system
      M = Sensor, Thermistor, or RTD
      P = Thermostat

APPENDIX B

Values of Trigonometric Functions

VALUES OF THE TRIGONOMETRIC FUNCTIONS

Angle	Sine	Cos	Tan
0°	0.0000	1.0000	0.0000
1°	0.0175	0.9998	0.0175
2°	0.0349	0.9994	0.0349
3°	0.0523	0.9986	0.0524
4°	0.0698	0.9976	0.0699
5°	0.0872	0.9962	0.0875
6°	0.1045	0.9945	0.1051
7°	0.1219	0.9925	0.1228
8°	0.1392	0.9903	0.1405
9°	0.1564	0.9877	0.1584
10°	0.1736	0.9848	0.1763
11°	0.1908	0.9816	0.1944
12°	0.2079	0.9781	0.2126
13°	0.2250	0.9744	0.2309
14°	0.2419	0.9703	0.2493
15°	0.2588	0.9659	0.2679
16°	0.2756	0.9613	0.2867
17°	0.2924	0.9563	0.3057
18°	0.3090	0.9511	0.3249
19°	0.3256	0.9455	0.3443
20°	0.3420	0.9397	0.3640
21°	0.3584	0.9336	0.3839
22°	0.3746	0.9272	0.4040
23°	0.3907	0.9205	0.4245
24°	0.4067	0.9135	0.4452
25°	0.4226	0.9063	0.4663
26°	0.4384	0.8988	0.4877
27°	0.4540	0.8910	0.5095
28°	0.4695	0.8829	0.5317
29°	0.4848	0.8746	0.5543
30°	0.5000	0.8660	0.5774

VALUES OF THE TRIGONOMETRIC FUNCTIONS

Angle	Sine	Cos	Tan
31°	0.5150	0.8572	0.6009
32°	0.5299	0.8480	0.6249
33°	0.5446	0.8387	0.6494
34°	0.5592	0.8290	0.6745
35°	0.5736	0.8192	0.7002
36°	0.5878	0.8090	0.7265
37°	0.6018	0.7986	0.7536
38°	0.6157	0.7880	0.7813
39°	0.6293	0.7771	0.8098
40°	0.6428	0.7660	0.8391
41°	0.6561	0.7547	0.8693
42°	0.6691	0.7431	0.9004
43°	0.6820	0.7314	0.9325
44°	0.6947	0.7193	0.9657
45°	0.7071	0.7071	1.0000
46°	0.7193	0.6947	1.0355
47°	0.7314	0.6820	1.0724
48°	0.7431	0.6691	1.1106
49°	0.7547	0.6561	1.1504
50°	0.7660	0.6428	1.1918
51°	0.7771	0.6293	1.2349
52°	0.7880	0.6157	1.2799
53°	0.7986	0.6018	1.3270
54°	0.8090	0.5878	1.3764
55°	0.8192	0.5736	1.4281
56°	0.8290	0.5592	1.4826
57°	0.8387	0.5446	1.5399
58°	0.8480	0.5299	1.6003
59°	0.8572	0.5150	1.6643
60°	0.8660	0.5000	1.7321

VALUES OF THE TRIGONOMETRIC FUNCTIONS

Angle	Sine	Cos	Tan
61°	0.8746	0.4848	1.8040
62°	0.8829	0.4695	1.8807
63°	0.8910	0.4540	1.9626
64°	0.8988	0.4384	2.0503
65°	0.9063	0.4226	2.1445
66°	0.9135	0.4067	2.2460
67°	0.9205	0.3907	2.3559
68°	0.9272	0.3746	2.4751
69°	0.9336	0.3584	2.6051
70°	0.9397	0.3420	2.7475
71°	0.9455	0.3256	2.9042
72°	0.9511	0.3090	3.0777
73°	0.9563	0.2924	3.2709
74°	0.9613	0.2756	3.4874
75°	0.9659	0.2588	3.7321

VALUES OF THE TRIGONOMETRIC FUNCTIONS

Angle	Sine	Cos	Tan
76°	0.9703	0.2419	4.0108
77°	0.9744	0.2250	4.3315
78°	0.9781	0.2079	4.7046
79°	0.9816	0.1908	5.1446
80°	0.9848	0.1736	5.6713
81°	0.9877	0.1564	6.3138
82°	0.9903	0.1392	7.1154
83°	0.9925	0.1219	8.1443
84°	0.9945	0.1045	9.5144
85°	0.9962	0.0872	11.4300
86°	0.9976	0.0698	14.3010
87°	0.9986	0.0523	19.0810
88°	0.9994	0.0349	28.6360
89°	0.9998	0.0175	57.2900
90°	1.0000	0.0000	∞

APPENDIX C

Answers to Questions

Chapter 1

1. A transformer is an electromagnetic device that transfers energy from one circuit to another through mutual inductance.
2. AC. It is more efficient.
3. The primary windings of a transformer are connected to the source voltage, and the secondary windings are connected to the loads.
4. "H" terminals indicate the high-voltage windings, and the "X" terminals indicate the low-voltage windings.
5. Transformers are rated in kVA. They may have to deliver current to loads which do not consume power (watts) such as inductive and capacitive loads.
6. Step-up. "X".
7. No. The output kVA must equal the input kVA.
8. Maximum power will be transferred when the impedance of the load is equal to the impedance of the transformer.
9. By the number system. H1 is in-phase with X1.
10. Copper and core losses.
11. Laminating the core material.
12. An increase in current flow in the transformer will not increase the number of flux lines.
13. Efficiency cannot be improved by the electrician. The electrician can ensure that the transformer will operate at its peak efficiency by good maintenance procedures.
14. Application, purpose, cooling method, number of phases, type of insulation, and method of mounting.
15. Overheating a transformer can destroy it. Reducing its temperature allows one to operate it above its capacity.
16. Class H.
17. Given on the nameplate.
18. It is the percent of change of the output voltage from no-load to full-load.
19. 2500 amperes.
20. Compensate for changes in the value of the primary voltage. All power must be removed before changing the tap.
21. Fuses or circuit breakers in the primary and secondary circuits. Lightning and surge arresters.
22. 600 volts; 83.36 amperes.
23. Autotransformers have a single winding common to both the primary and secondary circuit.
24. kVA X co-ratio.

25. Instrument transformers reduce the values of high voltages and currents to safer levels so that they can be measured.

Chapter 2

1. kVA rating and voltages.
2. The leads marked "H" are the high-voltage leads, and the ones marked "X" are the low-voltage leads.
3. See Figure 2–4.
4. Yes, if losses are not considered.
5. 6.25 amperes. 25 amperes.
6. See Figure 2–7.
7. Zero.
8. 2160 volts.
9. Subtractive.
10. Use an ohmmeter.
11. See Figure 2–10.
12. Primary: H1–H1, H2–H2. Secondary: X1–X1, X2–X2.
13. No. A circulating current would be established between the two transformers resulting in high loss. These currents should be limited to 10% of the full-load current.
14. The transformer with the smaller impedance will take a disproportionate part of the load.
15. Yes. See Figure 2–12.
16. 120 degrees.
17. 25%
18. Counterclockwise.
19. Zero.
20. 480 volts.
21. H2 terminals are wired together and taken to the neutral.
22. Easier to install; lower cost; higher efficiency; less space required; lower transportation costs.
23. They are equal.
24. Line voltage is equal to 1.73 times the winding voltage.
25. Line current equals 1.73 times the winding current.
26. Line voltage equals winding voltage.
27. Wye.
28. Delta.
29. Yes. Delta-delta, which cannot be wired parallel with delta-wye or wye-delta.
30. See Figure 2–20.
31. 62.28 kVA.
32. Wye-wye.
33. Yes. See Figure 2–23.
34. Three. 120/240/208 volts.
35. See Figure 2–28.
36. 58%
37. 15.5% greater.
38. They can add to or subtract from the line voltage.
39. Scott transformers.
40. Providing a neutral where none exists.

Chapter 3

1. The transformer should be inspected and damage reported.
2. Some transformers have gauges, and others have a level mark inside of the tank.
3. Put tank under pressure using an inert gas and use soapy water to locate the leak.
4. The coolant will expand when heated, building up pressure in sealed units and forcing liquid to overflow in units not sealed.
5. 0.06%
6. The temperature of the liquid can be elevated above the boiling point of water by using the short-circuit method.
7. Mineral oil.
8. Skin irritation; PCBs.
9. Highest.
10. 8 feet.
11. Minimum of 8 feet. Grounded. Warning sign.
12. Locked room or vault.
13. 4 inches.
14. Yes.
15. One that will remove the heat losses without causing a temperature rise in excess of the transformer's temperature rating. An exchange of 100 cubic feet per minute per kilowatt loss is recommended.
16. 62 decibels.

17. Select a transformer with a low level of noise.
18. Earth potential; the conductor between the system and ground; permanent joining of metallic parts.
19. Keep noncurrent-carrying metallic parts safe to work around. Limit the rise of excessive voltages on the system.
20. Overvoltage and overcurrent.
21. Lightning and surge arresters. Fuses and breakers.
22. Overload: The current is from one to six times the operating level but is confined to the current carrying conductors of the system. Short circuit: Currents are not confined to the system conductors and may be hundreds of times the normal operating current.
23. Volts, amperes, and interrupting capacity.
24. Usually only one link will open on overload. In the case of a short circuit, two or more links will open. There is usually more evidence of arcing under short circuit conditions.
25. Completely self-protected. The transformer has lightning and surge protection and breakers or fuses in the primary and secondary.
26. 225 amperes. 250 amperes.
27. Metal frame of the building.
28. Operating under an overload condition.
29. Primary and secondary voltages. Impedance. Primary and secondary currents. Where protection is provided. Whether there is thermal protection or not.
30. Megger.
31. 20,071 amperes.
32. The voltage is adjusted to the point that the rated primary current is obtained.

Chapter 4

1. Repairs and down-time cost money. Maintenance can be scheduled, whereas emergency repairs cannot.
2. Repair crews are inherently inefficient. They spend most of their time waiting for failures. This time could be better spent preventing the failures through routine maintenance.
3. Maintenance begins on a transformer when it is installed. The operating information collected at this time is the basis for future tests.
4. How often maintenance is scheduled on a transformer is largely due to the environment in which it operates and the record of failures.
5. Manufacturer's recommendations should be followed. Also use common sense about the environment in which the device is to be operated.
6. Damaged conductors are most often found around the pothead.
7. Contacts oxidize causing resistance. This in turn results in a hot spot in the transformer. Power should be removed when cleaning the tap changer.
8. 22 kilovolts.
9. The oil will be discolored.
10. Rubber has sulfur in it. The sulfur will be given off into the coolant and will attack the copper wire.
11. The oil should be filtered using a filter press.
12. The filters in the press are becoming clogged and should be changed.
13. A sudden rise in operating temperature.
14. A gas absorber should be part of the transformer. The gas given off by the coolants may be explosive and could explode in the event of an arc or fire.
15. A solution of hydrochloric acid is allowed to stand in the coils.
16. The acid should be poured into the water and not water into the acid.
17. The nitrogen exhausts all the air and prevents moisture from mixing with the coolant.
18. Monthly.
19. Maintenance periods can be extended when no problems are detected after several maintenance checks.
20. Heat, dirt, and moisture.
21. This statement is false. The best way is to hear the increased vibrations of the addi-

tional load and to feel the temperature increase.

22. Study, familiar.
23. Faults in high-power systems usually leave some physical observable evidence.
24. The transformer itself is more likely to be at fault.
25. b. short circuit.
26. The transformer can catch on fire.
27. a. lower. Life and property are put in jeopardy.
28. Megger.
29. No. The coolant may have sealed the fault by the time the test is made.
30. Yes. Hydrogen chloride gas is formed and when combined with water will form hydrochloric acid which will attack the transformer.

Chapter 5

1. Type of structure; present utilization and possible future uses; projected life of the structure; flexibility of the structure; location of service entrance equipment; load requirements and their locations; sources, quality, and continuity of power, switchgear, distribution equipment, and panels; installation methods. Cost is always a factor.
2. 18.65 kVA.
3. 22.46 amperes.
4. 601 volts to 15 kilovolts.
5. 480/277 volts.
6. A ground fault will not interrupt service.
7. Transient voltages may cause damage to wiring and equipment; fault may not be detected with repeated arcing resulting; insulation may be stressed at voltages above its rating; faults are difficult to locate.
8. 300 volts.
9. No limit provided conductors are rated for the value and protected by rigid steel conduit.
10. No.
11. Open the low-voltage feeder breaker supplying power to the defective unit.
12. Separate transformers are used at each load unit with high-voltage primary being supplied to each.
13. A defect in the feed r system will affect only one power cer...r.
14. Balance current load on each part of the feeder; allow for isolation of a defective power center for repair or maintenance; allow for isolation of defective feeder cable.
15. Power can be maintained to all loads even if a power center or a feeder cable becomes defective.
16. Limiters protect the system from ground or short circuit faults on the system.
17. Use a network system that has more than one primary supply with separate primary feeder breakers. Each feeder supplies one or only a few of the total power centers. Each power center is capable of supplying loads through a banked-secondary distribution system.
18. Each floor has a transformer to convert the 480/277 volt distribution voltage to 207/120 volt utilization voltage.
19. 100,000 amperes.
20. HVAC, busway, emergency lighting, and elevator.
21. Only the fuse or circuit breaker associated with the fault will trip.
22. According to cable size.
23. Voltage and current can be supplied from either end of the cable.

Chapter 6

1. Volta, chemical source, the battery.
2. Oersted.
3. Davenport.
4. So many are in use to take the drudgery out of work.
5. An electromagnetic device that converts electrical energy into mechanical energy.
6. 746.
7. Three.
8. Three.
9. DC, AC.
10. Frequency of power source, number of poles.

11. 1200 RPM.
12. Difference between the synchronous speed and actual speed of a motor.
13. When the motor has only two poles.
14. True.
15. 83%.
16. Turning force.
17. Dynamometer.
18. Measure motor speed.
19. 0.81 horsepower.
20. Windage, friction, core losses, stator loss, rotor loss, and stray load losses.
21. Air gap between stator and rotor. Motor has friction. Moving parts of a motor are resisted by the air.
22. The percent rotor loss is approximately equal to the percent slip.
23. Cost of electrical energy has increased very rapidly in the past few years.
24. The thermal protector is not UL approved. It may be either manual or automatic reset.
25. H.
26. This is the amount of time a motor can operate under full load without overheating.
27. The motor can be operated 15% over its rated capacity without causing damage.
28. Stator, rotor, and a means of support.
29. Wound rotor and squirrel cage rotor.
30. Sleeve and ball bearing types.
31. Watts, horsepower.
32. Difference in rate of expansion due to temperature change on two dissimilar metals.
33. The *north pole* of a magnet is the one that points to the north geographic pole of the earth.
34. Repel, attract.
35. Reluctance.
36. Rotating magnetic field.
37. Left-hand rule.
38. Left hand rule.
39. Right-hand rule.
40. Zero.
41. Convert AC to DC.
42. The attraction and repulsion of the magnetic fields between the field winding and the rotor.
43. Neutral.
44. Add additional loops in parallel on the rotor.
45. 90, 120.

Chapter 7

1. 115 volts, 230 volts, 115/230 volts.
2. Constant, adjustable, multispeed, varying, and adjustable-varying speed.
3. No. There must be some slip or no current will be induced into the armature.
4. Start windings are added to the stator. These windings have a higher resistance than the run windings. The result is that the magnetic field created by the current flowing through these windings leads the magnetic field for the run windings. This creates a rotating magnetic field.
5. 90.
6. Centrifugal switch.
7. T.
8. The lead is connected to the thermal protector.
9. Black, red.
10. Yes. Use an ohmmeter. The start windings will have a higher resistance than the run windings.
11. Yes. Reverse either the leads of the run windings or the start windings to the power source. Do not reverse both.
12. Series, yes.
13. When motor is operated at the higher voltage.
14. By changing the number of poles.
15. Run windings are on individual poles rather than pairs.
16. Higher starting torque.
17. The capacitor is wired in series with the start windings and the centrifugal switch.
18. Microfarads and V AC.
19. Increase.
20. They are identical.
21. No.

22. Yes.
23. No.
24. Oil-filled capacitor. An electrolytic capacitor would be destroyed if it remained in the circuit for long periods of time.
25. Provides higher run torque.
26. The capacitor is switched between the run windings and the start windings.
27. Very low. Shaded-pole motors have low starting torque.
28. Rotation is toward the shaded pole.
29. Repulsion motor.
30. Shorted.
31. Short, lift.
32. No.
33. Maximum torque is developed when the brushes are set at approximately 17° from hard neutral.
34. Repulsion-start, induction-run.
35. Repulsion-induction.
36. Repulsion.
37. Varying speed, high-starting torques, and ability to reverse direction of rotation.
38. CEMF.
39. Series DC motor.
40. Armature reaction, self-induction.
41. Torque, speed.
42. Change direction of current flow through either the armature or the field windings.
43. Decrease.
44. Series.
45. Compound.
46. Suicide.
47. Universal.
48. The loads are always applied through gears or direct coupling.
49. Compensating windings.
50. Motor operates at a high RPM.

Chapter 8

1. 90, 120
2. Three-phase systems deliver almost constant power to the load. Single-phase systems deliver zero power during alternations.
3. Stator, rotor.
4. The rotor's impedance is equal to the vector sum of its resistance and reactance.
5. Frequency of stator current times percent slip.
6. Rotor current, strength of the magnetic field, and the phase angle between the magnetic field and the rotor currents.
7. Wye, delta.
8. Delta.
9. T1, T2, and T3.
10. Three 3-wire circuits.
11. T5.
12. Parallel with each other.
13. Series.
14. No. Synchronous speed – slip.
15. Frequency, number of poles, percent slip.
16. No.
17. Resistance.
18. Interchange any two phases of the power to the windings.
19. Squirrel cage rotor type.
20. It requires a much larger magnetizing force to have the flux lines leave the iron in the stator to cross the airgap to cut the rotor windings.
21. 100 horsepower.
22. For starting high-inertia loads requiring large starting torque.
23. Additional resistance in the windings of the wound rotor motor during its run cycle results in reduced torque and the motor turning at a lower RPM.
24. Synchronous motor.
25. 1800 RPM.
26. Three-phase, AC for the stator, and DC for the rotor.
27. Speed, type of service, and power factor.
28. A synchronous motor may use a separate motor to bring it up to speed or it may have special windings to start it as an induction-type motor.

29. 1% to 5%.
30. Current is induced into these windings when the motor is running either faster or slower than the synchronous speed. This sets up a magnetic field which tends to bring the motor back in sync with the rotating magnetic field.
31. No. The exciter generator is sometimes mounted on the rotor shaft of the motor. Direct connections can be made between the exciter and the motor's rotor.
32. Lagging.
33. Delta, 240 volts. Wye, 416 volts.
34. Four. One 3-wire, and three 2-wire.
35. T6, T8.
36. Either may be used.
37. When breaking the current path to the coil, the meter should either be set to its highest voltage rating or removed from the circuit.
38. Line current can tell you if the motor is drawing its normal current for its horsepower rating. Also, the current should be about the same on each of the phases because the motor is a balanced load. If the load is not balanced and the current is either too high or low, the motor has a problem yet to be corrected.

Chapter 9

1. The motor should be checked to ensure it meets the specifications of the order.
2. The shaft can be bent.
3. System voltage and nameplate voltage must be within 10%.
4. Special paints and stainless steel hardware may be used.
5. Provided the service factor is above 1.0.
6. No. If anything, they should be overfilled.
7. 1782 volts.
8. 1,600,000 ohms.
9. The motor must be dried.
10. Raintight and dusttight.
11. The lubrication system may not perform in any direction and the lubricant may leak from the motor.
12. 10:1.
13. 400 RPM.
14. Close to the motor. Reduces stress on the bearings.
15. 1/8 inch sag for each foot of length of the belt.
16. 64.26 inches.
17. 2708.25 ft/min.
18. Chain drives eliminate slippage.
19. Reduces noise and protects gear box in case of a jam.
20. Mechanical.
21. 85 amperes.
22. Overloads.
23. 98 amperes.
24. 47.25 amperes.
25. True.
26. Six.
27. Starts, stops.
28. The contacts provide memory so that the motor will continue to run when the start button is released.
29. Closed.
30. No. FLC passes through the heaters of the overload which in turn will cause the overload contacts to open. The overload contacts are in series with the control circuit.
31. Thermostats, thermistors, RTDs, and thermocouples.
32. Loss of one phase of a three-phase system.
33. Motors may be protected from low and high voltages, excessive vibration, excessive bearing temperature, excessive speeds, reverse direction of rotation, line surges, and lightning strikes.
34. Ungrounded.
35. Yes. Dual-element-type fuses provide both.
36. The overcurrent device must have a high enough rating to take into account the starting current of the motor which may be six times the FLC.
37. For overcurrent protection, the FLC is found in the tables of Article 430 of the NEC. Nameplate FLC is used for determining overload protection.

38. 125%, 100%.
39. Overcurrent devices protect the conductors in a circuit. Overload devices protect the equipment.
40. The interrupting capacity of a fuse or breaker is the amount of short-circuit current the device can withstand safely while removing the load from the source.
41. Before starting a motor for the first time, the insulation resistance, lubricants and levels, freedom of rotation, alignment of motor with the load, proper and sufficient ventilation, and the power source all need to be checked. The motor's wiring scheme must also be confirmed with the wiring diagram.
42. If a motor is performing according to its specifications, the currents and voltages will be within tolerance. Temperature rise will not be excessive.
43. Power factor is the ratio of the watts used by the load to the volt-amperes delivered to the load. Mathematically this is expressed as PF = Watts/VA. Power factor is equal to the cosine of the angle.
44. 40 kW, 30 kVAR.
45. The power factor can be corrected to 95% by placing a capacitor rated about 16.85 CkVAR in parallel with the motor.
46. Utilities may penalize a plant by applying additional charges for low power factor. Low power factor also reduces the overall capacity for doing useful work with the available kVA capacity of the transformers supplying power to the plant. Conductors along with related overcurrent devices, controllers, and switches must be larger. It costs more to operate with poor power factor.
47. The primary cause of low poor factor is motors operating at less than their full horsepower capacity. This causes the current to the motors to lag the voltage in the system.
48. Correcting the power factor will allow for additional loads to be placed on the system. The voltage conditions within the plant will be improved, and power losses will be reduced.
49. Power factor may be improved by keeping motors fully loaded or synchronous motors may be used to correct the power factor.
50. When power factor is corrected at the motor, power factor is corrected throughout the plant. If the motor is moved, the power factor correction goes with it. When power factor is corrected at the main, it is cheaper than doing so at each individual motor. Power factor throughout the plant is not corrected.

Chapter 10

1. Motors. Motors have moving parts which cause friction. Friction causes rise in temperature which is one of the primary destroyers of electromagnetic devices.
2. Preventive and corrective maintenance.
3. Temperatures above the motor's rating, dirt, dust, debris, abrasive particles, chemicals, and moisture all can destroy a motor.
4. Continuously.
5. Changes in the motor's performance become obvious. Time between maintenance procedures is charted. Costs can be controlled.
6. Size of motor, usage, and environmental conditions.
7. Handle parts with care. Tag and mark parts. Store parts in an organized manner and in a safe place.
8. Use a block of wood and machinist hammer to break the seal between the bell housing and the motor's frame. Check for electrical connections between the bell housing and the motor's frame.
9. Dirt and grime restrict the transfer of heat from the motor causing the motor's temperature to rise. High temperatures destroy motors. Debris collecting on motors may have metallic particles that can cause shorts or chemicals that will attack the motor's finish or insulation of its electrical parts.
10. False. High-pressure air used in this way can cause particles to collect deep in crevices of the motor and result in later problems.
11. Yes. The motor must be completely dried and megger readings taken to insure the insulation quality.
12. Insulation quality of windings needs to be restored by varnishing the windings.
13. Radial, thrust, and angular loads.

14. Maintenance for large motors is needed more frequently.
15. The pressure applied by power grease guns may rupture the bearing seal causing grease to leak from the reservoir.
16. Bearings will overheat if there is none, too little, too much, or deteriorated lubricant. Bad bearings and misalignment of motor and load will also cause this problem. In some rare instances, shaft current may develop and cause this problem.
17. Increased friction will cause the motor to draw more current and overheat. The rotor may strike the stator poles or become frozen and not turn in severe cases.
18. Arbor and press. This method eliminates the accidental striking of the housing and does not damage the bearing.
19. Ball.
20. Cleanliness of commutator, good brushes that fit the curvature of the commutator, freedom of movement of the brush in its holder, proper pressure of brush against the commutator, proper position of the brush on the commutator circumference, concentricity of commutator, proper undercut mica, the formation of a uniform copper oxide–carbon film.
21. Brushes must be able to carry the load currents, act as a lubricant on the commutator, and withstand high levels of short-circuit currents and peaks.
22. High-grade carbon, electrographite, natural graphite, copper graphite. Copper graphite.
23. Sandpaper.
24. Causes brush damage and the need for replacements.
25. Spring balance and a strip of paper.
26. One-eighth.
27. Wear is different for the two polarities.
28. Opens, shorts, and ground faults.
29. Megger.
30. Yes. Use a clamp-on ammeter.
31. Temperatures above motor's rating. Environmental conditions such as moisture, chemicals, dust, and abrasive particles. Electrical spikes on the power line.
32. Use your senses to detect problems before beginning electrical tests.
33. Two. External and internal. External growlers are used to detect shorts and opens in the armature. Internal growlers are used to detect opens and shorts in the stator windings.
34. The winding's reaction would be zero when the ohmmeter is used because of the DC source voltage used by the meter. Only the DC resistance of the windings would be measured and a few shorted turns may not be noticeable in the meter reading. Impedance testing uses the line frequency for the measurement, and the reactance would be present to add with the resistance.
35. Multimeter.
36. Rotor windings are not shorted.
37. An ohmmeter uses a lower voltage than the rated voltage of the capacitor and will not indicate voltage breakdowns at higher voltage. Also, the ohmmeter will not indicate capacitance value. The capacitor checker overcomes both of these problems.
38. Capacitors are not repaired, they are replaced.
39. Yes. Capacitors can store a charge. Discharge them before attempting to make any measurements.
40. There is no problem. Because DC was used instead of AC, the capacitor charged to the value of the applied voltage, the ammeter deflected, and the current stopped flowing due to opposing voltages. The capacitor is probably good.
41. Power supply, controller, motor, and load.
42. Power supply.
43. Measure line current and compare it to the FLC rating of the motor.
44. Motor has a short to ground and is not grounded itself.
45. The power supply overcurrent device should open removing power from the motor.
46. Centrifugal switch, start capacitor, and start winding.
47. Look for another problem. The exciter voltage should not be applied until the rotor is at or near the synchronous speed.

48. The brushes are probably set at either hard or soft neutral. They need to be set at plus or minus 17° from hard neutral in the direction of desired rotation.

49. The friction causes an increase in the mechanical load. The motor must draw more current from the power source to overcome the load. The increased current though the motor windings causes a temperature rise.

50. The motor has thermal protection. The thermal protector automatically resets when it cools. The motor is probably overloaded and the line current should be measured when it is running.

51. Potential problems that can cause loss of production are often detected and corrected so that they do not occur at inopportune times. These programs increase the life of the machine, and in the long run, reduce costs.

Glossary

AC Generator: An electromagnetic device used to convert mechanical energy into electrical energy. The device has an alternating current output. The machine uses slip rings and brushes. On large generators, the armature is fixed (stator), and the field windings are rotated (rotor).

AC: Alternating current. The current reverses its direction of flow periodically twice each cycle.

Ambient temperature: The air temperature surrounding a transformer or motor.

Ampere: Unit of measurement for electrical current. An ampere is equal to 1 coulomb of electrons flowing by a given point in 1 second. Also known as *coulomb/second.*

Ampere turns: Unit of measure of magneto force. Force = 1.257 X current X number of turns.

Apparent power: Volts X amperes in an AC circuit.

Armature: A wound rotor with a commutator.

Armature reaction: The magnetic field created due to the current in the armature set up due to its movement in a magnetic field.

Askarel: A generic term for a group of nonflammable synthetic chlorinated hydrocarbons used as an insulating coolant in transformers. There are many types. Gases may be produced in askarels under arcing conditions. The principal gas is hydrogen chloride which is noncombustible, but small quantities of combustible gases may also be present.

Autotransformer: A transformer with common primary and secondary windings that can either step voltages up or down.

Breakdown torque: Maximum amount of torque a motor develops under load without an abrupt drop in speed and power.

Bridge circuit: A combination circuit with series and parallel components that are tied to a common bridge. These circuits are frequently used to make precision measurements of component values.

Capacitance: Ability to store an electrical charge. The unit of capacitance is the *farad.* Capacitance opposes changes in voltage.

Capacitive reactance: Opposition to current flow in an AC circuit due to the presence of capacitance. Capacitive reactance is measured in ohms. The symbol used for capacitive reactance is X_c. $X_c = 1/(2\pi fC)$.

Capacitor: A device that will store an electrical charge. A simple capacitor consists of two conductors separated by an insulator.

Capacitor-start motor: A version of the split-phase motor that has a capacitor in series with the start windings. This produces a greater phase displacement between the magnetic fields of the start and run windings resulting in higher starting torque.

Circuit breaker: An overcurrent device that operates on the principles of temperature rise or the strength of a magnetic field. Circuit breakers may use either principle or both at the same time.

Coefficient of coupling: Percentage of magnetic flux lines from a primary coil that cut the secondary coil of a transformer. The number is expressed as a decimal fraction.

Commutation: Process of reversing current so that current flow from an armature of a generator to a load is always in the same direction, and the current to the windings of the motor's armature is always the same.

Commutator: Group of bars providing connections from brushes to the armature windings. The commutator is a mechanical switch that maintains current in one direction.

Compound DC motor: A DC machine that has both shunt and series field windings.

Conductor: A material whose atoms have many free electrons and that allows current to pass though it easily. Conductors are generally restricted to those materials that have atoms with less than one-half the number of electrons required to fill the outer shell of the atom.

Copper losses: Heat losses in transformers and motors due to the resistance of the wire. These are sometimes referred to as I^2R losses.

Coulomb: A quantity of electrons. Various authorities differ on the exact number of electrons in a coulomb. An authority at the National Bureau of Standards, now the National Institute of Measurements and Technology, reported the number as *about* 6.15×10^{18}.

Counterelectromotive force (CEMF): Current induced into a conductor moving through a magnetic field which sets up a field that opposes the current that caused it.

Coupling: Percentage of mutual inductance between primary and secondary windings of a transformer. Coupling losses occur when all the flux lines generated by the primary windings do not cut the secondary windings.

CW/CCW: Clockwise/counterclockwise rotation.

Cycle: A set of events that reoccur in some sequence. A cycle is a complete set of values. One sine wave in a sequence of sine waves is a complete set of positive and negative values that reoccur in sequence and is a cycle.

Cycling: Motor shuts off due to excessive heat and then comes back on when cool. The condition continually repeats itself. Cycling is sometimes called *tripping*.

DC: Direct current. Current flows only in one direction in a circuit.

DC generator: An electromagnetic device used to convert mechanical energy into electrical energy. The output current flows in only one direction. Direct current generators have commutators and brushes.

Dielectric: Insulating material.

Dielectric constant: Ability of a material to withstand electric stress before breaking down and becoming a conductor. The dielectric strength of air is 1.

Dynamometer: Measuring device that converts the magnetic force between two coils into the physical units of pound foot (lb-ft) or ounce inch (oz-in).

Eddy currents: Losses due to currents being induced into the core materials of transformers and motors.

Efficiency: Ratio of power output to power input: P_{out}/P_{in}.

Electrolytic capacitor: Capacitors that are polarized with relatively high capacitance when compared to other types. The dielectric is a thin oxide coating on the aluminum foil conductors.

Enclosures of motors: Motor enclosures may be open type (OT) or totally enclosed (TE). Open-type motors may have ventilation openings in the end shields and/or the shell. Totally enclosed motor enclosures may be nonvent (TENV) with no external fan, fan cooled (TEFC), and explosive proof.

End-bell: Shield at each end of a motor that supports the bearings. The end-bell is sometimes called the end-plate.

Energy: Ability to do work. Measured in foot-pounds in the English system and joules in the metric system. A foot-pound equals about 1.356 joules.

Farad: Unit measurement of electrical capacitance. One farad is the ability of the device to store 1 coulomb of electrons for each volt applied. Farad = coulomb/volt.

FHP: Fractional-horsepower motor. Motors whose ratings are less than 1 horsepower.

FLA: Full-load amperes. The amount of current normally drawn by a motor when at rated load and voltage. Same as FLC.

FLC: Same as FLA.

Flux: Force field of a magnet.

Flux density: Number of magnetic flux lines per cross-sectional area. Unit of measurement is the gauss. One *gauss* equals one line of force per square centimeter.

Frame or frame size: Usually refers to the NEMA system of standardized motor mounting dimensions.

Frequency: The number of cycles that occur in 1 second (cycles/second). The unit of measure is the *hertz* (Hz).

Full-load torque: The amount of torque developed by a motor when it is running at full-load speed and at rated horsepower.

Fuse: An overcurrent device that operates on rise in temperature. The fuse link melts and removes the power source from the load.

Gilbert: Unit of measurement of magneto force. This is the force needed to establish one single line of force (maxwell) with one rel of reluctance.

Ground: Earth potential. Exposed metal parts of electrical equipment are sometimes maintained at ground potential to prevent electrical shock.

Ground fault: Any point on an ungrounded motor component where the resistance to the frame is 1 megohm or less.

Grounding conductor: The wire between the electrical device and earth.

Growler: A device used to locate opens and shorts in a motor's windings.

Harmonic frequency: This frequency is a multiple of the fundamental frequency of the power source. Harmonic frequencies of 60 Hz would be 120, 180, 240, and so on, hertz. Harmonic frequencies can cause problems with motor operation and power distribution systems.

Henry: Unit of inductance. One henry is present in a circuit when 1 volt is induced into the inductor for a change of 1 ampere in 1 second. $E = L(I/T)$.

HP: Horsepower. A measure of the rate of doing work. In electrical terms, 1 horsepower is equal to 746 watts. In physical terms, 1 horsepower is equal to 550 foot-pounds per second, or 33,000 foot-pounds per minute. Horsepower equals torque in ounce inches times RPM divided by one million.

Hysteresis: A property of magnetic materials that causes the magnetization of the material to lag behind the force that is producing it. Hysteresis is a basic loss in electromagnetic devices.

Hz: Hertz. Unit of measure of frequency in cycles per second. Alternating current power in the United States is usually 60 Hz.

IBEW: International Brotherhood of Electrical Workers.

Impedance: The total opposition to current flow in an AC circuit measured in ohms. Impedance is equal to the vector addition of resistance, inductive reactance, and capacitive reactance in a circuit.

Inductance: The ability of a changing or moving magnetic field to induce a current into a conductor. The unit of inductance is the *henry*. Inductance opposes the change of current in a circuit.

Induction motor: Motors that operate on the principles of magnetic rotating fields. No direct connection is made between the power source and the rotor. Rotor power is obtained through transformer action between stator windings and rotor windings.

Inductive reactance: Opposition to current flow in an AC circuit due to the presence of inductance. Inductive reactance is measured in ohms. The symbol used for inductive reactance is X_L. $X_L = 2\pi fL$.

Inductor: A component that processes the electrical quality of inductance. In circuit diagrams, L is used to indicate an inductor.

Insulation classes: Insulation is rated by its temperature capability for reasonable life of the material. The two most common classes are Class A, 105°C, and Class B, 130°C. These values are for the total temperature and not for temperature rise above the ambient.

Insulator: A material that opposes current flow. By definition, the element has more than one-half the required number of electrons in the outer orbit of its atom. A more precise definition is that it requires at least 1 electron-volt of energy to cause an electron to leave the outer orbit of the atom and become a current carrier.

Isolation transformer: A transformer with a 1:1 turns ratio used to isolate the load from the power source.

JATC: Joint Apprenticeship and Training Committee. In the electrical industry, these committees are composed of NECA and IBEW members whose purpose is to provide electrical training for apprentice electricians.

Joule: Metric measure of potential energy and work. It is equal to about 0.738 foot-pound. In electrical terms it is equal to 1 watt-second.

Kilo: Prefix meaning 1000 times the root word.

kVA: Kilovolt amperes. (Volts X amperes)/1000.

Laminates: Thin sheets of insulated steel sheets used in the cores of transformers, motors, and generators to limit eddy current losses.

Laws of magnetism: Like poles repel and unlike poles attract each other.

Lead: The conductor wire brought out from the windings of a transformer or motor.

Left-hand rule: A method used to predict the direction of current flow in a generator when the direction of motion and the magnetic field are known.

Lenz's law: Induced current in any circuit is always in such a direction as to oppose the field that caused it.

Magnetic circuit: Complete path of the magnetic field. All magnetic flux lines that leave the source must return to the source.

Magnetic line of force: An invisible line along which a compass needle will align itself.

Magnetic materials: All materials that are attracted by a magnet. Common metals are iron, steel, nickel, and cobalt.

Magnetic saturation: An increase in the magnetizing force produces little or no further increase in the flux density.

Magnetomotive force (F) (mmf): Force that produces flux in a magnetic circuit.

Maximum power transfer: This condition exists when the impedance of the load equals the impedance of the source:

Maxwell: One single line of magnetic flux.

Meg: Prefix meaning one million times the root word. Example: 1 megohm = 1,000,000 ohms.

Micro: Prefix meaning one-millionth of the root word (0.000001). a 1 microfarad (μF) capacitor is equal to 0.000001 farad.

Mil: One-thousandth of an inch (0.001 inch).

Milli: Prefix meaning one-thousandth of the root word (.001). 100 mH = 0.1 henry.

Motor: An electromagnetic device that converts electrical energy into mechanical energy.

Motor reaction: Opposing force to rotation developed in a generator caused by the load current.

Multimeter: Test instrument that combines the voltmeter, ammeter, and ohmmeter in the same device.

Mutual inductance (M): Two or more coils have the same magnetic circuit. Flux change in one coil will cause an EMF in the other(s).

NEC: National Electrical Code. This code is sponsored by The National Fire Protection Institute and is used by many government bodies to regulate building codes pertaining to the installation of electrical wiring, lighting, and appliances. The purpose of the code is to safeguard persons and property from the hazards arising from the improper use of electricity.

NECA: National Electrical Contractors Association.

NEMA: National Electrical Manufacturers Association.

No-load voltage: Output voltage of a source such as a transformer when no current is being drawn. This voltage will fall when a load is applied due to the current flowing through the internal impedance of the source.

Oersted: Unit of magnetic intensity equal to 1 gilbert per centimeter.

Ohm: Unit of measure for resistance(Ω). The value of the ohm has been internationally agreed to as being the amount of electrical resistance that a column of mercury, 106.3 centimeters long, 1 millimeter square, at 0°C, will present to current flow in a circuit. The mercury has a mass of 14.4521 grams. A common definition is that when 1 volt is applied to a circuit, and 1 ampere flows in the circuit, then 1 ohm of resistance is present.

Ohm's law: The relationship that exists among voltage, current, and resistance in a circuit. E = I X R. For AC circuits, the resistance is replaced by the impedance (Z) of the circuit.

Open circuit: Circuit broken or the load is removed. Load resistance is undefined in value.

Oscilloscope: Test instrument that displays amplitude, frequency, and wave shape on a cathode ray tube.

Ounce-inch: Measure of torque for a small motor. An ounce-inch is equal to 1 ounce of force applied at 1 inch from the centerline of the shaft.

Overcurrent protector: A device such as a fuse or circuit breaker that removes the load from the power source when a predetermined level of current is drawn by the circuit.

Overload protector: A temperature detecting device built into a motor circuit which disconnects the motor for the power source if the temperature rise becomes excessive.

PCB: Polychlorinated biphenyl. A chemical used in askarels which has been banned by the EPA due to its cancer-causing characteristics.

Period: The amount of time required for one cycle. The period can be determined by dividing the frequency into 1. T = 1/F.

Permanent magnet: Bars of steel and other materials that have been permanently magnetized.

Permeability: Relative ability of a material to allow the passage of magnetic flux lines as compared to air.

Permeance (P): Ability of a material to allow the passage of magnetic flux lines. It is equal to 1/reluctance.

Pico: Prefix that means one-trillionth (0.000000000001). 0.000000000001 farad = 1 picofarad (pF).

Polarity: Magnetic polarity has been defined so that the flux lines move from north to south. Circuit polarity is defined so that current flow in the external circuit is from negative to positive.

Poles: Number of stator poles in a motor.

Polyphase: Power system with two or more phases.

Potentiometer: A variable resistor used to control the voltage in a circuit. A potentiometer has three leads connected to it.

Power: Rate of doing work. Foot-pounds/second. Power in an electrical circuit is equal to amperes X volts for DC, and amperes X volts X $\cos\theta$ for AC. The unit of measure of electrical power is the *watt*. Substituting the definitions for ampere and volt, watt = (coulomb/second) X (joule/coulomb) = j/s.

Power factor: The ratio of true power to apparent power. Power factor is equal to the $\cos\theta$.

Primary winding: The coil of a transformer that is connected to the power source.

Reactance: Opposition to current flow due to the presence of inductance or capacitance in an AC circuit.

Relay: Magnetic switch.

Reluctance: Opposition to the passage of magnetic flux lines.

Repulsion-start motor: Single-phase motor that develops starting torque by interaction of rotor magnetic fields and stator magnetic fields.

Residual magnetism: The magnetism that remains in the material after the magnetizing force is removed.

Resistance: Opposition to current flow measured in ohms.

Resistor: An electrical component that opposes the flow of current. In circuit diagrams, resistors are designated with R.

Rheostat: A variable resistor used to control the flow of current in a circuit. A rheostat has two leads connected to it.

Right-hand rule: Using the right hand, the direction of rotation of a motor can be predicted if the direction of current flow and the magnetic field are known.

Rotor: The moving part of a motor or a generator.

RPM: Revolutions per minute. Speed of rotor rotations on a motor.

Running torque: The turning force as determined by the horsepower and speed of a motor at any given time during operation.

Secondary winding: The coil of a transformer that receives energy from the primary winding of a transformer and is connected to the load.

Series motor: Field windings are connected in series with the armature. The current is the same through both windings.

Shaded-pole motor: Each field pole is split to accommodate a short-circuit copper ring called a *shading coil*. The coil produces a rotating magnetic field in conjunction with the field pole to start and run the motor.

Short circuit: A direct electric connection of very low resistance across the power source or a component, usually resulting in fault currents.

Shunt: A connection in parallel with some component or circuit.

Shunt motor: Field windings are connected in parallel with the armature.

Sine wave: Graphic representation of a wave whose size is proportional to the sine of an angle that is a linear function of time and distance.

Single-phase motor: A motor that operates on single-phase alternating current.

Slip rings: Metal rings that are connected to the rotor of a motor or generator to connect the rotor windings to an external circuit using brushes.

Solenoid: Coil of wire wrapped around an iron core to produce an electromagnet.

Split-phase motor: A single-phase motor that has start and run windings. A rotating magnetic field is created by the two sets of windings to start the motor. Once the motor starts, a centrifugal switch removes the start windings from the circuit.

Squirrel cage rotor: A rotor consisting of bars placed in slots of the rotor core and joined together by end-rings. These rotors are used by single-phase and three-phase induction motors.

Stack: Thickness of the stator.

Starting torque: Amount of turning force produced by a motor when it begins to turn after being stopped. This is sometimes called the *locked-rotor torque*.

Stator: The stationary part of a motor or a generator on which coils are wound.

Step-down transformer: A transformer whose secondary voltage is less than the primary voltage.

Step-up transformer: A transformer whose secondary voltage is higher than the primary voltage.

Stroboscope: An electrical or mechanical device that can be used to measure the RPM of a motor. The stroboscope makes the rotor appear to stand still when the RPM matches the frequency of the stroboscope.

Switch: Device for directing or controlling the current flow in a circuit.

Synchronous motor: An electrical machine that runs at the synchronous speed.

Synchronous speed: This speed can be calculated by multiplying the frequency times 120 and dividing this value by the number of poles. Synchronous speed = F X 120/N.

Tachometer: A mechanical device used to measure the RPM of a motor.

Temperature rise: Amount of heat a transformer or motor generates above the ambient temperature.

Thermalcouple: Two dissimilar metals joined at a common junction point. The junction of two dissimilar metals when heated produces a current flow. As this junction changes in temperature, the thermal couple will provide different levels of currents.

Thermistor: A solid state device whose resistance changes with change in temperature.

Three-phase AC: Three separate but common voltages or currents that are displaced from each other by one-third cycle, or 120°.

Transient pulses: Unwanted electrical disturbances introduced into a system that may disrupt the operation or cause damage. Lightning strikes often cause transient pulses.

True power: Power dissipated in watts. Volts X Amperes X Cosθ. It is the actual power used by a circuit.

Turns ratio: Ratio of the number of turns in the primary of a transformer to the numbers of turns in the secondary.

Unity coupling: When all the flux lines created by the primary coil cut through the secondary coil of a transformer, unity coupling exists.

Universal motor: A series AC motor that also operates on DC.

VA: Volt-ampere. Volts X Current. In a DC circuit the VA is equal to the watts, or power dissipated. This is not true for AC circuits.

Vector: A straight line drawn to scale that shows magnitude and direction of a force.

Volt: Measure of potential energy. One volt is equal to 1 joule per coulomb. Volt = joule/coulomb.

Voltage drop: The difference between the source voltage and the voltage across a load supplied by the source.

Voltmeter: An electrical instrument used to measure the potential energy in a circuit.

Watt: Electrical unit of power. Joule/second.

Wattless power: Power that is not consumed in an AC circuit due to reactance.

Wattmeter: An electrical instrument that reads the true power in a circuit.

Bibliography

Adams, J.E. *Electrical Principles and Practices*, 2nd ed. New York: McGraw-Hill, 1951.

Alford, L.P., Bangs, J.R. *Production Handbook*. New York: Ronald Press, 1954.

American Petroleum Institute. *Form-Wound Squirrel-Cage Induction Motors—250 Horsepower and Larger*. Washington, DC: American Petroleum Institute, 1987.

Barks, H.B. *Electrical Apparatus*. "Public Vocational Education for Electric Motor Winders: How Good Is It?" 1981.

Bussman Division, McGraw-Edison Company. *Electrical Protection Handbook*. St. Louis, MO: Bussman Division, McGraw-Edison, 1984.

Clidero, R., Sharpe, K. *Construction Wiring*. Albany, NY: Delmar Publishers, 1982.

Cooke, N.M. *Basic Mathematics for Electronics*, 2nd ed. New York: McGraw-Hill, 1960.

Croft, T., Watt, J.H., Summers, W.I. *American Electricians' Handbook*, 10th ed. New York: McGraw-Hill, 1981.

Fasco®Industries, Inc. *Fasco Facts*. St. Louis: Fasco Industries, Inc., 1983.

General Electric Company. *AC Motor Selection and Application Guide*. Fort Wayne. IN: General Electric Co., 1987.

General Electric Company. *All New KinaMatic Direct Current Motors, 10–60 HP*. Erie, PA: General Electric Co., 1985.

General Electric Company. *Dry Type Transformers-Information Guide*. Fort Wayne, IN: General Electric Co., 1986.

General Electric Company. *GE Vertical Motors*. Fort Wayne, IN: General Electric Co., 1987.

General Electric Company. *How To Specify and Evaluate Energy-Efficient Motors*. Fort Wayne, IN: General Electric Co., 1985.

General Electric Company. *Motors—Commercial and Industrial Fractional Horsepower Motors*. Fort Wayne, IN: General Electric Co., 1986.

General Electric Company. *Motors—Energy $aver Motors and Normal Efficiency Motors*. Fort Wayne, IN: General Electric Co., 1986.

General Electric Company. *Transformer Connections*. Schenectady, NY: General Electric Co., 1960.

General Electric Company. *Vertical Induction Motors*. Fort Wayne, IN: General Electric Co., 1985.

Gerrish, H.H. *Electricity and Electronics*. Homewood, IL: Goodheart-Willcox Co., 1968.

Gingrich, H.W. *Electrical Machinery, Transformers, and Controls*. Englewood Cliffs, NJ: Prentice-Hall, 1979.

Herman, S.L., Alerich, W.N. *Industrial Motor Control*. Albany, NY: Delmar Publishers, 1985.

Joint Apprenticeship and Training Committee, Local #26, IBEW and NECA-Plant Engineering Training Systems. *Training Program for Maintenance Electricians*. Barrington, IL: Technical Publishing Co., 1971.

Kurtz, E.B., Shoemaker, T.M. *The Lineman's and Cableman's Handbook*, 5th ed. New York: McGraw-Hill, 1976.

Lloyd, T.C. *Electric Motors and Their Applications*. New York: John Wiley & Sons, 1969.

Marcus, A. *Electricity for Electricians*. Englewood Cliffs, NJ: Prentice-Hall, 1969.

Mileaf, H. *Electricity One-Seven*, 2nd ed. rev. Rochelle Park, NJ: Hayden, 1978.

Miller, H.N. *Nondestructive High Potential Testing*. New York: Hayden, 1964.

Moore, F.G. *Manufacturing Management*. Homewood, IL: Richard D Irwin, 1955.

Nailen, R.L. "Electrical Maintenance on a University Campus." *Electrical Apparatus*, 1986.

National Electrical Manufacturers Association. *Instructions for Handling, Installation, Operation and Maintenance of Busway Rated 600 Volts Or Less*. Washington, DC: National Electrical Manufacturers Association, 1986.

National Electrical Manufacturers Association. *MG 1, Motors and Generators*. Washington, DC: National Electrical Manufacturers Association, 1987.

National Fire Protection Association. *National Electrical Code—1987*. Quincy, MA: National Fire Protection Association, 1986.

Rawlplug Company, Inc. *Rawl Masonry Anchors*. New Rochelle, NY: Rawlplug Co., 1986.

Robertson, M.E. *Transformer Application Fundamentals*, 2nd ed. Albany, NY: Delmar Publishers, 1970.

Rockis, G. *Solid State Fundamentals for Electricians*. Alsip, IL: American Technical Publishers, 1985.

Rockis, G., Mazur, G. *Electrical Motor Controls*. Alsip, IL: American Technical Publishers, 1982

Rosenberg, R. *Electric Motor Repair*, 2nd ed. New York, NY: Holt, Rinehart, and Winston, 1970.

Schweitzer, G. *Basics of Fractional Horsepower Motors and Repair*. Rochelle Park, NJ: Hayden, 1960.

Singer, B.B., Forster, H. *Basic Mathematics for Electricity and Electronics*, 5th ed. New York, NY: McGraw-Hill, 1984.

Soares, E.C. *Grounding Electrical Distribution Systems for Safety*. Wayne, NJ: Marsh Publishing Co., Inc., 1966.

Sorgel Transformers, Square D Company. *Sorgel Dry-Type Transformer Study Course*. Milwaukee, WI: Sorgel Transformers, 1976.

Sprague Electric Company. *Power Factor Correction—A Guide for the Plant Engineer*. North Adams, MA: Sprague Electric Co., 1985.

Square D Company. *Motor Control Fundamentals*. Milwaukee, WI: Square D Co., 1978.

Square D Company. *Wiring Diagrams*. Milwaukee, WI: Square D Co., 1986.

Stallcup, J.G. *Motors and Transformers*. Homewood, IL: American Technical Publishers, 1987.

Steinmetz, C.P. *Report on Hysteresis Losses*. General Electric Co., 1892.

Traister, J.E. *Electrical Applications Guidebook*. Reston, VA: Prentice-Hall, 1979.

United States Army. *Electric Motor and Generator Repair*. Washington, DC: Headquarters, Department of the Army, 1972.

Westinghouse Electric Corporation. *Transformer Study Manual*. Westinghouse Electric Corp., 1951.

Westinghouse Electric Corporation. *Westinghouse Consulting Application Guide*. Westinghouse Electric Corp., 1987.

Wildi, T., DeVito, M.J. *Investigations in Electric Power Technology, Instructor's Edition*, 1st ed. Farmingdale, NJ: Buck Engineering Co., 1971.

Index